A **Visual Analogy Guide** to

Human Anatomy & Physiology

Third Edition

D0864996

Paul A. Krieger

Grand Rapids Community College

MORTON
PUBLISHING

925 W. Kenyon Avenue, Unit 12
Englewood, CO 80110
800-348-3777

www.morton-pub.com

Book Team

President	David M. Ferguson
Senior Acquisitions Editor	Marta R. Martins
Associate Project Editor	Sarah D. Thomas
Assistant Editor	Trina Lambert
Interior Design and Composition	Sarah Bailey
Production Manager, Cover Design	Will Kelley
Illustrations	Paul A. Krieger

Printed in the United States of America

10 9 8 7 6 8 4 3 2 1

ISBN 10: 1-61731-626-1
ISBN 13: 978-1-61731-626-5

Library of Congress Control Number: 2016956402

For my wife, Lily

For your love, constant support, and endless encouragement.

Without your help this book would never have been written.

Acknowledgments

Many people contributed to this book. First, I acknowledge David Ferguson, president of Morton Publishing, for signing me to my first book contract while he was working as an editor. David graciously gave me the extra time and autonomy I needed to complete the book the way I originally envisioned it. Doug Morton, founder of Morton Publishing, for agreeing to take a chance on my new book concept and publish it. Working as an author for Morton Publishing has always been a pleasure because everyone treats you with respect and works as a team to create the best book possible. It certainly makes me proud that they remain true to their mission of developing very high-quality learning materials at sensible prices. Marta Martins, senior acquisitions editor, did a wonderful job compiling all the reviewer comments in a spreadsheet and promptly responded to all my inquiries. To Dan Matusiak, for his diligence in faithfully creating the original glossary. I would also like to offer a special thank you to the reviewers listed below for their valuable feedback during the writing process. The production design team at Morton—Will Kelley, Sarah Bailey, and Joanne Saliger—deserves special recognition for the terrific work they have consistently done in producing all of the *Visual Analogy Guides*. Will Kelley, production manager, designed the dazzling new book cover and added new features such as the illustrations on the section dividers. Trina Lambert, assistant editor, and Sarah Thomas, associate project editor, did a fantastic job handling all the changes and corrections while copy editing and proofreading the manuscript. To all the dedicated sales representatives at Morton for introducing my books to students all over the country. Without them, there would not be a new edition. To indexer Doug Easton for carefully creating the helpful index. My fabulous wife, Lily, kindly accepted all my long working hours and offered valuable feedback. Without her help and encouragement this book would not have been written. Thanks to veteran author Mike Timmons for helping me crystallize my initial ideas about the visual analogy guides to turn them into a tangible reality. To my fellow author and friend, Kevin Patton, for graciously responding to all my annoying authoring questions along the way. Thanks to Jim Gibson for teaching me how to improve the graphic layout and page design. To my father, Eugene Krieger, and my mother, Norma Krieger, for emphasizing the need for a good education and for giving me the freedom to pursue my own path. Finally, for my students, colleagues, friends in the Human Anatomy & Physiology Society (HAPS) and the Michigan Community College Biologists (MCCB), and everyone else who offered suggestions, support, and encouragement—thanks to all of you. I am truly grateful.

Reviewers

Steve Bassett, Southeast Community College (first edition)

Neelima Bhavsar, Ph.D., Saint Louis Community College–Florissant Valley (second edition)

David Canoy, Chemeketa Community College (first edition)

John E. Fishback, Ozarks Technical Community College (second edition)

Cynthia Herbrandson, Kellogg Community College (first edition)

Dr. Tony Lamanna, Paradise Valley Community College (second edition)

Darren Mattone, Muskegon Community College (second edition)

Suzanne Pundt, University of Texas–Tyler (second edition)

Dr. Andrea Scollard, Baton Rouge Community College (second edition)

Marilyn Shopper, Ed.D., Johnson Community College (second edition)

Alan J. Spindler, M.D., D.C., A.R.N.P., Brevard Community College–Melbourne (second edition)

Chris Sullivan, California State University–Sacramento (first edition)

Faith Vruggink, Kellogg Community College (first edition)

Contents

How to Use This Book

Purpose

This book was written primarily for students of human anatomy and physiology; however, it will be useful for teachers or anyone else with an interest in this topic. It was designed to be used in conjunction with any of the major anatomy or anatomy and physiology textbooks. What makes it unique, creative, and fun is the **visual analogy** learning system. This will be explained later. The modular format allows you to focus on one key concept at a time. Each module has a text page on the left with corresponding illustrations on the facing page. Most illustrations are unlabeled so that you can quiz yourself on the structures. A handy key to the illustration is provided on the text page. Although this book covers most of the major organ systems, the topics are weighted more toward areas that typically give students difficulty. It uses a variety of learning activities such as labeling, coloring, and mnemonics to help instruct. In addition, it offers special study tips for mastering difficult topics.

What Are Visual Analogies?

A visual analogy is a helpful way to learn new material based on what you already know from everyday life. It compares an anatomical structure to something familiar such as an animal or a common object. For example, the vertebral column has three different types of vertebrae. One type looks like a giraffe. Comparing the vertebra to a giraffe allows you to mentally correlate the *unknown* (*vertebra*) with the *known* (*giraffe*). Doing this accomplishes several things.

1. It reduces your anxiety about learning the material and helps you focus on the task at hand.

2. It forces you to observe anatomical structures more carefully. After all, being a good observer is the first step to becoming a good anatomist—or any type of scientist!

3. It makes the learning more fun, relevant, and meaningful so you can better retain the information.

Whenever a visual analogy is used in this book, a small picture of it appears in the upper right-hand corner of the illustration page for easy reference. This allows you to quickly reference a page visually by simply flipping through the pages.

Icons Used

The following icons are used throughout this book:

 Microscope icon—indicates any illustration that is microscopic.

 Crayon icon—indicates illustrations that were specially made for coloring. Even though students may color any of the illustrations to enhance their learning, they may benefit more by referring to this icon. In some cases, written instructions appear next to this icon with directions about exactly what to color or how to color it.

 Scissors icon—indicates that something is either cut or broken. For example, it may be used to show that a chemical bond between two molecules is broken.

 Three-dimensional icon—Indicates a three-dimensional view of an anatomical structure.

 Two-dimensional icon—Indicates a two-dimensional view of an anatomical structure.

 Dashed lines usually are used to indicate a structure that is behind another structure.

Abbreviations and Symbols

The following abbreviations are commonly used throughout this book:

l.   = left
r.   = right
a.   = artery
v.   = vein
m. = muscle

n.   = nerve
ex.   = example
sing. = singular
pl.   = plural

I sincerely hope that this book will be a fun, helpful tool for anyone interested in learning human physiology. Many of the visual analogies used in the book have been tested with students in both lab and lecture settings to ensure that they are useful.

Enjoy learning with visual analogies!

Visual Analogy Index

Visual Analogy Index

Visual Analogy Index

Visual Analogy Index

Visual Analogy Index

TOPIC	ANALOGY	ICON(S)	PAGE NO.
51. Tissues: Smooth Muscle	Smooth muscle stacks in a sheet like **bricks in a wall**		133
52. Tissues: Neural Tissue	A multipolar neuron looks like an **octopus**		135
53. Skeletal System: Long Bone Structure	Spongy bone looks like a **sponge**		149
54. Skeletal System: Skull	The coronal suture is like a **tiara (crown)**; the sutures on the posterior aspect of the skull look like a **modified peace sign**		154
55. Skeletal System: Skull	The sella turcica of the sphenoid bone looks like a **horse's saddle**		155
56. Skeletal System: Temporal Bone	A temporal bone resembles a **rooster's head**		161
57. Skeletal System: Ethmoid Bone	Superior view looks like a **door hinge**; crista galli looks like a **shark fin**		163
58. Skeletal System: Sphenoid Bone	Sphenoid resembles a **bat**		165
59. Skeletal System: Sphenoidal Foramina	Remember the sphenoidal foramina with **Ros the cowboy**		167
60. Skeletal System: Palatine Bones	Palatine bones are like **two letter "L"s**—one the mirror image of the other		168
61. Skeletal System: Numbers of Vertebrae	Total number of each type of vertebrae corresponds to **mealtimes**; coccyx— **rattlesnake tail**		171
62. Skeletal System: Atlas and Axis	Atlas, the **turtle**; Axis, the **football player**		173

Visual Analogy Index

Visual Analogy Index

TOPIC	ANALOGY	ICON(S)	PAGE NO.
74. Nervous System: Multipolar Neuron	Axon of a neuron is like an **electrical cord**		241
75. Nervous System: Nerve Impulse Conduction	Membrane potential is like a **battery**; graded potential is like a **small, local ripple effect**; an action potential is like a **wave of electric current**; synaptic transmission is like a **speedboat crossing a channel** and pulling into its dock		243
76. Nervous System: Nerve Impulse Conduction	Action potential is like a **wave of electric current**		247
77. Nervous System: Continuous Conduction	Continuous conduction is like a **domino effect**		249
78. Nervous System: Continuous Conduction	Saltatory conduction is like **skipping a stone**		251
79. Nervous System: Peripheral Nerve	A peripheral nerve is like **tubes within tubes**		255
80. Nervous System: Spinal Cord	The gray matter in the spinal cord is shaped like a **butterfly**		263
81. Nervous System: Reflexes	A reflex arc involves input, processing, and output, like **typing a letter on your laptop**		265
82. Nervous System: Brain Ventricles	The ventricular system can be compared to the **neck, head, and horns of a ram**		277
83. Nervous System: Sympathetic Division of ANS	Sympathetic division is like **stepping on the gas**		291
84. Nervous System: Parasympathetic Division of ANS	Parasympathetic division is like **stepping on the brake**		293
85. Special Senses: Eye—External Structure	The rectus muscles are positioned like **points on a compass (north, south, east, and west).**		323

Visual Analogy Index

Visual Analogy Index

Visual Analogy Index

LANGUAGE OF ANATOMY

Planes of the Body

Description

An essential skill in anatomy is being able to visualize a sliced section of a tissue, organ, or region of the human body. This requires you to mentally jump from the three-dimensional to the two-dimensional. The prerequisite to developing this skill is being able to visualize the different ways an object can be sliced. The three basic planes that can pass through an object to section it are:

- **Sagittal** (*median*) **plane**—This plane slices an object down the middle, making a left half and a right half.

- **Frontal plane**—This is also called the coronal plane. The term *coronal* means "crown." This plane splits an object into a front half and a back half.

- **Transverse plane**—This plane divides the body into an upper half and a lower half.

Key to Illustration		
1. Frontal plane	2. Sagittal *(median)* plane	3. Transverse plane

Planes of the Body
(through a 2-year-old boy)

1. _____

Where's the toy truck they promised me for posing like this?

3. _____

2. _____

Color each plane a different color.

Description   Directional terms are part of the working language of anatomy. They are commonly used to describe the position of an anatomical structure or the position of one body part in relation to another. Some of the most common terms for body orientation/direction are given below, along with their meanings and examples of how to use them correctly in a sentence.

Superior: above
Inferior: below

> *ex.:* The brain is *superior* to the lungs.
> The lungs are *inferior* to the brain.

Medial: toward the midline of the body
Lateral: away from the midline of the body

> *ex.:* The nose is *medial* to the external ear.
> The external ear is *lateral* to the nose.

Proximal: nearer the trunk of the body or the reference point
Distal: farther from the trunk of the body or the reference point

> *ex.:* The shoulder is *proximal* to the wrist.
> The wrist is *distal* to the shoulder.

Superficial: toward the body surface
Deep: away from the body surface

> *ex.:* The skin is *superficial* to the muscle.
> The muscles are *deep* to the skin.

Anterior/ventral: toward the front
Posterior/dorsal: toward the back

> *ex.:* The heart is *anterior* to the spinal cord.
> The spinal cord is *posterior* to the heart.

Key to Illustration			
1. Superior	3. Anterior *(ventral)*	5. Proximal	7. Medial
2. Inferior	4. Posterior *(dorsal)*	6. Distal	8. Lateral

Directional Terms

1. _____

2. _____

3. _____

4. _____

5. _____

6. _____

7. _____

8. _____

2-year-old boy

LANGUAGE OF ANATOMY

Regional Terms

Description

Regional terms are part of the working language of anatomy. Many anatomical structures are named after the region of the body in which they are found. For example, there are the *brachial* artery, the *femoral* nerve, and the *deltoid* muscle. Memorizing these terms now will serve you well throughout your study of anatomy.

The table below gives the key to the numbers on the illustration as well as the definition of each regional term.

Regional Term	Description	Regional Term	Description
1. Cranial	Skull	23. Digital	Fingers
2. Facial	Face	24. Pelvic	Pelvis
3. Cephalic	Head	25. Inguinal	Groin
4. Frontal	Forehead	26. Coxal	Hip
5. Orbital	Eye	27. Pubic	Anterior region of pelvis
6. Nasal	Nose	28. Femoral	Thigh
7. Buccal	Cheek	29. Patellar	Patella
8. Oral	Mouth	30. Crural	Leg, from knee to ankle
9. Mental	Chin	31. Tarsal	Ankle
10. Cervical	Neck	32. Pes (Pedal)	Foot
11. Deltoid	Shoulder	33. Occipital	Back of head
12. Pectoral	Region of chest covered by pectoralis major muscles	34. Otic	Ear
		35. Scapular	Scapula
13. Sternal	Sternum	36. Vertebral	Vertebral column
14. Mammary	Breast	37. Lumbar	Lower back
15. Thoracic	Thorax or chest	38. Olecranon	Elbow
16. Axillary	Armpit	39. Gluteal	Buttocks
17. Brachial	Arm, from shoulder to elbow	40. Manual	Hand
18. Antecubital	Front of elbow	41. Perineum	Region between anus and genitals
19. Abdominal	Abdomen		
20. Antebrachial	Forearm	42. Popliteal	Back of knee
21. Carpal	Wrist	43. Sural	Calf
22. Palmar	Palm	44. Calcaneal	Heel

LEFT OR RIGHT?

For paired regions or structures, it is important to designate each as left or right. In anatomy, left and right always refers to the individual's left and right. For example, there is both a left femoral region and right femoral region.

Right femoral region

Left femoral region

Anterior

I hope these numbers wash off!

Posterior

Color each different regional area a different color.

Note: dashed lines indicate borders of regional areas.

Dorsum of the hand

Dorsum of the foot

Body Cavities

1. **Dorsal body cavity**—consists of the cranial and vertebral cavities

2.	Cranial cavity	contains the brain
3.	Vertebral cavity	contains the spinal cord

4. **Ventral body cavity**—consists of the thoracic and abdominopelvic cavities

5.	Thoracic cavity	contains the heart and the lungs
6.	Mediastinum	median compartment of the thoracic cavity
7.	Pleural cavity	fluid-filled space around the lungs
8.	Pericardial cavity	fluid-filled space around the heart
9.	Abdominopelvic cavity	consists of the abdominal and pelvic cavities
10.	Abdominal cavity	contains the digestive organs, kidneys, and ureters
11.	Pelvic cavity	contains the urinary bladder, internal reproductive organs, and the rectum

Membranes

Serous membranes are double-layered, fluid-filled sacs that surround organs like the heart and the lungs.

Analogy

To visualize how a serous membrane surrounds an organ, imagine a fist pushed into a partially filled water balloon. The fist is like the **organ** and the balloon is like the **serous membrane**. The **inner layer** of the balloon that touches the fist is like the **visceral layer**. The **outer wall** of the balloon is like the **parietal layer**. The balloon's **inner space**, filled with water, is like the **serous cavity**, which is normally filled with a lubricant called serous fluid. This fluid is secreted by cells within the serous membrane.

12.	Parietal pericardium	outermost layer of the serous membrane around the heart
13.	Visceral pericardium	innermost layer of the serous membrane around the heart
14.	Pericardial cavity	narrow, fluid-filled space between the parietal and visceral layers
15.	Parietal pleura	outermost layer of the serous membrane around the lungs
16.	Visceral pleura	innermost layer of the serous membrane around the lungs
17.	Pleural cavity	narrow, fluid-filled space between the parietal and visceral layers
18.	Parietal peritoneum	outermost layer of the serous membrane that lines the abdominopelvic cavity
19.	Visceral peritoneum	innermost layer of the serous membrane that lines most digestive organs
20.	Peritoneal cavity	narrow, fluid-filled space between the parietal and visceral layers

BODY CAVITIES

I have this EMPTY feeling inside

Color the body cavities different colors.

Diaphragm (separates the thoracic cavity from the abdominopelvic cavity)

Midsagittal view

Frontal view

MEMBRANES

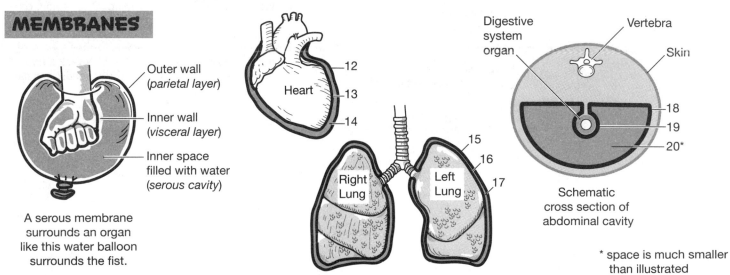

Outer wall (*parietal layer*)

Inner wall (*visceral layer*)

Inner space filled with water (*serous cavity*)

A serous membrane surrounds an organ like this water balloon surrounds the fist.

Heart

Right Lung

Left Lung

Digestive system organ

Vertebra

Skin

Schematic cross section of abdominal cavity

* space is much smaller than illustrated

Abdominopelvic quadrants

Medical professionals often divide the **abdominopelvic region** into the following four areas to help locate pain sites due to injuries and other medical problems. Note that the terms "left" and "right" refer to the subject's left and right.

RUQ	=	Right upper quadrant
LUQ	=	Left upper quadrant
RLQ	=	Right lower quadrant
LLQ	=	Left lower quadrant

Abdominopelvic regions

Anatomists often prefer to divide the abdominopelvic region into a more detailed nine-square grid. The table below explains each region.

Symbol	Regional Name	Description
E	Epigastric region (*epi* = above; *gastric* = stomach)	Located at the top of the middle column; contains the duodenum and parts of the following: liver, stomach, and pancreas
U	Umbilical region (navel)	Located in the center of the grid; contains parts of both the transverse colon and the small intestine
H	Hypogastric region (*hypo* = below; *gastric* = stomach)	Located at the bottom of the middle column; contains the urinary bladder, sigmoid colon, and part of the small intestine
RH	Right hypochondriac region (*hypo* = below; *chondro* = cartilage)	Located to the subject's right of the epigastric region; contains the gallbladder and parts of both the right kidney and the liver
LH	Left hypochondriac region (*hypo* = below; *chondro* = cartilage)	Located to the subject's left of the epigastric region; contains the spleen and parts of the following: stomach, left kidney, and large intestine
RL	Right lumbar region (*lumbus* = loin)	Located to the subject's right of the umbilical region; contains parts of the following: large intestine, small intestine, and right kidney
LL	Left lumbar region (*lumbus* = loin)	Located to the subject's left of the umbilical region; contains parts of the following: large intestine, small intestine, and left kidney
RI	Right iliac region (*iliac* = largest part of hip bone)	Located to the subject's right of the hypogastric region; contains the bottom of the cecum, the appendix, and part of the small intestine
LI	Left iliac region (*iliac* = largest part of hip bone)	Located to the subject's left of the hypogastric region; contains parts of both the large and small intestine

Color the quadrants different colors.

RUQ   LUQ

RLQ   LLQ

The **Four** Abdominopelvic Quadrants

Color the regions different colors.

RH   E   LH
Liver   Stomach

Gallbladder

Pancreas

Small intestine

Large intestine

RL   U   LL

Cecum

RI   H   LI

Appendix

Urinary bladder

The **Nine** Abdominopelvic Regions

PHYSIOLOGY OVERVIEW

SET
POINT

LO   HI

Description

The human body has different levels of structural organization. **Atoms** (*ex: oxygen*) combine to form **molecules** (*ex: water*). Very large molecules called **macromolecules** (*ex: carbohydrates, lipids, proteins, nucleic acids*) are the building blocks used to form any **cell** (*ex: cardiac muscle cell*). A group of similar cells working together is called a **tissue** (*ex: cardiac muscle tissue*). A variety of different tissues working together forms an **organ** (*ex: the heart*), and a group of organs working together forms a complex **organ system** (*ex: cardiovascular system = heart and the blood vessels*). These complex organ systems all working together form an **organism** (*ex: human organism*). For convenience, teachers and textbook authors sometimes discuss each organ system independent of the others. But it should be emphasized that no organ system exists as an island unto itself, because they are all interdependent. Substances such as water, nutrients, and waste products have to move between organ systems to maintain **homeostasis** (see pp. 16–17) for the entire organism. Keep in mind that when studying physiology it is common to jump back and forth between these different levels of structure when learning about physiological mechanisms.

The following 10 organ systems are shown in the illustration on the facing page: **integumentary system, musculoskeletal system, nervous** and **endocrine systems, cardiovascular system, lymphatic/immune system, respiratory system, digestive system, urinary system,** and **reproductive systems.** This schematic illustration attempts to show the interrelationships between organ systems. The white rectangular band around the periphery represents the **integumentary system.** This includes the skin and all its components, such as hair follicles, oil glands, and a host of other structures. Our skin functions as a thin mantle to protect us from our external environment. The **musculoskeletal system** provides an internal framework for support and allows for body movements. Consider that movement would be impossible without muscles acting on bones as a kind of lever system.

Two organ systems work together to act as the regulators of body function—the **nervous** and **endocrine** systems. Because they both serve a similar purpose, they were placed together in the same box. Without the nervous system we would have no way to detect stimuli from either our external or internal environments, no way to process this information, and no way to respond to it. The endocrine system is composed of glands (*ex: pancreas*) that release chemical messengers called **hormones** (*ex. insulin*) that travel through the bloodstream to target a particular structure such as an organ to induce a response. Working together, these two systems help maintain the vital activities that keep us alive.

The **cardiovascular system** consists of the heart and all the blood vessels (*ex. arteries, veins, arterioles, venules,* and *capillaries*). The heart functions as a pump to constantly circulate blood throughout the body. The blood vessels must connect to all the other organ systems to deliver nutrients (*ex. oxygen*) to all the body cells via the bloodstream.

The following four organ systems contain organs/structures that are hollow: **respiratory system, digestive system, urinary system,** and **reproductive systems.** Structurally speaking, all of these systems have an internal chamber called a **lumen** that extends to the external environment. Any substance within this lumen is still considered part of the external environment until it crosses the wall of that structure and enters body tissues. The major function of the **respiratory system** is to exchange the respiratory gases, oxygen and carbon dioxide, between the body and the external environment. Oxygen is brought in from the external environment, transferred to the blood, and delivered to body cells. Carbon dioxide is a normal waste product made by body cells. It is transferred to the blood, then sent back to the respiratory system, where some of it is released to the external environment.

The **digestive system** functions to take in nutrients, break them down to their simplest components, and absorb them into the blood. Then they are delivered to cells. In addition, the digestive system eliminates waste products from the body. The **urinary system** constantly filters and processes the blood to eventually form urine, which contains nitrogen wastes.

The **lymphatic** and **immune systems** typically are grouped together (*even though they are not illustrated that way*). One major function of the **lymphatic system** is to take the **interstitial fluid** found between cells and to cleanse it of debris and possible pathogens, then to deliver it back to the bloodstream. The **immune system** functions to protect the body from foreign pathogens such as bacteria and viruses. For example, when a break in the skin occurs, pathogens may enter the body and the immune system must launch a response akin to soldiers going to war. It accomplishes this by making antibodies and other substances that help fight off these invaders.

Trace the pathway oxygen follows until it enters a body cell. How many barriers must it cross?

O_2 (Oxygen)

CO_2 (Carbon dioxide)

Food

Integumentary system

Respiratory system

Musculoskeletal system

Digestive system

(Gray arrows indicate direction of blood flow)

CO_2

O_2

Heart

n u t r i e n t s

Lymphatic system

Cardiovascular system

Body cell

Color the different organ systems different colors.

Lymph node

Nervous and endocrine systems

CO_2

Interstitial fluid

O_2

Sperm (male); egg (female)

Urinary system

Reproductive system

Immune system

Urine containing nitrogen wastes

Feces, undigested materials

Description

Homeostasis is defined as maintaining a stable, constant internal environment within an organism. As changes constantly occur in the external environment, organisms must have control systems so they can detect these changes and respond to them. Anything that must be maintained in the body within a normal range must have a control system. For example, body temperature, blood pressure, and blood glucose levels all have to be regulated.

A **control system** consists of the following four parts:

1. **Variable:** the item to be regulated. Consider the variable like a teeter-totter. If it is perfectly balanced in the horizontal position, it is at the normal value called the **set point**. Just as a single person resting on the teeter-totter can cause an imbalance, various stimuli can cause the variable to either rise above normal levels or fall below normal levels.

2. **Receptor:** senses the change in the variable; provides input to the control center.

3. **Control center:** determines the value for the set point.

4. **Effector:** provides output to influence the stimulus; uses feedback to return the system to the original set point.

Feedback mechanisms may be either positive or negative. **Negative feedback** mechanisms are the more common in the human body. As a city works, they regulate the day-to-day activities of the body. The output in this system is used to turn off the stimulus. A good example of this is explained below, with the cooling system in a room used to regulate room temperature. **Positive feedback** mechanisms, in contrast, are like the emergency responders. They are not as common, and they use the responses to intensify the original stimulus. A biological example of this is labor contractions during childbirth. A hormone called oxytocin stimulates the muscle in the wall of the woman's uterus to strengthen contractions so they become progressively more forceful until the baby is born.

Analogy

Let's use the analogy of regulating room temperature as an example of a control system. If we set the room temperature at 68°F, then it falls below normal, the change is detected and triggers the furnace to heat the room back to the set point. Alternatively, if the room temperature increases above this level, it also is detected and triggers the central air conditioning to turn on to cool the room temperature back to normal. The illustration on the facing page shows the example of a room getting warmer on a hot summer day. Let's identify each of the four elements in this control system.

- **Variable:** Room temperature. Set point? 68°F.
- **Receptor:** Thermometer within the thermostat.
- **Control center:** Thermostat.
- **Effector:** Central air conditioning.

Similarly, the body must regulate its own body temperature. Our normal set point is 98.6°F. When we are overheating our body responds in numerous ways in an effort to cool itself, such as sweating, increasing our breathing rate, and increasing our heart rate. Alternatively, if our body temperature is falling, we stimulate processes that conserve or generate heat, such as shivering, decreasing our breathing rate, and decreasing our heart rate.

- **Variable:** Body temperature. Set point? 98.6°F.
- **Receptor:** Hot and cold temperature sensors in the skin.
- **Control center:** Hypothalamus in the brain.
- **Effector:** Skeletal muscles—shivering.

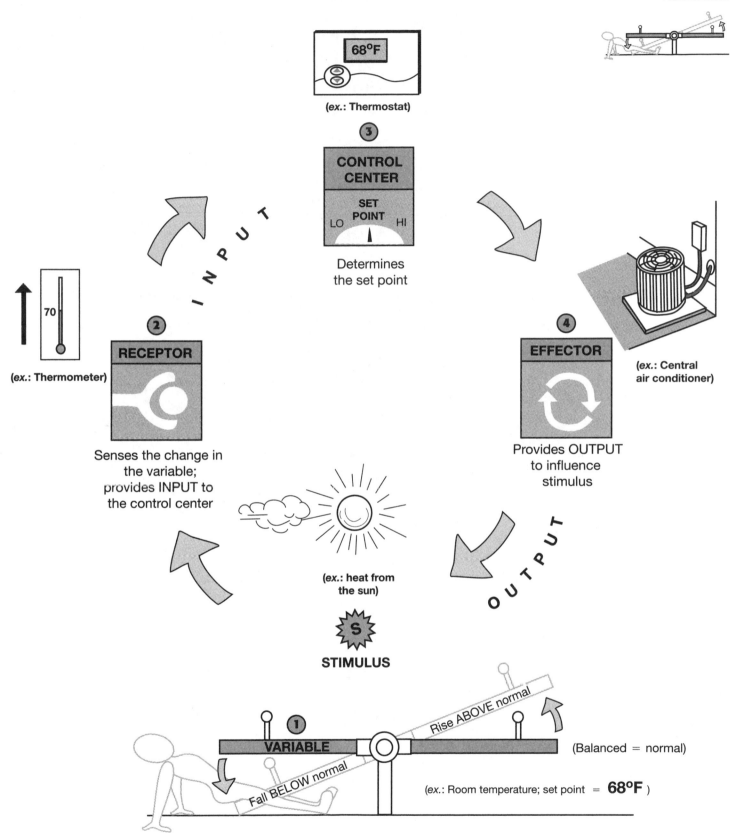

(ex.: Thermostat)

③ CONTROL CENTER

SET POINT
LO   HI

Determines the set point

INPUT

70

(ex.: Thermometer)

② RECEPTOR

Senses the change in the variable; provides INPUT to the control center

68°F

④ EFFECTOR

(ex.: Central air conditioner)

Provides OUTPUT to influence stimulus

OUTPUT

(ex.: heat from the sun)

S

STIMULUS

① VARIABLE

Rise ABOVE normal

(Balanced = normal)

Fall BELOW normal

(ex.: Room temperature; set point = **68°F**)

Control system: This illustration shows all the features of a control system. As an analogy, regulating room temperature on a hot day is shown.

Notes

CHEMISTRY

Atomic Structure

The illustration on the facing page shows a cylinder of **helium** gas to represent the element helium and graphite in a pencil to represent the element **carbon**. Helium gas is used to fill balloons, and we all know how pencils are used. Each of these elements, like all others, is made of many identical **atoms**. With the exception of hydrogen, each atom contains three subatomic particles: **protons, neutrons**, and **electrons**. The protons and neutrons are packed tightly together at the center of the atom called the **nucleus**. Electrons are particles found moving at high speed in a spherical region around the nucleus called the **electron cloud**. Here is a summary of the properties of each subatomic particle:

- **Proton**—a particle that has a **positive** (+) charge (remember the phrase "Protons are Positive")

- **Neutron**—a particle that is **neutral** (no charge) (remember the phrase "Neutrons are Neutral")

- **Electron**—a small, almost massless particle with a **negative** charge

Atomic Number

Atomic number = number of protons in an atom

Every element on the periodic table has a different atomic number that indicates the number of protons in a single atom of that element. The illustrations show helium as atomic number 2, telling us that it has only 2 protons (+2); carbon, with atomic number 6, has 6 protons (+6). The net charge on any atom is always zero, which means that the number of protons and electrons is always equal. Notice that the helium atom has 1 proton and 1 electron, and the carbon atom has 6 protons and 6 electrons. Because protons are always positive, and electrons are always negative, the net charge of both atoms is zero.

Practice Problems

Determine the number of protons in an atom of the following elements:

| 1 | 19 **K** 39.10 Potassium | 2 | 30 **Zn** 65.40 Zinc | 3 | 7 **N** 14.01 Nitrogen | 4 | 35 **Br** 79.90 Bromine |

5 Are the total number of protons and electrons in an atom the same or different? Explain.

6 Which subatomic particle has no charge?

Answers

1 19   2 30   3 7   4 35   5 They are the same. The net charge on any atom is always zero, which requires the number of protons and electrons to be equal.   6 neutron

20

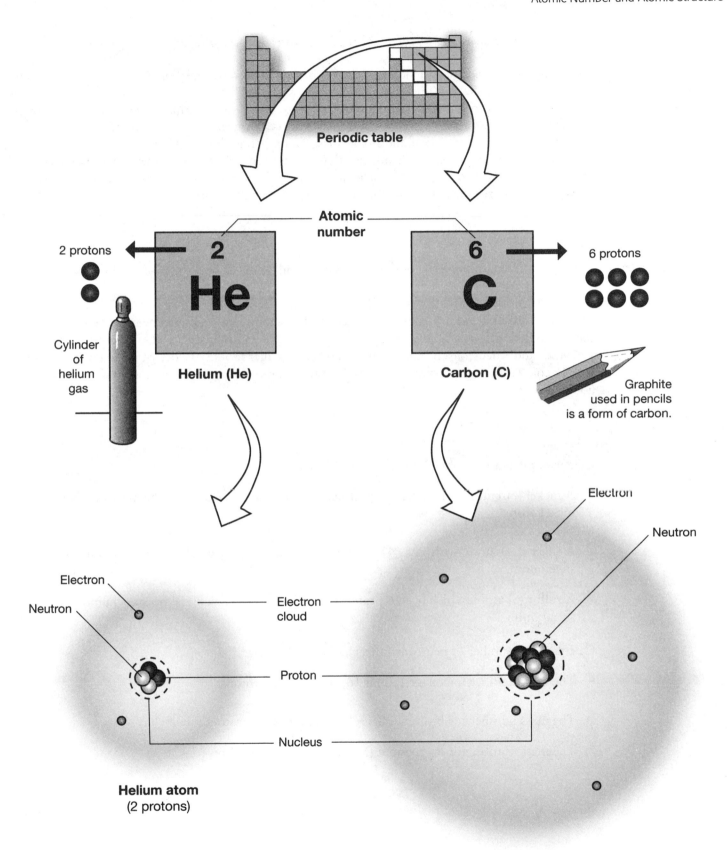

Periodic table

Atomic number

2 protons

He

Helium (He)

Cylinder of helium gas

6 protons

C

Carbon (C)

Graphite used in pencils is a form of carbon.

Electron

Neutron

Electron cloud

Neutron

Electron

Proton

Nucleus

Helium atom
(2 protons)

Carbon atom
(6 protons)

Atomic Mass

Atomic mass = average mass of all the naturally occurring isotopes of that element

On the periodic table, the atomic mass number is found below the symbol of the element. For example, on the facing page, helium (He) has an atomic mass of **4.003**. The units used for measuring very small particles are called atomic mass units (amu). These represent the atomic mass expressed in grams (g). Why isn't the atomic mass a whole number? It is calculated as an *average* of all the naturally occurring **isotopes** of that element. By comparison, the atomic mass for carbon is given as **12.011**. Notice that it also is not a whole number for the reason already mentioned. The atomic mass tells us the mass of the nucleus of an atom, which contains both protons and neutrons. Why not consider the electrons? They have a negligible mass, so they contribute almost nothing to the overall mass of the atom.

Mass Number

Mass number = number of **protons and neutrons** in the nucleus of an atom

Unlike the atomic mass, the mass number is always a whole number because it tells us the total number of protons and neutrons in the nucleus of an atom. Let's examine the mass number for each of the illustrated atoms. As the second element on the periodic table, the helium atom has only 2 protons and 2 neutrons. By adding these numbers together (2 + 2) we determine that the mass number is 4. In comparison, the carbon atom has 6 protons and 6 neutrons, giving it a mass number of 12 (by adding 6 + 6).

Practice Problems

Calculate the **mass number** for each of the following:

1 A nickel (Ni) atom, which has 28 protons and 31 neutrons. Calculate the **mass number** for this nickel atom.

2 A silver (Ag) atom, which has 47 protons and 61 neutrons. Calculate the **mass number** for this silver atom.

3 A sulfur (S) atom, which has 16 protons and 16 neutrons. Calculate the **mass number** for this sulfur atom.

Calculate the number of **neutrons** for each of the following:

4 The mass number for oxygen (O) is 16, and it has 8 protons.

5 The mass number for bromine (Br) is 80, and it has 35 protons.

6 The mass number for magnesium (Mg) is 24, and it has 12 protons.

Answers

1 59 2 108 3 32 4 8 5 45 6 12

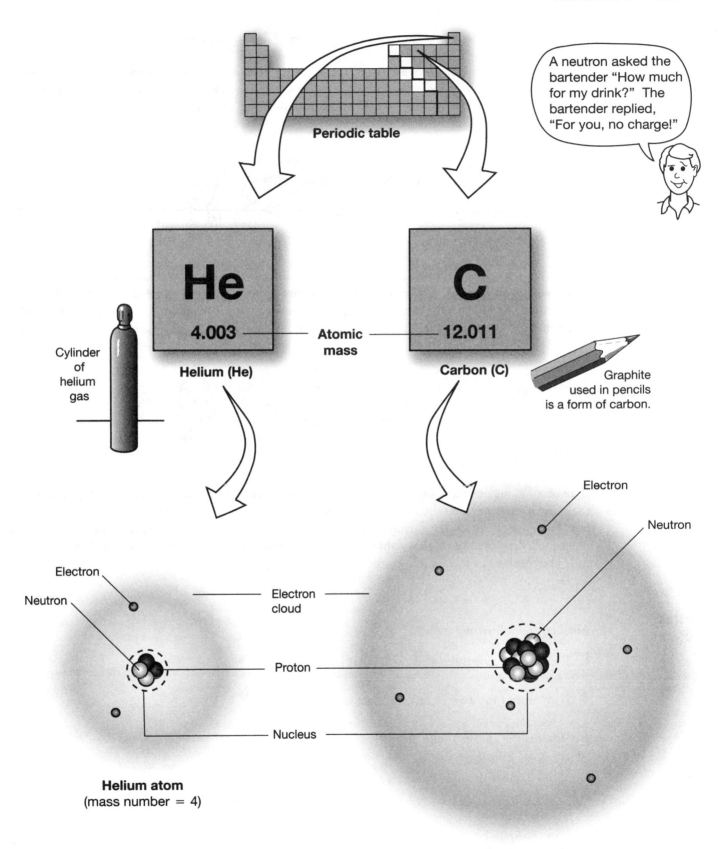

Periodic table

A neutron asked the bartender "How much for my drink?" The bartender replied, "For you, no charge!"

Cylinder of helium gas

He
4.003
Helium (He)

Atomic mass

C
12.011
Carbon (C)

Graphite used in pencils is a form of carbon.

Electron

Neutron

Electron
cloud

Electron

Neutron

Proton

Nucleus

Helium atom
(mass number = 4)

Carbon atom
(mass number = 12)

Ions

Ion = atom that has either *gained* or *lost* one or more electrons

When an atom gains or loses electrons, it is called an **ion** instead of an atom. Recall that atoms are electrically neutral, so they have no charge. Atoms that lose one or more electrons become positively charged and are called **cations**. When an atom gains one or more electrons, it becomes negatively charged and is called an **anion**.

Positive (+) Ions (Cations)

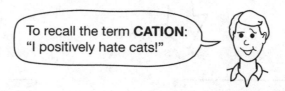

To recall the term **CATION**:
"I positively hate cats!"

If an atom loses one electron, its charge is +1, if it loses two electrons, its charge is +2, and so on. Electrons are always negatively charged, so the *loss* of this negative charge results in a positively (+) charged ion. The illustration on the facing page shows an atom of sodium (Na). Sodium has only one electron in its outermost shell, so it tends to give it up quite easily to become an ion with a +1 charge.

Negative (–) Ions (Anions)

To recall the term **ANION**:
"A Negative ION!"

If an atom gains one electron, its charge is –1, if it gains two electrons, its charge is –2, and so on. Electrons are always negatively charged, so the *gain* of a negative charge results in a negatively (–) charged ion. The illustration on the facing page shows an atom of chlorine. It has seven electrons in its outermost shell. A full shell contains eight electrons, so chlorine tends to readily accept an electron to become full. The result is an ion with a –1 charge.

Practice Problems

Answer the following:

1 If an atom *gains* three electrons, what is the charge?

2 If an atom *loses* two electrons, what is the charge?

3 What is the charge on an atom?

4 What is the difference between a **cation** and an **anion**?

Answers

1 –3   2 +2   3 No charge   4 Cations are positively (+) charged and anions are negatively (–) charged.

Positive (+) Ions
(cations)

An atom walks into a bar
and tells the bartender,
"I think I lost my electron."
The bartender asks,
"Are you sure?"
The atom replies "I'm positive!"

11 e⁻

LOSES one
valence
electron

11 p⁺

10 e⁻

11 p⁺

Sodium ATOM
(no charge)

Sodium ION
(charge of + 1)

$+1$

Negative (–) Ions
(anions)

17 e⁻

GAINS one
valence
electron

17 p⁺

18 e⁻

17 p⁺

Chlorine ATOM
(no charge)

Chloride ION
(charge of – 1)

-1

Octet Rule and Exceptions

Matter always wants to be in its most stable state. But what determines this stability? One measure is an atom's total number of valence electrons. The **octet rule** describes the *general* rule for stability for an atom: 8 electrons in its outermost shell (or 8 valence electrons). Why 8? This is the typical maximum number of electrons that makes most electron shells "full." To remember the octet rule, think of the eight ball used in playing pool.

EIGHT BALL = octet rule

For example, neon (Ne) has 8 valence electrons, so it is very stable.

Are there exceptions to this rule? Yes. For example, hydrogen is the simplest element on the periodic table and has one valence electron. Because it has only one electron shell located close to the nucleus, that shell is considered full when it contains 2 electrons. As a result, hydrogen's "magic number" for stability is 2, not 8. The same is true for helium, which contains 2 electrons in its only electron shell.

Analogy

This octet rule is a driving force for chemical bonding. Unstable atoms combine with other unstable atoms to form a more stable product. As an analogy, let's classify a lonely, single guy looking for a spouse as "unstable." Ideally speaking, if he bonds with a suitable companion, and they work hard at their relationship, they have formed a more "stable" couple. Obviously, there are LOTS of exceptions to this rule, but you get the point, right?

unstable   unstable   **STABLE!**
(in an ideal world)

Let's apply this analogy to the concept of chemical bonding. Atoms bond with other atoms to form more stable products in several different ways. If they share a pair of electrons, they form a **covalent bond**, but if one atom donates an electron and the other receives it, an **ionic bond** is formed. Both of these types of bonds will be discussed as separate topics.

Example of Octet Rule

Let's use the example of an ionic bond in the formation of table salt to understand this further.

Word description

A sodium atom (Na) plus a chlorine atom (Cl) forms sodium chloride salt (NaCl)

Chemical equation

Na + Cl → NaCl

Visual equation

Sodium becomes a cation (+)

Chlorine becomes an anion (−)

Sodium is unstable because it has only one valence electron; it donates its electron to chlorine

Chlorine is also unstable because it has seven valence electrons; it receives the electron from sodium

This results in the atoms becoming more stable ions. The chlorine ion now has eight valence electrons, so it is more stable. After donating, the sodium ion is still left with no valence electrons, however it does have eight electrons in a lower shell.

unstable   unstable   STABLE!

Practice Problems

On a separate sheet of paper, answer the following:

1 Alkali metals (1A) such as lithium and sodium each have one valence electron. Based on what you know about chemical bonding, predict if they are more chemically reactive or less chemically reactive. Explain.

2 The noble gases (8A) such as neon and argon all have eight valence electrons. Based on what you know about chemical bonding, predict if they are more reactive or less reactive. Explain.

3 Calcium has two valence electrons. If it formed an ionic bond, would it more likely be an electron donor or a recipient? Explain.

4 Oxygen has six valence electrons. How many more does it need to have a full outermost shell?

Answers

1 Alkali metals need eight valence electrons to be stable, but have only one, so they are unstable. This makes them more chemically reactive because they have to lose an electron to become more stable.   2 Noble gases have a full outermost shell with eight valence electrons, so they are very stable and less chemically reactive. They do not have to bond to become stable.   3 Calcium has only two valence electrons but needs eight, so it would be easier for it to donate two than receive six. Thus, it would be an electron donor.   4 Oxygen needs eight valence electrons to have a full outermost shell. It already has six, so it would need two more. Thus, it would be an electron acceptor.

CHEMISTRY: Chemical Bonds

The Ionic Bond and the Covalent Bond

Description   Atoms bond to be in a more stable state. We're going to examine two common types of bonds: **ionic bonds** and **covalent bonds**.

Ionic bonds = one atom *donates* its electron; the other *receives* that same electron	The **key concept** is that positive and negative ions are formed; opposite charges attract.

Ex.: formation of sodium chloride salt

Word description: A sodium atom plus a chlorine atom form sodium chloride salt.

Chemical equation: Na   +   Cl   ⟶   NaCl

Sodium has become a **cation (+)**

Chlorine has become an **anion (−)**

Explanation: Na· + ·Cl: ⟶ Na⁺ :Cl:⁻

Sodium atom *donates* its electron to chlorine

Chlorine atom *receives* the electron from sodium

This results in the atoms becoming ions; opposite charges attract

Visual: Electron   Electron cloud

+1   −1

Na   Cl   Na⁺¹   Cl⁻¹

Opposite charges attract

Covalent bonds = **shared pair** of electrons (like a couple in love, the word *love* is scrambled in the word *covalent*)

 The **key concept** is that a covalent bond links two atoms like two people locking arms.

Word description: A hydrogen atom plus another hydrogen atom forms hydrogen gas.

Chemical equation:

$$H \quad + \quad H \quad \longrightarrow \quad H_2$$

Explanation:

Shared pair

Line is a symbol for a covalent bond

$$H\cdot \quad + \quad \cdot H \quad \longrightarrow \quad H \vdots H \quad = \quad H_2 \quad = \quad H{-}H$$

Hydrogen atom *shares* its valence electron with ...

... another hydrogen atom.

This results in a shared pair of electrons or a covalent bond.

Visual:

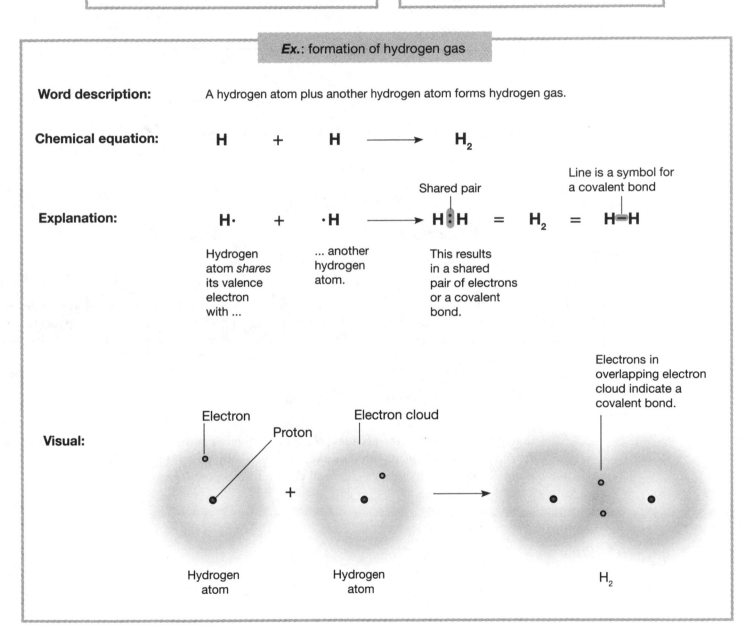

Electron

Proton

Electron cloud

Electrons in overlapping electron cloud indicate a covalent bond.

Hydrogen atom

+

Hydrogen atom

H_2

Description

Knowing the terms and symbols used in chemical equations is essential to understanding chemistry. Let's use making pancakes as an analogy. Mixing ingredients together in a recipe is similar to what occurs in some chemical reactions. By combining two substances, then heating the batter, you form something new.*

Description in Words

Water   plus   pancake mix   forms   pancakes

Equation

Water   +   pancake mix   ⟶   pancakes

Visual

Reactants   Product

In the above example, the *water* and the *pancake mix* are the **reactants**. They always are written to the left of the arrow. After the batter is heated in a frying pan, the final **product**, pancakes, is formed. The product always is written to the right of the arrow. Knowing this, let's write the general formula for any chemical equation:

Reactant #1   +   Reactant #2   ⟶   Product(s)

* This analogy works only in the broadest sense of combining two substances to make something new. When water is combined with pancake mix, the resulting batter is actually a mixture. When the batter is heated, it undergoes a chemical change to form pancakes.

Understanding Chemical Equations

Practice Problems

Identify the REACTANTS in the following chemical equations:

1 $H_2CO_3 \longrightarrow H_2O + CO_2$

2 $BaCl_2 + Na_2SO_4 \longrightarrow 2NaCl + BaSO_4$

3 $C + O_2 \longrightarrow CO_2$

Identify the PRODUCTS in the following chemical equations:

4 $CaCO_3 + 2HCl \longrightarrow CaCl_2 + CO_2 + H_2O$

5 $2KClO_3 \longrightarrow 2KCl + 3O_2$

6 $CH_4 + 2O_2 \longrightarrow CO_2 + 2H_2O$

Change the following word descriptions into chemical equations:

7 Hydrochloric acid plus sodium hydroxide forms sodium chloride plus water.

8 Glucose plus oxygen forms carbon dioxide plus water.

9 Chlorine plus sodium bromide forms sodium chloride plus bromine.

Answers

1 H_2CO_3 = carbonic acid

2 $BaCl_2, Na_2SO_4$ = barium chloride, sodium sulfate

3 C, O_2 = carbon, oxygen

4 $CaCl_2, CO_2, H_2O$ = calcium chloride, carbon dioxide, water

5 KCl, O_2 = potassium chloride, oxygen

6 CO_2, H_2O = carbon dioxide, water

7 $HCl_{(aq)} + NaOH \longrightarrow NaCl + H_2O$

8 $C_6H_{12}O_6 + 6O_2 \longrightarrow 6CO_2 + 6H_2O$

9 $Cl_2 + 2NaBr \longrightarrow 2NaC + Br_2$

CHEMISTRY: Chemical Reactions

Types of Chemical Reactions

Description Chemical reactions have predictable patterns depending on their type. Let's examine three basic types of chemical reactions: *combination reactions*, *decomposition reactions*, and *replacement reactions*.

Combination Reactions Two or more reactants combine to form a **single** product—the opposite of a decomposition reaction.

Analogy: cup   +   cake   ⟶   cupcake

(Combining the word cup with the word cake forms a new compound word, cupcake)

Ex.: H_2O + CO_2 ⟶ H_2CO_3

[Water (H_2O) plus carbon dioxide gas (CO_2) forms carbonic acid (H_2CO_3)]

Decomposition Reactions A compound breaks down into its component parts—the opposite of a combination reaction.

Analogy: cupcake   ⟶   cup   +   cake

(Decomposing the compound word cupcake results in forming the words cup and cake)

Ex.: H_2CO_3 ⟶ H_2O + CO_2

[Carbonic acid (H_2CO_3) is broken down into water (H_2O) and carbon dioxide (CO_2)]

Replacement Reactions

Like changing dance partners, components recombine to form new products. There are two types: **single replacement** and **double replacement**.

Single replacement

May I cut in?

Sure!

Forget "B" and stay with me!

OK

Darn!

A + B C ⟶ A C + B

these components combine

Ex.: Zn + CuCl$_2$ ⟶ ZnCl$_2$ + Cu

[Zinc (Zn) plus copper chloride (CuCl$_2$) forms zinc chloride (ZnCl$_2$) plus copper (Cu)]

Double replacement

"outside" components combine

You dance better than "B."

"D" stepped on my foot!'

A B + C D ⟶ A D + B C

"inside" components combine

Ex.: CaCl$_2$ + K$_2$CO$_3$ ⟶ CaCO$_3$ + KCl

[Calcium chloride (CaCl$_2$) plus potassium carbonate (K$_2$CO$_3$) forms calcium carbonate (CaCO$_3$) plus potassium chloride salt]

Classify each of the following reactions as *combination, decomposition, single replacement,* or *double replacement*:

1 $2Mg + O_2 \longrightarrow 2MgO$

2 $NaCl + AgNO_3 \longrightarrow NaNO_3 + AgCl$

3 $Cl_2 + 2NaBr \longrightarrow 2NaCl + Br_2$

4 $SO_2 + H_2O \longrightarrow H_2SO_3$

Answers

1 Combination 2 Double replacement 3 Single replacement 4 Combination

33

Description

A **solution** is a homogeneous mixture consisting of a substance (solute) dissolved and uniformly distributed in another (solvent). A common example is a **saline** (salt) **solution**. Water is the **solvent** that dissolves sodium chloride (NaCl), the **solute**.

| Formula: | SOLVENT | + | SOLUTE | = | SOLUTION |

Ex.: **Water** H_2O + **NaCl** (sodium chloride) = **Saline** (salt) **solution**

Liquid Solutions

Water serves as the solvent for most **liquid solutions**. These solutions are called aqueous. Solutes can be gases (*ex.:* CO_2), solids (*ex.:* NaCl), and/or liquids (*ex.:* acetic acid) dissolved in water. Wine, vinegar, and seawater are all examples of liquid solutions.

Intravenous (I.V.) solution

Solvent = water
Solutes = NaCl (solid), KCl (solid), and other salts

Coffee solution

Solvent = water
Solutes = caffeine (solid) and other substances from coffee beans

Beer solution

Solvent = water
Solutes = alcohol (liquid), CO_2 (gas), and malt (solid)

Gas and Solid Solutions

In additon to liquid solutions, there are also **gas solutions** (the solvent and the solute are both gases) and **solid solutions** (the solvent and the solute are both solids). The solvent is the substance present in a greater amount and surrounds the solute.

Gas solution

The **air** in the atmosphere; more nitrogen gas is present than oxygen gas.

Solvent = nitrogen (N_2)
Solute = oxygen (O_2)

Solid solution

Steel pipes: more iron (Fe) than carbon (C) is present.

Solvent = iron (Fe)
Solute = carbon (C)

Dental amalgam: more silver than mercury is present.

Solvent = silver (Ag)
Solute = mercury (Hg)

Water as a Solvent

Water is a common solvent. Each water molecule (H₂O) contains two hydrogens (H) covalently bonded to one oxygen (O) atom. Like a magnet, the hydrogen ends of each O–H bond possess a partially positive charge ($\delta+$) and the oxygen end has a partially negative charge ($\delta-$). This partial charge makes these bonds polar. The slightly positive H of one molecule is attracted to the slightly negative O of another molecule to form a hydrogen bond. Because of water's shape and its polar O–H bonds, water is a polar molecule.

The chemical rule for dissolving is: "Like dissolves like." Water is polar, so it dissolves other polar compounds but not nonpolar compounds such as corn oil. When crystals of NaCl are added to water, the water molecules are attracted to their charge. The oxygen ends are attracted to the sodium ions (Na⁺), and the hydrogen ends are attracted to the chloride ions (Cl⁻). The final result is that the ionic bond between NaCl salt is broken, the ions are separated, and water molecules surround each ion. This action (solvation) is what you observe as you add sodium chloride crystals to water, and they seem to "disappear" as you stir them into solution.

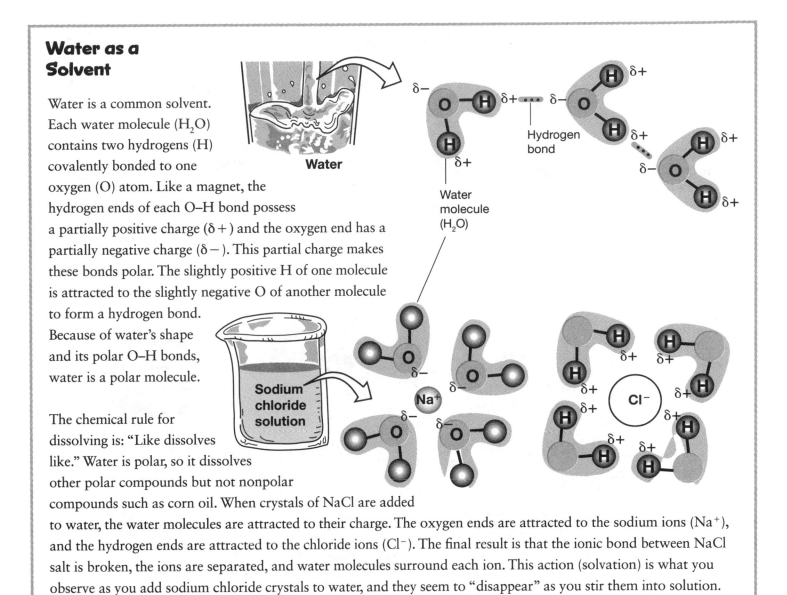

Water

Water molecule (H₂O)

Hydrogen bond

Sodium chloride solution

Na⁺

Cl⁻

Practice Problems

You make a sugar solution to put in your hummingbird feeder. Answer the following:

1 What is the **solvent** in your sugar solution?

2 What is the **solute** in your sugar solution?

3 What type of solution did you make? Explain.

Answers

1 Water 2 Sugar 3 It's an aqueous solution, a type of liquid solution, because the solid solute (sugar) was dissolved in a liquid solvent (water).

COLLOIDS

Definition and Properties

Large solute (1–1,000 nm)

Dispersing agent

Definition:
A **colloid** is a heterogeneous mixture containing either large solute particles or solid, liquid, or gas aggregates scattered within a dispersing agent such as a gas, a liquid, or a solid.

Properties of a colloid:

• **Particle size**—solute particle is large, 1 to 1,000 nm

• **Turbidity**—scatters light so it appears very turbid, or cloudy

• **Settling**—solute particles do NOT settle

• **Separation**—solute particles CANNOT be separated by filters

General Examples of Colloids

Colloids can be found in the kitchen and other common places. It's important to note that they can involve substances in different states such as a gas in a liquid or a gas in a solid.

Whipped cream

Gas in a liquid

(Nitrous oxide [N_2O] gas is dispersed in liquid cream.)

Marshmallows

Gas in a solid

(Air is dispersed in a solid, gelatinous sugar mixture.)

Milk

Liquid in a liquid

(Lipid globules are dispersed in water.)

Biological Examples

The **cytosol** is the liquid portion of the cell (excluding the organelles), and it is both a solution and a colloid. It's a solution because it contains many different ions dissolved in water. But it also contains large organic substances like proteins that are dispersed in the water, which makes it a colloid.

Body cell

Plasma

Red blood cells

Tube of centrifuged blood

The **plasma** is an example of a colloid. When whole blood is spun in a centrifuge it separates into its liquid portion—the plasma—and its solid portion—mostly red blood cells. Various proteins are scattered in water, which gives the plasma a pale yellow color.

SUSPENSIONS

Definition and Properties

Definition:
A **suspension** is a mixture containing very large solute particles evenly distributed in a dispersing agent by mechanical means that will eventually settle out.

Very large solute (> 1,000 nm)

Dispersing agent

Properties of a suspension:
• **Particle size**—solute particle is very large, > 1,000 nm
• **Turbidity**—scatters light so it appears very turbid, or cloudy
• **Settling**—solute particles settle quickly
• **Separation**—solute particles can be separated by filters

General Examples of Suspensions

Flour in water
(flour suspended in water)

Clouds in air
(ice crystals and dust suspended in air)

Mud
(soil suspended in water)

Biological Examples

Cytoplasm is composed of the liquid portion inside the cell called the cytosol and all the very large organelles, such as the mitochondrion. These organelles are suspended in the cytosol, making the cytoplasm a suspension.

Whole blood is a suspension because red and white blood cells are suspended in the liquid portion of the blood called the plasma. Cell fragments called platelets are another large solute in this suspension.

Mitochondrion

Body cell

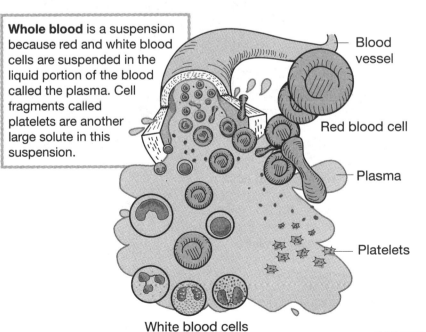

Blood vessel

Red blood cell

Plasma

Platelets

White blood cells

ACIDS

Definition and Common Substances Containing Acids

Soda pop

Oranges

Coffee

Definition:

Acids are substances that donate hydrogen ions and/or ionize in water to produce hydrogen ions (H^+).

ACIDS
(H^+ donors) → H^+

Soda pop, oranges, and coffee all contain acids.

Hydrogen Ion (H^+) = Proton

Electron

Proton

Hydrogen atom

Hydrogen ion (H^+) or a proton

A hydrogen atom in an acid contains one proton and one electron. When the hydrogen loses its electron to become a hydrogen ion (H^+), all that remains is one proton. Therefore, H^+ = **one proton.** Many references use these terms interchangeably.

Chemical Equations for Acids

HYDROCHLORIC ACID

$$HCl \xrightarrow{H_2O} H^+ + Cl^-$$

Hydrogen chloride (HCl) is a gas, but it ionizes into hydrogen ions (H^+) and chloride ions (Cl^-) when it is mixed with water.

$$H^+ + H_2O \longrightarrow H_3O^+$$

The hydrogen ions do not remain as H^+; instead, they react with water molecules to form **hydronium ions (H_3O^+).**

$$HCl + H_2O \longrightarrow H_3O^+ + Cl^-$$

In summary, when hydrogen chloride (HCl) reacts with water (H_2O), it completely ionizes into hydronium ions (H_3O^+) and chloride ions (Cl^-).

Hydrochloric acid (HCl) is an example of an acid. It is produced by cells in your stomach and is a component of your gastric juice.

BASES

Definition and Common Substances Containing Bases

Oven cleaner

Antacid tablets

Surface cleaner (with ammonia)

Definition:

Bases are substances that accept hydrogen ions and/or ionize in water to produce hydroxide ions (OH^-).

H^+ → **BASES** (H^+ acceptors)

Oven cleaner, antacid tablets, and any surface cleaner containing ammonia all contain bases.

Chemical Equations for Bases

SODIUM HYDROXIDE

Sodium hydroxide (NaOH) is an example of a base that is a metal hydroxide.

$$NaOH \xrightarrow{H_2O} Na^+ + \boxed{OH^-}$$

Sodium hydroxide (NaOH) is a solid, but it ionizes to form sodium ions and hydroxide ions (OH^-) when it is mixed with water. The H_2O above the arrow indicates that water is involved in the ionization process but does not serve as a chemical reactant.

Many bases are metal hydroxides, such as NaOH, KOH, and $Mg(OH)_2$, but some are not.

Let's look at an example of another base that is not a metal hydroxide—namely, **ammonia (NH_3)**.

AMMONIA

Ammonia (NH_3) is an example of a base that is *not* a metal hydroxide.

$$NH_3 + H_2O \underset{\longrightarrow}{\longleftarrow} NH_4^+ + \boxed{OH^-}$$

Ammonia (NH_3) is a gas that forms ammonium ions (NH_4^+) and hydroxide ions (OH^-) when it reacts with liquid water. NH_3 accepts H^+ from a water molecule to form NH_4^+. This leaves behind an OH^- in the process.

This perfectly fits the definition of a base as being both a substance that accepts hydrogen ions (H^+) and produces hydroxide ions (OH^-).

Description

The pH scale stands for "potential hydrogen" scale because it measures the molar concentration of hydrogen ions (H^+) in solution. Brackets [ ] are used to indicate molar concentration, so the phrase "molar concentration of hydrogen ions in solution" can be abbreviated as [H^+]. But hydrogen ions immediately bind to water molecules to form hydronium ions (H_3O^+), so pH is really measuring [H_3O^+] in solution. To avoid confusion, because a hydrogen ion is a proton, it's important to note that all three of these are equivalent when discussing hydrogen ions in solution:

$$\text{hydrogen ion } (H^+) = \text{proton} = \text{hydronium ion } (H_3O^+)$$

The pH values identify solutions as being acidic, basic (alkaline), or neutral.

pH Scale Facts

Range:	0–14
Acid:	< 7.0
Base:	> 7.0
Neutral point:	7.0; this is where [H_3O^+] = [OH^-].
Increments:	Each number represents a $10\times$ (tenfold) change in [H_3O^+]
Units:	[H_3O^+] per liter of solution

1 pH is mathematically calculated as: $\mathbf{pH = -log\,[\,H_3O^+]}$.

2 CpH is a logarithmic scale like the Richter scale used to measure the intensity of earthquakes. An earthquake at 6 is 1,000 times more intense than an earthquake measured at 3.

3 Changes in pH reflect exponential changes in the [H_3O^+] of the solution. Each unit on the scale represents a $10\times$ (tenfold) change in [H_3O^+], so changes occur as 10^x, where x = the number of units changed on the scale.

For example, a solution with a pH of 5 is moderately acidic, like coffee. If we add acid and the pH decreases from 5 to 3, how much of a change is it? Well, $5 - 3 = 2$, so the solution changes 2 pH units. Therefore, $10^2 = 10 \times 10 = 100$, so there is a $100\times$ change in [H_3O^+]. But is it an increase or a decrease in [H_3O^+]? If the pH value decreases, the solution is more acidic. Likewise, if the pH value of a solution increases, it is becoming more basic and therefore less acidic.

Now, if you add a base to your pH = 3 solution, and the pH increases to 7, you are decreasing the [H_3O^+] by 4 pH units. Similarly, $3 - 7 = -4$, so $10^{-4} = 10^{-1} \times 10^{-1} \times 10^{-1} \times 10^{-1} = 1/10,000$. At a pH of 7, the solution has 1/10,000 the [H_3O^+] as it did at pH = 3.

So if you are hired as a lab technician and are asked to make a solution with a pH of 5, but you are a little sloppy and make a solution with a pH of 6 instead, you can see that you have made a big error. Your solution has a [H_3O^+] 10 times lower than it is supposed to be. Let's hope you don't get fired!

The **pH scale** showing the pH values of some common substances

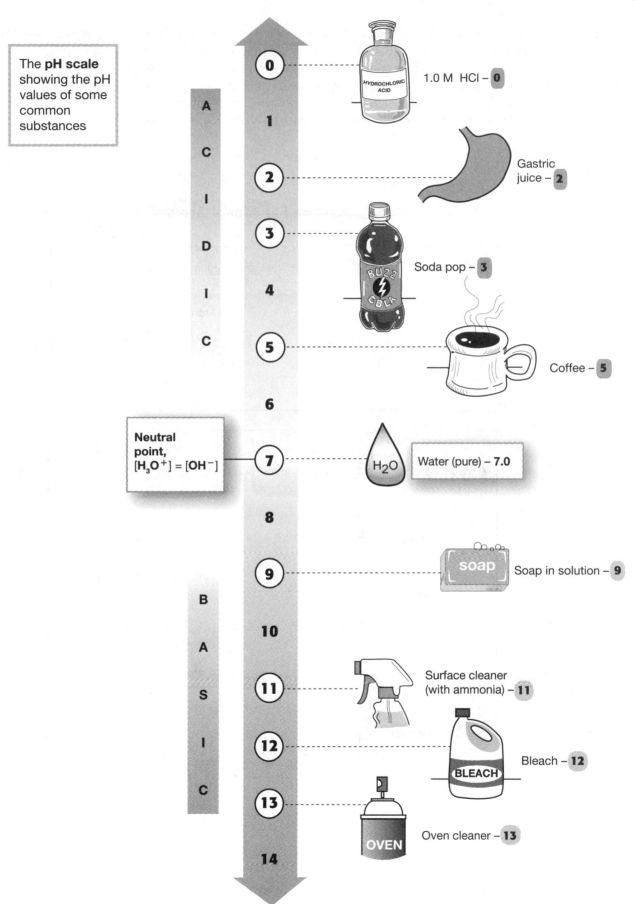

A
C
I
D
I
C

Neutral point, $[H_3O^+] = [OH^-]$

B
A
S
I
C

0 — 1.0 M HCl – **0**

2 — Gastric juice – **2**

3 — Soda pop – **3**

5 — Coffee – **5**

7 — Water (pure) – **7.0**

9 — Soap in solution – **9**

11 — Surface cleaner (with ammonia) – **11**

12 — Bleach – **12**

13 — Oven cleaner – **13**

Description
Organic compounds contain the element carbon (C). They typically also contain hydrogen (H) and also may contain other elements such as oxygen (O) and nitrogen (N). Within all organic compounds are groups of atoms called **functional groups** that determine many of the chemical and physical properties of that compound. Functional groups allow us to predict chemical behaviors and thus serve as a basis for classification of the different types of organic compounds. We will examine the following four common types of functional groups: hydroxyl, carbonyl, carboxyl, and amino groups.

Four Common Functional Groups

Hydroxyl group —OH   oxygen bonded to hydrogen

Ex.:

Alcohol   — C — OH

Hydroxyl group poem

*Do not stumble,
Do not fall,
Hydroxyls are found
In alcohols.*

Carbonyl group   carbon double bonded to oxygen

Ex.:

Aldehyde

Ketone

Amide

Carbonyl group poem

*Carbonyls are common,
They fail to hide,
They're in aldehydes, ketones,
And amides.*

Carboxyl group
(*carbonyl* group +
hydro*xyl* group)   a carbonyl group bonded
to a hydroxyl group

Ex.:

Carboxylic
acid

Carboxyl group poem

*Take one carbonyl,
Plus one hydroxyl group,
Combine 'em together,
Carboxyl soup!*

Practice Problems

1 What type of functional group is found in a **ketone**?

2 What type of functional group is found in an **alcohol**?

3 Identify the two different types of functional groups found in an **amide**?

4 What type of functional group is found in a **carboxylic acid**?

Answers

1 carbonyl group   2 hydroxyl group   3 carbonyl group and amino group   4 carboxyl group

CHEMISTRY: Organic Chemistry

Polymers: Synthetic and Natural

Description Polymers are large, long-chain molecules formed by linking building blocks called **monomers**. Think of a children's necklace formed by pushing plastic beads together. In this analogy, the beads are the monomers, and the necklace is the polymer. Some polymers are made from the same monomer repeated again and again. This is indicated here with beads having the same color. Other polymers are made from more than one type of monomer, as indicated with beads of different colors. Synthetic polymers, derived from alkenes, are the foundation of the plastics industry. In contrast, our bodies have the ability to make natural polymers, and they also are found in many of the foods we ingest. Let's examine both synthetic and natural polymers.

Analogy

Natural Polymers

MONOMER → (only one kind of monomer in polymer)

MONOMER

POLYMER

COMMON SOURCES

Cellulose
- straight-chain polymer that does **not** form coils
- found in plants

Glucose

Amylose (type of starch)
- straight-chain polymer that forms coils
- produced and stored in plants

Glycogen
- polymer with **branches** coming off main chain
- produced and stored in animal tissues

- Wood
- Cotton
- Paper

- Baked potato
- Rice
- Corn

Stored in:
- Liver
- Muscle

(different monomers in polymer)

Amino acids

Protein
- chain of amino acids

- Poultry
- Fish
- Beans

Nucleotide
(see p. 64)

twist ⇌ untwist

Nucleic acid (DNA)
- chains of nucleotides

- Mushrooms
- Spinach
- Nuts
(found in most foods)

Description

Recall that **polymers** are large, long-chain molecules formed by linking building blocks called **monomers**. **Polypeptides** are natural polymers formed from **amino acid** monomers. It's like how a children's necklace is formed by pushing plastic beads together. In this analogy, the beads are the amino acid monomers, and the necklace is the polypeptide polymer.

Analogy 1

Overview: Building the Chain

When one amino acid links to another it forms a C-N covalent bond called a **peptide bond**. Two amino acids linked together are called a **dipeptide**, and three linked together are called a **tripeptide**. As the chain grows in length, it becomes a **polypeptide**. A chain with 50 or more amino acids *and* biological function is called a **protein**. (*Note:* There is no clear consensus on the exact number of amino acids to define a protein).

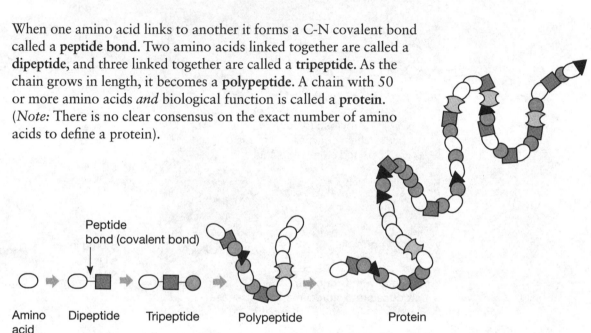

Analogy 2

Think of the 20 different amino acids as an alphabet with 20 letters. Just as different combinations of letters are used to make hundreds of thousands of different words, various amino acids are linked in different sequences to form different peptides and proteins.

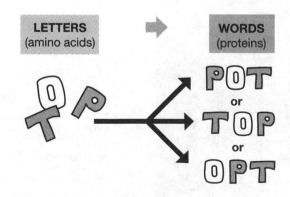

The same three letters—O, T, and P—can be used to form three different words: POT or TOP or OPT. Similarly, different sequences of amino acids form different peptides and proteins.

Body cell

Forming the Peptide Bond

The process of creating proteins from amino acids occurs in body cells and is called **protein synthesis**. As a preview, let's look at the formation of a peptide bond between two amino acids. Using the same icons on the facing page, imagine we are forming a dipeptide. When one amino acid links to another, it forms a dipeptide, and one water molecule is produced as a product:

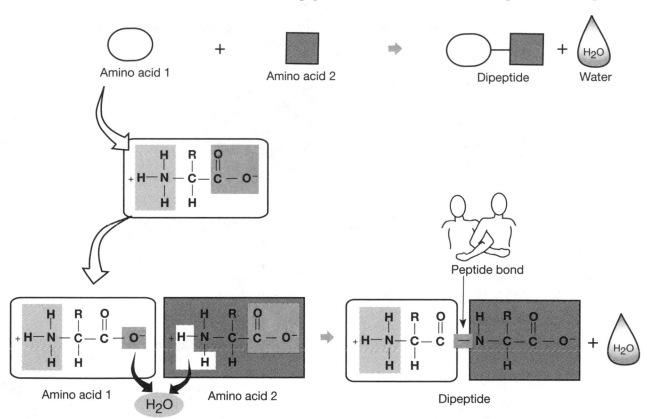

The oxygen is removed from amino acid 1, which combines with two hydrogens from amino acid 2 to form H_2O. The carbon and nitrogen form a peptide bond.

Notice that the peptide bond is a covalent bond between the carbon (C) of amino acid 1 and the nitrogen (N) of amino acid 2. Because all covalent bonds are a shared pair of electrons, it is like two people locked arm in arm.

To form a tripeptide and then a polypeptide, the process simply repeats itself.

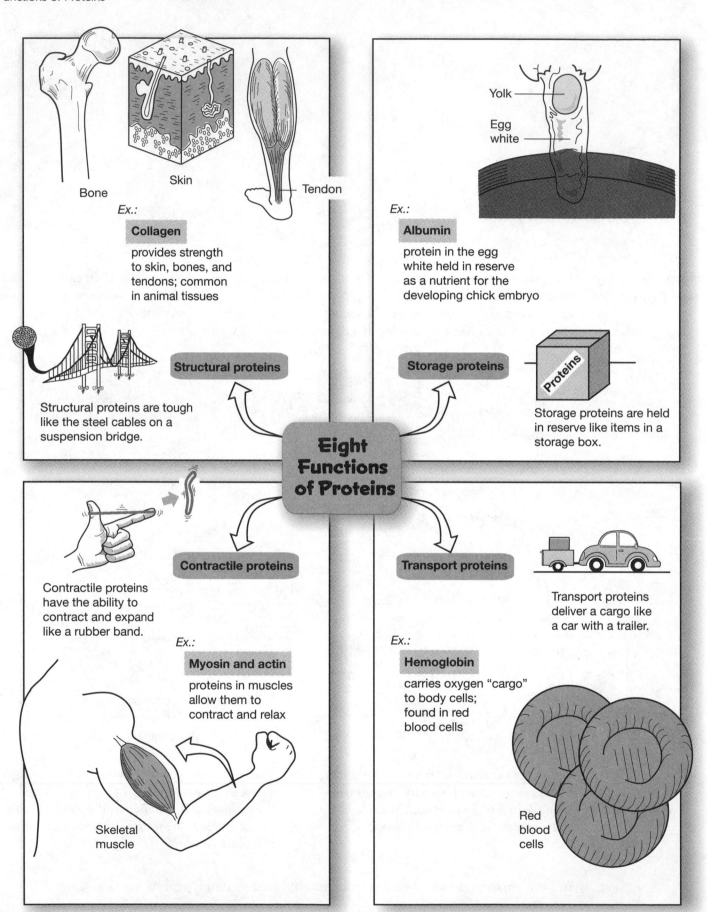

Bone

Skin

Tendon

Ex.:

Collagen

provides strength
to skin, bones, and
tendons; common
in animal tissues

Structural proteins

Structural proteins are tough
like the steel cables on a
suspension bridge.

Yolk

Egg
white

Ex.:

Albumin

protein in the egg
white held in reserve
as a nutrient for the
developing chick embryo

Storage proteins

Proteins

Storage proteins are held
in reserve like items in a
storage box.

**Eight
Functions
of Proteins**

Contractile proteins

Contractile proteins
have the ability to
contract and expand
like a rubber band.

Ex.:

Myosin and actin

proteins in muscles
allow them to
contract and relax

Skeletal
muscle

Transport proteins

Transport proteins
deliver a cargo like
a car with a trailer.

Ex.:

Hemoglobin

carries oxygen "cargo"
to body cells;
found in red
blood cells

Red
blood
cells

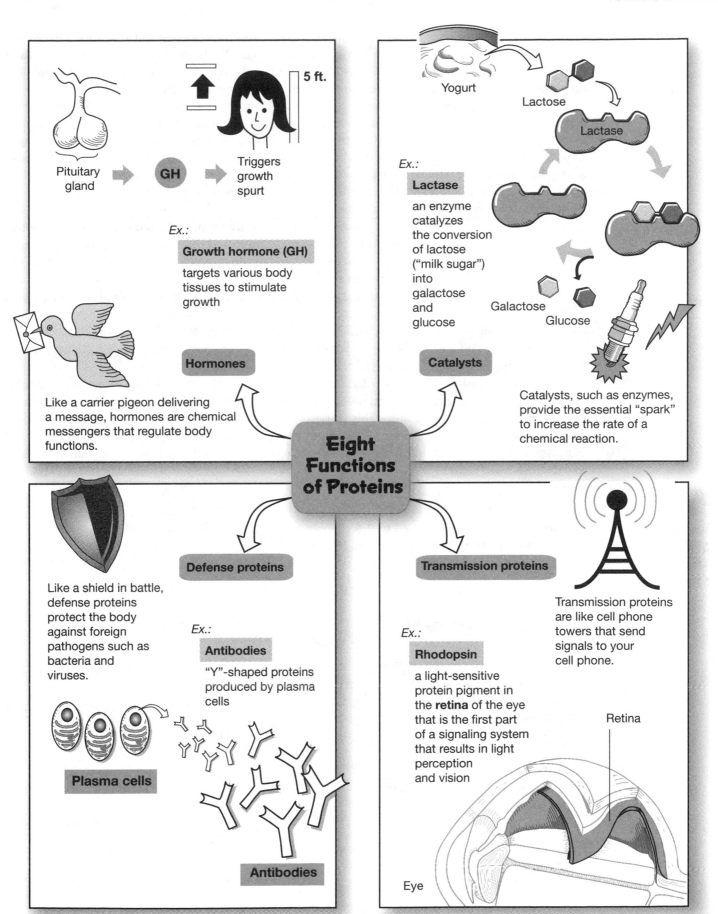

Pituitary gland

GH

Triggers growth spurt

5 ft.

Ex.:

Growth hormone (GH)

targets various body tissues to stimulate growth

Hormones

Like a carrier pigeon delivering a message, hormones are chemical messengers that regulate body functions.

Yogurt

Lactose

Lactase

Ex.:

Lactase

an enzyme catalyzes the conversion of lactose ("milk sugar") into galactose and glucose

Galactose

Glucose

Catalysts

Catalysts, such as enzymes, provide the essential "spark" to increase the rate of a chemical reaction.

Eight Functions of Proteins

Defense proteins

Like a shield in battle, defense proteins protect the body against foreign pathogens such as bacteria and viruses.

Ex.:

Antibodies

"Y"-shaped proteins produced by plasma cells

Plasma cells

Antibodies

Transmission proteins

Transmission proteins are like cell phone towers that send signals to your cell phone.

Ex.:

Rhodopsin

a light-sensitive protein pigment in the **retina** of the eye that is the first part of a signaling system that results in light perception and vision

Retina

Eye

Description

Enzymes are proteins that act as **catalysts** to increase the rate of chemical reactions. Astonishingly, some raise the reaction rate by as much as 1 million times! Without enzymes, the chemical reactions needed to sustain life would not be possible. Consequently, they are found throughout your body in such places as the plasma, the digestive tract, and inside your body cells.

The three-dimensional shape of an enzyme is vital to its function. Every enzyme has a cleft or depression called the **active site**. This is where it binds the **substrate**, the substance upon which the enzyme acts. Every enzyme has a shape that specifically fits with only one kind of substrate. Digestive enzymes are common, so they are a good example to use. Every time you eat food, enzyme-producing cells either lining your digestive tract or within glands secrete many different digestive enzymes to assist in breaking down food into smaller substances that can be used by body cells.

Lactose is a disaccharide, or "double sugar," that breaks down into its two component sugars, galactose and glucose. Imagine that you just ate some yogurt for breakfast. It contains many nutrients, but let's focus only on the lactose. After the yogurt is swallowed, it moves down your esophagus, into your stomach, then into your small intestine. Enzyme-producing cells lining the small intestine secrete the enzyme **lactase**. As lactase mixes with the food inside the intestine, the lactase comes in contact with the lactose, and the two bind together. This is called the **enzyme-substrate complex**. Next, a type of chemical reaction called a **hydrolysis** reaction occurs in which a water molecule helps to break the covalent bond between the two sugars. The two sugars, glucose and galactose, are then released from the enzyme, absorbed into the blood, and delivered to body cells. Once inside body cells, they can either be used as a fuel immediately or stored for later use.

Have you heard of lactose intolerance? This is the condition in which a person is unable to digest lactose. As you might suspect, the cause of this disorder is a deficiency in the production of the enzyme lactase. To compensate for this problem, lactase can be purchased from stores and added to dairy products prior to ingestion. Some dairy products such as milk claim to be "lactose-free." This indicates that milk already has been treated with lactase and that the lactose already has been digested to galactose and glucose.

Two Models for Enzyme– Substrate Interaction

Two models commonly are used to describe how the substrate binds to the enzyme:
(1) **lock and key model**, and (2) **induced fit model**.

Lock and key model

The substrate fits in the active site on the enzyme like a lock and key.

Induced fit model

Like a baseball glove changes its shape to catch a baseball, the active site changes its shape to bind the substrate.

Galactose

Glucose

Substrate (S)
(lactose)

yogurt

Active
site

1 **Lactose** is the sugar
found in dairy products
such as milk and yogurt.
We break it down in our
small intestine into its
two component sugars—
galactose and glucose.

2 **Enzyme (E)**
(lactase)

The enzyme lactase
is made by cells in
your small intestine
and digests *only* the
substrate lactose.

ENZYME FUNCTION
(*Ex.:* digesting the
sugar lactose
with the enzyme
lactase)

H_2O

Covalent
bond

5 **Free enzyme**

The lactase enzyme
is now free to bind
another lactose molecule
and repeat the cycle.

Galactose   Glucose

3 **Enzyme-substrate (E-S)
complex**

Inside the small intestine,
lactose binds to the active
site on lactase. Then, with
the help of a water molecule,
the covalent bond shown is
broken, forming the products
glucose and galactose.

4 **Products formed**

The products galactose and
glucose are released from
the enzyme. These sugars
are small enough to be
absorbed into the blood
and delivered to body cells
where they can be used as
fuel or stored.

CHEMISTRY: Carbohydrates

Description

When runners talk about "carb loading" or advertisers talk about a "low-carb diet," they are referring to carbohydrates. In simplest terms, carbohydrates are "sugars," or polymers made from sugars, that often have names ending in the suffix *–ose*, such as glucose. They are composed of the elements carbon, hydrogen, and oxygen. From simplest to most complex, carbohydrates are divided into the following three major types:

- **Monosaccharides** (single sugars)
  Ex.: glucose (blood sugar)
- **Disaccharides** (double sugars)
  Ex.: maltose (grain sugar)
- **Polysaccharides** (many sugars)
  Ex.: amylose (plant starch)

The hexagon is used as a symbol to represent a monosaccharide.

Carbohydrates have several functions vital to all living organisms. As an overview, let's put them into three functional categories: energy, stored energy, and structural components.

Major Functional Types

These three functional categories are presented visually on the facing page.

① **Energy:** Like gasoline for your car, monosaccharides are used as energy for body cells.

Ex.:

- **Glucose** ("blood sugar"): The oxidation of the monosaccharide glucose within body cells produces carbon dioxide and water. During this process, free energy is released that can be used to power cellular work such as muscle contraction.

② **Structural components:** Some serve as markers on cell surfaces; others are key parts of molecular structures.

Ex.:

- **Cellulose:** This polysaccharide is found in plants and is composed of long, straight chains of glucose molecules. It makes the cell walls around plant cells rigid.

- **DNA:** The backbone of the DNA double helix is made of deoxyribose sugars and phosphate groups.

- **Antigens for ABO blood groups:** Your blood type (A, B, AB, or O) is determined by proteins on the surface of red blood cells. Attached to these proteins are short-chain carbohydrates.

③ **Stored energy:** Carbohydrates are held in reserve like items in a storage box.

Ex.:

- **Glycogen:** This polysaccharide is found in animal tissues (such as liver and muscle cells) and is composed of long chains of glucose molecules. When we ingest excess glucose not immediately used for energy, glycogen is produced.

- **Amylose** (type of starch): This polysaccharide is found in plants and is composed of a long chain of glucose molecules that functions as an important energy storage molecule.

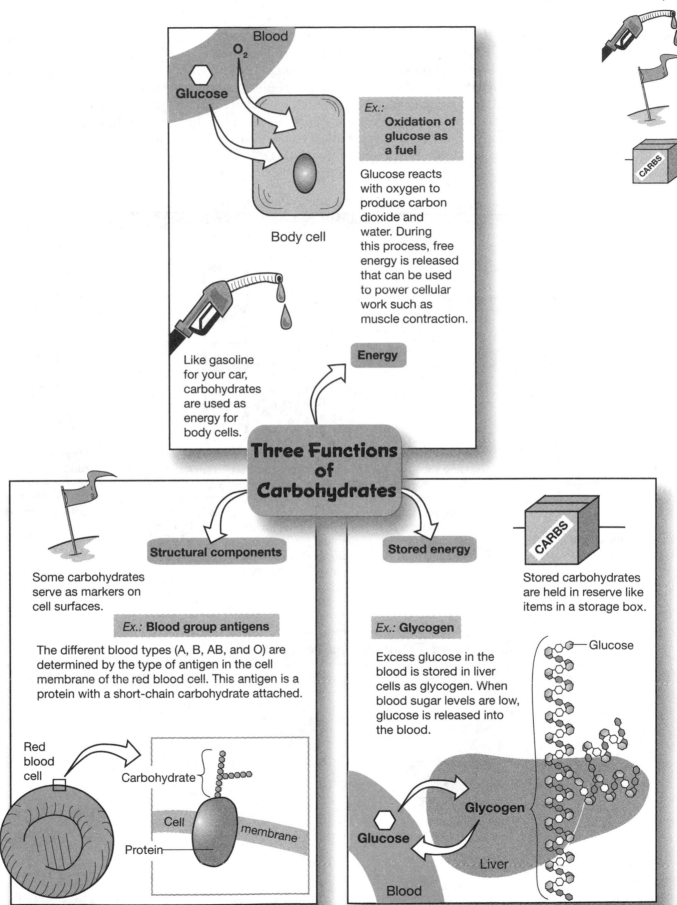

Three Functions of Carbohydrates

Like gasoline for your car, carbohydrates are used as energy for body cells.

Energy

Ex.:
Oxidation of glucose as a fuel

Glucose reacts with oxygen to produce carbon dioxide and water. During this process, free energy is released that can be used to power cellular work such as muscle contraction.

Blood

O₂

Glucose

Body cell

Structural components

Some carbohydrates serve as markers on cell surfaces.

Ex.: **Blood group antigens**

The different blood types (A, B, AB, and O) are determined by the type of antigen in the cell membrane of the red blood cell. This antigen is a protein with a short-chain carbohydrate attached.

Red blood cell

Carbohydrate

Cell membrane

Protein

Stored energy

Stored carbohydrates are held in reserve like items in a storage box.

CARBS

Ex.: **Glycogen**

Excess glucose in the blood is stored in liver cells as glycogen. When blood sugar levels are low, glucose is released into the blood.

Glucose

Glycogen

Glucose

Liver

Blood

Description

Nucleotides are the monomers, or building blocks, for making long-chained nucleic acid polymers such as **DNA** and **RNA**. If you ate a spinach salad for lunch, you ingested DNA from inside the nucleus of the plant cells. In your digestive system, DNA is broken down into nucleotides, which then are transported through the blood to all your cells. Once inside your cells, the free nucleotides float around like raw materials at a construction site. These nucleotides are used to synthesize new DNA in your cells. A DNA nucleotide from a plant or an animal is chemically the same, so the original source doesn't matter. It just reminds you that you really are what you eat! Let's examine in more detail how a DNA nucleotide is used as a building block for making DNA.

① OVERVIEW: Nucleotides as building blocks for DNA

Phosphodiester bond (covalent bond)

twist

untwist

This schematic is used as a symbol to represent a **nucleotide**. It contains three parts:

• **Sugar (S)**
• **Phosphate group (P)**
• **Base (B)**

When two nucleotides bind together, a covalent bond called a **phosphodiester bond** forms between the sugar of one and the phosphate group of another nucleotide. Binding more nucleotides allows long chains to form.

This ladder is a schematic of a section of a DNA molecule. Notice that the sugars and phosphates form the backbone of the ladder, and the bases form the steps or rungs. Imagine this ladder is made of rope, allowing you to twist and untwist it.

By twisting the rope ladder, you form a spiral staircase called the DNA **double helix**.

② Two different schematics for a nucleotide

P

S B

=

Phosphate group

Base

O

Sugar

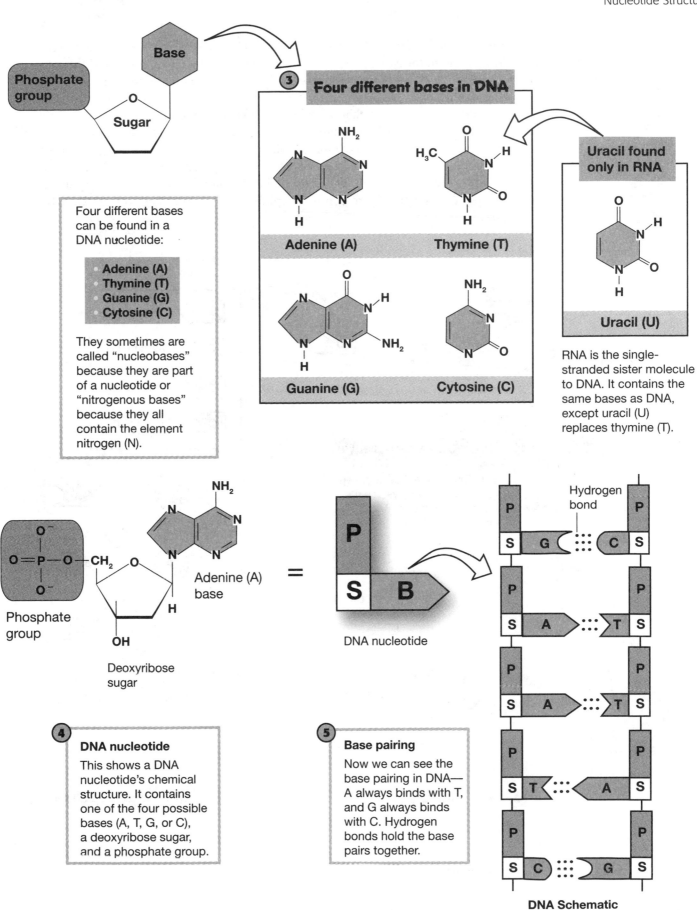

Base

Phosphate group

Sugar

3 **Four different bases in DNA**

Four different bases can be found in a DNA nucleotide:

- **Adenine (A)**
- **Thymine (T)**
- **Guanine (G)**
- **Cytosine (C)**

They sometimes are called "nucleobases" because they are part of a nucleotide or "nitrogenous bases" because they all contain the element nitrogen (N).

Adenine (A)

Thymine (T)

Guanine (G)

Cytosine (C)

Uracil found only in RNA

Uracil (U)

RNA is the single-stranded sister molecule to DNA. It contains the same bases as DNA, except uracil (U) replaces thymine (T).

Phosphate group

Adenine (A) base

Deoxyribose sugar

=

P

S **B**

DNA nucleotide

Hydrogen bond

4 **DNA nucleotide**

This shows a DNA nucleotide's chemical structure. It contains one of the four possible bases (A, T, G, or C), a deoxyribose sugar, and a phosphate group.

5 **Base pairing**

Now we can see the base pairing in DNA—A always binds with T, and G always binds with C. Hydrogen bonds hold the base pairs together.

DNA Schematic

Description

The most common, stable form of DNA—the double helix—resembles a spiral staircase. In terms of its structure, the railing of the staircase is made of a sugar-phosphate backbone, and the steps are made of base pairs. Imagine that you cut the staircase straight down the middle through the steps. Now you have two halves of the spiral staircase. This more accurately reflects the structure of the double helix because the two halves are held together by hydrogen bonds between the base pairs.

DNA is a nucleic acid—a natural polymer made from long chains of nucleotides. It is found in the nucleus of your cells and functions as the master blueprint for making all the different proteins found throughout the body. It is tightly coiled and condensed into structures called chromosomes. Imagine that you took all the DNA in one cell and unraveled it. If you were microscopically small, guess how many steps you would have to take to climb the spiral staircase from end to end? Roughly 3 billion! I'm tired just thinking about it.

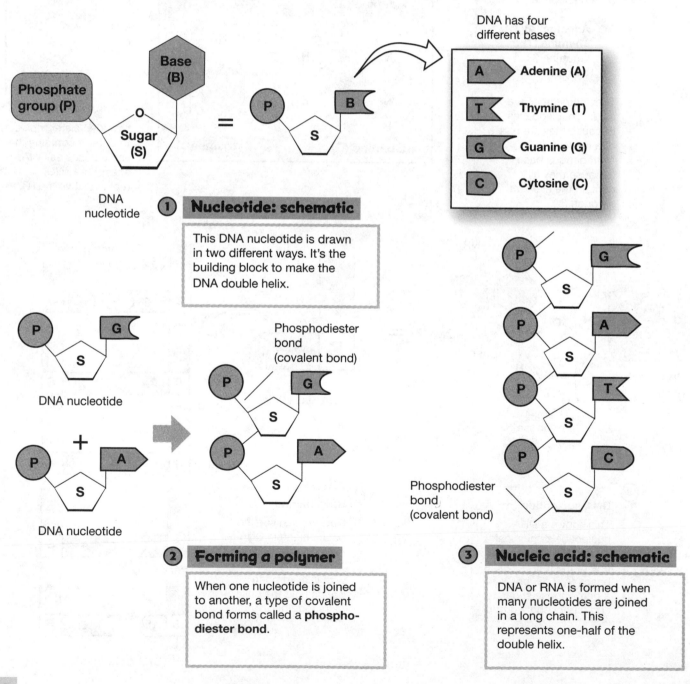

DNA has four different bases

A — Adenine (A)
T — Thymine (T)
G — Guanine (G)
C — Cytosine (C)

① Nucleotide: schematic

This DNA nucleotide is drawn in two different ways. It's the building block to make the DNA double helix.

② Forming a polymer

When one nucleotide is joined to another, a type of covalent bond forms called a **phosphodiester bond**.

③ Nucleic acid: schematic

DNA or RNA is formed when many nucleotides are joined in a long chain. This represents one-half of the double helix.

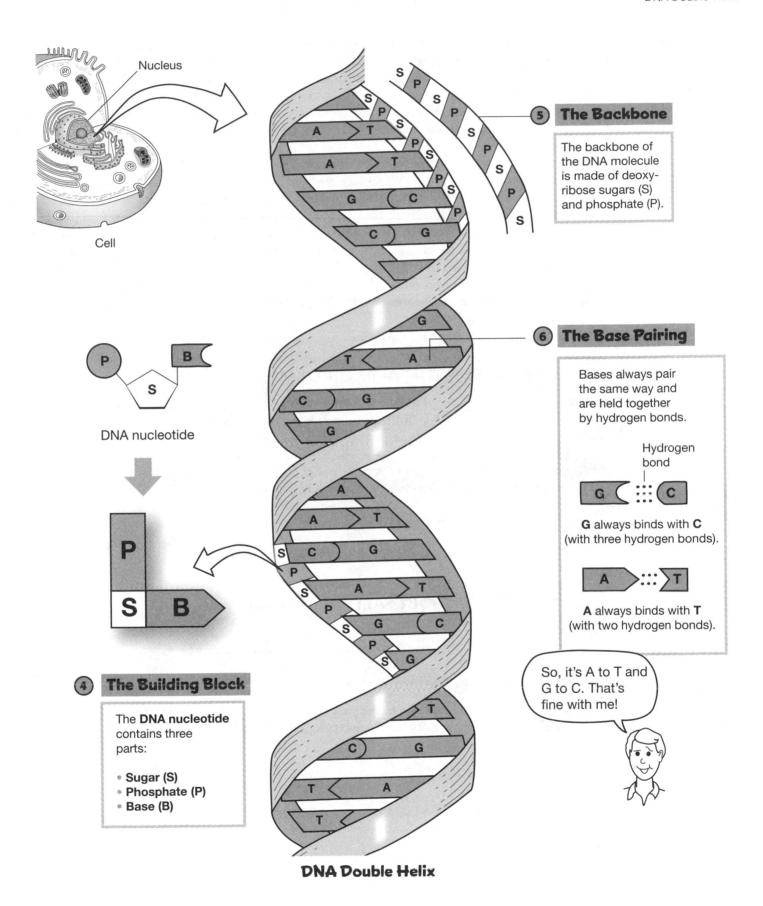

Nucleus

Cell

DNA nucleotide

5 The Backbone

The backbone of the DNA molecule is made of deoxy-ribose sugars (S) and phosphate (P).

6 The Base Pairing

Bases always pair the same way and are held together by hydrogen bonds.

Hydrogen bond

G always binds with **C** (with three hydrogen bonds).

A always binds with **T** (with two hydrogen bonds).

So, it's A to T and G to C. That's fine with me!

4 The Building Block

The **DNA nucleotide** contains three parts:

• **Sugar (S)**
• **Phosphate (P)**
• **Base (B)**

DNA Double Helix

Description

Lipids (*lipos* = fat) consist of a broad category of organic compounds that include waxes, fats and oils, phospholipids, and steroids. They are made of carbon, hydrogen, and oxygen and are not soluble in water. Most lipids, except steroids, contain fatty acid chains. The fatty acid chains are composed of carbon and hydrogen ranging from 10 to 20 carbons in length. The total number of fatty acid chains in a lipid molecule ranges from one to three, depending on the type of lipid. For example, phospholipids contain two fatty acid chains per molecule, and triglycerides (in fats and oils) contain three fatty acid chains per molecule.

Let's consider some practical examples. Shoe polish is a type of wax. The same is true of waxes for your car and your wood furniture. Waxes also are found in candles, soap, and some cosmetics. Fats and oils contain lipids called triglycerides (triacylglycerols). Most fats we consume are derived from animals, and oils are derived from plants. For example, when cooking, you could choose either lard (fat) or corn oil (oil) to "grease" the frying pan. The healthier choice would be corn oil because it contains unsaturated fatty acids, which are less likely to lead to coronary artery disease. Consuming large amounts of cholesterol is often reported to be bad for your health. But what isn't mentioned is that this steroid is made in small amounts by liver cells and is a vital component in your plasma membranes. To promote good health, your multivitamin contains other lipids—the fat-soluble vitamins (A, D, E, and K). Lipids are common in everyday life.

As an overview, let's compare the major classes of lipids with and without fatty acid chains.

LIPIDS WITH FATTY ACID CHAINS

Waxes	Found in commercial products such as shoe polish, furniture polish, car wax, candles, soaps, and cosmetics
Fats and oils (contain triglycerides)	Found in commercial products such as lard, corn oil, olive oil, and canola oil
Phospholipids	Found in plasma membranes

LIPIDS WITHOUT FATTY ACID CHAINS

Steroids	Examples include sex hormones such as progesterone, estrogen, and testosterone

As an introduction to lipids, the facing page illustrates and compares the structure and function of three classes of lipids: triglycerides (triacylglycerols), phospholipids, and steroids.

Comparison of three different classes of lipids: triglycerides, phospholipids, and steroids

LIPIDS

Triglyceride
(triacylglycerol)

Phospholipid

Steroid

Glycerol — Fatty acid
Glycerol — Fatty acid
Glycerol — Fatty acid

Glycerol — Fatty acid
Glycerol — Fatty acid
Alcohol — PO_3^{-1}

Phospholipid molecule

Cholesterol molecule

Triglycerides (triacylglycerols) are stored in fat cells below the skin.

Phospholipids are a major component of cell membranes.

Cholesterol, a steroid, is also a structural component of cell membranes.

Plasma (*cell*) membrane

Cell

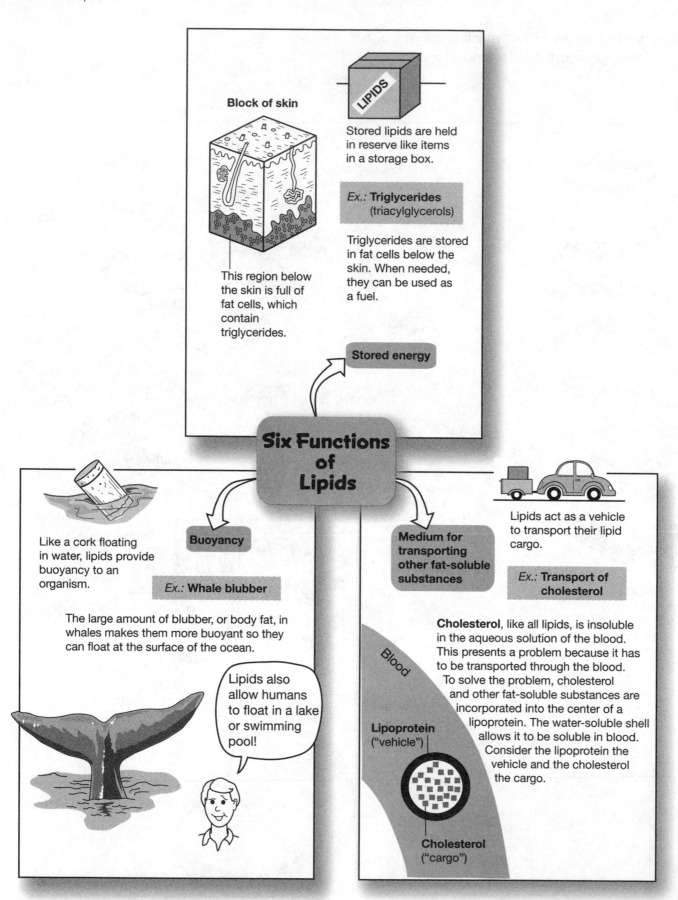

Block of skin

LIPIDS

Stored lipids are held in reserve like items in a storage box.

Ex.: **Triglycerides** (triacylglycerols)

Triglycerides are stored in fat cells below the skin. When needed, they can be used as a fuel.

This region below the skin is full of fat cells, which contain triglycerides.

Stored energy

Six Functions of Lipids

Buoyancy

Like a cork floating in water, lipids provide buoyancy to an organism.

Ex.: **Whale blubber**

The large amount of blubber, or body fat, in whales makes them more buoyant so they can float at the surface of the ocean.

Lipids also allow humans to float in a lake or swimming pool!

Medium for transporting other fat-soluble substances

Lipids act as a vehicle to transport their lipid cargo.

Ex.: **Transport of cholesterol**

Cholesterol, like all lipids, is insoluble in the aqueous solution of the blood. This presents a problem because it has to be transported through the blood. To solve the problem, cholesterol and other fat-soluble substances are incorporated into the center of a lipoprotein. The water-soluble shell allows it to be soluble in blood. Consider the lipoprotein the vehicle and the cholesterol the cargo.

Blood

Lipoprotein ("vehicle")

Cholesterol ("cargo")

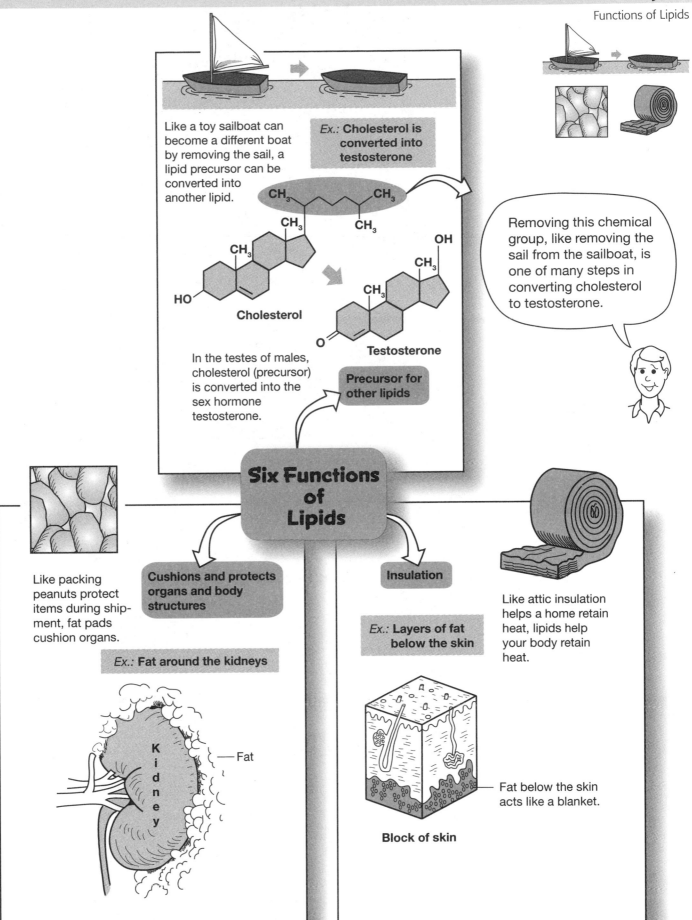

Like a toy sailboat can become a different boat by removing the sail, a lipid precursor can be converted into another lipid.

Ex.: **Cholesterol is converted into testosterone**

Cholesterol

Testosterone

In the testes of males, cholesterol (precursor) is converted into the sex hormone testosterone.

Precursor for other lipids

Removing this chemical group, like removing the sail from the sailboat, is one of many steps in converting cholesterol to testosterone.

Six Functions of Lipids

Cushions and protects organs and body structures

Like packing peanuts protect items during shipment, fat pads cushion organs.

Ex.: **Fat around the kidneys**

Kidney — Fat

Insulation

Like attic insulation helps a home retain heat, lipids help your body retain heat.

Ex.: **Layers of fat below the skin**

Fat below the skin acts like a blanket.

Block of skin

61

Notes

CELLS:
STRUCTURE AND FUNCTION

CELLS: Structure and Function

Description

The **cell** is the basic unit of life, and all cells have many common structures. Each cell is surrounded by a **plasma** (*cell*) **membrane**—the *gatekeeper* that only allows certain substances to cross it. In various epithelial cells, there are two types of possible extensions of the plasma membrane—either shorter **microvilli** used to absorb nutrients or much longer **cilia** used to move materials such as mucus over the cell surface. Beneath the plasma membrane is the **cytoskeleton**—a complex network of protein filaments that provide strength and shape to the cell. Many **organelles** (*small organs*) are scattered within the cell, with the largest one being the **nucleus**. **Membranous organelles** are covered by a membrane, and **nonmembranous organelles** are not. The region between the plasma membrane and the nucleus is called the **cytoplasm**, which is composed of the fluid portion called the **cytosol** and the **organelles**.

NONMEMBRANOUS ORGANELLES

Organelle	Function
Centrosome	Composed of two centrioles that each move to opposite ends of the cell during cell division
Microfilaments	Thin protein fibers commonly found at the periphery of the cell; anchor the cytoskeleton to the plasma membrane
Microtubules	Hollow tubes of protein that can act as tracks along which organelles move
Ribosome	*Protein factory*; composed of two subunits that function in protein synthesis

MEMBRANOUS ORGANELLES

Organelle	Function
Endoplasmic reticulum (ER)—(two types) Smooth ER (has no ribosomes attached) Rough ER (has ribosomes attached)	Smooth ER: lipid and carbohydrate synthesis Rough ER: modification and packaging of newly made proteins
Golgi complex (Golgi body; Golgi apparatus)	*Mailroom*; receives proteins from the rough ER, then delivers them to the cytosol in a vesicle or secretes them
Lysosome	*Digestion center*; vesicles containing digestive enzymes to remove pathogens and broken organelles
Mitochondrion	*Powerhouse*; site of aerobic cellular respiration; produces ATP for the cell
Nucleus • nucleolus (region within nucleus containing DNA and RNA)	*Control center*; contains DNA—the genetic material for making proteins
Peroxisome	*Detox center*; vesicles containing enzymes to break down substances such as hydrogen peroxide, fatty acids, amino acids; detoxifies toxic substances

Key to Illustration

1. Microvilli
2. Plasma (*cell*) membrane
3. Smooth endoplasmic reticulum
4. Nuclear pores
5. Peroxisome
6. Nucleus
7. Nucleoplasm
8. Nucleolus
9. Ribosomes
10. Rough endoplasmic reticulum
11. Microfilament
12. Mitochondrion
13. Lysosome
14. Vacuole
15. Cytoskeleton
16. Golgi complex
17. Cytosol
18. Cilia
19. Microtubule
20. Centrioles

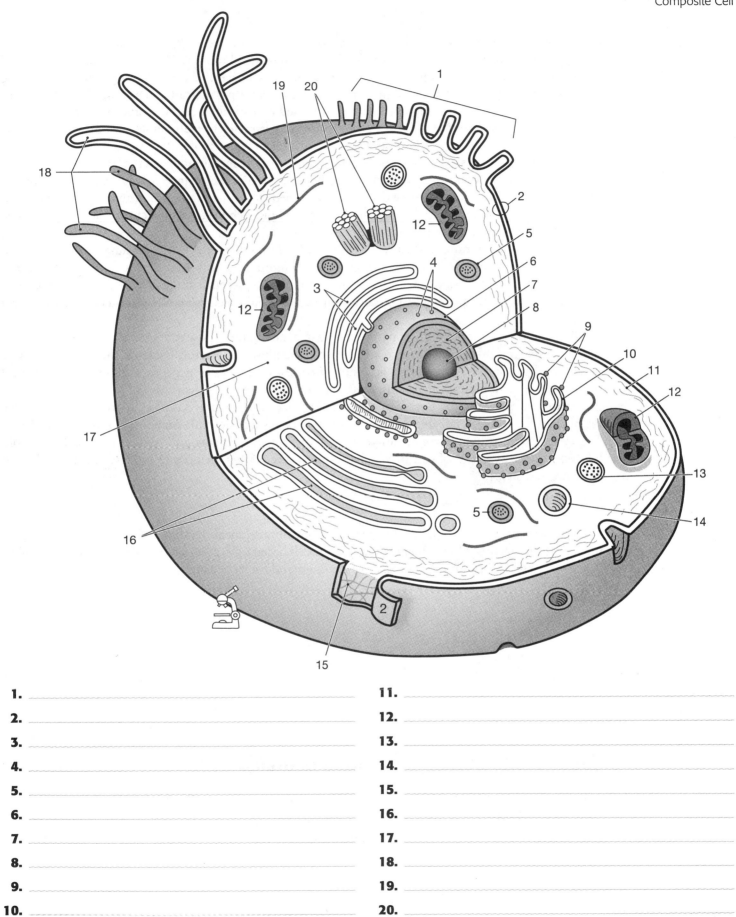

1. _____

2. _____

3. _____

4. _____

5. _____

6. _____

7. _____

8. _____

9. _____

10. _____

11. _____

12. _____

13. _____

14. _____

15. _____

16. _____

17. _____

18. _____

19. _____

20. _____

CELLS: Structure and Function

Plasma (*Cell*) Membrane Structure

Description   Every cell in your body is enclosed by an envelope called a **plasma (*cell*) membrane**. It is the gateway through which substances enter or exit any cell. A plasma membrane is a double layer of lipids referred to as the **phospholipid bilayer** or, more simply, the **lipid bilayer**. The **phospholipid molecule** is the repeating unit within the cell membrane. Each molecule has a **polar head** and a **nonpolar tail** made of two fatty acid chains—typically one chain is saturated and straight, and the other is unsaturated and bent. The polar head is hydrophilic (*water loving*), so it is attracted to water molecules, and the nonpolar tail is hydrophobic (*water fearing*), so it is not attracted to water molecules. As a result, the nonpolar tails position themselves away from water in the middle of the membrane, and the polar heads position themselves toward polar water molecules on the outer and inner surfaces of the plasma membrane. Some phospholipids have a carbohydrate chain attached to them and are called **glycolipids**. Embedded and scattered within the lipid bilayer is another type of lipid—**cholesterol molecules**. They provide some rigidity for the plasma membrane.

The plasma membrane is often referred to as a **fluid mosaic model**—*fluid* because the phospholipid molecules move from side to side due to their bent, unsaturated fatty acid chains, and *mosaic* because it is made of more than one type of macromolecule (lipids, proteins, and carbohydrates). If you were microscopically small and could fall on its surface, it would feel more like a waterbed than a hard wooden floor.

Proteins are another important part of the plasma membrane. Those that span the entire lipid bilayer are called **integral proteins** and serve a variety of functions. Some act as *channels* through which only particular types of ions pass. Others act as *receptors* for hormones or neurotransmitters. Some integral proteins have carbohydrate chains attached to them and are called **glycoproteins**.

Peripheral proteins do not span the bilayer. Instead, they only connect to one surface of the membrane. Some act as enzymes, and others may serve as a structural component of the cytoskeleton. The **cytoskeleton** is composed of different types of protein filaments and is located beneath the lipid bilayer. It serves as a kind of scaffolding that supports the plasma membrane and gives the cell a defined structure and shape.

Function   The plasma membrane is a selectively permeable membrane, meaning that not all substances can pass through it. It controls what can enter and exit the cell based on factors such as size (molecular weight), charge, lipid solubility, and the presence of channels and transporters.

Key to Illustration

Plasma (*Cell*) Membrane
1. Polysaccharide chain
2. Glycoprotein
3. Phospholipid monolayer
4. Cholesterol molecule
5. Filaments of cytoskeleton

6. Peripheral protein
7. Integral protein
8. Glycolipid
9. Phospholipid bilayer

Phospholipid Molecule
10. Headgroup of phospholipid
11. Tailgroup of phospholipid

Cell

Plasma membrane
(OUT)
(IN)

Imagine you were a microscopically small person, and you could flop on a cell membrane. It would feel more like falling onto a water bed than landing on a hard surface like the floor.

(ECF)

(OUT)

(ICF)

Phospholipid molecule

(IN)

1. _____
2. _____
3. _____
4. _____
5. _____
6. _____
7. _____
8. _____
9. _____
10. _____
11. _____

KEY
(ICF) = intracellular fluid
(ECF) = extracellular fluid

CELLS: Structure and Function

Description
The life cycle of a cell, called the **cell cycle**, is divided into two major parts: interphase and mitosis. Interphase is subdivided into G_1, S, and G_2, and mitosis is subdivided into prophase, metaphase, anaphase, and telophase. The duration of the cell cycle varies widely from one cell type to another. Assuming a total cell cycle time of 24 hours, interphase may last roughly 22 hours and mitosis about 2 hours.

Interphase involves lots of activity. First, during G_1 (Gap 1), the longest phase, the cell is growing and synthesizing proteins. Next, in the S phase (synthesis), DNA replication occurs in preparation for cell division, so that each new cell has an identical set of DNA in its chromosomes. The final phase, G_2 (Gap 2) is relatively short in duration. Enzymes and other proteins needed for cell division are produced.

After interphase cells prepare to undergo cell division to form two identical daughter cells. Mitosis specifically refers to the duplication, separation, and reorganization of chromosomes into two identical nuclei. During mitosis a related process called cytokinesis occurs in which the cytoplasm is divided. Cytokinesis can begin in late anaphase, continues through telophase, and ends after mitosis is completed.

A few key events to identify each stage of the cell cycle are listed in the table below:

Phase of Cell Cycle	Key Event
Interphase	Nucleus remains intact; cell growth; DNA replication; protein synthesis
Prophase	Nuclear membrane disappears; duplicated chromosomes called **sister chromatids** become visible; centrioles move to opposite ends of the cell; **spindle fibers** appear; **kinetochore** of each chromatid attaches to a spindle fiber
Metaphase	Sister chromatids align at the middle of the cell
Anaphase	Spindle fibers pull sister chromatids apart and move them to opposite poles of the cell; early cleavage furrow forms
Telophase	Cleavage furrow deepens; cell pinches into two and divides cytoplasm (**cytokinesis**); nuclear membranes re-form

Study Tips
To recall the phases of the cell cycle, use this mnemonic:

IPMAT = "I Passed My Anatomy Test"

Key to Illustration

Stages of Cell Cycle
1. Interphase
2. Prophase
3. Metaphase
4. Anaphase
5. Telophase (*and cytokinesis*)

Final Products
6. Daughter cells

Significant Structures
a. Plasma (*cell*) membrane
b. Nucleolus
c. Nucleus
d. Centriole
e. Sister chromatids
f. Kinetochore
g. Spindle fibers
h. Cleavage furrow
i. Nuclear membrane (still forming)

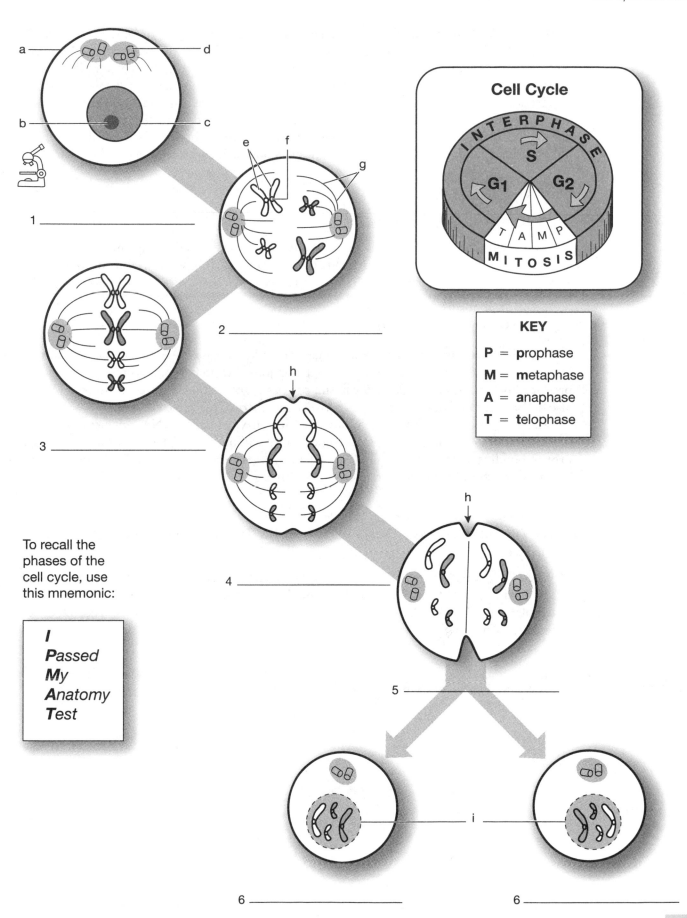

Cell Cycle

INTERPHASE

S

G1   G2

T A M P

MITOSIS

KEY

P = **p**rophase

M = **m**etaphase

A = **a**naphase

T = **t**elophase

a

d

b

c

1 _____

e   f

g

2 _____

3 _____

h

4 _____

5 _____

i

6 _____   6 _____

To recall the phases of the cell cycle, use this mnemonic:

I
Passed
My
Anatomy
Test

Description

Because DNA (deoxyribonucleic acid) is THE genetic material, it has to be replicated in preparation for normal cell division so that a copy of it can be placed in each new daughter cell (see p. 68). This replication occurs during the S phase of the cell cycle. Structurally, DNA is a complex macromolecule located in the cell nucleus in a double helix form resembling an extremely tall spiral staircase.

To discuss its structure, we will simplify it. Imagine that the spiral staircase is flexible so it can be untwisted into a tall ladder. Then cut the ladder down the middle of the rungs so it is in two equal halves. Each half-ladder is bonded to the other by hydrogen bonds. Each half-ladder is created by linking a long chain of building blocks called **nucleotides**. Each nucleotide has three parts:

1. **Sugar** (five-carbon)
2. **Phosphate group**
3. **Base** (containing nitrogen).

The backbone of the ladder is made of alternating sugars and phosphate groups bonded to one another. The rungs on the ladder are made of base pairs. Each base can be one of four different types: adenine (A), thymine (T), guanine (G), or cytosine (C). The base pairing is as follows:

- A bonds to **T** (or **T** to **A**)
- G bonds to **C** (or **C** to **G**)

This base pairing is the basis for the genetic code. This message directs the creation of proteins by specifying the sequence of amino acids to be joined to make that protein. The type of proteins produced determines the type of organism, such as a goldfish, maple tree, or human. These proteins also account for individual differences between members of the same species.

Replication occurs with the help of specialized enzymes called **DNA polymerases**. They attach to one section of DNA, and it unzips. Free nucleotides within the cytoplasmic soup are paired with their proper bonding partners (A to T, G to C) to synthesize a new strand. Each new DNA molecule contains one strand from the parent DNA molecule.

Analogies

- The DNA **double helix** is like a spiral **staircase**. The **DNA molecule** (simplified version) is like a **ladder**.

- When DNA prepares to replicate itself, it is said to **UNZIP**. This is an accurate description. When you **unzip** your coat, each **half of the zipper** is like one-half of the DNA molecule, called the **DNA template**, used to synthesize a new strand. As the new strand is synthesized, DNA **ZIPS** back together again.

Location

DNA is found in the nucleus of all body cells.

Cell

untwist
twist

cut down
the middle

Twisted rope ladder
model of DNA
double helix

Rope ladder
untwisted

Nucleotide

Q: Why replicate DNA?

A: In preparation for cell division.
Each new cell needs to have a
copy of the DNA.

Original DNA strand

Cell

New DNA strand

DNA replicating

Replicated
DNA

Cell divides

Daughter cells
(each new cell has the same DNA)

P

S   B

Nucleotide (detail)

KEY

P = phosphate group
S = sugar (five carbons)
B = nitrogen base
(can be one of four types:
adenine (A), thymine (T),
guanine (G), or cytosine (C))

Base pairing in DNA

A > T

A always binds with T.

G < C

G always binds with C.

Nucleotide

Newly synthesized
DNA strand

DNA "unzips" to
replicate by breaking
the hydrogen bonds
between the base
pairs. Each separated
strand of DNA then
serves as a template
for making a new
strand of DNA.

Color each nucleotide,
sugar, and phosphate
group a different color
to match your textbook.

Description

Protein synthesis is a major function in most cells. It requires that DNA first be transcribed into a similar molecule called messenger RNA, or mRNA. Then the mRNA must be translated from the nucleotide "language" of RNA into the amino acid "language" of proteins.

(**DNA** = master blueprint)   (**mRNA** = copy of master blueprint)   (**protein** = new office building)

Analogy

The process of protein synthesis in a cell is like constructing a new office building. The cytoplasm is like the construction site, and the DNA is like the master blueprint. The master blueprint for the office building is stored in the trailer on the construction site, just as DNA is stored in the nucleus (because it is too large to leave). Just as copies can be made of the original blueprint, a copy of DNA is made in the form of mRNA. The mRNA is small enough to leave the nucleus, enter the cytoplasm, and bind to the ribosomes, which are like the construction workers. The raw materials for the office building are steel, concrete, and glass, to name a few. To build a protein, the needed raw materials are amino acids.

Basic Steps in Protein Synthesis

① **Transcription:** The enzyme RNA polymerase travels along an "unzipped" section of DNA for the purpose of making mRNA. The building blocks for making mRNA are called RNA nucleotides. The enzyme uses one side of the DNA molecule as a template to match the base of the RNA nucleotide to the corresponding base in DNA. As the RNA nucleotides are lined up next to each other, they are covalently bonded to eventually form a short, single-stranded mRNA molecule.

② **RNA Processing:** Like old film being spliced together during editing, the single-stranded mRNA molecule also is processed. Small segments called introns are removed from the molecule.

③ The single-stranded mRNA is small enough to leave the nucleus through the nuclear pore and move into the cytoplasm, where it binds to the small subunit of a ribosome.

④ A transfer RNA (tRNA) molecule has an amino acid binding site section at one end and a three-base section at the other end called the anticodon. Just as a taxi transports passengers, the tRNA transports the proper amino acid to the ribosome. With the help of an enzyme and free energy from ATP hydrolysis, one specific amino acid is bonded to one specific tRNA molecule based on its anticodon sequence.

⑤ The mRNA message is read in groups of three bases. Each triplet group is called a codon. The tRNA with the amino acid called methionine always binds to the beginning of the message called the start codon. Next, the large ribosomal subunit binds to the small ribosomal subunit to activate the ribosome. The large subunit contains two binding sites for tRNAs called the P site and the A site.

⑥ The initial tRNA binds to the P site. One by one, amino acids are positioned next to each other, then bonded together. This results in a growing polypeptide chain. Here is how it works:

⑦ A tRNA with the anticodon matching the mRNA codon arrives at the A site with the next amino acid to be added to the chain. The growing polypeptide chain at the P site is transferred to the tRNA at the A site as it binds to the next amino acid to be added to the chain.

⑧ As the ribosome moves to the right, the tRNA moves from the A site over to the P site. Then the tRNA with the next amino acid to be added to the chain arrives at the A site, and the process repeats itself.

⑨ The process of growing the polypeptide chain ends when a special codon called the stop codon is reached at the A site. The final polypeptide is released from the ribosome along with the last tRNA. Finally, the ribosome separates into its large and small subunits. This marks the end of protein synthesis.

1 direction RNA polymerase moves

RNA polymerase

Nucleus   Cytoplasm

DNA

Intron

2

mRNA

RNA nucleotide
DNA base

RNA base

3

Nuclear pore

Nuclear membrane

Amino acid

Large subunit

P site   A site

Large subunit binds to small subunit

Methionine

tRNA

Anticodor

Codon

mRNA

Start codon   Small subunit

5

Amino acids

Enzyme

ATP

tRNA   **4**   Anticodon

Methionine

Ribosome

A site

mRNA

6

Growing polypeptide chain

Next amino acid to be added to chain

mRNA

7

Large subunit

Polypeptide is released

Stop codon

mRNA

Large ribosomal subunit separates from small subunit

Small subunit

9

P site

mRNA

Ribosome moves right

8

Description

Adenosine triphosphate (ATP) is the universal energy currency for all cells. Look at the illustration of ATP on the facing page to understand its structure. It is classified as a nucleotide and consists of three parts: (1) **adenine base**, (2) **ribose sugar**, and (3) **phosphate groups**—three total. The term "adenosine" refers to the adenine base and ribose sugar bonded together, and the term "triphosphate" comes from the three phosphate groups covalently bonded to each other. There is a net negative charge on each of the phosphate groups, so they repel each other and make ATP relatively unstable.

Cells are highly organized structures constantly doing work to maintain their physical structure and carry out their general functions. This cycle of regular work requires energy. In biological systems, it is common to couple a spontaneous reaction with a nonspontaneous reaction. A metal rusting when exposed to moist air is an example of a spontaneous reaction (it occurs all by itself). A muscle cell attempting to contract is an example of a nonspontaneous reaction (it occurs only with the input of additional energy). ATP hydrolysis occurs as a spontaneous reaction and is represented by the following chemical equation:

$$\text{ATP} + \text{H}_2\text{O} \xrightarrow{\text{ATPase}} \text{ADP} + \text{P}_i \text{ (inorganic phosphate group)} + \textbf{free energy}$$

This reaction requires the assistance of an enzyme called an **ATPase** that binds an ATP molecule to itself with a shape-specific fit like a lock and key. In any hydrolysis ("water-splitting") reaction, a water molecule is used to cleave a single covalent bond. In this case, it cleaves the covalent bond linking the terminal phosphate group to the second phosphate group. During this process, water is split into a **hydroxyl group** (OH$^-$) and a **hydrogen ion** (H$^+$). The hydroxyl group binds to the phosphorus atom (P) in the terminal phosphate group, and the hydrogen binds to the oxygen atom on the second phosphate group (see illustration). As a result, a more stable molecule called **adenosine diphosphate** (**ADP**) is formed as a result of the decreased repulsion between the phosphate groups. The free phosphate group is often transferred to another substrate or to an enzyme. In addition, some free energy is released in the process. The free energy is used to drive processes that must occur in cells, such as muscle contraction, active transport of substances across cell membranes, movement of a sperm cell's tail, manufacturing a hormone, or anything else a cell has to do.

ATP is manufactured inside your cells via the energy derived from foods ingested. During this process, called cellular respiration, ATP is formed by bonding ADP and P$_i$ through a series of **oxidation-reduction** reactions.

Analogies

ATP hydrolysis is also like an **investment in a profitable stock or mutual fund**. Though you must make an initial investment of your own money, you may reap a greater reward (dividend) in the end. Similarly, it takes an initial investment of energy to break the covalent bond in ATP, but the result is a larger amount of free energy when the new bonds in the final products are formed.

Study Tips

Unfortunately, the function of ATP and the process of ATP hydrolysis often are explained incorrectly in many textbooks. Let's correct some common misconceptions:

Misconception 1: **Breaking chemical bonds releases energy.**
Actually, just the opposite is true. During a chemical reaction, the formation of new chemical bonds in a more stable product results in the release of energy.

Misconception 2: **ATP has a special "high-energy phosphate bond" between the second phosphate group and the terminal phosphate group.**
The bond here is actually a covalent bond. This statement gives the false impression that this bond is ready to fly apart like a "jack-in-the-box." In fact, bonds are a force that hold atoms together so an input of energy actually is required to break a chemical bond. Instead, the three phosphate groups in ATP are all linked by relatively strong bonds called covalent bonds that represent shared pairs of electrons. It takes energy to break these bonds.

Adenosine Triphosphate (ATP) Chemical Structure

ATP

ATP is found inside body cells

Phosphate group (P)

Negative charge

Covalent bond to be broken

Triphosphate (three phosphate groups)

NH_2

Adenine base

Ribose sugar

Adenosine (A)

ATP Hydrolysis Equation

| ATP | + | H_2O | $\xrightarrow{\text{ATPase}}$ | ADP | + | P_i | + | free energy |

| Adenosine triphosphate | plus | water | and ATPase enzyme forms | Adenosine diphosphate | plus | inorganic phosphate ion | plus | free energy |

covalent bond to be broken

ATPase

OR

free energy

The free energy can be used for any activity the cell has to do, such as contraction of a muscle cell.

Investment Analogy

$$$

$$$$

ATP hydrolysis is like a good investment. Though it requires some seed money to invest in a good stock, it will return more money in its dividend. Similarly, an initial investment of energy is required to break the covalent bond in ATP. But the amount of energy released upon formation of the new bonds in the products is greater than the initial investment. In summary, like a net profit, there is a net release of energy in ATP hydrolysis.

Description

Any substance able to either enter or leave the cell must do so by passing through the plasma membrane. In general, nutrients such as oxygen have to pass into cells, and waste products such as carbon dioxide have to pass out of them. Because the cell membrane is selectively permeable, it allows some substances to pass through but prevents others from doing so. Factors that determine whether a substance can pass through include: (1) size (molecular weight), (2) lipid solubility, (3) charge, and (4) presence of channels and transporters. Therefore, if a substance is smaller, more lipid-soluble, has no charge, and has the appropriate protein channels or transporters available, it is more likely to be able to cross the membrane. The smaller the size, the easier it is for a molecule to cross. The more lipid-soluble the better, because the cell membrane is made of phospholipids. As for charge, uncharged substances have an easier time crossing directly through the membrane (without using a membrane protein).

Transport of a substance through the membrane is categorized as either passive or active. **Passive transport** refers to processes that occur spontaneously without any energy investment by the cell via ATP hydrolysis. **Active transport** refers to any transport process that occurs only when the cell invests energy from ATP hydrolysis to force transport to occur.

Passive Transport

The illustration on the facing page shows three types of passive transport through the plasma membrane. All of them deal with the diffusion process, in which a solute particle spontaneously moves from an area of higher solute concentration to an area of lower solute concentration. Each is described below:

(1) In **simple diffusion** a small, nonpolar, uncharged particle diffuses directly through the phospholipid bilayer of the plasma membrane. Oxygen gas enters our cells by simple diffusion, and carbon dioxide gas leaves our cells by the same process. Other substances that diffuse across plasma membranes include fatty acids, steroids, and fat-soluble vitamins (A, D, E, and K).

(2) **Simple diffusion** also can be a channel-mediated-type **protein channel**. This is how small substances such as ions diffuse through a plasma membrane. For example, sodium ions diffuse through sodium channels to enter a cell. Similarly, potassium ions pass through potassium channels to leave the cell.

(3) **Facilitated diffusion** differs from simple diffusion in the following ways: (1) involves the transport of larger solutes, (2) uses a protein carrier that has a shape-specific fit for a specific solute, (3) involves a shape change in the protein carrier. The solute must bind to the shape-specific binding site. This puts a limitation on how quickly diffusion can occur. The action of the binding of the solute to its binding site induces a shape change in the protein carrier. This allows the solute to be released on the opposite side of the plasma membrane. Glucose enters cells via facilitated diffusion.

Active Transport

The illustration on the facing page shows three types of active transport processes through the plasma membrane. Each is described below:

(4) **Primary active transport** refers to the movement of a specific substance against its concentration gradient. In other words, a solute particle is moved from an area of low solute concentration to an area of high solute concentration. Like paddling against the current in a river, this requires energy. The protein used in this process is generally referred to as a pump, which appropriately implies the need for energy. Free energy from ATP hydrolysis is the fuel that keeps the pump working. This protein pump always contains a binding site for the substance to be transported. For example, the iodine pump actively transports iodine out of the blood and into the cells in our thyroid gland. The thyroid requires iodine as a raw material for manufacturing the hormones it normally produces.

Unlike primary active transport, **secondary active transport** does not directly use free energy from ATP hydrolysis. Instead, it gets its energy indirectly from the established concentration gradients from either sodium ions (Na^+) or hydrogen ions (H^+). These ion gradients were established by primary active transport. Movement of these ions down their concentration gradients is then coupled to the simultaneous transport of another substance via either symport or antiport.

(5) **Cotransport** refers to the simultaneous transport of two substances in the same direction across the membrane with the aid of a specific type of membrane protein. One example is the sodium-glucose symporter located in the epithelial cells in the intestinal mucosa. This transport process helps glucose get absorbed into the blood from the digestive tract.

(6) **Countertransport** refers to the simultaneous transport of two substances in the opposite direction across the membrane with the aid of a specific type of membrane protein. One example is the sodium-calcium antiporter, which keeps calcium levels low inside cells by pumping it out.

PASSIVE TRANSPORT

ACTIVE TRANSPORT

Solute particle

1

(ECF)

(ICF)

Solute particle

2

Protein channel

Plasma membrane

Calcium

6

Sodium

ATP

Glucose

3 Glucose

5

A.

B.

A.

Iodide ion

4

ATP

1 Simple diffusion: no membrane protein involved (*ex.*: oxygen diffusing into a cell)

2 Simple diffusion: channel-mediated type (*ex.*: sodium ions diffuse into cells through sodium channels)

3 Facilitated diffusion (*ex.*: glucose enters cells by facilitated diffusion)

6 Counter-transport (*ex.*: sodium-calcium antiporter)

5 Cotransport (*ex.*: sodium-glucose symporter)

4 Primary active transport (*ex.*: iodide pump)

A.

B.

A.

B.

ATP

Outline for Items on This Page

I. Passive transport
    A. Diffusion
        1. Simple diffusion
        2. Facilitated diffusion

II. Active transport
    A. Primary active transport
    B. Secondary active transport
        1. Cotransport
        2. Countertransport

Color the different proteins in the cell membrane and the solutes crossing the cell membrane. Color the ATP yellow.

KEY

ICF = intracellular fluid

ECF = extracellular fluid

77

Description

The plasma membrane is selectively permeable because it controls which substances enter and exit the cell. Nutrients and waste products both must pass through this structure, but not all substances are able to cross the plasma membrane. Factors that determine whether a substance can pass through include: (1) size (molecular weight), (2) lipid solubility, (3) charge, and (4) presence of channels and transporters. Therefore, if a substance is smaller, more lipid-soluble, has no charge, and has the appropriate protein channels or transporters available, it is more likely to be able to cross the membrane.

Transport of a substance is classified as either **passive transport** or **active transport**. In passive transport, the cell does not have to expend energy for the process to occur. Active transport requires the use of cellular energy. Typically, this energy is in the form of the free energy liberated from ATP hydrolysis.

Simple diffusion falls under the category of **passive transport**. Typically, it deals with the movement of a solute particle. For example, this process allows oxygen to enter the cells and carbon dioxide to exit them. In diffusion, substances tend to move from areas where they are in higher concentration to areas of lower concentration. The driving force for simple diffusion is the movement toward a state of dynamic equilibrium, in which concentrations of the designated substance are in equal concentrations on both sides of the plasma membrane. This **driving force** is like a **room full of crowded people at a party who all need their personal space**. As the illustration shows, this *need* induces people to move from the crowded room A over to room B.

Facilitated diffusion is also a **passive transport** process, but it typically deals with larger solute particles than those in simple diffusion. It differs from simple diffusion primarily in that it is a **carrier-mediated** process. This means that a protein carrier in the plasma membrane is needed to assist in the diffusion process. The carrier protein has a shape-specific binding site for the specific solute to be transported. The solute binds to this site temporarily before being released on the other side of the membrane.

Analogy

In the illustration, **a person** represents **a solute particle.** The **wall** between the rooms is like the **plasma membrane. Room A** is like a **solution** on one side of the plasma membrane, and **room B** is like the **other solution** on the other side of the plasma membrane. In the "Before" illustration, the **group of crowded people in room A** represents a **high-concentration gradient** of solute particles relative to room A. The **need** for people to have their personal space is like the **driving force** to achieve dynamic equilibrium. In the "After" illustration, the **equal distribution of people** in both rooms A and B is like the **state of dynamic equilibrium.**

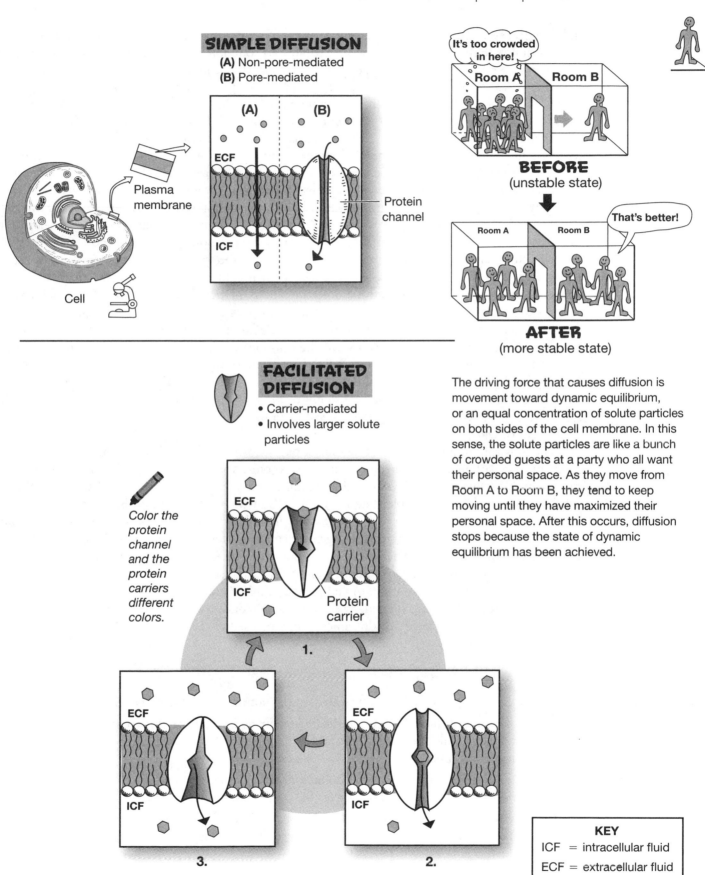

SIMPLE DIFFUSION

(A) Non-pore-mediated
(B) Pore-mediated

ECF

ICF

(A)   (B)

Protein channel

Cell

Plasma membrane

It's too crowded in here!

Room A   Room B

BEFORE
(unstable state)

Room A   Room B

That's better!

AFTER
(more stable state)

FACILITATED DIFFUSION

• Carrier-mediated
• Involves larger solute particles

Color the protein channel and the protein carriers different colors.

ECF

ICF

Protein carrier

1.

ECF

ICF

3.

ECF

ICF

2.

The driving force that causes diffusion is movement toward dynamic equilibrium, or an equal concentration of solute particles on both sides of the cell membrane. In this sense, the solute particles are like a bunch of crowded guests at a party who all want their personal space. As they move from Room A to Room B, they tend to keep moving until they have maximized their personal space. After this occurs, diffusion stops because the state of dynamic equilibrium has been achieved.

KEY	
ICF =	intracellular fluid
ECF =	extracellular fluid

Description

In biological systems, **osmosis** is defined as the flow of **water** across a semipermeable membrane toward the solution with the higher solute concentration. This common, important process occurs when water moves in and out of plant and animal cells by crossing their cell membranes. The driving force is the tendency toward achieving dynamic equilibrium, in which there is an equal concentration of solute on both sides of the cell membrane. When examining osmosis, the solute in question is impermeable to the cell membrane. It is also essential to realize that the solution inside the cell (intracellular solution) is always compared to the solution outside the cell (extracellular solution). For example, if you told me I was tall, I might ask, "Compared to what?" Compared to a toddler, I may be tall but not when compared to a giraffe. Similarly, if you were to tell me that a solution was concentrated, I could also ask, "Compared to what?" The word "concentrated" makes sense only in the context of a relative comparison. Although your body has regulatory mechanisms to keep the concentration of the intracellular and extracellular solutions relatively stable, the extracellular solution is more likely to change in its solute concentration because of external influences. Sometimes it is more dilute, and other times it is more concentrated.

The illustration of the hollow, glass, U-shaped tube at the top of the facing page demonstrates the process of osmosis. An artificial, semipermeable membrane separates two solutions of different concentrations—one on side A and the other on side B. Notice that the solute in both solutions is a protein that is impermeable because it is too large to cross the membrane. The solution on side B is more concentrated than the solution on side A. Any time the two solutions are unequal in concentration, water always moves toward the solution with the higher solute concentration. In this case, water will move from side A to side B. Why? Think of this as water's attempt to dilute the more concentrated solution to reach dynamic equilibrium. As water moves, the volume of solution on side B increases, and the volume of solution on side A decreases. Water stops moving when the two solutions have equal solute concentrations—the state of dynamic equilibrium. At dynamic equilibrium, the net movement of water across the membrane is zero.

The illustration in the middle of the facing page shows a red blood cell in three different types of solutions: (1) **isotonic**, (2) **hypertonic**, and (3) **hypotonic**. Your body's extracellular solutions are normally **isotonic** (iso = equal) **solutions** with the same concentration of impermeable solutes as the intracellular solution. This is a stable state for body cells. A cell in this solution has already achieved dynamic equilibrium, so the net movement of water in and out of the cell is zero. A **hypertonic solution** (hyper = more, greater) is one that has a greater concentration of impermeable solutes relative to the intracellular solution. A cell placed in this solution will shrink (crenate) because of water leaving the cell. A **hypotonic** (hypo = less) **solution** is one that has a lesser concentration of impermeable solutes relative to the intracellular solution. A cell placed in this solution will swell and possibly burst (lyse) because of water rushing into the cell. By convention, biologists typically use these terms to describe the extracellular solution, though they technically can be used to describe either the intracellular or the extracellular solution as a comparison.

Applications

Three applications for osmosis are illustrated on the bottom of the facing page:

1. Using an isotonic solution is illustrated by giving a patient an intravenous (IV) saline solution at the hospital. The saline solution has to be isotonic to your red blood cells so they neither shrink nor burst.

2. Creating a hypertonic solution is illustrated by adding salt to the slimy film coating the surface of the slug. As a result, water leaves the cells of the slug and can cause it to shrivel up and die.

3. Using a hypotonic solution is illustrated by placing wilted lettuce in a bowl of water in the refrigerator overnight to refresh it. Water in the bowl enters the cells of the lettuce leaf, making it firm again.

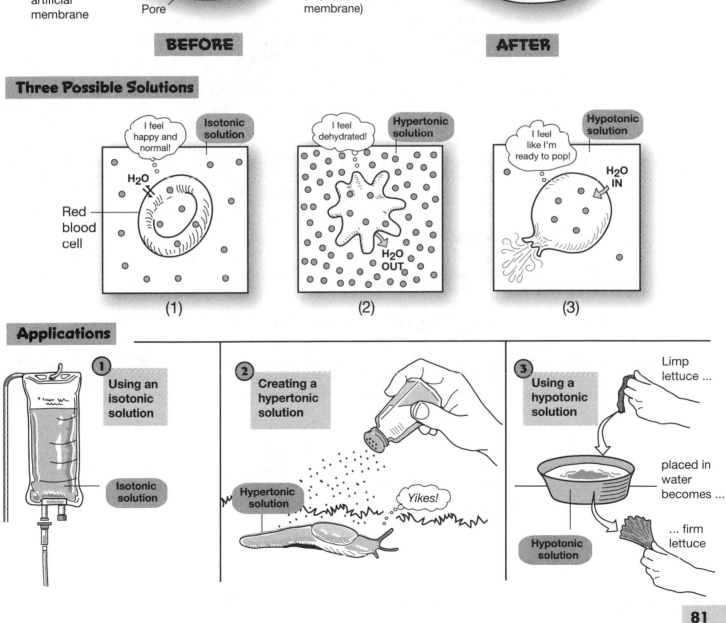

Description

Filtration falls under the category of **passive transport** and has the following features:

- Requires a force (*blood pressure*) to make it happen
- Requires a pressure gradient.
- Separates liquids (*water* and *small solutes*) from solids (*plasma proteins, red blood cells, white blood cells,* and *platelets*).
- Produces a filtrate (*final filtered solution*)

Filtration is a common process that occurs constantly through capillary beds that link arteries to veins. Just as narrow, single-lane roads lead to larger streets and then to highways, capillaries eventually lead to the body's highways—large arteries and large veins. Capillaries are found throughout the body and are the most microscopic of the blood vessels. In fact, they are so small that red blood cells must pass through them single-file. Through filtration, they provide direct contact with body cells. Their structure is simple: The wall of a capillary is made of a single, flat layer of cells called **simple squamous epithelium**. These thin cells allow for easy filtration. Surrounding this epithelial layer is a thin **basement membrane** composed of extracellular materials, such as protein fibers, secreted by epithelial cells.

Almost all tissues have an ample supply of capillary beds—a branching network of capillaries with a higher-pressure arteriole end and a lower-pressure venule end. Like a one-way street, blood always moves from the arteriole end to the venule end. Because filtration is completely dependent on pressure, most of it occurs nearer the arteriole end. The blood pressure is the force that makes it all happen by creating a pressure gradient. As a result of the higher pressure within the arteriole end, the liquid plasma is forced through the wall of the capillary, so water and small solutes leave the blood to form a **filtrate** called **interstitial fluid**. This fluid fills the spaces within tissues called **interstitial spaces**, which include gaps between body cells as well as those between blood vessels and body cells. Simultaneously, large solids such as blood cells and large molecules such as plasma proteins remain inside the capillary, because they are too large to pass through it.

Filtration is responsible for producing all of the following:

- **interstitial fluid** (the only way we have to get fluid to tissues)
- **urine** (processed blood plasma that contains mostly water)
- **cerebrospinal fluid (CSF)**—(constantly circulates around the brain and spinal cord and has many functions: provides a protective cushion, distributes nutrients, and removes waste products)

There are three different types of capillaries (see p. 362). Some are more permeable than others, which determines how easily filtration occurs. For example, specialized capillaries in the kidneys called **glomeruli** (sing. *glomerulus*) (see p. 492) are highly permeable to allow for easy filtration of the plasma to form urine, and capillaries in the brain are the least permeable in the body to protect the brain's precious neurons from being damaged by toxic substances that may have entered the bloodstream.

Analogy

The process of filtration is the same, generally, as a coffeemaker making coffee. Compared with filtration through a capillary, the **force** is **gravity** instead of **blood pressure**. The **coffee filter** is like the **simple squamous epithelium**. The **coffee grounds** are like the **solid** materials in the blood (**red blood cells, white blood cells, plasma proteins**, etc.). The **water** is like the **plasma**. The **coffee solution** is called the *filtrate*, which is like the **interstitial fluid**.

Capillary network

Arteriole end

Venule end

Artery

Vein

Black arrows indicate the direction of blood flow.

Coffee filter

Coffee grounds

Water

Filtrate

Coffeemaker (*cutaway view*)

A coffeemaker uses the same filtration process that occurs in your body.

Basement membrane

Color the simple squamous epithelium and the coffee filter the same color, because they are analogous structures.

White arrows indicate filtration.

Simple squamous epithelium

Interstitial fluid (filtrate)

Small solute particles in filtrate

Body cell

Red blood cell in lumen

Body cell

Nucleus of body cell

Interstitial space

Body cell

Interstitial space

Description

The plasma membrane is like the gatekeeper of the cell, because it controls the substances that enter and exit. Plasma membranes selectively allow only certain substances (*like nutrients*) in and other substances (*like waste products*) out. Not all substances are able to cross the plasma membrane. Generally, they have to be small and lipid-soluble to cross. If they meet these criteria, they can be transported by two different methods: (1) **passive transport** or (2) **active transport**. In passive transport, the cells do not have to expend any energy for the process to occur. Active transport requires that cellular energy be used. Typically, this energy is in the form of the free energy liberated from ATP hydrolysis.

The process of **active transport** falls under the category of the same name—**active transport**. It moves a solute particle against a concentration gradient with the help of a protein transporter within the plasma membrane. This protein uses the free energy from ATP hydrolysis to move the solute particle(s) against the gradient. An example is the sodium-potassium pump present in all cells (see p. 90).

Analogy

The process of **active transport** is like the action of a **sump pump** in a homeowner's basement, which prevents flooding inside the home. Both require energy to allow the pump to transport a substance against a concentration gradient. The **protein pump** uses the **free energy** from ATP hydrolysis, and the **sump pump** uses **electrical energy**. The protein pump transports a solute particle from a region of lower concentration to a region of higher concentration. Similarly, the sump pump transports *water* from a region of lower *water* concentration inside the basement to a region of higher *water* concentration outside the home.

Note: The limitation of this analogy is that active transport always involves a solute particle rather than a solvent such as water.

Study Tips

- Students often confuse the process of **active transport** and the category of **active transport** because they have the same name. Don't make this mistake. The former refers to the specific "sump pump-like process" described above, and the latter refers to the general category for any process that uses the free energy from ATP hydrolysis to transport substances into or out of the cell.

- If a transport process is active, this means that it uses the free energy from ATP hydrolysis to perform a task. To distinguish the different types of active transport (exocytosis, endocytosis, active transport, etc.), ask yourself: "What is the free energy specifically used for in each case?"

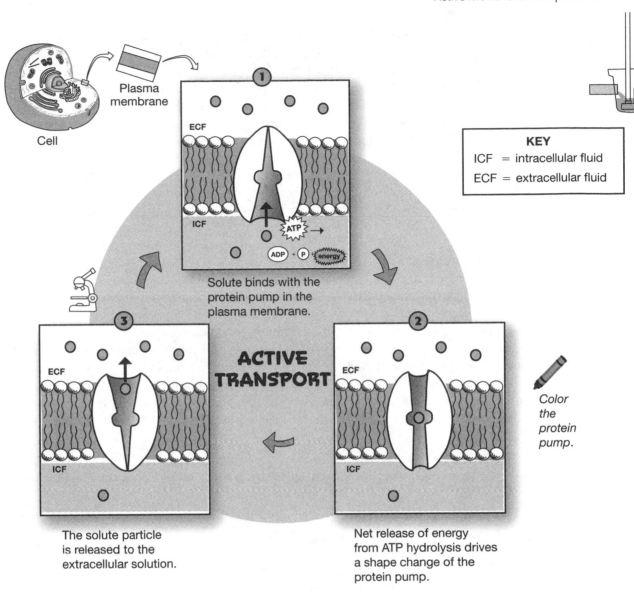

Cell

Plasma membrane

KEY

ICF = intracellular fluid

ECF = extracellular fluid

① Solute binds with the protein pump in the plasma membrane.

ACTIVE TRANSPORT

② Net release of energy from ATP hydrolysis drives a shape change of the protein pump.

③ The solute particle is released to the extracellular solution.

Color the protein pump.

Electrical outlet

energy

Rain

Outlet pipe

Soil saturated with water

Basement floor

Sump pump

Perforated drain tube

Water

Perforated drain tube

Sump Pump Analogy

Active transport is like a sump pump. Both work to move a substance against its concentration gradient from a region of lower concentration to a region of higher concentration. Both also require energy to power a pump to move the substance against its concentration gradient.

Description

The plasma membrane is like the gatekeeper of the cell because it controls which substances enter and exit. Plasma membranes selectively allow only certain substances (*like nutrients*) in and other substances (*like waste products*) out. Not all substances are able to cross the plasma membrane. Generally, they have to be small and lipid-soluble to cross. If they meet these criteria, they can be transported by two different methods: (1) **passive transport** or (2) **active transport**. In passive transport, the cell doesn't have to transport any energy for the process to occur. Active transport requires cellular energy be used. Typically, this energy is in the form of the free energy liberated from ATP hydrolysis.

Exocytosis is a type of active transport. This is the process by which substances made by the cell (e.g., a hormone) are concentrated within a vesicle. Special regulatory proteins associated with the plasma membrane use cellular energy to move the vesicle toward the plasma membrane, so the two can merge. Then the contents of the vesicle are released to the extracellular solution. This is how cells within your pancreas release the hormone insulin, and for example, how neurons release neurotransmitters such as acetylcholine.

Endocytosis is another type of active transport that is actually the reverse of exocytosis. In this process the plasma membrane invaginates to trap a solid particle (*bacteria, virus*) or some liquid (*extracellular solution*) in a small, pouch-like structure. This pouch uses cellular energy to gradually close, pinch itself off from the plasma membrane, and become a vesicle within the cell.

There are two types of endocytosis: (1) **pinocytosis** and (2) **phagocytosis**. If the material in the pouch is a liquid solution such as the extracellular solution, it is called **pinocytosis**. Cells use this process to obtain nutrients from the extracellular solution. If the material is a solid object such as a bacterium, it is called **phagocytosis**. Macrophages are a type of white blood cell that attacks foreign pathogens. Because they normally are engaged in phagocytosis, they are called *phagocytic* cells.

Study Tips

Knowing common roots, prefixes, and suffixes is important to decode the meaning of scientific terms (see p. 547).

> *exo-* = *outer, outside*   **cyto-** = *cell*   **-sis** = *condition*
>
> *exocytosis* = *to bring outside the cell*
>
> *endo-* = *within*   **cyto-** = *cell*   **-sis** = *condition*
>
> *endocytosis* = *to bring within the cell*

Cell   Plasma membrane

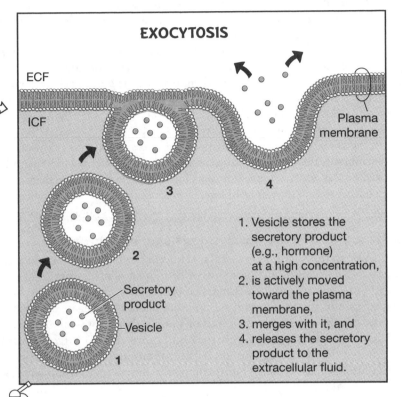

EXOCYTOSIS

ECF

ICF

3

4

Plasma membrane

2

Secretory product

Vesicle

1

1. Vesicle stores the secretory product (e.g., hormone) at a high concentration,
2. is actively moved toward the plasma membrane,
3. merges with it, and
4. releases the secretory product to the extracellular fluid.

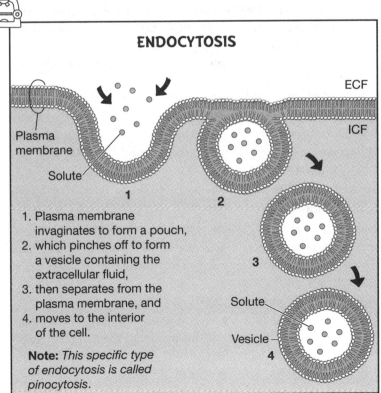

ENDOCYTOSIS

ECF

ICF

Plasma membrane

Solute

1

2

3

Solute

Vesicle

4

1. Plasma membrane invaginates to form a pouch,
2. which pinches off to form a vesicle containing the extracellular fluid,
3. then separates from the plasma membrane, and
4. moves to the interior of the cell.

Note: *This specific type of endocytosis is called pinocytosis.*

KEY

ICF = intracellular fluid

ECF = extracellular fluid

Description A **membrane** (*resting*) **potential** (MP) is a separation of positive (+) and negative (−) charges across a **plasma membrane** that results in an *electrical potential* or a *voltage*. This voltage is measured in units called *millivolts* (mV)—much smaller than that produced by a battery. Typically, the outside of the plasma membrane is positively charged, and the inside is negatively charged. Most human body cells have an MP, but the exact value is different for different cell types and ranges between −5 mV and −100 mV. The negative sign means that the inside of the membrane is negative relative to the outside.

Let's compare the MP to a 1.5 V battery, which also has an electrical potential because of its positive (+) pole at one end and negative (−) pole at the other. The battery has **potential energy** because of its *potential* to do work.

Potential energy is like water behind a dam. Because the *position* of the water level is higher on one side of the dam and lower on the other, it also has the *potential* to do work. When the floodgate opens, this potential energy gets converted into kinetic energy or energy in motion. As the water flows through the gate, it can turn a turbine that generates electricity—hydroelectric power.

Similarly, the potential energy in the battery can be converted into a flow of electricity to power a flashlight. For the **plasma membrane**, this potential energy is stored not as water behind a dam but as a cation (+) concentration gradient. Based on the rule for simple diffusion, ions passively flow down their concentration gradients from regions of higher ion concentration to regions of lower ion concentration. This cation concentration gradient is the potential energy source for cells to do work. For example, in a neuron this gradient is used to generate a nerve impulse.

What is the source of these positive and negative charges?

The short answer to this question is that the *extracellular fluid* (ECF) and *intracellular fluid* (ICF) are electrically neutral because they contain equal numbers of *cations* (+ ions) and *anions* (− ions). The only exceptions are at the inner and outer surfaces of the **plasma membrane**. Here, there is an accumulation of positive (+) charges on the outside surface of the membrane and an accumulation of negative (−) charges on the inside. The positive charges on the outside are mainly the result of **sodium** (Na^+) ions—the most dominant cations in the **ECF**. The negative charges on the inside are mostly the result of negatively charged **proteins** too large to leave the cell. But it's also because of the loss of the most dominant cations in the ICF—**potassium** (K^+) ions.

The real situation is actually more complex (surprise, surprise!). Let's summarize the key factors that contribute to the MP:

- **Ion concentration gradients for Na^+ and K^+ across the plasma membrane**
  - **Na^+** has a greater concentration in the ECF than the ICF, so it could diffuse *into* the cell.
  - **K^+** has a greater concentration in the ICF than the ECF, so it could diffuse *out* of the cell.
  - Na^+ and K^+ can pass through gated channels only in the membranes that are usually closed but can be stimulated to open.
  - These ion concentration gradients are a source of potential energy to do work.
- **Differing permeabilities for Na^+ and K^+**
  - Membrane proteins called *leakage channels* allow both Na^+ and K^+ to diffuse across the membrane.
  - Plasma membranes are *much* more permeable to K^+ than Na^+.
    - **(1)** K^+ leaves the cell as it diffuses down its concentration gradient. The result? This loss of positive charge makes the interior of the cell more negative (*see illustration*).
- **Opposite charges attract**
  - As previously mentioned, as K^+ diffuses out of the cell, this loss of positive charge makes the interior more negative.
    - **(2)** Because opposite charges attract, some of the K^+ ions are drawn back into the negatively charged interior. The result? This relatively small gain in positive charge makes the interior less negative than it might be otherwise (*see illustration*).
- **Action of the $Na^+ − K^+$ pump** (see p. 90)
    - **(3)** With each cycle, the **$Na^+ − K^+$ pump** actively transports three Na^+ ions out for every two K^+ ions it brings into the cell. The result? This maintains the MP by preserving the charge imbalance across the membrane. How? More positive charges are moved out of the cell (+3), and fewer positive charges are brought into the cell (+2) (*see illustration*).

Study Tips To distinguish *cations* (+ ions) from *anions* (− ions) remember this phrase: "I *positively* hate *cats*."

Cell

Plasma membrane

Membrane potential

ECF

Plasma membrane

ICF

Positive pole

Negative pole

SUPER VOLT

Analogy

Color the positive (+) and negative (−) signs different colors.

Both a plasma membrane and a battery have a positive (+) end and a negative (−) end. By separating positive and negative charges, both are able to store **potential energy**.

POTENTIAL ENERGY...

is like water behind a dam. It is ready to do work.

Water (*ion gradient*)

Dam (*plasma membrane*)

Floodgate (*closed*)

For a battery, the energy can be used to generate electricity for a flashlight.

ENERGY CONVERSION

For a **plasma membrane**, the energy can be used to generate a **nerve impulse**.

Neuron

Where do the positive and negative charges come from?

Color the K+ ions.

ECF

K+
(3 Na+ OUT)
2 K+

① Leakage channel ②

③ Na+ - K+ pump

Plasma membrane

Proteins

ICF

K+
(2 K+ IN)
3 Na+

Description
The **sodium-potassium pump** is a protein found in the plasma membrane of all animal cells. It functions as an ion exchanger to actively transport sodium for potassium across the membrane. The net release of energy from the hydrolysis of one ATP molecule is used to transport three sodium ions *out* of the cell and two potassium ions *into* the cell. Since a single cell has many of these protein pumps, the net effect of all of them working together is to maintain a gradient for both sodium and potassium.

The concentration of potassium ions is normally 10 to 20 times greater inside the cells than out. The reverse is true for sodium. Because of this difference in concentration, sodium's tendency is to diffuse into cells, and potassium's tendency is to diffuse out of cells. This occurs through leakage channels in the membrane.

A significant amount of the cell's resting energy—about 33 percent—is used to keep the pump working. The activity of the pump depends on the concentration of sodium in the cytosol. The greater the concentration, the more active the pump; the lesser the concentration, the less active. If the pump were to stop working, the sodium concentration in the cytosol would increase, and nothing could prevent water from entering the cell via osmosis (see p. 80).

Cycle
In the illustration on the facing page, the sodium-potassium pump is shown going through a four-step cycle:

(1) **Sodium binding:** three sodium ions in the cytosol bind at their respective binding sites on the pump.

(2) **Shape change:** The binding of sodium causes the hydrolysis of one ATP molecule into ADP, a phosphate group and a net release of free energy. The free energy is used to bind the phosphate group to the pump, which induces a shape change in the pump. This shape change both enables the release of the three sodium ions and makes it easier for two potassium ions to bind from the extracellular solution.

(3) **Potassium binding:** The binding of the two potassium ions causes the phosphate group to be released from the pump, which, in turn, causes the shape of the protein to change again.

(4) **Potassium released:** The restoring of the pump to its original shape releases the two potassium ions into the cytosol. The pump is now ready to bind three new sodium ions, and the cycle repeats itself.

Analogy
Functionally, the sodium-potassium pump is like a revolving door. It takes energy to move the revolving door, just as it takes ATP energy to drive the pump. Each cycle of the pump results in three sodium ions moving *out* of the cell and two potassium ions moving *into* the cell.

REVOLVING DOOR ANALOGY

BEFORE   AFTER

(OUT)   K+   (OUT)   Na+ → Na+ Na+
                K+   (3 sodium OUT)

(IN)   Na+   (IN)
       Na+   Na+
         Na+   K+ K+
                               (2 potassium IN)

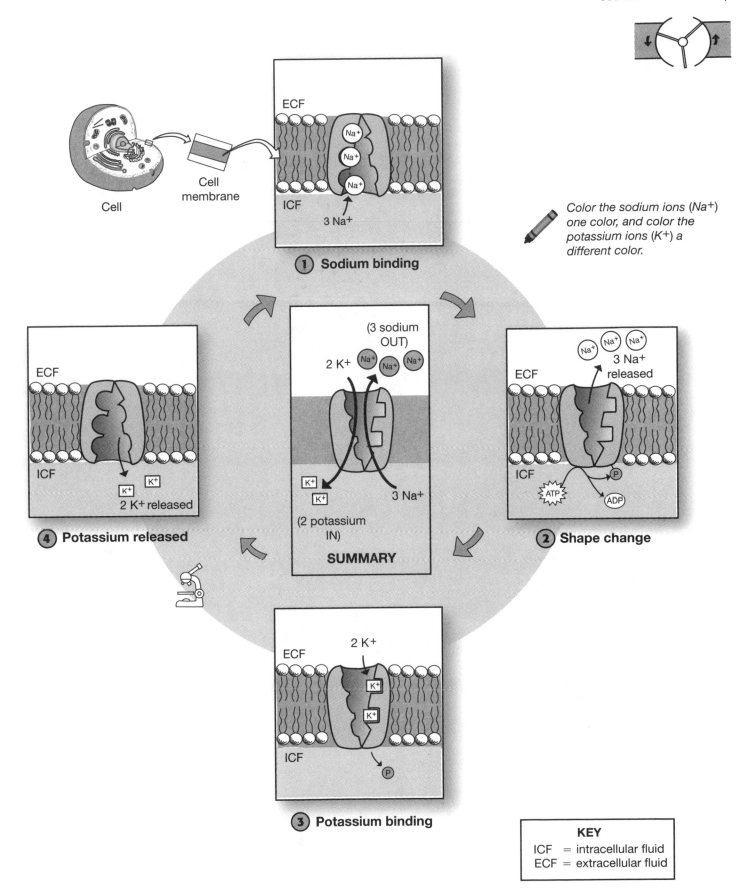

Color the sodium ions (Na⁺) one color, and color the potassium ions (K⁺) a different color.

Cell

Cell membrane

ECF
ICF
3 Na⁺
① **Sodium binding**

(3 sodium OUT)
2 K⁺
3 Na⁺
K⁺
(2 potassium IN)
SUMMARY

ECF
ICF
2 K⁺ released
K⁺   K⁺
④ **Potassium released**

3 Na⁺ released
ECF
ICF
ATP   ADP   P
② **Shape change**

2 K⁺
ECF
K⁺
K⁺
ICF
P
③ **Potassium binding**

KEY
ICF = intracellular fluid
ECF = extracellular fluid

91

Notes

TISSUES

Description

Tissues are a group of similar cells working together for a common purpose. The four different types of tissues are: (1) **epithelial**, (2) **connective**, (3) **muscular**, and (4) **nervous**. Let's consider only the first tissue type—**epithelial tissues**. There are many different kinds of epithelial tissues, and they are often named after their cell shape. They are found lining internal body cavities and passageways and covering body surfaces. Epithelial tissues are composed mostly of cells that rest on a thin **basement membrane**. In physiology, epithelial cells are an important theme because substances have to pass through them before they can enter the blood. Every organ system contains epithelial tissues. For example, they line (*are inside of*) the following structures: blood vessels, digestive organs, the urinary bladder, and the microscopic air sacs in the lungs. Moreover, they cover the surface of the skin. In short, they are everywhere in the body.

Each cell in an epithelial tissue has three different surfaces that serve as reference points on the cell. Each surface represents a different side of the cell and is composed of the **plasma membrane**. The top of the cell, which faces the lumen or body surface, is called the **apical surface**. The sides of the cell are called the **lateral surfaces**. The bottom of the cell, which rests on the basement membrane and is closer to the blood, is called the **basal surface**. The typical pathway a substance follows through an epithelial cell is as follows: **apical surface** to **cytosol** to **basal surface** to **basement membrane** to **capillary**. (**Note:** A good way to distinguish the apical surface from the basal surface is to remember the alliteration "**B**asal, **B**asement, **B**lood." These three structures are in close proximity to each other.)

Intercellular Junctions

The lateral surfaces of adjacent epithelial cells are connected one to another by different kinds of junctions. The four major types of junctions are: (1) **tight junctions**, (2) **adhering junctions**, (3) **desmosomes**, and (4) **gap junctions**. Each is summarized below.

① Tight junctions loop around the whole cell and are located near the apical surface. They look like lines of rivets that form a seal between adjacent cells by stapling their plasma membranes together. Each rivet-like structure is actually a **transmembrane protein**. Tight junctions prevent substances from passing between epithelial cells. This forces substances to move through cells instead of squeezing between them. For example, tight junctions in the digestive tract keep digestive enzymes inside the intestine and prevent them from entering the blood.

② Adhering junctions function like seam welds. They are typically located below tight junctions, so they are also near the apical surface. They look like a belt that wraps around the whole cell and contain band-like proteins called **plaques**. Running along the length of the plaques are thin, contractile proteins called **microfilaments**. Together, the microfilaments and plaques form a structure called an **adhesion belt**. Adhering junctions prevent lateral surfaces from separating while providing a space for substances to enter. For example, a substance that already passed through the apical surface can enter this space as it continues on to the basal surface.

③ Desmosomes are like spot welds. They consist of disklike proteins called **plaques** held together and spaced apart by **linker proteins**. Instead of surrounding the entire cell, desmosomes are found at discrete points on the cell. They function as structural reinforcements at specific stress points along the cell. **Intermediate filaments** are long, strong cable-like proteins that extend into the cytosol and connect the plaque of one desmosome to the plaque of another desmosome on the opposite side of the cell. This helps maintain the structural integrity of the cell and the tissue. Desmosomes are common in the cells within the epidermis of the skin.

Half-desmosomes or **hemidesmosomes** are located on the basal surface, where they serve to anchor this surface to the basement membrane.

④ Gap junctions allow for small substances to be transported between adjacent cells. Each gap junction consists of a small group of tubular structures. Each tubular structure is composed of a cluster of six proteins called **connexins** and has a fluid-filled **channel** running down its center. This channel allows small substances such as ions and glucose to travel between adjacent cells. In ciliated epithelial cells, this may help to coordinate the movement of the cilia. Gap junctions are also found in other tissues such as cardiac and smooth muscle.

Nucleus

Basement membrane

Epithelial tissue

Metal plates

Plasma membrane

Rivet

Transmembrane protein

① Tight junction

② Adhering junction

Adhesion belt

Plaque

Microfilament

Black arrows show two different pathways a substance may follow through an epithelial cell.

Substance

Apical surface *(lumen side)*

Tight junction

Desmosome

Nucleus

Nucleus

Lateral surface

Gap junction

Hemidesmosome

Basal surface *(blood side)*

Basal surface

Basement membrane

Color the capillary red.

Capillary

Linker protein

Plaque

Intermediate filament

Intercellular space

③ Desmosome

Channel between cells

Connexin

Plasma membranes

④ Gap junction

Color the basement membrane and apical, lateral, and basal surfaces using different colors for each.

Simple Squamous Epithelium

Description

Epithelial tissues line internal cavities and passageways and cover external body surfaces. They are composed mostly of cells that rest on a thin **basement membrane** made of protein fibers. No blood vessels are present. One method of classifying epithelial tissues is by the number of layers of cells. *Simple* epithelia have a single layer of cells, and *stratified* epithelia have multiple layers of cells. Classification is also based on the following cell shapes: *squamous* (thin, flat), *cuboidal* (cube-shaped), and *columnar* (column-shaped).

Simple squamous epithelium is a single row of thin, flat cells.

Analogy

Each **simple squamous epithelial cell** compares with a **fried egg** because both are flat with an irregular border. The nucleus of the cell is like the yolk.

Location

Lines internal surface of ventral body cavities, blood vessels, and heart; parts of kidney tubules; alveoli of the lungs.

Function

Flat shape allows substances to either diffuse easily through the cell or be filtered through it; secretion; reduces friction.

Study Tips

This tissue is often shown in two different views: (1) superior view, (2) side view. In a superior view, it looks like a fried egg, and in a side view, it looks like a flat cell. This is often confusing for students. To identify this tissue *in a side view*, on either a microscope slide or a photograph, look for the following:

- Flat cells.
- Flat nuclei.

Key to Illustration		
1. Simple cuboidal epithelial cell	2. Nucleus of simple squamous epithelial cell	3. Alveoli

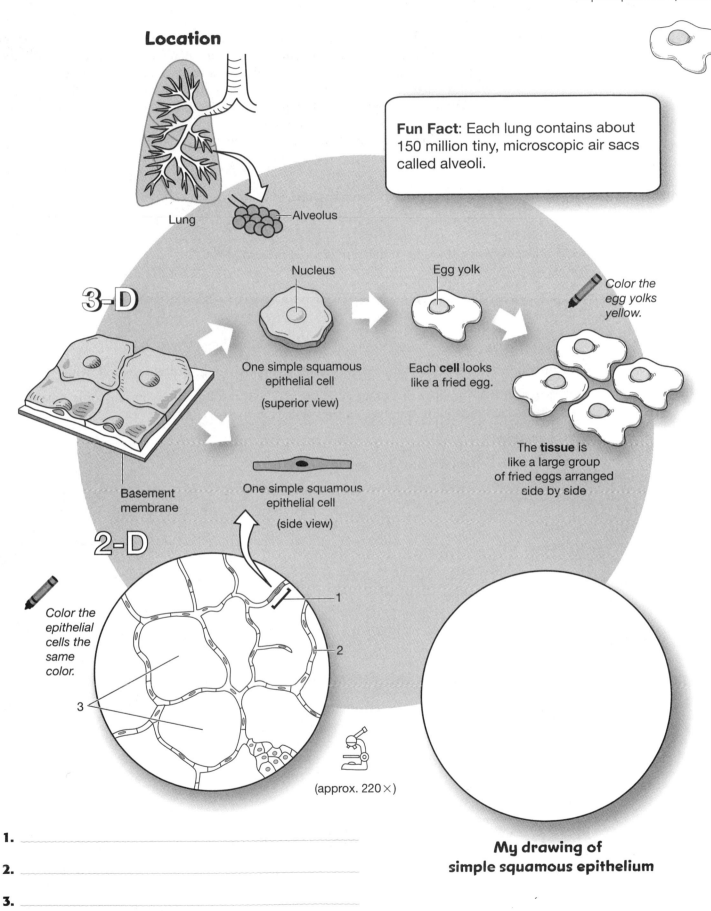

Location

Lung

Alveolus

Fun Fact: Each lung contains about 150 million tiny, microscopic air sacs called alveoli.

3-D

Nucleus

One simple squamous epithelial cell

(superior view)

Egg yolk

Each **cell** looks like a fried egg.

Color the egg yolks yellow.

The **tissue** is like a large group of fried eggs arranged side by side

Basement membrane

One simple squamous epithelial cell

(side view)

2-D

Color the epithelial cells the same color.

1

2

3

(approx. 220×)

1. _____

2. _____

3. _____

My drawing of simple squamous epithelium

TISSUES: Epithelial

Description Epithelial tissues line internal cavities and passageways and cover external body surfaces. They are composed mostly of cells that rest on a thin **basement membrane** made of protein fibers. No blood vessels are present. One method of classifying epithelial tissues is by the number of layers of cells. *Simple* epithelia have a single layer of cells, and *stratified* epithelia have multiple layers of cells. Classification is also based on the following cell shapes: *squamous* (thin, flat), *cuboidal* (cube-shaped), and *columnar* (column-shaped).

 Simple cuboidal epithelium is a single row of cube-shaped cells.

Analogy Each **simple cuboidal cell** is shaped like an **ice cube**.

Location Ducts of glands; parts of kidney tubules; follicles of thyroid gland.

Function Secretion; absorption.

Study Tips To identify this tissue on either a microscope slide or a photograph, look for the following:

- Ring of cells (because this tissue often forms the wall of a hollow microscopic tube such as the duct of a gland, it appears as a ring of cells in cross section).

- Square cells.

- Nucleus is relatively large and round.

Key to Illustration		
1. Simple cuboidal epithelial cell	2. Nucleus of simple cuboidal epithelial cell	3. Connective tissue

Location

Kidney

Fun Fact: On average, an adult's kidneys process about 165 liters of filtered blood plasma every day, but only 1 percent is excreted as urine.

3-D

Basement membrane

One simple cuboidal epithelial cell

Nucleus

Each cell is shaped like an ice cube.

2-D

Color the epithelial cells the same color.

(approx. 275×)

1
2
3

1. _____

2. _____

3. _____

My drawing of simple cuboidal epithelium

TISSUES: Epithelial

Description   **Epithelial tissues** line internal cavities and passageways and cover external body surfaces. They are composed mostly of cells that rest on a thin **basement membrane** made of protein fibers. No blood vessels are present. One method of classifying epithelial tissues is by the number of layers of cells. *Simple* epithelia have a single layer of cells, and *stratified* epithelia have multiple layers of cells. Classification is also based on the following cell shapes: *squamous* (thin, flat), *cuboidal* (cube-shaped), and *columnar* (column-shaped).

    Simple columnar epithelium appears as a single layer of tall, column-shaped cells with oblong nuclei. They are of two types: *ciliated* and *nonciliated*. **Cilia** are extensions of the plasma membrane that contain microtubules and move back and forth to propel fluids or secretions across cell surfaces.

Analogy   Each simple columnar cell in this tissue looks like a column.

Location

Ciliated type
- uterine tubes

Nonciliated type
- stomach
- small intestine
- large intestine
- gallbladder

Function   The primary function is absorption; secretion of mucus, enzymes, and other substances; movement of mucus by cilia.

Study Tips   To identify this tissue on either a microscope slide or a photograph, look for the following:

- Tall rectangular cells.
- Oblong-shaped nucleus.
- Nucleus usually located in the lower half of the cell.

Key to Illustration		
1. Microvilli	3. Nucleus of simple columnar epithelial cell	5. Connective tissue
2. Simple columnar epithelial cell (nonciliated type)	4. Basement membrane	

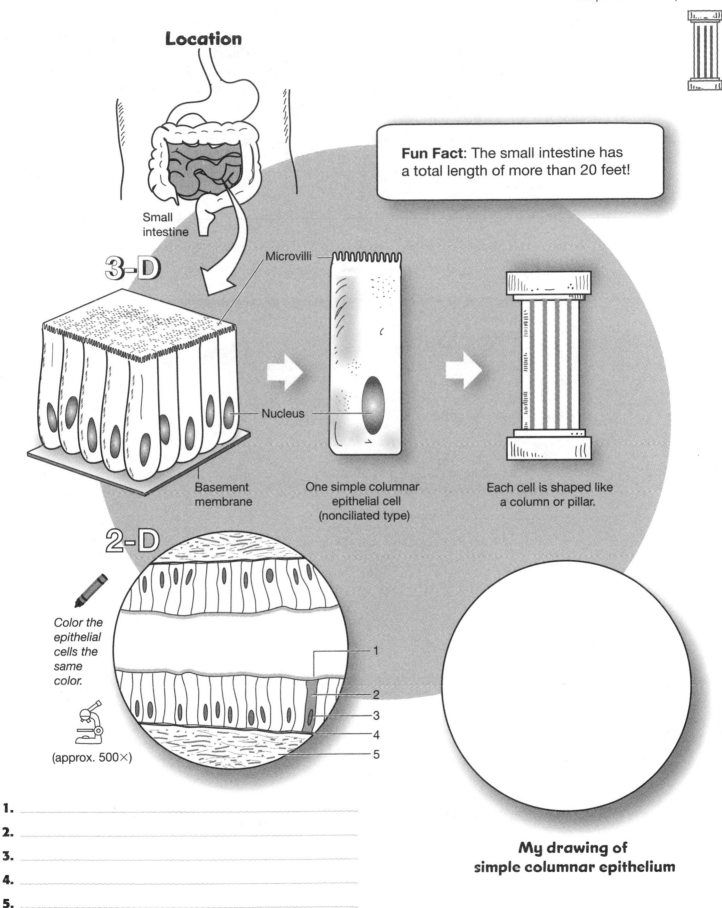

Location

Small
intestine

Fun Fact: The small intestine has
a total length of more than 20 feet!

3-D

Microvilli

Nucleus

Basement
membrane

One simple columnar
epithelial cell
(nonciliated type)

Each cell is shaped like
a column or pillar.

2-D

*Color the
epithelial
cells the
same
color.*

(approx. 500×)

1
2
3
4
5

**My drawing of
simple columnar epithelium**

1. _____

2. _____

3. _____

4. _____

5. _____

TISSUES: Epithelial

Description

Epithelial tissues line internal cavities and passageways and cover external body surfaces. They are composed mostly of cells that rest on a thin **basement membrane** made of protein fibers. No blood vessels are present. One method of classifying epithelial tissues is by the number of layers of cells. *Simple* epithelia have a single layer of cells, and *stratified* epithelia have multiple layers of cells. Classification is also based on the following cell shapes: *squamous* (thin, flat), *cuboidal* (cube-shaped), and *columnar* (column-shaped).

Pseudostratified columnar epithelium consists of a single row of cells. Most cells have a columnar shape, but other shorter cells may look cuboidal. The term "pseudostratified" literally means "falsely stratified." In other words, it looks as if it has multiple layers but actually has only one layer, because the cells are of differing heights.

Location

The two types of pseudostratified columnar epithelium are ciliated and nonciliated.

Ciliated	Nonciliated
• nasal cavity • trachea • bronchi	• ducts of male reproductive tract

Function

Protection; secretion; movement of mucus by **cilia**.

Study Tips

Under the microscope at higher magnifications, you can use the following landmarks to distinguish pseudostratified columnar epithelium cells:

- Cells have differing heights.

- Nuclei are not in an organized row, but are more staggered.

Key to Illustration

1. Cilia
2. Nucleus of pseudostratified ciliated columnar epithelial cell
3. Basement membrane
4. Connective tissue

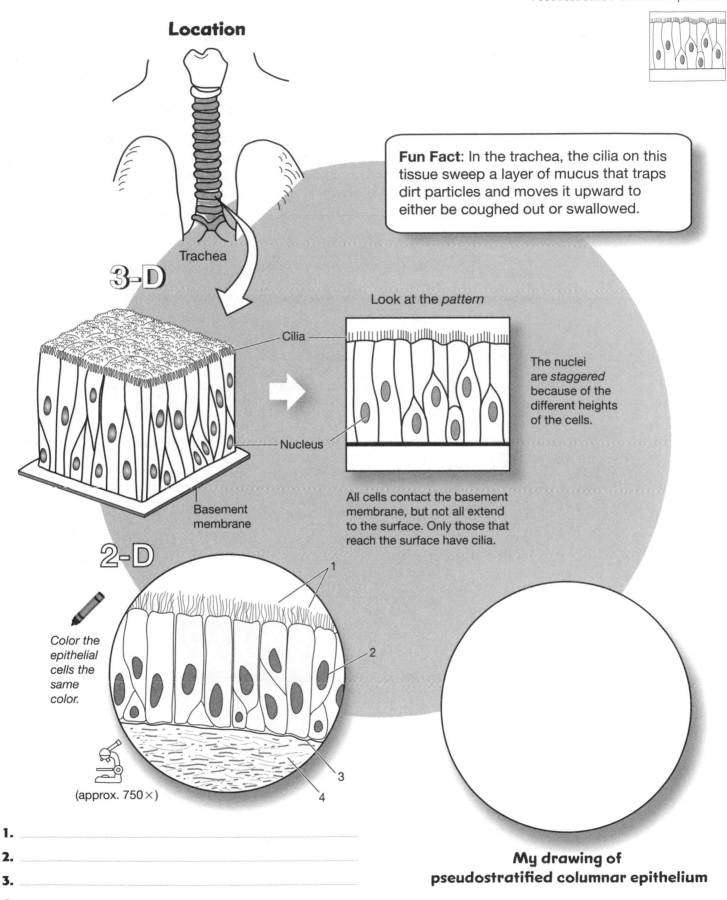

Location

Trachea

Fun Fact: In the trachea, the cilia on this tissue sweep a layer of mucus that traps dirt particles and moves it upward to either be coughed out or swallowed.

3-D

Look at the *pattern*

Cilia

Nucleus

The nuclei are *staggered* because of the different heights of the cells.

Basement membrane

All cells contact the basement membrane, but not all extend to the surface. Only those that reach the surface have cilia.

2-D

Color the epithelial cells the same color.

(approx. 750×)

My drawing of
pseudostratified columnar epithelium

1. _____

2. _____

3. _____

4. _____

Stratified Squamous Epithelium

Description

Epithelial tissues line internal cavities and passageways and cover external body surfaces. They are composed mostly of cells that rest on a thin **basement membrane** made of protein fibers. No blood vessels are present. One method of classifying epithelial tissues is by the number of layers of cells. *Simple* epithelia have a single layer of cells, and *stratified* epithelia have multiple layers of cells. Classification is also based on the following cell shapes: *squamous* (thin, flat), *cuboidal* (cube-shaped), and *columnar* (column-shaped).

Stratified squamous epithelium is of two different types—keratinized and nonkeratinized. The bottom layer in either type is composed of cuboidal or columnar cells active in cell division. New cells are pushed upward toward the surface. In a process called **keratinization**, the new cells in the keratinized type fill with a protein called **keratin**. The result is that the outer surface of this tissue is tough and water-resistant.

Location

Keratinized	Nonkeratinized
• epidermis of skin	• lining of mouth, pharynx, esophagus, anus, and vagina

Function

Provides physical protection to underlying tissues against abrasion and pathogens.

Study Tip

Pattern: Transition in cell shape from cuboidal or columnar cells in the lower region to flat cells on the top.

Key to Illustration	
1. Stratified squamous epithelium	3. Connective tissue
2. Basement membrane	4. Nuclei

Stratified Squamous Epithelium

Location

Epidermis
of skin

3-D

Fun Fact: On average, the epidermis of your skin regenerates about every 27 days.

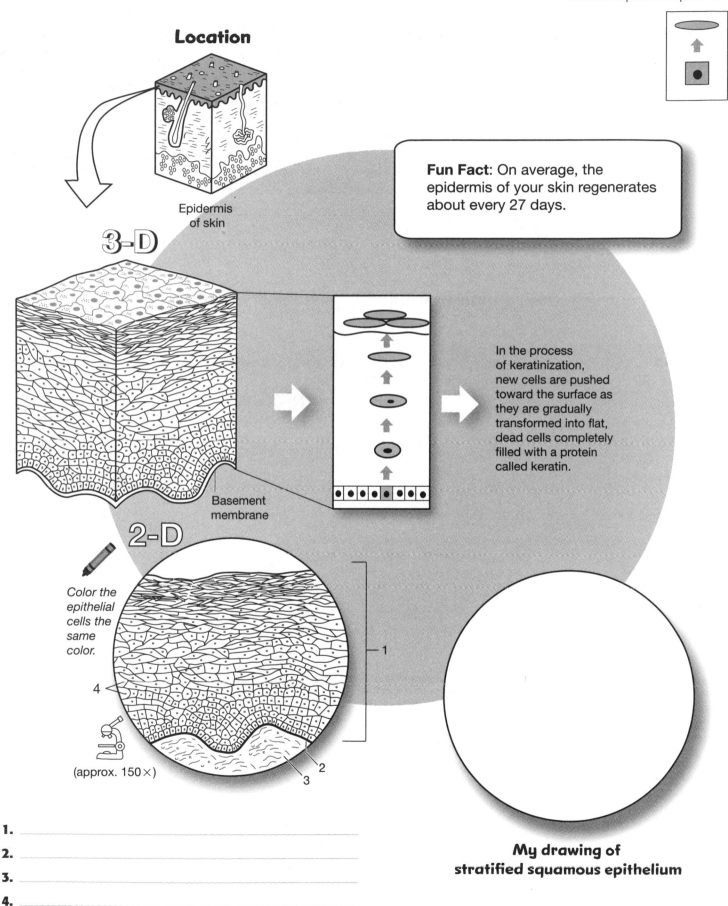

Basement
membrane

In the process of keratinization, new cells are pushed toward the surface as they are gradually transformed into flat, dead cells completely filled with a protein called keratin.

2-D

Color the epithelial cells the same color.

4

1

2

3

(approx. 150×)

1. _____

2. _____

3. _____

4. _____

My drawing of stratified squamous epithelium

Stratified Transitional Epithelium

Description

Epithelial tissues line internal cavities and passageways and cover external body surfaces. They are composed mostly of cells that rest on a thin **basement membrane** made of protein fibers. No blood vessels are present. One method of classifying epithelial tissues is by the number of layers of cells. *Simple* epithelia have a single layer of cells, and *stratified* epithelia have multiple layers of cells. Classification is also based on the following cell shapes: *squamous* (thin, flat), *cuboidal* (cube-shaped), and *columnar* (column-shaped).

Transitional epithelium is able to stretch and recoil, so it can be illustrated in either a stretched or a relaxed state. In the relaxed state it appears to be composed of a variety of cell shapes. On the bottom it may contain cuboidal or columnar cells, and at the surface are large, dome-shaped cells that transform into a squamous shape when stretched.

Location

Lines ureters, urinary bladder, urethra, and renal pelvis.

Function

Easily allows stretching and recoiling.

Study Tip

Pattern: In the relaxed state, look for the cell pattern of cuboidal cells near the bottom, columnar in the middle, and large, dome-shaped cells on top.

Key to Illustration		
1. Stratified transitional epithelium	2. Basement membrane	3. Connective tissue

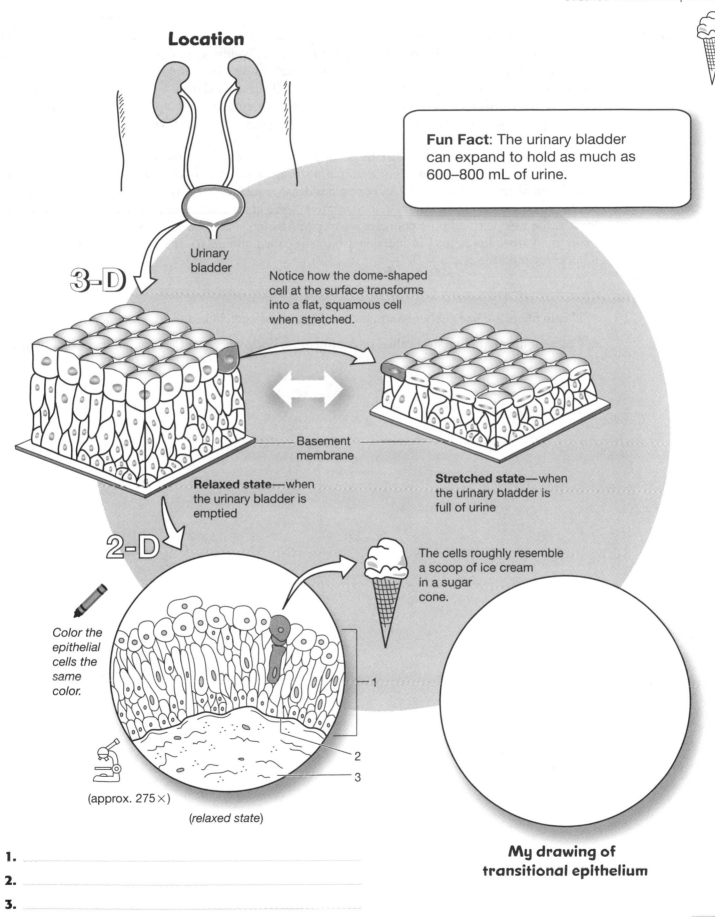

Location

Urinary bladder

Fun Fact: The urinary bladder can expand to hold as much as 600–800 mL of urine.

3-D

Notice how the dome-shaped cell at the surface transforms into a flat, squamous cell when stretched.

Basement membrane

Relaxed state—when the urinary bladder is emptied

Stretched state—when the urinary bladder is full of urine

2-D

The cells roughly resemble a scoop of ice cream in a sugar cone.

Color the epithelial cells the same color.

1

2

3

(approx. 275×)

(relaxed state)

My drawing of transitional epithelium

1. _____

2. _____

3. _____

Variations in Connective Tissues

Description

Connective tissues primarily give structural support to other tissues and organs in the body. Though there are a wide variety of types, all are composed of cells, fibers, and ground substance. The most common cell type is called a **fibroblast,** which manufactures fibers and other extracellular material. The two most common types of protein fibers are **collagen** and **elastin.** Collagen is for strength, and elastin is for elasticity. The cells and the fibers are both embedded in a gel-like material called the **ground substance.** The ground substance varies in its consistency from gelatin-like to a much more rigid material.

The physical traits of connective tissue are determined mainly by the varying proportion of cells, fibers, and the ground substance. For example, a strong connective tissue requires a greater proportion of collagen fibers and fewer cells. An example is dense regular connective tissue, found in tendons that anchor muscle to bone. In contrast, a connective tissue composed mostly of cells is not very strong. Such is the case with adipose connective tissue (*fat tissue*). The main purpose of adipose connective tissue is to store lipids (*fat*) in individual fat cells called **adipocytes.** This tissue contains numerous adipocytes and little else.

Analogy

- **Elastin fibers** are like **rubber bands,** because they allow stretching and recoiling in a tissue.

- **Collagen fibers** are like **steel cables on a suspension bridge,** because they give strength to a tissue.

All connective tissues contain the following basic components:

Cells + Fibers + Ground substance

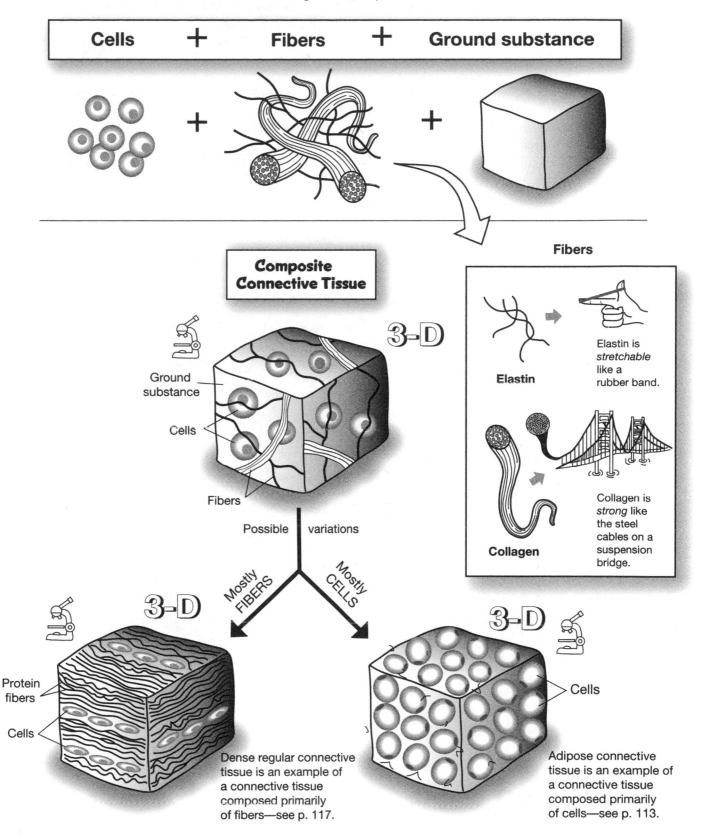

Composite Connective Tissue

3-D

Ground substance

Cells

Fibers

Possible   variations

Fibers

Elastin is *stretchable* like a rubber band.

Elastin

Collagen is *strong* like the steel cables on a suspension bridge.

Collagen

Mostly FIBERS

Mostly CELLS

3-D

Protein fibers

Cells

Dense regular connective tissue is an example of a connective tissue composed primarily of fibers—see p. 117.

3-D

Cells

Adipose connective tissue is an example of a connective tissue composed primarily of cells—see p. 113.

Description

Connective tissues primarily give structural support to other tissues and organs in the body. Though there are a wide variety of types, all are composed of cells, fibers, and ground substance. The most common cell type is called a **fibroblast**, which manufactures the fibers and other extracellular material. The two most common types of protein fibers produced are **collagen** and **elastin**. Collagen is for strength, and elastin is for elasticity. The cells and the fibers are both embedded in a gel-like material called the **ground substance**. The ground substance varies in consistency from being gelatin-like to a much more rigid material.

Loose connective tissues have fewer fibers than other connective tissues and serve as a protective padding in the body. The three tissues classified as loose connective tissues are: *areolar connective tissue*, *adipose connective tissue*, and *reticular connective tissue*.

Areolar connective tissue has a random arrangement of cells, fibers, and ground substance. It contains all the basic components of any connective tissue without being specialized.

Location

Beneath epithelial tissues all over the body; between skin and skeletal muscles; surrounding blood vessels; within skin; around organs; around joints.

Function

Cushions and protects organs; its phagocytes protect against pathogens; holds tissue fluid.

Key to Illustration		
1. Collagen fibers	**2.** Elastin fibers	**3.** Fibroblast nuclei

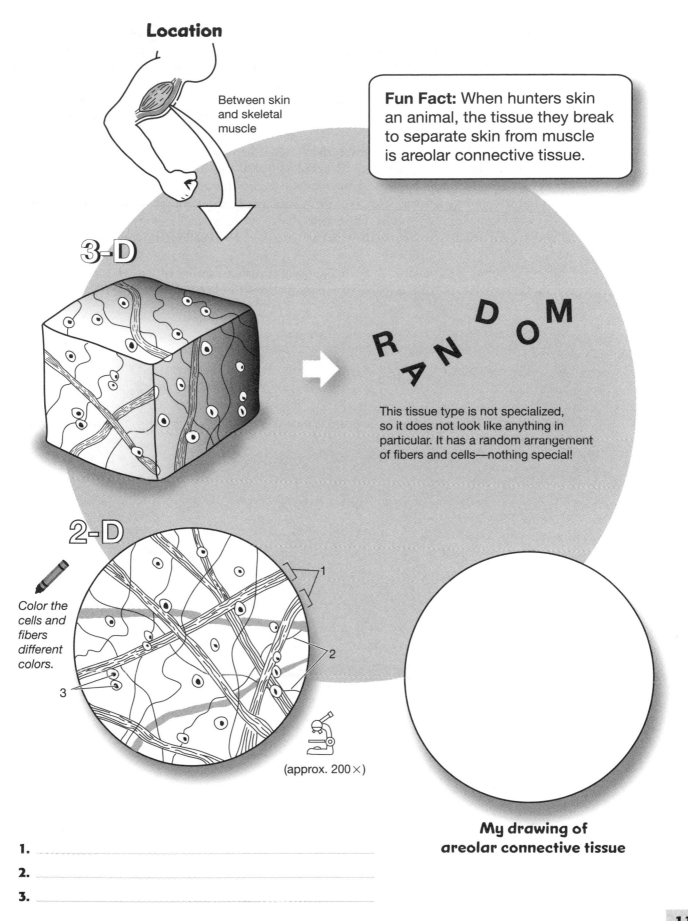

Location

Between skin and skeletal muscle

Fun Fact: When hunters skin an animal, the tissue they break to separate skin from muscle is areolar connective tissue.

3-D

R A N D O M

This tissue type is not specialized, so it does not look like anything in particular. It has a random arrangement of fibers and cells—nothing special!

2-D

Color the cells and fibers different colors.

(approx. 200×)

1

2

3

My drawing of areolar connective tissue

1. _____

2. _____

3. _____

Adipose Connective Tissue

Description

Connective tissues primarily give structural support to other tissues and organs in the body. Though there are a wide variety of types, all are composed of cells, fibers, and ground substance. The most common cell type, called a **fibroblast**, manufactures fibers and other extracellular material. The two most common types of protein fibers produced are **collagen** and **elastin**. Collagen is for strength, and elastin is for elasticity. The cells and the fibers are both embedded in a gel-like material called the **ground substance**. The ground substance varies in its consistency from being gelatin-like to a much more rigid material.

Loose connective tissues have fewer fibers than other connective tissues and serve as a protective padding in the body. There are three tissues classified as loose connective tissues: *areolar connective tissue*, *adipose connective tissue*, and *reticular connective tissue*.

Adipose connective tissue is fat tissue. It is composed almost entirely of fat cells called **adipocytes**, along with some blood vessels. These cells have a large vacuole to store lipids (*fats*). Though adipocytes are not able to divide, they do change in size by expanding or shrinking depending on the amount of lipid stored inside their vacuoles. For example, as a person loses weight, the amount of lipid in the adipocyte's vacuole decreases, causing the cell to shrink in size. Unfortunately, if a person regains that weight, the cells are able to expand back to their original size.

Location

Under all skin but especially in abdomen, buttocks, and breasts; around some organs such as eyeballs and kidneys.

Function

Protects certain organs and other structures; insulates against heat loss through the skin; stores energy as a reserve fuel.

Key to Illustration		
1. Blood vessel	3. Vacuole for lipid storage	4. Plasma membrane of adipocyte
2. Nuclei of adipocytes (*fat cells*)		(*fat cell*)

Location

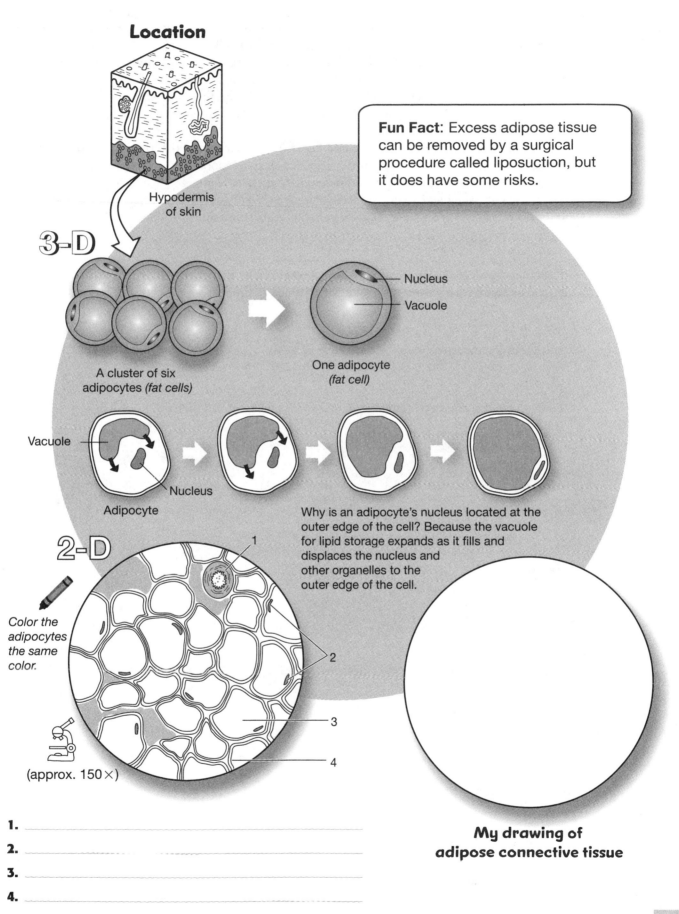

Hypodermis
of skin

Fun Fact: Excess adipose tissue can be removed by a surgical procedure called liposuction, but it does have some risks.

3-D

A cluster of six adipocytes *(fat cells)*

One adipocyte *(fat cell)*

Nucleus

Vacuole

Vacuole

Nucleus

Adipocyte

Why is an adipocyte's nucleus located at the outer edge of the cell? Because the vacuole for lipid storage expands as it fills and displaces the nucleus and other organelles to the outer edge of the cell.

2-D

Color the adipocytes the same color.

1

2

3

4

(approx. 150×)

My drawing of adipose connective tissue

1. _____

2. _____

3. _____

4. _____

Description   Connective tissues primarily give structural support to other tissues and organs in the body. Though there are a wide variety of types, all are composed of cells, fibers, and ground substance. The most common cell type, called a **fibroblast**, manufactures fibers and other extracellular material. The two most common types of protein fibers produced are **collagen** and **elastin**. Collagen is for strength, and elastin is for elasticity. The cells and the fibers are both embedded in a gel-like material called the **ground substance**. The ground substance varies in its consistency from being gelatin-like to a much more rigid material.

Loose connective tissues have fewer fibers than other connective tissues and serve as a protective padding in the body. There are three tissues classified as loose connective tissues: *areolar connective tissue*, *adipose connective tissue*, and *reticular connective tissue*.

Reticular (*reticulata* = net) **connective tissue** primarily consists of a network of reticular fibers. The most common cell type is the reticular cell, but it also contains fibroblasts and macrophages.

Analogy   **Reticular connective tissue** is like **many cobwebs**. The **cobweb** itself is like the **network of reticular fibers** scattered throughout the tissue that physically support a variety of cell types.

Location   Spleen; bone marrow; lymph nodes; liver; kidney.

Function   Fibers form a supportive netlike structure for a variety of cell types.

Key to Illustration		
1. Reticular fibers	**2.** Cell	**3.** Ground substance

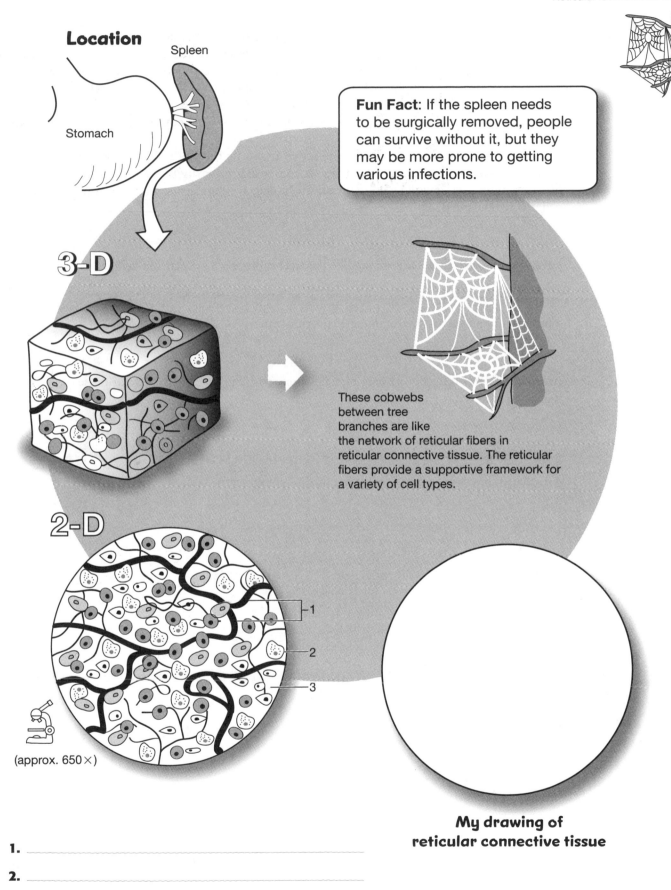

Location

Spleen

Stomach

Fun Fact: If the spleen needs to be surgically removed, people can survive without it, but they may be more prone to getting various infections.

3-D

These cobwebs between tree branches are like the network of reticular fibers in reticular connective tissue. The reticular fibers provide a supportive framework for a variety of cell types.

2-D

1

2

3

(approx. 650×)

My drawing of reticular connective tissue

1. _____

2. _____

3. _____

TISSUES: Connective

Dense (*Fibrous*) Regular Connective Tissue

Description

Connective tissues primarily give structural support to other tissues and organs in the body. Though there are a wide variety of types, all are composed of cells, fibers, and ground substance. The most common cell type is called a **fibroblast**, which manufactures fibers and other extracellular material. The two most common types of protein fibers produced are **collagen** and **elastin**. Collagen is for strength, and elastin is for elasticity. The cells and fibers are both embedded in a gel-like material called the **ground substance**. The ground substance varies in its consistency from being gelatin-like to a much more rigid material.

There are two types of dense connective tissue in the body: (1) dense *regular* and (2) dense *irregular*. **Dense regular connective tissue** is composed primarily of collagen fibers, so it is also called *fibrous connective* or *collagenous tissue*. The collagen fibers are stacked on top of each other in an orderly arrangement.

Analogy

Layers of **collagen fibers** are strong like the **steel cables on a suspension bridge**.

Location

Tendons and aponeuroses; ligaments; covering around skeletal muscles.

Function

Anchors skeletal muscle to bone; attaches bone to bone; packages skeletal muscles; stabilizes bones within a joint.

Study Tips

- Fibroblasts are in rows sandwiched between collagen fibers.
- Collagen fibers are layered in an organized arrangement.

Key to Illustration	
1. Collagen fibers	2. Nuclei of fibroblasts

Dense (*Fibrous*) Regular Connective Tissue

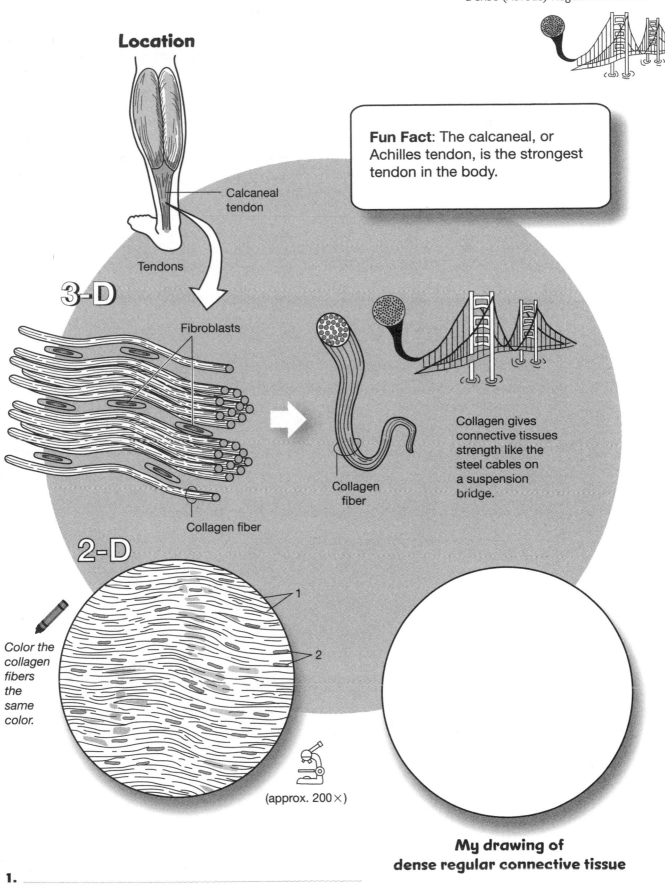

Location

Calcaneal tendon

Tendons

Fun Fact: The calcaneal, or Achilles tendon, is the strongest tendon in the body.

3-D

Fibroblasts

Collagen fiber

Collagen fiber

Collagen gives connective tissues strength like the steel cables on a suspension bridge.

2-D

Color the collagen fibers the same color.

1

2

(approx. 200×)

My drawing of dense regular connective tissue

1. _____

2. _____

TISSUES: Connective

Dense Irregular Connective Tissue

Description   Connective tissues primarily give structural support to other tissues and organs in the body. Though they are of a wide variety of types, all are composed of cells, fibers, and ground substance. The most common cell type is called a **fibroblast**, which manufactures the fibers and other extracellular material. The two most common types of protein fibers produced are **collagen** and **elastin**. Collagen is for strength, and elastin is for elasticity. The cells and the fibers are both embedded in a gel-like material called the **ground substance**. The ground substance varies in its consistency from being gelatin-like to a much more rigid material.

There are two types of dense connective tissue in the body: (1) dense *regular* and (2) dense *irregular*. **Dense irregular connective tissue** is characterized by a random arrangement of collagen fibers and a greater proportion of ground substance.

Location   Dermis of the skin; periosteum; visceral organ capsules; around muscles.

Function   Resists stresses applied in many different directions.

Study Tips
- Fibroblasts are more scattered throughout the tissue.
- Collagen fibers are **not** stacked on top of each other, randomly arranged.

Key to Illustration

1. Nucleus of fibroblast	2. Collagen fibers	3. Ground substance

Location

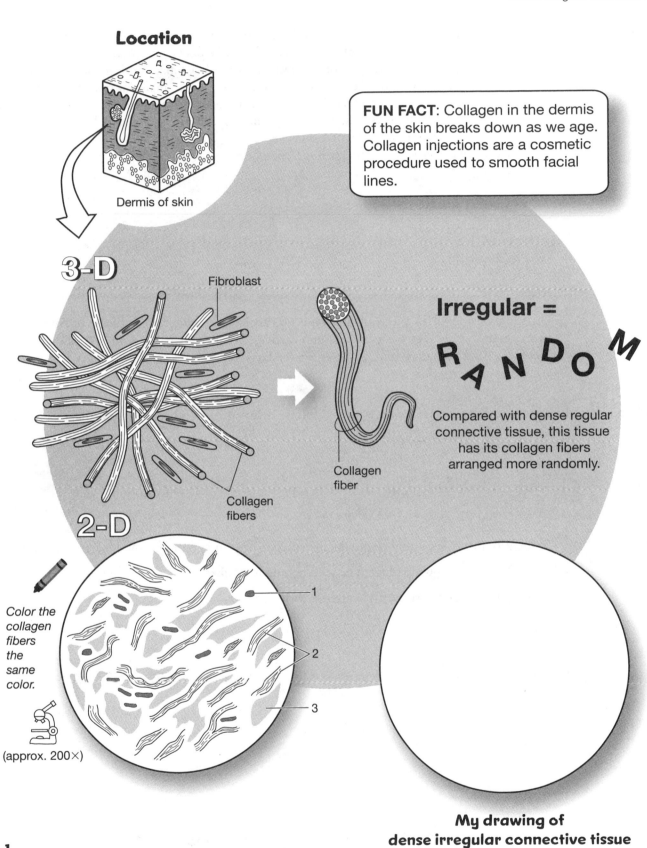

Dermis of skin

FUN FACT: Collagen in the dermis of the skin breaks down as we age. Collagen injections are a cosmetic procedure used to smooth facial lines.

3-D

Fibroblast

Collagen fibers

Irregular =

R A N D O M

Collagen fiber

Compared with dense regular connective tissue, this tissue has its collagen fibers arranged more randomly.

2-D

Color the collagen fibers the same color.

(approx. 200×)

My drawing of dense irregular connective tissue

1. _____

2. _____

3. _____

Description

Connective tissues primarily give structural support to other tissues and organs in the body. Though there are a wide variety of types, all are composed of cells, fibers, and matrix.

Cartilage is a specialized type of connective tissue. It is characterized by three traits: **lacunae**, **chondrocytes**, and a rigid **matrix**. The **matrix** is a firm, gel-like material that contains protein fibers and other substances. Within the matrix are small cavities called **lacunae**. Within the lacunae are living cartilage cells called **chondrocytes**. Because cartilage lacks blood vessels, chondrocytes rely on the diffusion of nutrients into the matrix to survive.

The three basic types of cartilage in the body are:

- **Hyaline cartilage**
- **Elastic cartilage**
- **Fibrous cartilage**

Hyaline cartilage is the most common type of cartilage.

Analogy

Three-dimensionally, a piece of **any type of cartilage** is similar to a **block of Swiss cheese** in its structure and general consistency. Though cartilage is much stronger, both are solid and flexible. The **cheese** itself is the **matrix**, and the **holes** are the **lacunae**.

Location

Covers ends of long bones in synovial joints; between ribs and sternum; cartilages of nose, trachea, larynx, and bronchi; most portions of embryonic skeleton.

Function

Structural reinforcement, slightly flexible support; reduces friction within joints.

Study Tips

To identify this tissue either under the microscope or from a photograph, look for the following:

- This is the **only cartilage type with no apparent fibers** (*they are present but do not stain well*).
- Chondrocytes are evenly scattered within matrix.

Key to Illustration	
1. Matrix	3. Chondrocyte (*cartilage cell*)
2. Lacunae	4. Nucleus of chondrocyte

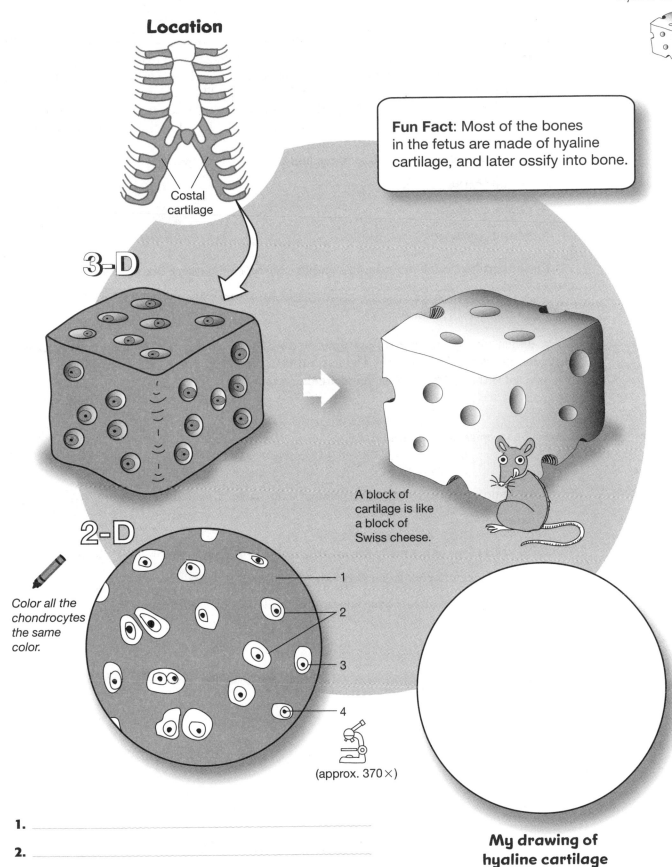

Location

Costal cartilage

Fun Fact: Most of the bones in the fetus are made of hyaline cartilage, and later ossify into bone.

3-D

2-D

Color all the chondrocytes the same color.

A block of cartilage is like a block of Swiss cheese.

1

2

3

4

(approx. 370×)

My drawing of hyaline cartilage

1. _____

2. _____

3. _____

4. _____

Description

Connective tissues primarily give structural support to other tissues and organs in the body. Though there are a wide variety of types, all are composed of cells, fibers, and matrix.

Cartilage is a specialized type of connective tissue. It is characterized by three traits: **lacunae**, **chondrocytes**, and a rigid **matrix**. The **matrix** is a firm, gel-like material that contains protein fibers and other substances. Within the matrix are small cavities called **lacunae**. Within the lacunae are living cartilage cells called **chondrocytes**. Because cartilage lacks blood vessels, chondrocytes rely on the diffusion of nutrients into the matrix to survive.

The three basic types of cartilage in the body are:

- **Hyaline cartilage**
- **Elastic cartilage**
- **Fibrous cartilage**

Elastic cartilage is the most durable and flexible type of cartilage because of the presence of many elastic fibers.

Analogy

Three-dimensionally, a piece of **any type of cartilage** is similar to a **block of Swiss cheese** in its structure and general consistency. Though cartilage is much stronger, both are solid and flexible. The **cheese** itself is the **matrix**, and the **holes** are the **lacunae**.

Location

External ear; epiglottis; auditory canal.

Function

Provides support and easily returns to original shape when distorted.

Study Tips

To identify this tissue either under the microscope or from a photograph, look for the following:

- Chondrocytes appear larger than other cartilages.
- Numerous elastic fibers have appearance of plant roots branching in the soil.

Key to Illustration		
1. Matrix	3. Chondrocyte (*cartilage cell*)	5. Nucleus of chondrocyte
2. Lacunae	4. Elastin fiber	

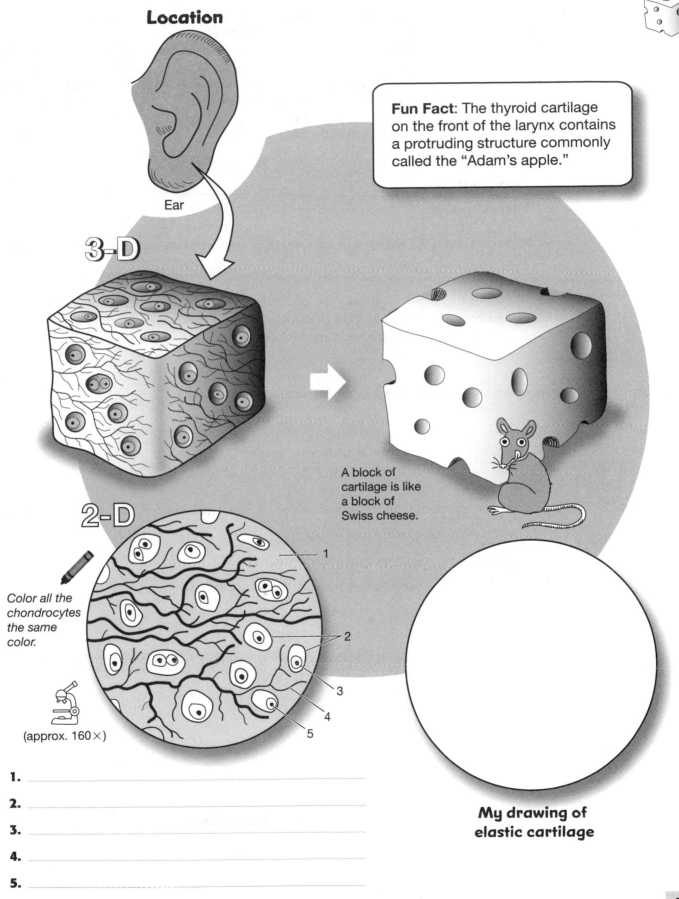

Location

Ear

Fun Fact: The thyroid cartilage on the front of the larynx contains a protruding structure commonly called the "Adam's apple."

3-D

2-D

Color all the chondrocytes the same color.

(approx. 160×)

A block of cartilage is like a block of Swiss cheese.

1.
2.
3.
4.
5.

My drawing of elastic cartilage

TISSUES: Connective

Description   Connective tissues primarily give structural support to other tissues and organs in the body. Though there are a wide variety of types, all are composed of cells, fibers, and matrix.

Cartilage is a specialized type of connective tissue. It is characterized by three traits: **lacunae**, **chondrocytes**, and a rigid **matrix**. The **matrix** is a firm, gel-like material that contains protein fibers and other substances. Within the matrix are small cavities called **lacunae**. Within the lacunae are living cartilage cells called **chondrocytes**. Because cartilage lacks blood vessels, chondrocytes rely on the diffusion of nutrients into the matrix to survive.

The three basic types of cartilage in the body are:

- **Hyaline cartilage**
- **Elastic cartilage**
- **Fibrous cartilage**

Fibrous cartilage is the strongest of the three types because of the presence of many **collagen** fibers.

Analogy   Three-dimensionally, a piece of **any type of cartilage** is similar to a **block of Swiss cheese** in its structure and general consistency. Though cartilage is much stronger, both are solid and flexible. The **cheese** itself is the **matrix**, and the **holes** are the **lacunae**.

Location   Intervertebral disks; pubic symphysis; pads within knee joint.

Function   Shock absorber in a joint; resists compression.

Study Tips   To identify this tissue either under the microscope or from a photograph, look for the following:

- Has the most collagen fibers of any cartilage.
- Collagen fibers often appear in a wavy pattern.
- Chondrocytes are often seen in rows and/or small clusters.

Key to Illustration

1. Lacuna	**3.** Nucleus of chondrocyte	**4.** Collagen fibers in matrix
2. Chondrocyte		

Fibrous Cartilage

Location

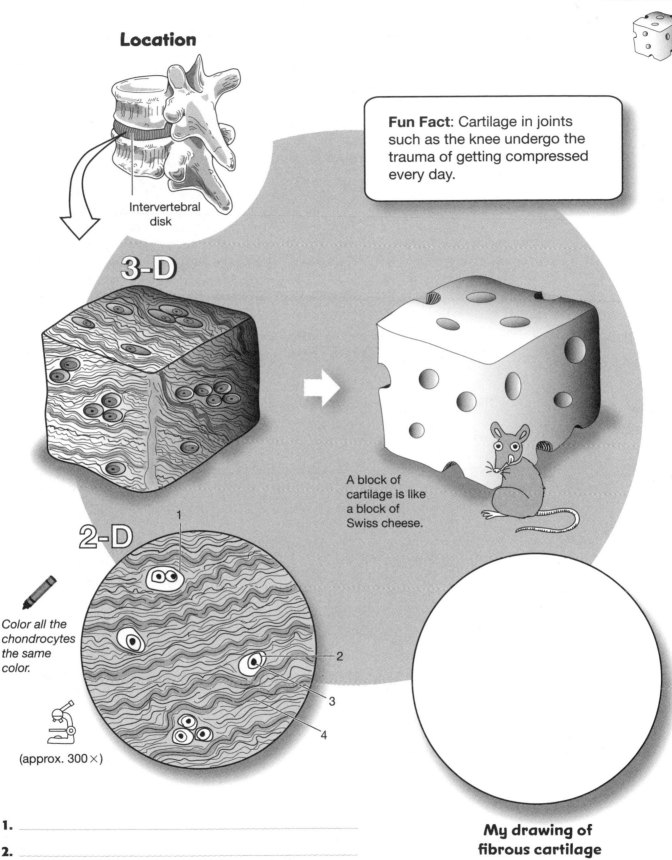

Intervertebral disk

Fun Fact: Cartilage in joints such as the knee undergo the trauma of getting compressed every day.

3-D

A block of cartilage is like a block of Swiss cheese.

2-D

Color all the chondrocytes the same color.

(approx. 300×)

1. _____

2. _____

3. _____

4. _____

My drawing of fibrous cartilage

Bone (*Osseous*) Tissue

Description

Connective tissues primarily give structural support to other tissues and organs in the body. Though there are a wide variety of types, all are composed of cells, fibers, and matrix.

Bone is a specialized type of connective tissue that has calcified into a hard substance. It is composed of organic and inorganic substances. The inorganic portion that constitutes about two-thirds of bone mass is made of modified calcium phosphate compounds called **hydroxyapatite**, and the organic portion is composed of **collagen** fibers. The two general types of bone are **spongy** and **compact**. Spongy bone is less organized and is found in the ends of long bones and other places. Compact bone is more complex and orderly in structure and is found in the shaft of long bones and other locations.

Let's examine compact bone in more detail. The individual units in compact bone are tall, cylindrical towers called **osteons** (*Haversian systems*). In the middle of each osteon is a **central canal** that serves as a passageway for blood vessels. Around this canal are concentric rings of bony tissue called **lamellae**. Along each of these rings at regular intervals are small spaces called **lacunae** that contain a mature bone cell, or **osteocyte**. Branching between individual lacunae are smaller passageways called **canaliculi** that allow fluid with dissolved nutrients to travel to osteocytes.

Analogy

In a transverse cut, each **surface of an osteon** looks like a **tree stump**. Both structures are made of hard, dense materials. Like the **growth rings** in a tree, the osteon has concentric rings called **lamellae**.

Location

Bones.

Function

Bone supports the body and protects vital organs; provides attachments for muscle to form a lever system for movement; stores calcium compounds and fat. Marrow contains stem cells that produce all blood cell types.

Key to Illustration		
1. Osteon	2. Central canal	3. Osteocytes inside lacunae

Bone (*Osseous*) Tissue

Location

Fun Fact: Bone is stronger than concrete and nearly as strong as steel.

3-D

Wedge of compact bone

Wedge of compact bone (detailed)

The surface of one osteon

From a superior view, each osteon looks like a tree stump.

2-D

Color all the osteons the same color.

1

2

3

(approx. 120×)

1. _____

2. _____

3. _____

My drawing of compact bone

Description

There are three different types of muscle tissue:

- **Skeletal muscle**
- **Cardiac muscle**
- **Smooth muscle**

Skeletal muscle is under conscious control, so it is also referred to as *voluntary* muscle. Each skeletal muscle cell is a long cylinder with a banding pattern, and each band is called a **striation**. Most body cells have only one nucleus per cell, but skeletal muscle has multiple nuclei in each cell—a unique feature.

Analogy

Under the microscope, **skeletal muscle** appears as a bunch of **stacked logs**. Each log is equivalent to one **skeletal muscle cell**. Consider them birch logs that have a striped pattern on them. These **stripes** are the **striations**. Note that the log doesn't show us the entire tree, just as the image under the microscope doesn't show us the entire cell. This is because the cells are very long.

Location

All the major muscles of the body are composed of skeletal muscle. Examples of skeletal muscle are the biceps brachii, gluteus maximus, and pectoralis major.

Function

Contraction of muscles (*conscious control*).

Study Tips

Under the microscope at higher magnifications, you can use the following landmarks to distinguish skeletal muscle tissue:

- Striations.
- Multiple nuclei per cell.
- Long cells. (Each cell is so long that you cannot see the ends of it under high magnification.)

Key to Illustration		
1. Nuclei within one skeletal muscle cell	2. One skeletal muscle cell	3. Striation

Location

Skeletal muscle

Fun Fact: The human body is composed of more than 600 skeletal muscles. Together, they constitute about 40 percent of body mass.

3-D

Nucleus
Striations
Sarcolemma
(*plasma membrane*)

Skeletal muscle cell

Each skeletal muscle cell (or fiber) is like a birch tree log because it is a long cylinder with a striped pattern.

2-D

1

2

3

Color the muscle cells the same color.

(approx. 150×)

1. _____

2. _____

3. _____

My drawing of skeletal muscle

Description

There are three different types of muscle tissue:

- Skeletal muscle
- Cardiac muscle
- Smooth muscle

Cardiac muscle is under our unconscious control. Each cell has a somewhat cylindrical shape and a single nucleus per cell. One cell connects with another to form a union called an **intercalated disk**. This structure can be seen with a compound microscope. Vertical bands run up and down each cell to form a striped pattern in which each stripe is a **striation**.

Analogy

The **intercalated disks** are like **two pieces of a jigsaw puzzle** fitting together.

Location

Cardiac muscle is found *only* in the heart.

Function

Contraction of muscles (*unconscious control*).

Study Tips

Under the microscope at higher magnifications, you can use the following landmarks to distinguish cardiac muscle tissue:

- Striations.
- Single nucleus per cell.
- Nucleus appears oval or rounded.
- Intercalated disks.
- Forking or branching pattern.

Key to Illustration		
1. Nucleus of one cardiac muscle cell	2. Striation 3. Intercalated disk	4. Individual cardiac muscle cell

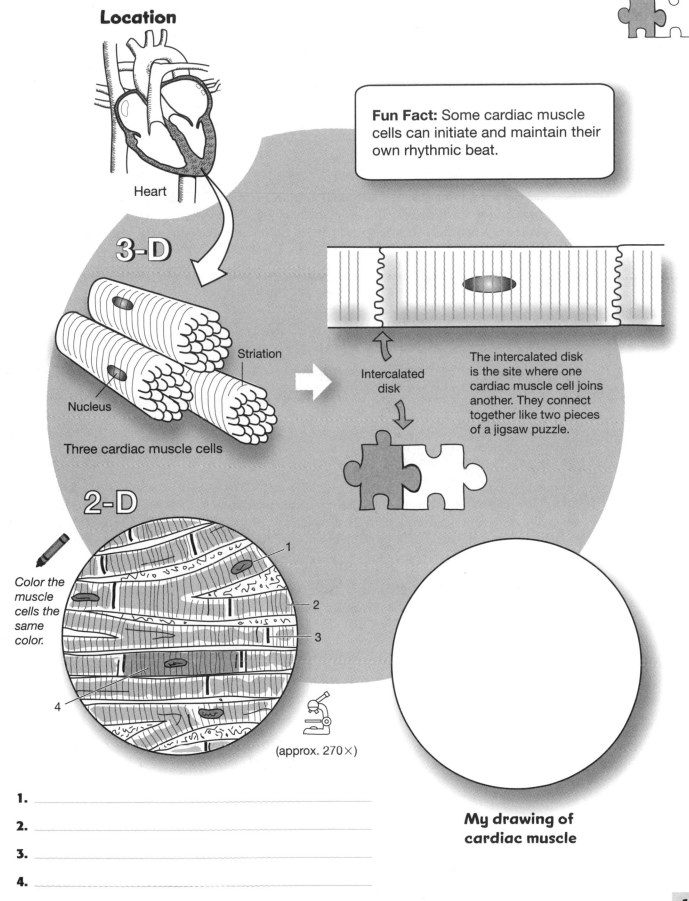

Location

Heart

Fun Fact: Some cardiac muscle cells can initiate and maintain their own rhythmic beat.

3-D

Striation

Nucleus

Three cardiac muscle cells

Intercalated disk

The intercalated disk is the site where one cardiac muscle cell joins another. They connect together like two pieces of a jigsaw puzzle.

2-D

Color the muscle cells the same color.

1
2
3
4

(approx. 270×)

1. _____

2. _____

3. _____

4. _____

My drawing of cardiac muscle

Description

There are three different types of muscle tissue:

- Skeletal muscle
- Cardiac muscle
- Smooth muscle

Smooth muscle is under unconscious control. It *lacks the striations* found in the other two types of muscle tissue, and each cell has only one nucleus.

Analogy

In a **sheet of smooth muscle** the **individual cells** are stacked one on top of the other and staggered in their appearance. Each short cell is shaped like elongated ravioli, because it is thicker in the middle and tapered on each end. This staggered pattern is similar to the pattern of **bricks in a wall**.

Location

The two types of smooth muscle are **visceral** and **multiunit**:

Type	Location
Visceral	Walls of hollow organs (stomach, intestines, urinary bladder, etc.)
Multiunit	Walls of large arteries; trachea; muscles in the iris and ciliary body of the eye; arrector pili muscles, which attach to hair follicles

Function

Contraction of muscles (*unconscious control*).

Study Tips

Under the microscope, it will be difficult to see the plasma membrane of individual cells. Instead, you will have to rely on the overall pattern of staggered cells stacked on top of one another to identify this tissue. The general pattern will show up in how the nuclei are arranged with respect to each other. You can also use the following landmarks:

- Single nucleus per cell.
- No striations (this is why it is called *smooth*).

Key to Illustration		
1. Individual smooth muscle cell	2. Nucleus of one smooth muscle cell	3. Plasma membrane

Location

Wall of digestive tract

Fun Fact: The drug pitocin is a hormone used to induce labor. It causes the smooth muscle in the uterus to contract.

3-D

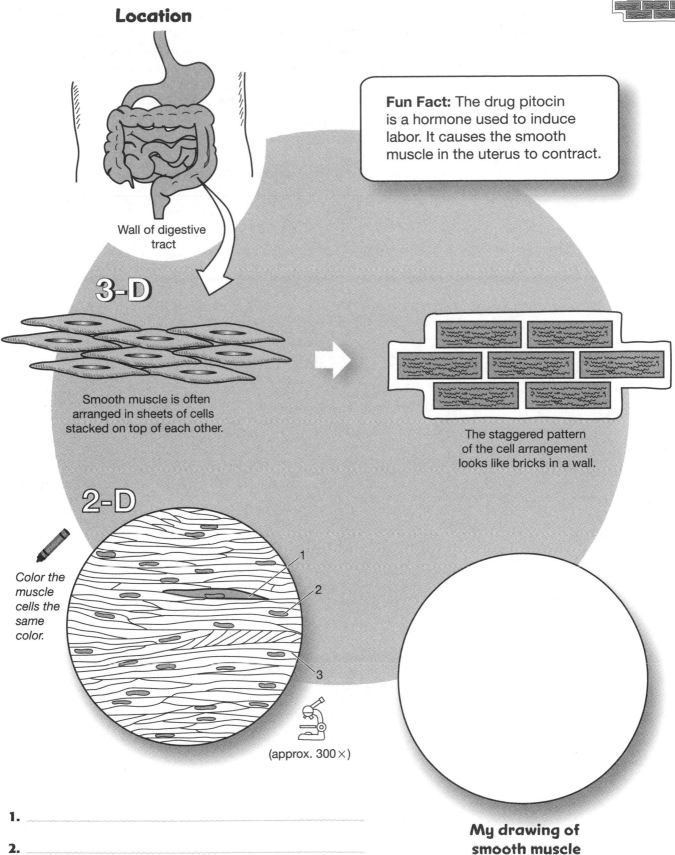

Smooth muscle is often arranged in sheets of cells stacked on top of each other.

The staggered pattern of the cell arrangement looks like bricks in a wall.

2-D

Color the muscle cells the same color.

1
2
3

(approx. 300×)

My drawing of smooth muscle

1. _____

2. _____

3. _____

Neural Tissue

Description

Neurons, or nerve cells, are some of the fundamental cells in neural tissue. Of the variety of types of neurons, all share certain features. Surrounding the nucleus of every neuron is a region called the **cell body**. Most of the organelles are found here. Branching out from the cell body is one of two types of processes—**dendrites** or an **axon**. Each neuron has only one axon per cell but may have one or more dendrites. At the end of every axon is the **synaptic knob**, which defines the end of the cell. Surrounding the neurons are various smaller cells that constitute the **neuroglia**, which offer structural support and protection.

Analogy

Under the microscope, a **neuron** resembles an **octopus**. The **head** of the octopus is the **cell body**, and the **tentacles** are the **cellular processes** (dendrites or axon).

Location

Brain; spinal cord; peripheral nerves.

Function

Conduct nervous (electrical) impulses to other neurons, muscles, or glands to regulate their function.

Study Tips

Under the microscope at higher magnifications, you can use the following landmarks to distinguish nervous tissue:

- Cellular processes (unique to nervous tissue).
- Cell body.
- Rounded nucleus.
- Nucleolus (may or may not be visible depending on slide quality and magnification).

Key to Illustration		
1. Dendrites	3. Nucleus	5. Axon
2. Cell body	4. Nucleolus	6. Neuroglia

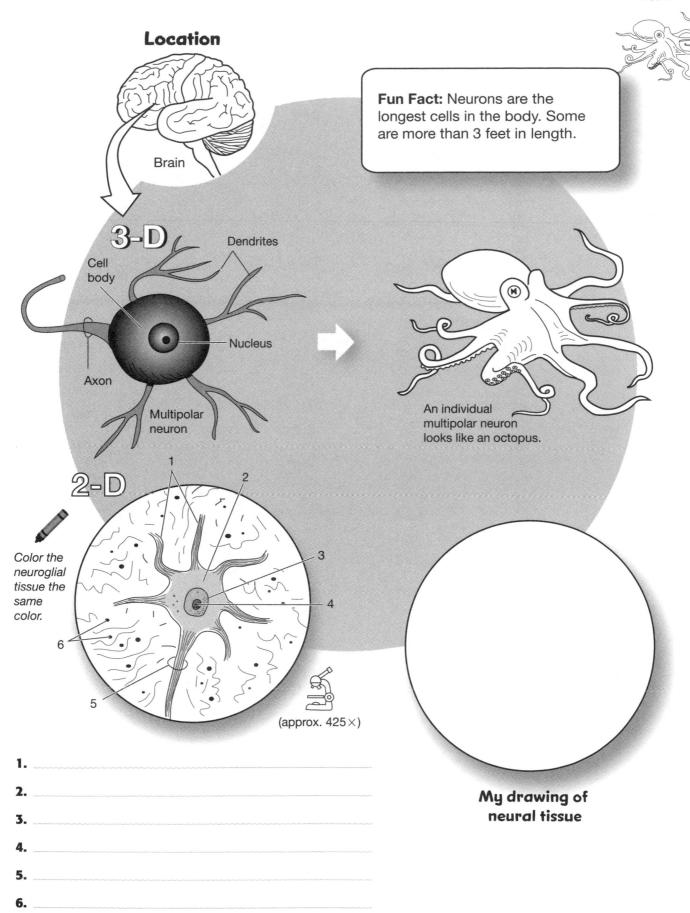

Location

Brain

Fun Fact: Neurons are the longest cells in the body. Some are more than 3 feet in length.

3-D

Cell body

Dendrites

Nucleus

Axon

Multipolar neuron

An individual multipolar neuron looks like an octopus.

2-D

Color the neuroglial tissue the same color.

(approx. 425×)

My drawing of neural tissue

1. _____

2. _____

3. _____

4. _____

5. _____

6. _____

INTEGUMENTARY SYSTEM

Description

The skin, or **integumentary system**, is the largest organ in the body by total surface area. In the average adult it measures about 21 sq. ft. It is divided into three regional areas: **epidermis**, **dermis**, and **hypodermis** (*subcutaneous region*). The epidermis is the outermost layer and is composed of layers of cells. The dermis lies directly beneath and is subdivided into the upper **papillary region** and the lower **reticular region**. The papillary region consists of loose connective tissue, and the reticular region consists of dense irregular connective tissue. The dermis houses many of the glands and hair follicles within the skin. The hypodermis is below the dermis and is composed of both areolar and adipose connective tissues. Though not a true part of the skin, it is the foundation on which the skin rests. The upper region contains arteries and veins. When a *subcutaneous injection* is given with a *hypodermic* needle, it penetrates into these vessels so the drug can be delivered directly into the bloodstream.

Accessory structures include hair follicles, hair shafts, glands, and sensory receptors. **Hair follicles** produce **hair shafts**. An **arrector pili** muscle is a small cluster of smooth muscle cells connected to hair follicles. When they contract, they pull the hair follicle up, causing "goose bumps." Typical glands are **sebaceous glands** and **sweat glands**. Sebaceous (oil) glands are often connected to hair follicles. They produce **sebum**—an oily substance that lubricates hair shafts.

There are two types of sweat glands: **merocrine** and **apocrine**. Merocrine glands are more common and widely distributed throughout the skin. They secrete a watery secretion called sweat into a duct and release it directly onto the skin surface. This watery film absorbs heat, then evaporates, resulting in cooling the body. Apocrine glands are less common and are connected to hair follicles. They are located in the armpits, around nipples, and in the groin. Their secretion contains both lipids and proteins. When bacteria act on this secretion, the result is body odor.

Sensory receptors include **tactile corpuscles**, **free nerve endings**, and **lamellated corpuscles**. Tactile corpuscles detect mostly light touch, free nerve endings detect mostly pain and temperature changes, and lamellated corpuscles respond to changes in deep pressure.

Function

Physical protection from environment, regulation of body temperature, synthesis of vitamin D, excretion of waste products, prevention of water loss, prevention of invasion by pathogens.

Study Tips

Layers of Epidermis

- To remember the layers of the epidermis from the outermost to innermost layer, use the following mnemonic: "Can Lucy Give Some Blood?"

  (**Note:** This works only with thick skin samples such as in the sole of the foot because the *stratum lucidum* is not present in thin skin.)

- You can remember that the stratum **b**asale is at the **b**ottom of the epidermis because *basale* and *basement* both begin with the letter **b**.

Key To Illustration		
Layers of Epidermis (***Thick Skin***)	**Regional Areas of the Skin**	**Accessory Structures**
1. Stratum corneum	6. Epidermis	11. Dermal papillae
2. Stratum lucidum	7. Dermis	12. Tactile corpuscle
3. Stratum granulosum	8. Papillary region	13. Free nerve ending
4. Stratum spinosum	9. Reticular region	14. Sebaceous (*oil*) gland
5. Stratum basale	10. Hypodermis (*subcutaneous region*)	15. Hair shaft
		16. Hair follicle
		17. Lamellated corpuscle
		18. Arrector pili muscle
		19. Sweat gland duct
		20. Merocrine sweat gland
		21. Apocrine sweat gland

Layers of Epidermis

1. _____
2. _____
3. _____
4. _____
5. _____

Regional Areas of the Skin

6. _____
7. _____
8. _____
9. _____
10. _____

Epidermis

Sebaceous gland

2-D

3-D

Adipocyte

Accessory Structures

11. _____
12. _____
13. _____
14. _____
15. _____
16. _____

17. _____
18. _____
19. _____
20. _____
21. _____

Hair Follicles, Hairs, and Sebaceous Glands

Description

Hair follicles, hairs, and sebaceous glands are all related structures. The hair follicles produce the hairs and the sebaceous glands lubricate them. Locate the epidermis, composed of epithelial cells, on the illustration on the facing page. Place your finger on the bottom layer of the epidermis called the **stratum basale** (labeled on the left side of the illustration). As you follow it to the right, you first loop around a glandular structure—the **sebaceous gland**. As you continue moving downward, you are now on the wall of the **hair follicle**. Notice that the tissue here is similar to that of the epidermis. When you hit the bottom you have arrived at a structure called the **hair bulb**. Now your finger is going up the right side of the follicle wall, and you finally end up at the bottom of the epidermis once again. The long structure in the middle of the hair follicle that extends above the epidermis is the **hair**. Now let's examine the details.

Hair follicle: Structure composed of epithelium and connective tissue that produce hairs through a process called **keratinization**. This process transforms epithelial cells into dead cells filled with a tough fibrous protein called **keratin**. Two general types of keratin are produced, namely, the more flexible, **soft keratin**, and the tougher, more rigid **hard keratin**. Hair follicles are divided into the following four parts listed from innermost to outermost:

- **Internal root sheath**—contains epithelial cells that separate the hair from the external root sheath
- **External root sheath**—contains many layers of epithelial cells, thins as it nears hair bulb
- **Glassy membrane**—thickened basement membrane layer
- **Connective tissue sheath**—protective coating of fibrous connective tissue

Hair bulb: The structure located at the base of the hair follicle, deep in the dermis or hypodermis. It contains the following:

- **Hair matrix**—contains cells from the **stratum basale** responsible for growth of existing hairs
- **Hair papilla**—region containing areolar connective tissue and many blood capillaries
- **Melanocyte**—a cell that produces a pigment called **melanin** that gives hair its color
- **Blood vessel**—blood capillaries deliver nutrients to cells in the follicle

Hair: A long cylindrical structure produced by hair follicles that contain keratinized cells. The portion deep within the hair follicle is called the *root* while the portion that extends above the skin is called the *shaft*. Both the root and the shaft contain the following three different layers of cells:

- **Cuticle**—tough, outermost cellular layer; cells contain hard keratin
- **Cortex**—middle and thickest cellular layer; contains numerous cells filled with hard keratin
- **Medulla**—innermost cellular layer; cells are filled with soft keratin

Sebaceous gland (*oil gland*)**:** Classified as a holocrine gland, the sebaceous gland contains secretory cells that produce a secretory product. During the secretion process, these cells increase in size and then burst, releasing both the product as well as the ruptured cells. In this case, the secretory product is an oily substance called **sebum**. As shown in the illustration, after being released, the sebum is discharged through a passageway and onto the hair surface to lubricate it. The reason that the gland never runs out of secretory cells is due to stem cells that line the inside of the gland. They continuously replace themselves through cell division, and many of the new cells transform into new secretory cells. Individuals who have oily hair have very active sebaceous glands.

Key To Illustration		
Hair structure	**Hair follicle structure**	**Hair bulb structures**
1. Cuticle	4. Internal root sheath	8. Hair matrix
2. Cortex	5. External root sheath	9. Hair papilla
3. Medulla	6. Glassy membrane	10. Melanocyte
	7. Connective tissue	11. Blood vessel

Hair Follicles, Hairs, and Sebaceous Glands

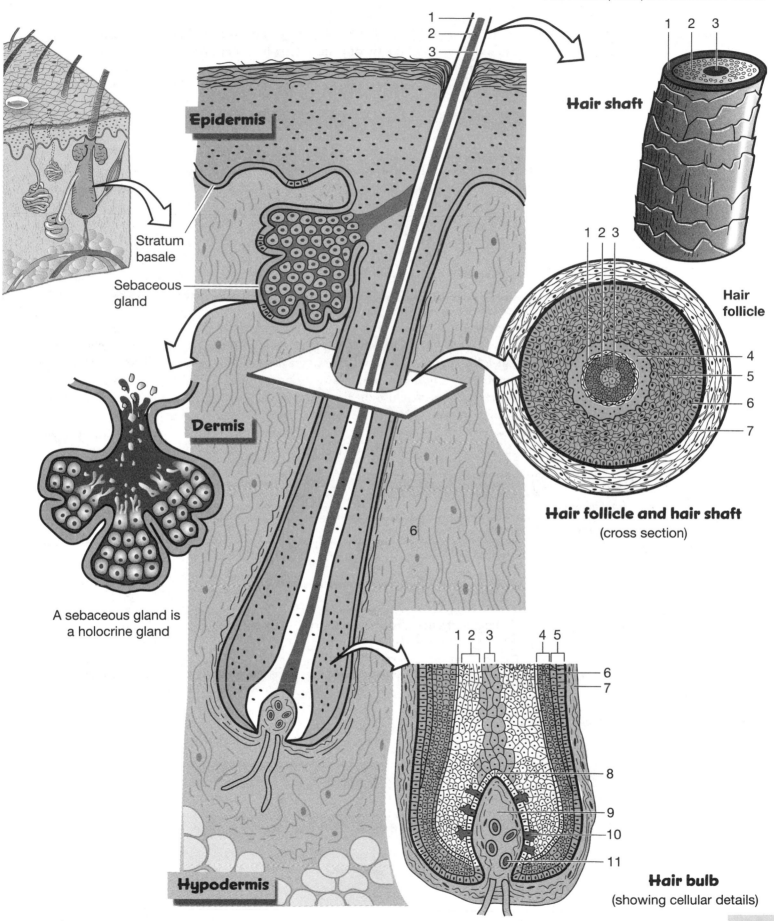

Hair shaft

Epidermis

Stratum basale

Sebaceous gland

Dermis

A sebaceous gland is a holocrine gland

Hypodermis

Hair follicle

Hair follicle and hair shaft
(cross section)

Hair bulb
(showing cellular details)

Description

Scattered throughout the human skin are millions of sweat glands that fall into two categories: **merocrine** (*eccrine*) **sweat glands** and **apocrine sweat glands**. First let's focus on what both types of sweat glands have in common. Both glands are tubular structures coiled into a ball where the sweat (*perspiration*) production occurs. The ball connects to a single excretory duct that terminates in an opening called the sweat pore where the sweat is released. Secretory cells within the tubule use exocytosis (see p. 86) to produce sweat and release it into the tubule's lumen. The nervous system stimulates contractile cells surrounding secretory cells to contract and squeeze the gland, thereby moving the collective sweat through the excretory duct.

Merocrine sweat glands are widely distributed throughout most all the skin and are most plentiful in the palms, soles, and forehead. Their excretory ducts always go through the epidermis so their sweat pore is located on the skin surface. With total number estimates at 2 to 5 million, they are very numerous. The sweat produced is a thin, watery secretion that has a salty taste due to the presence of sodium (Na^+) and chloride (Cl^-) ions. Other components within sweat include urea, amino acids, and glucose. These glands begin functioning in the newborn and their primary function is thermoregulation or cooling the body through evaporation. As sweat is produced and it covers the skin surface, the water in it absorbs heat from the skin's surface which leads to evaporation. Excretion of water, electrolytes (ex: Na^+, Cl^-), and wastes (ex: urea) is a secondary function.

Apocrine sweat glands are far fewer in number and are concentrated in specific skin regions such as the armpits, the area around the nipples, and the pubic region. Unlike merocrine sweat glands, they are associated with hair follicles and are found deeper in the dermis or hypodermis. The sweat produced contains the same components found in merocine sweat glands, but it is thicker due to the addition of lipids and proteins. Bacteria on the skin use the lipids and proteins as nutrients and break it down into other substances, resulting in body odor. These glands become active at puberty and are stimulated by emotional stress or sexual excitement. While their specific function is unknown, they may be related to the sexual scent glands found in other mammals.

Feature	Merocrine Sweat Glands	Apocrine Sweat Glands
Total number	2-5 million	Approximately 2,000
Overall size	Smaller	Larger
Location in skin	Dermis	Deep dermis or hypodermis
Distribution	Widely distributed throughout most of the skin	Concentrated in the armpits, the area around the nipples, and the pubic region
Secretion	Sweat—a thin, watery secretion containing sodium and chloride ions, urea, amino acids, and glucose	Sweat—a thicker secretion containing all the same components as merocrine sweat glands, as well as lipids and proteins
Function	Primary: Thermoregulation Secondary: Water, electrolyte, and waste removal	Unknown—may be related to sexual scent glands in other mammals; stimulated by emotional stress and sexual excitement
Site of sweat excretion	Skin surface (on epidermis)	Hair follicle
Beginning of function	Newborn	Puberty
Process used for sweat production	Exocytosis	Exocytosis

Key To Illustration

1. Droplet of sweat
2. Sweat pore
3. Duct of sweat gland
4. Merocrine gland cell
5. Apocrine gland cell
6. Connective tissue
7. Lumen

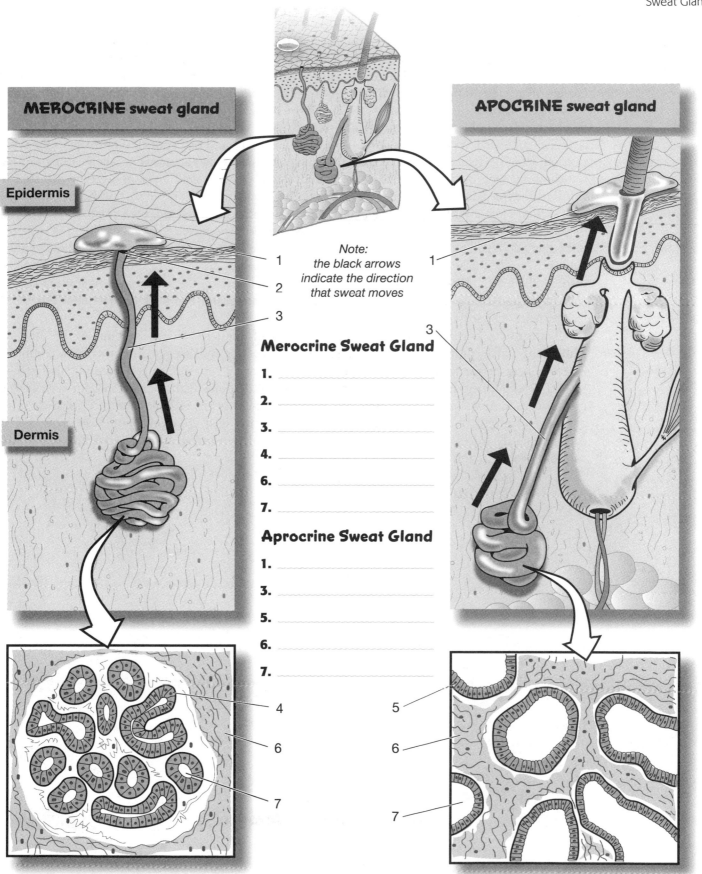

MEROCRINE sweat gland

Epidermis

Dermis

APOCRINE sweat gland

Note:
the black arrows
indicate the direction
that sweat moves

1

2

3

1

3

Merocrine Sweat Gland

1. _____

2. _____

3. _____

4. _____

6. _____

7. _____

Aprocrine Sweat Gland

1. _____

3. _____

5. _____

6. _____

7. _____

4

6

7

5

6

7

A section through a merocrine sweat gland.
Notice the narrower diameter tubes.

A section through an apocrine sweat gland.
Notice the larger diameter tubes.

Description

Nails, hair, and the epidermis of the skin are all formed by a process called **keratinization**. During this process, living cells are transformed into packed layers of dead cells completely filled with a tough, fibrous protein called **keratin**. The reason nails are harder than hair or epidermis is that they contain a harder form of keratin. In humans, nails cover the posterior surface of the ends of the fingers and toes just like claws or hoofs in other animals. They protect the digits from normal wear and tear and allow us to grasp small objects and scratch oursleves.

The nail is divided into three parts—**free edge**, **nail body**, and **nail root**. The **free edge** is the whitest portion, which extends beyond the end of the digit and is also the part we trim when cutting our nails. On the underside of the free edge is the **hyponychium**—a thickened layer of stratum corneum that connects the free edge to the fingertip. The **nail body** is the remaining visible part of the nail, which rests on a section of epidermis called the **nail bed**. Protecting the sides of the nail body are folds of skin called the **lateral nail folds**. Most of the nail body appears pinkish because of the many blood vessels in the underlying dermis. The whitish crescent shaped part at the bottom of the nail bed is called the **lunula**. Its color is from the thickened layer of epidermis beneath it, which prevents the blood vessels from showing through. Below the lunula where the skin meets the nail body is a thickened layer of stratum corneum called the **eponychium** or **cuticle**. This is the structure that gets pushed back during a manicure. Lastly, hidden below the surface in a fold of skin is the **nail root**.

Nail growth begins at the **nail matrix**—a mass of epithelial tissue surrounding the nail root. Mitotically active epithelial cells go through the process of keratinization to add many new compressed layers of keratin-filled cells onto the nail root. As new layers continue to be added, the nail grows in length. Normal, healthy nails have a regular growth rate along with a fairly smooth surface and a uniform color. Any changes in growth rate, texture, or color of the nail may indicate a disease state in a variety of organ systems. For example, yellow nails that lack an eponychium are associated with certain lung disorders. White nails can occur with arsenic poisoning, heart disease, and renal failure. Thyroid problems and aging can lead to brittle nails.

Key to Illustration

1. Free edge of nail body	5. Nail bed	9. Hyponychium
2. Nail body	6. Lateral nail fold	10. Nail root
3. Lunula	7. Epidermis	11. Nail matrix
4. Eponychium (cuticle)	8. Dermis	

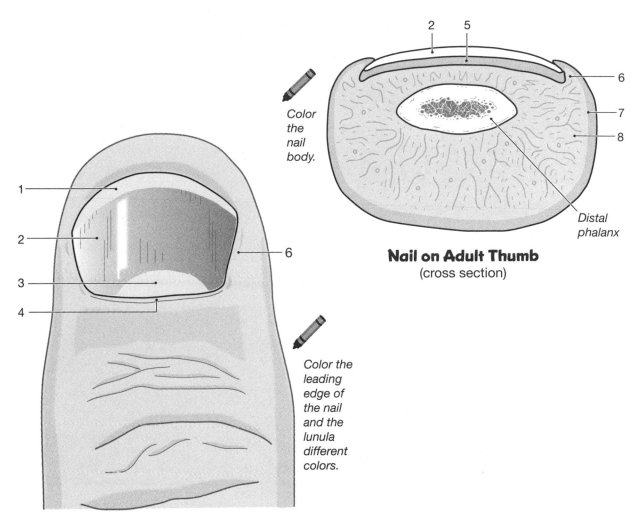

Color the nail body.

Nail on Adult Thumb
(cross section)

Distal phalanx

Color the leading edge of the nail and the lunula different colors.

Nail on Adult Thumb
(posterior view)

Color the nail body.

Distal phalanx

Proximal phalanx

Nail on Adult Thumb
(sagittal view)

1. _____
2. _____
3. _____
4. _____
5. _____
6. _____
7. _____
8. _____
9. _____
10. _____
11. _____

Notes

SKELETAL SYSTEM

Description

The shaft of the long bone is called the **diaphysis**, and each end is called an **epiphysis**. The region between the diaphysis and the epiphysis is called the **metaphysis**. Covering the diaphysis is a sheath of fibrous connective tissue called **periosteum** that aids in the attachment of muscles to bone. Covering each epiphysis is a smooth layer of hyaline cartilage, more generally referred to as **articular cartilage**, because it is used to form joints. This smooth surface helps reduce friction within the joint. Inside the diaphysis is a hollow chamber called the **medullary cavity**. It contains **yellow marrow**, which consists mostly of fatty tissue that acts as a reserve fuel supply for the body. Lining the inside of the medullary cavity is a thin cellular layer called the **endosteum**, which contains both **osteoblasts** and **osteoclasts**.

There are two types of bone within the body—**spongy** and **compact**. Spongy bone can be found within the epiphyses and lining the medullary cavity. Like a sponge, it is more porous, less organized, and contains many open spaces within it. By contrast, compact bone is much more organized and dense. It is much stronger than spongy bone, and it constitutes the wall of the diaphysis. In adults, a thin layer of compact bone is also found at the **epiphyseal line**. This marks where the epiphyseal growth plate used to be located before it ossified.

Analogy

Spongy bone looks similar to a sponge.

Study Tip

Epiphysis is the end of a long bone—"Epiphysis" and "End" both begin with the letter e.

Key to Illustration

1. Proximal epiphysis	6. Spongy bone	11. Yellow marrow
2. Metaphysis	7. Epiphyseal line	12. Compact bone
3. Diaphysis	8. Medullary cavity	13. Wedge of compact bone
4. Distal epiphysis	9. Periosteum	
5. Articular cartilage	10. Endosteum	

Long Bone Structure

Spongy bone
looks like
a sponge

The humerus is
an example
of a long bone.

*Color the
different
structures
different
colors.*

1.
2.
3.
2.
4.
5

Right Humerus
(posterior view)

6
7
8
10
11
12
9
13
(see p. 151)

1. _____   8. _____

2. _____   9. _____

3. _____   10. _____

4. _____   11. _____

5. _____   12. _____

6. _____   13. _____

7. _____

Description

There are two types of bone within the body—**spongy** and **compact**. Spongy bone can be found within the epiphyses and lining the medullary cavity. Like a sponge, it is more porous, less organized, and contains many open spaces within it. By contrast, compact bone is much more organized and dense. It is much stronger than spongy bone, and it constitutes the wall of the diaphysis.

Let's examine compact bone in more detail. The individual units in compact bone are tall, cylindrical towers called **osteons** (*Haversian systems*). In the middle of each osteon is a **central canal** that serves as a passageway for blood vessels. Around this canal are concentric rings of bony tissue called **lamellae**. Along each of these rings at regular intervals are small spaces called **lacunae** that contain a mature bone cell or **osteocyte**. Branching between individual lacunae are smaller passageways called **canaliculi** that allow fluid with dissolved nutrients to travel to osteocytes.

Analogy

Each **surface of an osteon** looks like a **tree stump**. Both structures are made of hard, dense materials. Like the **growth rings** in a tree, the osteon has concentric rings called **lamellae**.

Key to Illustration		
1. Osteon (*Haversian system*)	5. Spongy bone	9. Osteocyte
2. Lamellae	6. Communicating canal	10. Canaliculi
3. Collagen fibers	7. Central canal	
4. Periosteum	8. Lacuna	

Femur

Wedge of compact bone

The surface of an osteon looks like a tree stump (see p. 127).

Osteon (superior view)

1. _____
2. _____
3. _____
4. _____
5. _____
6. _____
7. _____
8. _____
9. _____
10. _____

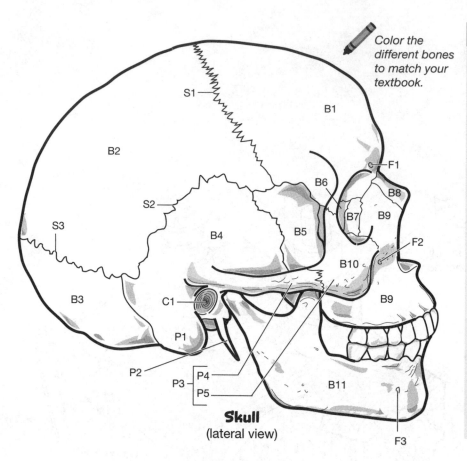

Color the different bones to match your textbook.

Skull
(lateral view)

Key to Lateral View

Bones (B)

B1	Frontal bone
B2	Parietal bone
B3	Occipital bone
B4	Temporal bone
B5	Sphenoid
B6	Ethmoid
B7	Lacrimal bone
B8	Nasal bone
B9	Maxilla
B10	Zygomatic bone
B11	Mandible

Sutures (S)

S1	Coronal suture
S2	Squamous suture
S3	Lambdoid suture

Processes, Projections (P)

P1	Mastoid process
P2	Styloid process
P3	Zygomatic arch
P4	Zygomatic process of temporal bone
P5	Temporal process of zygomatic bone

Foramina (F)

F1	Supraorbital foramen
F2	Infraorbital foramen
F3	Mental foramen

Canal (C)

C1	External acoustic meatus (*canal*)

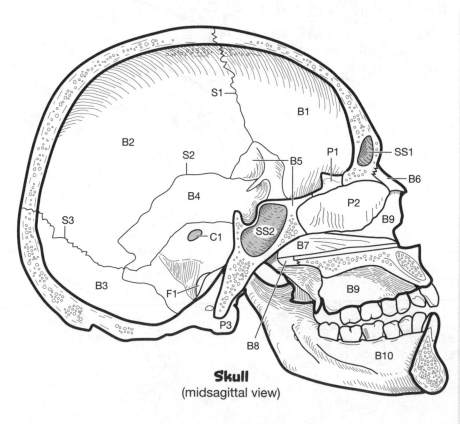

Skull
(midsagittal view)

Key to Midsagittal View

Bones (B)

B1	Frontal bone
B2	Parietal bone
B3	Occipital bone
B4	Temporal bone
B5	Sphenoid
B6	Nasal bone
B7	Vomer
B8	Palatine bone
B9	Maxilla
B10	Mandible

Sutures (S)

S1	Coronal suture
S2	Squamous suture
S3	Lambdoid suture

Processes, Projections (P)

P1	Crista galli of ethmoid
P2	Perpendicular plate of ethmoid
P3	Occipital condyle

Foramina (F)

F1	Jugular foramen

Sinuses (SS)

SS1	Frontal sinus
SS2	Sphenoidal sinus

Canal (C)

C1	Internal acoustic meatus (*canal*)

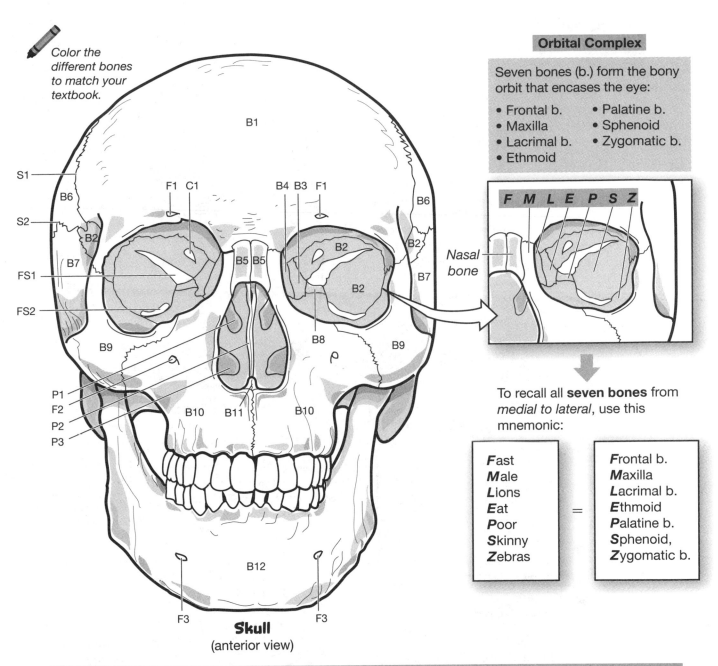

Color the different bones to match your textbook.

Orbital Complex

Seven bones (b.) form the bony orbit that encases the eye:

- Frontal b.
- Maxilla
- Lacrimal b.
- Ethmoid
- Palatine b.
- Sphenoid
- Zygomatic b.

F M L E P S Z

Nasal bone

To recall all **seven bones** from *medial to lateral*, use this mnemonic:

| **F**ast **M**ale **L**ions **E**at **P**oor **S**kinny **Z**ebras | = | **F**rontal b. **M**axilla **L**acrimal b. **E**thmoid **P**alatine b. **S**phenoid, **Z**ygomatic b. |

Skull
(anterior view)

Key to Anterior View

Bones (B)

B1 Frontal bone
B2 Sphenoid
B3 Ethmoid
B4 Lacrimal bone
B5 Nasal bone
B6 Parietal bone
B7 Temporal bone
B8 Palatine bone
B9 Zygomatic bone

B10 Maxilla
B11 Vomer
B12 Mandible

Sutures (S)

S1 Coronal suture
S2 Squamous suture

Foramina (F)

F1 Supraorbital foramen
F2 Infraorbital foramen
F3 Mental foramen

Canal (C)

C1 Optic canal

Processes, Projections (P)

P1 Middle nasal concha
P2 Perpendicular plate of ethmoid
P3 Inferior nasal concha

Fissures (FS)

FS1 Superior orbital fissure
FS2 Inferior orbital fissure

Color the different bones to match your textbook.

The term "coronal" means "crown." A tiara (*crown*) on the head of a beauty queen follows the same direction as the coronal suture.

This suture is named after the **sagittal plane** that it follows.

Skullcap
(superior view)

"*Lambdoid*" means resembling the Greek letter lambda, Λ

Skull
(posterior view)

In the posterior view, the skull looks like a modified "peace sign."

P = Parietal bone
O = Occipital bone

Key to Illustration		
Bones (B)	**Sutures (S)**	**Processes, Projections (P)**
B1 Frontal bone	S1 Coronal suture	P1 Mastoid process
B2 Parietal bone	S2 Sagittal suture	
B3 Occipital bone	S3 Lambdoid suture	

Simplify what you see by observing the pattern made by the foramina, and use the *sella turcica* and *foramen magnum* as key landmarks.

Sella turcica

Foramen magnum (largest hole)

From the lateral view, the **sella turcica** resembles a horse's saddle. It houses the pituitary gland.

The term **"Sella turcica"** means "Turkish saddle." It describes its shape as a saddle-like structure.

Key to Illustration

Bones (B)

B1 Frontal bone
B2 Ethmoid
B3 Sphenoid
B4 Temporal bone
B5 Parietal bone
B6 Occipital bone

Processes, Projections (P)

P1 Crista galli
P2 Cribriform plate

Canal (C)

C1 Internal acoustic meatus

Foramina (F)

F1 Foramen rotundum
F2 Foramen ovale
F3 Foramen spinosum
F4 Foramen lacerum
F5 Jugular foramen

Palatine bones

This inferior view of the skull offers a good view of the **palatine bones**.

Vomer

Foramen magnum

Use the largest foramen—the **foramen magnum**—as a landmark to locate the other foramina.

Color the different bones to match your textbook.

Key to Illustration

Bones (B)
- **B1** Palatine bone
- **B2** Vomer
- **B3** Maxilla
- **B4** Zygomatic bone
- **B5** Sphenoid
- **B6** Temporal bone
- **B7** Parietal bone
- **B8** Occipital bone

Sutures (S)
- **S1** Occipitomastoid suture
- **S2** Lambdoid suture

Foramina (F)
- **F1** Foramen lacerum
- **F2** Foramen ovale
- **F3** Foramen spinosum
- **F4** Jugular foramen

Processes, Projections (P)
- **P1** Palatine process of maxilla
- **P2** Temporal process of zygomatic bone
- **P3** Zygomatic process of temporal bone
- **P4** Styloid process
- **P5** Mastoid process
- **P6** Occipital condyle

Canal (C)
- **C1** External acoustic meatus *(canal)*
- **C2** Carotid canal

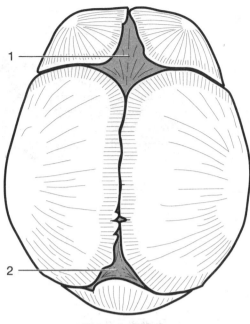

Fetal Skull
(superior view)

Fontanels are flat areas of fibrous connective tissue between bones of the skull that have not completed the ossification process. There are four major fontanels in the fetal skull. What is referred to as the "soft spot" on an infant's skull is actually the **anterior fontanel**, which typically closes by 2 years of age.

Color the different bones to match your textbook.

Fetal Skull
(lateral view)

Fetal Skull
(frontal view)

Fontanels

1. _____

2. _____

3. _____

4. _____

Key to Illustration

1. Anterior fontanel

2. Posterior fontanel

3. Mastoid fontanel

4. Sphenoid fontanel

Upper and Lower Jaws

The upper jaw consists of two **maxillae** (plural; *maxilla* = singular) that fuse together during development. These bones contain the largest sinuses—the maxillary sinuses.

* this rough surface has been cut away to reveal spongy bone

Right maxilla
(anterolateral view)

The lower jaw consists of a single bone called the **mandible** which is the strongest bone in the face.

Mandible
(lateral view)

Key to Illustration

Maxilla	Mandible		
1. Zygomatic process	1. Ramus	5. Coronoid process	9. Body
2. Infraorbital foramen	2. Condylar process	6. Mandibular foramen	10. Angle
3. Alveolar process	3. Mandibular notch	7. Alveolar process	
	4. Head	8. Mental foramen	

Bones Associated with the Skull

The single **hyoid bone** and the six **ear ossicles** are associated with the skull.

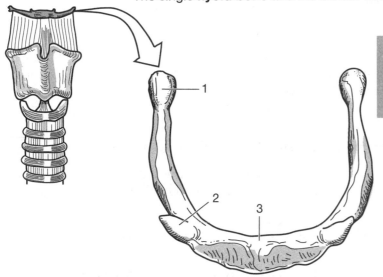

The **hyoid bone** is located above the larynx and serves as an attachment site for muscles of the tongue, larynx, and pharynx, which function in speech and swallowing.

Hyoid Bone
(superior view)

Hyoid Bone
(lateral view)

Ear Ossicles

These six bones—three in the middle ear of each temporal bone—are the smallest bones in the human body.
For more details see p. 326.

Malleus ➡ Hammer

Incus ➡ Anvil

Stapes ➡ Stirrup

Use this mnemonic for the ear ossicles:

*M*y		*M*alleus
*I*llustration	=	*I*ncus
*S*tinks		*S*tapes

Key to Illustration

Maxilla
1. Greater horn
2. Lesser horn
3. Body

Temporal Bone: Visualizing the Lateral View

Description

The temporal bone is located on the side of the skull. It articulates with the sphenoid, parietal, and occipital bones. The zygomatic process of the temporal bone articulates with the temporal process of the zygomatic bone to form the zygomatic arch.

Analogy

The **temporal bone in the lateral view** is like the **head of a rooster**. The **squamous portion** (*squama*) is the **rooster's comb**, the **external acoustic meatus** is the **rooster's eye**, the **styloid process** is the **rooster's beak**, and the **mastoid process** is the **rooster's wattles**.

Study Tips

Palpate (*touch*): Feel behind your ear to locate the large bump—the **mastoid process.**

- The **styloid process** is often broken off of a real skull because it is a delicate structure. Do not be surprised if you cannot locate it on a *real* skull, but a good-quality plastic skull will have this structure.

- Squamous means "scalelike," which accurately describes the flat, *squamous* portion of the temporal bone.

- Processes and other structures are sometimes named after the bones with which they articulate (e.g., zygomatic process of temporal bone and temporal process of zygomatic bone).

Key to Illustration		
1. Squamous part (*squama*)	3. External acoustic meatus	5. Styloid process
2. Zygomatic process	4. Mastoid process	

In the lateral view, a temporal bone looks like a rooster's head.

Temporal Bone
(lateral view)

Comb

Beak

Wattles

1. _____

2. _____

3. _____

4. _____

5. _____

Description

The **ethmoid** is embedded in the skull and located behind the bridge of the nose. The entire bone is not visible. Only specific portions of it can be seen, which makes it difficult to visualize how it fits into the skull. On its superior aspect are several important structures. A flat plate of bone called the **cribriform plate** has many small holes in it called **olfactory foramina** that allow olfactory nerves to pass from the olfactory organ to the brain. A partition called the **crista galli** separates the cribriform plate into left and right halves. In the anterior view, a long vertical plate of bone called the **perpendicular plate** forms part of the nasal septum. On either side of this plate are tiny, curved, bony projections called the **superior** and **middle nasal concha**. The sides of the bone consist of the lateral masses. These structures contain a network of interconnected, hollow chambers called **ethmoid air cells**. Together, these air cells constitute the **ethmoid sinuses**.

Analogy

- The entire bone can be compared to an iceberg floating in the water. Like the iceberg, the entire ethmoid bone cannot be seen, because much of it is embedded in the skull. The tip of the iceberg that can be seen is the superior aspect of the ethmoid bone, which looks like a door hinge folded flat. The hinge itself is the crista galli, the metal plate is the cribriform plate, and the screw holes in the plate are the olfactory foramina.

- The crista galli more closely resembles a shark's dorsal fin.

Location

Skull.

Key to Illustration

1. Crista galli	4. Superior nasal concha	6. Middle nasal concha
2. Cribriform plate	5. Ethmoid air cells	7. Perpendicular plate
3. Olfactory foramina		

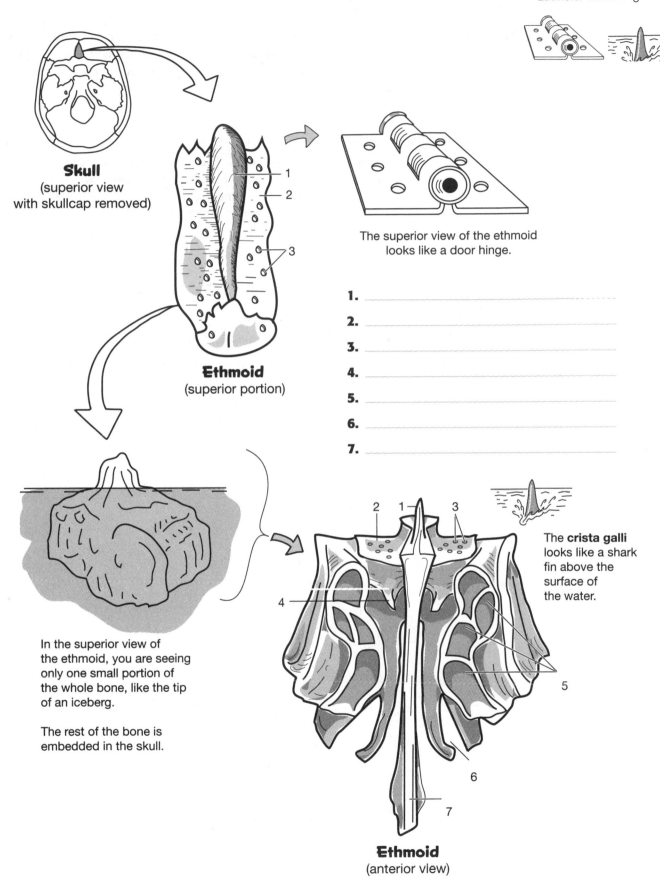

Skull
(superior view
with skullcap removed)

Ethmoid
(superior portion)

The superior view of the ethmoid
looks like a door hinge.

1. _____
2. _____
3. _____
4. _____
5. _____
6. _____
7. _____

In the superior view of
the ethmoid, you are seeing
only one small portion of
the whole bone, like the tip
of an iceberg.

The rest of the bone is
embedded in the skull.

The **crista galli**
looks like a shark
fin above the
surface of
the water.

Ethmoid
(anterior view)

Description
The **sphenoid** (*sphenoid bone*) is embedded within the skull, so it can be difficult to visualize. To be seen in its entirety, it must be removed from the skull. It articulates with the frontal, parietal, occipital, ethmoid, and temporal bones of the cranium and the palatine bones, zygomatic bones, vomer, and maxillae of the facial bones.

Analogy
The **entire sphenoid** resembles a **big-eared bat in flight**. The **legs of the bat** are the **pterygoid processes** and can only be seen from an inferior view of the skull. The **wing of the bat** is the **greater wing**, and the **bat's body** represents the **sella turcica**. The big **ears of the bat** are the **lesser wings**.

The **sella turcica** resembles a **horse's saddle**. This structure protects the pituitary gland.

Location
Skull.

Study Tips
Good landmarks:

- **Sella turcica.** This saddle-like structure is unique in its shape.
- **Foramen ovale.** This hole is usually oval-shaped, which makes it easier to distinguish from other foramina.

Key to Illustration		
1. Greater wing	4. Optic canal *(foramen)*	7. Foramen ovale
2. Lesser wing	5. Sella turcica	8. Foramen spinosum
3. Superior orbital fissure	6. Foramen rotundum	9. Pterygoid process

Skull
(superior view
with skullcap removed)

Sphenoid
(posterior view)

Sphenoid
(anterior view)

1. _____

2. _____

3. _____

4. _____

5. _____

6. _____

7. _____

8. _____

9. _____

Description

A **foramen** (sing.; *foramina* = plural) is a hole in bone through which a structure such as a blood vessel or nerve passes. The **sphenoid** bone contains many foramina. At first glance they appear like many small, indistinguishable holes. The challenge is to be able to differentiate one from the other.

Analogy

This analogy will link five different sphenoidal foramina. The **sella turcica** looks like a **horse's saddle**. Imagine a miniature cowboy sitting on the sella turcica. He is named **ROS** for the three foramina (F.) on either side of him. From medial to lateral, they are as follows: F. **Rotundum**, F. **Ovale**, and F. **Spinosum**. His **legs** go through foramen **lacerum** and his arms go through the **optic** foramen. Because the optic nerve passes through this opening, imagine ROS holding onto the optic nerves as if they were the reins of a horse.

Function

Foramen (F.)	Blood Vessel / Nerve Passing Through
Optic F.	Cranial nerve II (optic) and ophthalmic artery
F. rotundum	Maxillary branch of cranial nerve V (trigeminal)
F. ovale	Mandibular branch of cranial nerve V (trigeminal)
F. spinosum	Blood vessels to membranes around central nervous system
F. lacerum	Branch of ascending pharyngeal artery

Study Tips

- The **F. ovale** tends to be shaped like an oval.
- The **F. spinosum** is the smallest of the foramina.

Key to Illustration		
1. Optic foramen	3. Foramen ovale	5. Foramen lacerum
2. Foramen rotundum	4. Foramen spinosum	6. Sella turcica

Skull
(superior view
with scullcap removed)

Fun Fact: The skull is composed of 22 bones: 8 in the skullcap and 14 in the face.

Sphenoid Bone
(in skull)

Optic nerve

Ros the cowboy sits on the saddle (**sella turcica**) of the sphenoid bone.

1. _____

2. _____

3. _____

4. _____

5. _____

6. _____

Description

The skull has two small **palatine bones**. Two major parts of this bone are the **horizontal plate** and the **perpendicular plate**. The horizontal plate articulates with the maxillae to form the posterior portion of the hard palate in the roof of the mouth. When the horizontal plates of both bones touch, they form a narrow ridge called the **nasal crest** that articulates with the vomer. The perpendicular plate runs vertically and articulates with the maxillae, ethmoid, sphenoid, and inferior nasal concha. The **orbital process** is located on top of the perpendicular plate to mark the most superior part of the bone. It forms a small part of the posterior portion of the orbit.

Analogy

In the anterior view, each **palatine bone** looks like the **letter L. Both bones fused together** look like two mirror-image Ls touching each other.

Location

Skull.

Key to Illustration	
1. Horizontal plate	**3.** Perpendicular process
2. Nasal crest	**4.** Orbital process

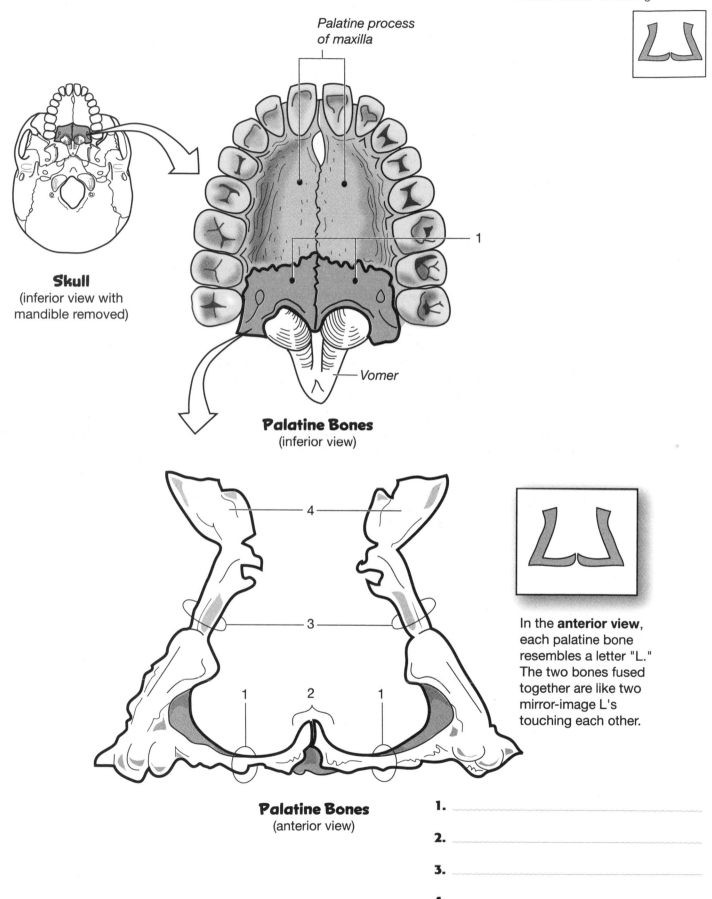

Palatine process
of maxilla

Skull
(inferior view with
mandible removed)

Vomer

Palatine Bones
(inferior view)

In the **anterior view**,
each palatine bone
resembles a letter "L."
The two bones fused
together are like two
mirror-image L's
touching each other.

Palatine Bones
(anterior view)

1. _____

2. _____

3. _____

4. _____

Description

The vertebral column contains a total of **24** vertebrae of **three** different types:

- **Cervical** (7)
- **Thoracic** (12)
- **Lumbar** (5)

In total there are 7 cervical, **12** thoracic, and 5 lumbar vertebrae. To remember the total number of each type of vertebrae, think of mealtimes:

- **Breakfast** at 7:00 a.m.
- **Lunch** at 12:00 noon
- **Dinner** at 5:00 p.m.

To remember the **sacrum** and **coccyx**, think of having two snacks in the evening:

- First snack = sacrum
- Second snack = coccyx

The **sacrum** results from the fusion of five vertebrae. On the top, in the anterior view, is a ridge of bone called the **sacral promontory** (*prominence*). This is an important landmark for a female's pelvic exam. The series of holes running through the bone are called the **sacral foramina**—nerves of the sacral plexus pass through them. On the posterior surface, an opening called the **sacral hiatus** leads into a long passageway called the **sacral canal**. Nerves from the spinal cord run through this canal.

The **coccyx** (*tailbone*) is located inferior to the sacrum and consists of three to five bones. It serves as an anchor point for muscles, tendons, and ligaments.

Total Number of Each Type of Vertebrae

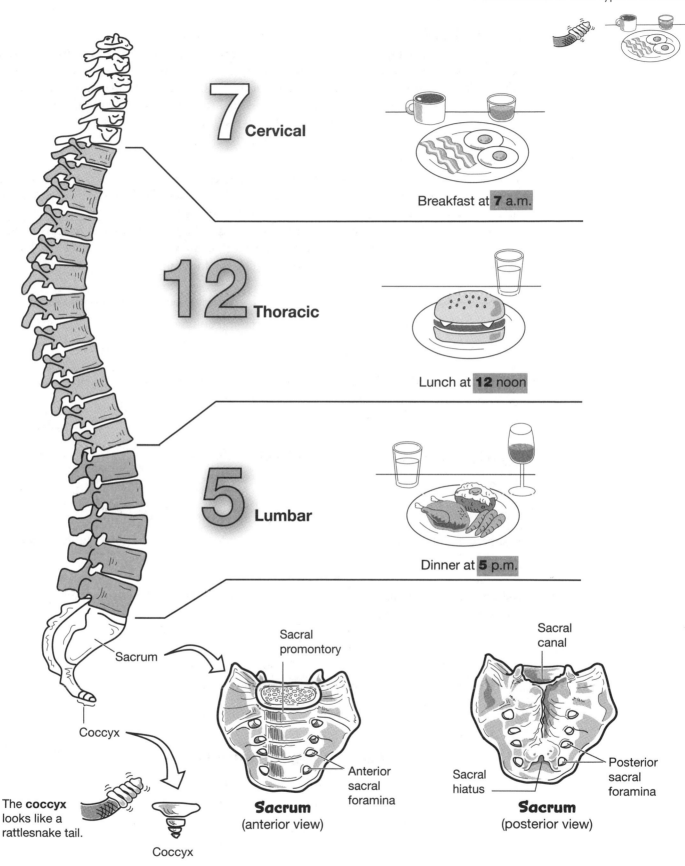

7 Cervical

Breakfast at **7** a.m.

12 Thoracic

Lunch at **12** noon

5 Lumbar

Dinner at **5** p.m.

Sacrum

Coccyx

The **coccyx** looks like a rattlesnake tail.

Coccyx

Sacral promontory

Anterior sacral foramina

Sacrum
(anterior view)

Sacral canal

Sacral hiatus

Posterior sacral foramina

Sacrum
(posterior view)

Description

The vertebral column contains a total of 24 vertebrae of three different types: cervical (7), thoracic (12), and lumbar (5). The first two cervical vertebrae at the top of the vertebral column are referred to as the **atlas** (*cervical 1* or *C1*) and the **axis** (*cervical 2* or *C2*). As the Greek god Atlas held up the Earth, the atlas vertebra is positioned at the base of the globe-like skull. The atlas is designed to pivot on the axis, which allows for rotational movement such as when you turn your head from side to side. Roughly half of all cervical rotation occurs between the atlas and the axis.

Analogy

The **atlas** (C1) resembles a **turtle's head with eyeglasses**. The **anterior arch** is the handle of the eyeglasses, the **superior articular facets** are the lenses of the glasses, the **transverse process** is the arm of the eyeglasses, and the **posterior arch** is the smile on the turtle's face.

The **axis** (C2) resembles a **football player grasping a football**. The **odontoid process** is the football player's helmet, the **superior articular facet** is the football player's shoulder pad, the **lamina** is the **forearm**, and the **spinous process** is the **hands grasping the football**.

Location

The first two cervical vertebrae at the top of the vertebral column.

Function

The atlas and axis together form a pivot joint. When you turn your head from side to side, the atlas is rotating on the more stationary axis.

Study Tips

The following are good landmarks for these bones:

Atlas

- transverse foramen
- **large** superior articular facets
- **large** inferior articular facets

Axis

- transverse foramen
- odontoid process (*dens*) (unique to axis only)

Note that *only* cervical vertebrae have a **transverse foramen**. This makes them easy to distinguish from thoracic or lumbar vertebrae.

Key to Illustration	
Atlas	**Axis**
1. Anterior arch	6. Odontoid process (*dens*)
2. Superior articular facet	7. Superior articular facet
3. Transverse process	8. Lamina
4. Posterior arch	9. Bifid spinous process
5. Transverse foramen	10. Transverse foramen

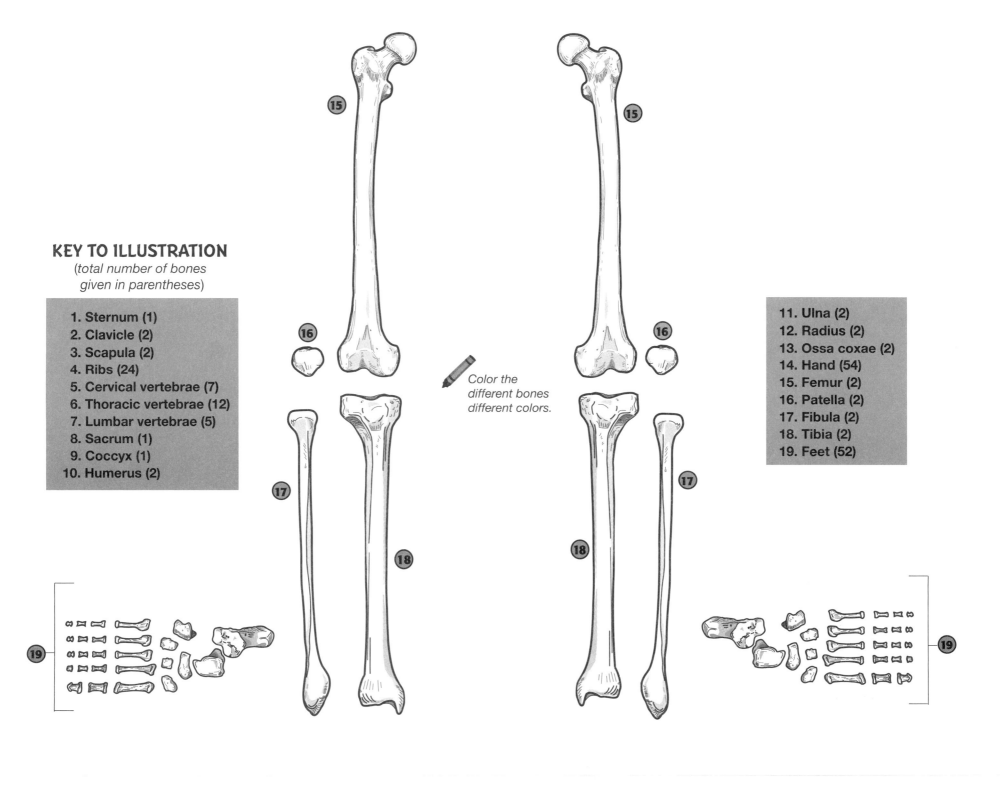

KEY TO ILLUSTRATION
(total number of bones given in parentheses)

1. Sternum (1)
2. Clavicle (2)
3. Scapula (2)
4. Ribs (24)
5. Cervical vertebrae (7)
6. Thoracic vertebrae (12)
7. Lumbar vertebrae (5)
8. Sacrum (1)
9. Coccyx (1)
10. Humerus (2)

11. Ulna (2)
12. Radius (2)
13. Ossa coxae (2)
14. Hand (54)
15. Femur (2)
16. Patella (2)
17. Fibula (2)
18. Tibia (2)
19. Feet (52)

Color the different bones different colors.

BONES OF THE SKELETON

The adult skeleton typically contains a total of

206

individual bones. Slightly more than half of these are found in the

HANDS

and

FEET.

SKULL
(*see back side*)

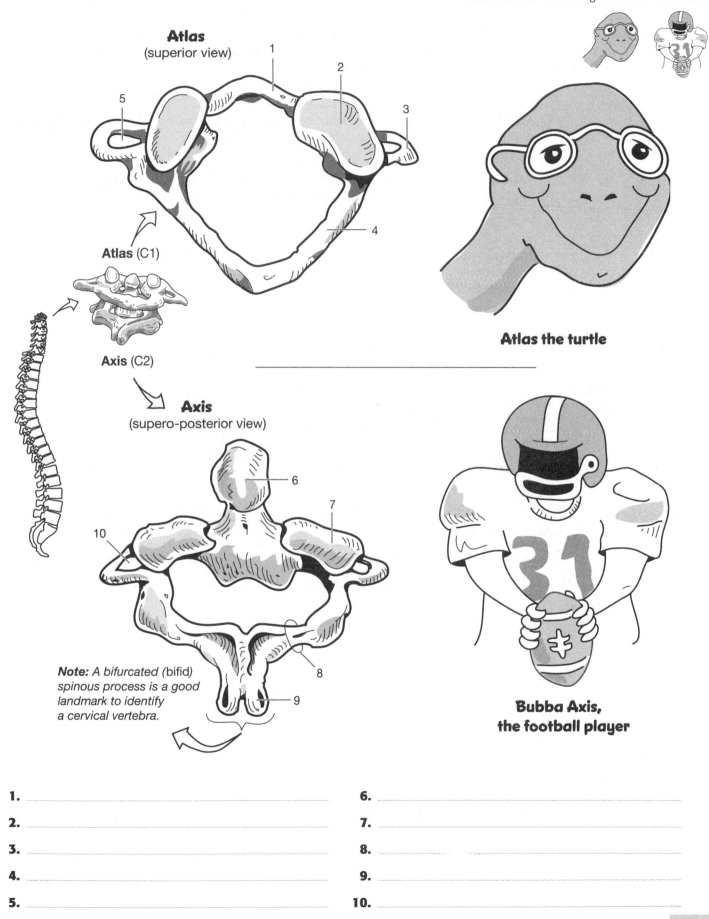

Atlas
(superior view)

Atlas (C1)

Axis (C2)

Axis
(supero-posterior view)

Note: *A bifurcated (bifid) spinous process is a good landmark to identify a cervical vertebra.*

Atlas the turtle

Bubba Axis, the football player

1. _____
2. _____
3. _____
4. _____
5. _____

6. _____
7. _____
8. _____
9. _____
10. _____

Lumbar Versus Thoracic Vertebrae: Distinguishing the Difference 1

Description

The vertebral column contains three different types of vertebrae: cervical (7), thoracic (12), and lumbar (5). Each type has its own unique features to distinguish one from another, yet all of the types have three basic features in common:

1. **Body**—bears weight and increases in size as one moves down the vertebral column.

2. **Vertebral arch**—structure that contains a **vertebral foramen, pedicles, lamina, spinous process,** and **transverse processes.** A defective fusion of the vertebral arch is associated with the birth defect spina bifida.

3. **Articular processes (two)**—the **superior articular process** and the **inferior articular process,** which join one vertebra to another.

Analogy

The posterior portion of a **thoracic vertebra** looks like a **goose with wings** *arched forward*. The posterior portion of a **lumbar vertebra** looks like a goose with wings *horizontal*. The **spinous process** is the **head and neck of a goose,** and the **transverse process** is the **wing of a goose.**

Location

Vertebral column.

Distinguishing Features

Feature	Thoracic	Lumbar
Location	Chest	Lower back
Body of vertebra	Medium-sized, heart-shaped; facets for ribs	Largest diameter; thicker; oval-shaped
Vertebral foramen	Medium-sized	Smaller-sized
Spinous process	Long, slender; points inferiorly	Broad; flat; blunt
Transverse process	10 of 12 have facets for rib articulations	Short; narrower; no articular facets or transverse foramina

Key to Illustration		
1. Spinous process	4. Transverse process	7. Body
2. Lamina	5. Superior articular facet	8. Pedicle
3. Vertebral foramen	6. Superior articular process	9. Facet for rib articulation

Fun Fact: A human neck has the same number of vertebrae as a giraffe's.

Location

HORACIC

LUMBAR

Thoracic vertebra
(superior view)

Note: *Most—but not all—thoracic vertebrae have an angle of <90° between the spinous process and the transverse process. Be sure to use all the distinguishing features in the table on p. 174.*

HORACIC

LUMBAR

Lumbar vertebra
(superior view)

1. _____
2. _____
3. _____
4. _____
5. _____

6. _____
7. _____
8. _____
*9. _____
(*found only on thoracic vertebrae)

Lumbar Versus Thoracic Vertebrae: Distinguishing the Difference 2

Description

The vertebral column contains three different types of vertebrae: cervical (7), thoracic (12), and lumbar (5). Each type has its own unique features to distinguish one from another, yet all of the types have three basic features in common:

1. **Body**—bears weight and increases in size as one moves down the vertebral column.

2. **Vertebral arch**—structure that contains a **vertebral foramen, pedicles, lamina, spinous process,** and **transverse processes.** A defective fusion of the vertebral arch is associated with the birth defect spina bifida.

3. **Articular processes (two)**—the **superior articular process** and the **inferior articular process,** which join one vertebra to another.

Analogy

In the posterolateral view, the **thoracic vertebra** looks like a **giraffe.** The **giraffe's snout** is the **spinous process,** the **giraffe's ears** are the **transverse processes,** and the **giraffe's horns** are the **superior articulating processes.** The **giraffe's cheek** is the **inferior articulating process.**

In the lateral view, a **lumbar vertebra** looks like the **head of a moose.** The **moose's snout** is the **spinous process,** the **moose's horns** are the **superior articulating processes,** and the **moose's bell** is the **inferior articulating process.**

Location

Vertebral column.

Key to Illustration		
1. Body	**4.** Transverse process	**7.** Inferior articular facet
2. Superior articular process	**5.** Spinous process	
3. Superior articular facet	**6.** Inferior articular process	

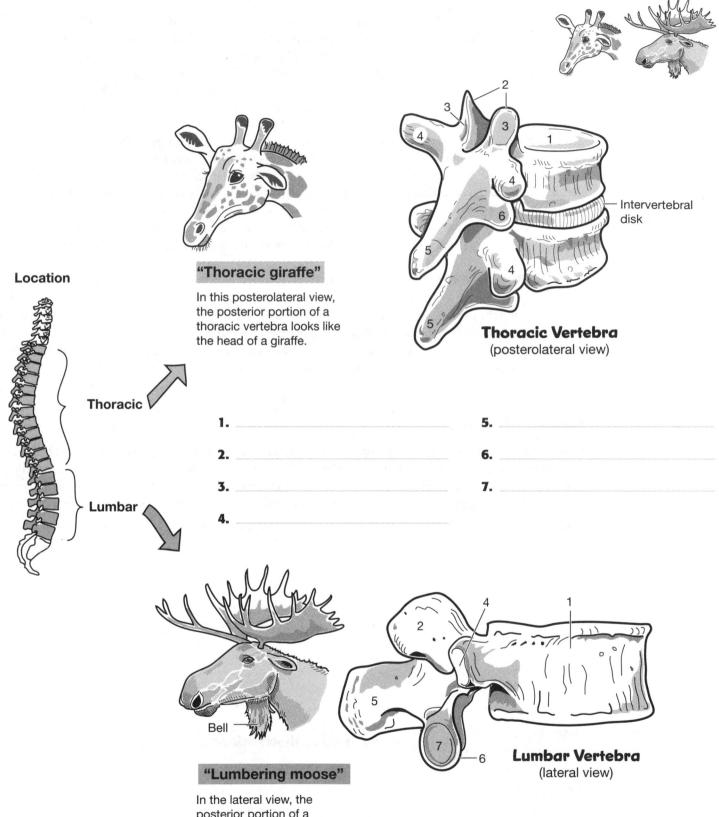

Location

Thoracic

Lumbar

"Thoracic giraffe"

In this posterolateral view, the posterior portion of a thoracic vertebra looks like the head of a giraffe.

Thoracic Vertebra
(posterolateral view)

Intervertebral disk

1. _____

2. _____

3. _____

4. _____

5. _____

6. _____

7. _____

Bell

"Lumbering moose"

In the lateral view, the posterior portion of a lumbar vertebra looks like the head of a moose.

Lumbar Vertebra
(lateral view)

Description

The **sternum** is classified as a flat bone, commonly called the breastbone, and is divided into three parts: (1) manubrium, (2) body, and (3) xiphoid process. The **manubrium** articulates with the clavicles and the costal cartilages of the first pair of ribs. The elongated **body** is the largest part of the sternum and articulates with the ribs through the costal cartilages. The **xiphoid process** is the smallest part and does not directly articulate with another bone. The term *xiphoid* means swordlike, which accurately describes its pointed tip. This cartilaginous structure is typically the last part of the sternum to ossify in the adult.

There are 12 pairs of **ribs**, commonly divided into two groups based on whether they attach directly to the sternum through the costal cartilages. The first seven ribs that do directly attach are called **true ribs** (1–7), and the last five ribs that do not directly attach are called **false ribs** (8–12). A more anatomically correct term for the true ribs is the **vertebrosternal ribs** (1–7). The false ribs are more correctly divided into two groups: **vertebrochondral ribs** (8–10) and **floating ribs** (11–12). The vertebrochondral ribs connect indirectly to the sternum via one common section of costal cartilage. The floating ribs have no connection to the sternum and are also called **vertebral ribs** because they connect to only the last two thoracic vertebrae (T11 and T12).

Each **rib**, like the sternum and the scapulae, is classified as a flat bone. The longest portion is called the **body** or shaft. The two ends of a rib look quite different. The flattened end is called the **sternal end** because it attaches to the sternum through the costal cartilages. The more bumpy end is called the **head** and it has two flat surfaces called articular facets that attach to a vertebra. Below the head is a short segment called the **neck**. Adjacent to the neck is a small, rounded projection called a **tubercle** that has an articular facet that attaches to the transverse process of a thoracic vertebra.

Study Tips

Palpate (*feel by touch*):

- **Sternum:** You can easily feel the middle of the **manubrium** and **body** of your sternum. The small tip at the end marks the **xiphoid process**. During CPR training, students are instructed to avoid doing compressions on the xiphoid process because it can break off easily and cause serious damage to the liver, diaphragm, or lungs. This is especially problematic for elderly females with osteoporosis.

- **Ribs:** The body of the ribs can be felt on the lateral surface of the thoracic cage.

Key to Illustration	
Sternum (S)	**Rib (R)**
S1 Jugular notch	R1 Head
S2 Clavicular notch	R2 Neck
S3 Manubrium	R3 Tubercle
S4 Body	R4 Body (*shaft*)
S5 Xiphoid process	R5 Sternal end

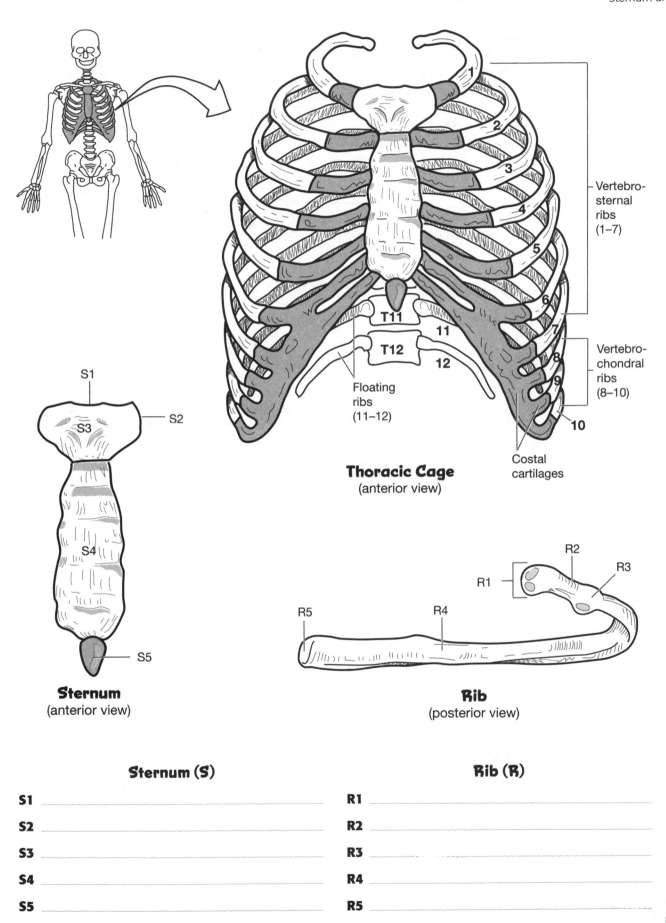

Thoracic Cage
(anterior view)

Vertebro-sternal ribs (1–7)

Vertebro-chondral ribs (8–10)

Costal cartilages

Floating ribs (11–12)

Sternum
(anterior view)

Rib
(posterior view)

Sternum (S)		Rib (R)	
S1	_____	R1	_____
S2	_____	R2	_____
S3	_____	R3	_____
S4	_____	R4	_____
S5	_____	R5	_____

Description

The **scapula** is commonly called the shoulder blade and is rarely fractured. It articulates with the clavicle and the humerus. The **glenoid cavity** (*fossa*) receives the head of the humerus, and the **acromion process** articulates with the acromial (lateral) end of the clavicle. The large, flat **body** of the scapula serves as an attachment for muscles, tendons, and ligaments.

The **clavicle** is commonly called the collarbone. At its sternal (*medial*) end it articulates with the clavicular notch of the sternum, and its acromial (*lateral*) end articulates with the acromion process of the scapula. Sports injuries and auto accidents often cause clavicle fractures—one of the most commonly fractured bones.

Location

- Scapula—acromial region (shoulder) and back.
- Clavicle—superior border of the pectoral region.

Study Tips

- To position the clavicle: The more rounded end is the sternal end, and the flatter end is the acromial end.

Palpate (*feel by touch*):

- **Clavicle:** You can feel the details of this bone.

- Feel your clavicle, and follow it to its lateral end; the bump it attaches to is the **acromion process.** The **spine** of the scapula can be felt by gliding your fingers along the back of the shoulder. The body of the scapula cannot be felt, because it is covered by muscle.

Key to Illustration		
Scapula (S)	6. Inferior angle	**Clavicle (C)**
1. Acromion process	7. Infraspinous fossa	C1. Acromial (*lateral*) end
2. Coracoid process	8. Medial border	C2. Sternal (*medial*) end
3. Spine of scapula	9. Supraspinous fossa	
4. Glenoid cavity (*fossa*)	10. Superior border	
5. Lateral border	11. Suprascapular notch	

Simplify what you see…

Body

Right Scapula
(posterior surface)

Right Scapula
(lateral view)

Right Clavicle
(superior view)

Description

The **humerus** is the only bone in the brachial region. The **head** of the humerus articulates with the glenoid cavity in the scapula to form the shoulder joint. At the distal end the **trochlea** articulates with the ulna, and the **capitulum** articulates with the head of the radius. The **olecranon fossa** articulates with the olecranon process of the ulna, and the **coronoid fossa** articulates with the coronoid process of the ulna.

Analogy

The **distal end of the humerus** resembles a **hitchhiker's hand**. The thumb is the medial epicondyle. The first two fingers adjacent to the thumb are the **trochlea**, and the last two fingers are the **capitulum**. The depression in the middle of the palm is the **coronoid fossa**. Note that the thumb always points medially.

Location

Brachial region (between shoulder and elbow).

Study Tips

- To distinguish between the two condyles at the distal end of the humerus: the Trochlea has a more pointed Tip or edge, and the Capitulum is simply Curved and not as pointed.

- The **olecranon fossa** is a good landmark to identify the posterior view, because it is the deepest depression on the bone.

- Note that the **head** of the humerus and **medial epicondyle** always point *medially*. This helps to distinguish a *left* humerus from a *right* humerus.

Key to Illustration

1. Greater tubercle	5. Deltoid tuberosity	9. Trochlea
2. Intertubercular groove	6. Coronoid fossa	10. Medial epicondyle
3. Lesser tubercle	7. Lateral epicondyle	11. Olecranon fossa
4. Head	8. Capitulum	

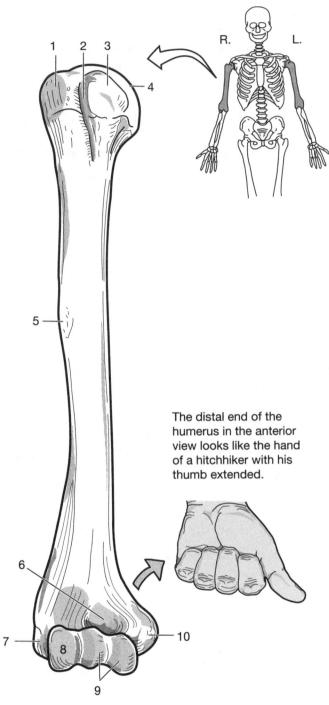

The distal end of the humerus in the anterior view looks like the hand of a hitchhiker with his thumb extended.

Right Humerus
(anterior view)

Right Humerus
(posterior view)

1. _____
2. _____
3. _____
4. _____
5. _____
6. _____

7. _____
8. _____
9. _____
10. _____
11. _____

Description
The **radius** and the **ulna** articulate with each other at both their proximal and distal ends. The distal end of the radius is on the thumb side of the arm, and the distal end of the ulna is nearer the fifth digit. At the proximal end, the head of the radius pivots on the radial notch of the ulna. The **olecranon process** of the ulna is the elbow, and it articulates with the olecranon fossa of the humerus. The **coronoid process** of the ulna articulates with the coronoid fossa of the humerus. A fibrous sheet of connective tissue connects the **diaphyses** of both bones.

The distal end of the radius articulates with the **carpal bones** in the wrist. The **olecranon process** of the ulna articulates with the olecranon fossa of the humerus, and the **coronoid process** of the ulna articulates with the coronoid fossa of the humerus. The head of the radius articulates with the capitulum of the humerus.

Analogy
- The **ulna** resembles a **crescent wrench**.
- The **head of the radius** resembles a **warped hockey puck**.

Location
The radius and ulna are located in the antebrachial region (forearm).

Study Tips
- The proximal end of the ulna has a u-shaped structure called the trochlear notch that identifies it as the Ulna because **u** is the first letter in the word ulna.

- The head of the **R**adius is **R**ounded. Use the alliteration "**Rounded Radius**" to distinguish the radius from the ulna.

- Notches and other structures are sometimes named after the bones with which they articulate:

  —radial notch of ulna

  —ulnar notch of radius

  —olecranon *process* of the ulna hooks into the olecranon *fossa* of the humerus

  —coronoid *process* of the ulna hooks into the coronoid *fossa* of the humerus

Key to Illustration		
Radius	**Ulna**	
1. Neck	5. Styloid process (of ulna)	10. Head of ulna
2. Radial tuberosity	6. Coronoid process	11. Radial notch of ulna
3. Styloid process	7. Olecranon process	
4. Ulnar notch of radius (not fully visible in illustration)	8. Trochlear notch	
5. Styloid process (of radius)	9. Ulnar tuberosity	

R. L.

Head

7

8

6

1

2

9

The head of the **R**adius is **R**ounded. It looks like a slightly warped hockey puck.

7

8

6

11

9

The trochlear notch and coronoid process form a "U" shape, identifying this bone as the ulna.

Radius

Ulna

Ulna

The ulna looks like a crescent wrench in the lateral view.

4

10

3

5

Right Radius and Ulna
(anterior view)

Right Ulna
(lateral view)

Radius

1. _____

2. _____

3. _____

4. _____

5. _____

Ulna

5. _____

6. _____

7. _____

8. _____

9. _____

10. _____

11. _____

Description

Each hand contains a total of 27 bones and is divided into three groups of bones: **carpals** (8), **metacarpals** (5), and **phalanges** (14). The carpals are the small bones of the wrist. The palm of the hand contains the metacarpals, and the phalanges are located in the fingers or digits.

Study Tips

The most difficult part of the hand to learn is the *carpals*. These eight bones are small and clustered together in the wrist. Simplify it by viewing them as two equal rows of bones stacked one on top of the other. Each row has four bones in it. The first row is nearer the radius and ulna. The second row is next to the metacarpals. Beginning on the thumb side of the first row, the proper order is: **scaphoid, lunate, triquetrum, pisiform**. Following the same pattern, the proper order for the second row is: **trapezium, trapezoid, capitate, hamate**.

Clinically, the carpals are important to know, because they are commonly fractured. When a carpal is fractured, it is referred to by its specific name. The scaphoid is the most commonly fractured carpal bone, often caused by falling on an outstretched arm. The lunate is the most commonly dislocated carpal bone.

The *hamate* (*hamatum,* or hooked) *bone* has a hook-like process on it (*visible only in the anterior view*). You can better recall this bone because the words *hamate* and *hook* both begin with the letter *h*.

The *thumb* has only two phalanges instead of three. It has a proximal and distal phalanx but lacks a middle phalanx. This is the same pattern as the great toe of the foot.

- On many plastic models, the **triquetrum** and pisiform bones are often fused together to appear as one bone. This causes some confusion for students.

Key to Illustration

Carpals
1. Scaphoid
2. Lunate
3. Triquetrum (*triangular bone*)
4. Pisiform

5. Trapezium
6. Trapezoid
7. Capitate
8. Hamate

Phalanges
9. Proximal phalanx
10. Middle phalanx
11. Distal phalanx

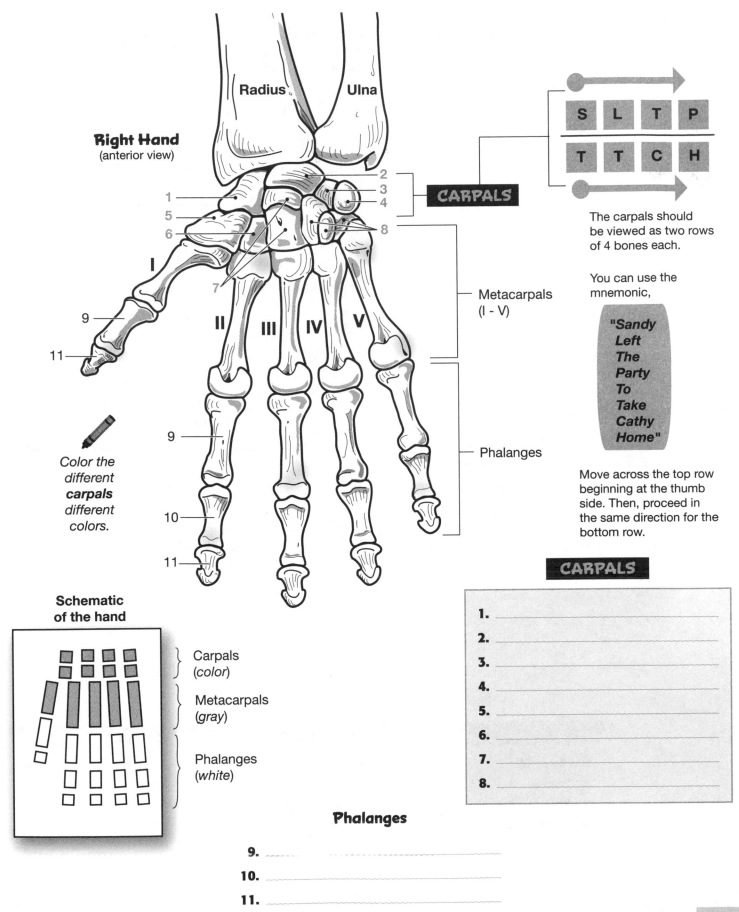

Radius **Ulna**

Right Hand
(anterior view)

1
5
6

7

I

9

11

II III IV V

9

10

11

**Schematic
of the hand**

Carpals
(*color*)

Metacarpals
(*gray*)

Phalanges
(*white*)

*Color the
different **carpals**
different
colors.*

2
3
4

8

CARPALS

S	L	T	P
T	T	C	H

The carpals should
be viewed as two rows
of 4 bones each.

You can use the
mnemonic,

*"Sandy
Left
The
Party
To
Take
Cathy
Home"*

Move across the top row
beginning at the thumb
side. Then, proceed in
the same direction for the
bottom row.

Metacarpals
(I - V)

Phalanges

CARPALS

1. _____
2. _____
3. _____
4. _____
5. _____
6. _____
7. _____
8. _____

Phalanges

9. _____
10. _____
11. _____

187

Description

The **pelvis** is composed of four bones: (1) left hip bone, (2) right hip bone, (3) sacrum, and (4) coccyx. The anatomical term for a hip bone is os coxae (plural, *ossa coxae*). Each hip bone results from the fusion of three separate bones: **ilium**, **ischium**, and **pubis** bones. These three meet together in the **acetabulum** (*means "cup of vinegar"*) to form what looks like a peace sign. This deep depression is where the head of the femur forms a ball-and-socket joint with the hip bone. The largest of the three hip bones is the ilium. The broad surface of this bone is a major attachment for muscles, tendons, and ligaments. The strongest bone of the three is the ischium. A large, rough projection—the **ischial tuberosity**—is a major landmark on this bone. It bears the body weight when seated. The pubis has the least bone mass and contains roughly one-half of the largest foramen in the skeletal system—the **obturator foramen** (the other half is part of the ischium).

Analogy

The **two pubis bones** in the anterior aspect of the pelvis together resemble a **mask** worn by someone. The mask itself represents both pubis bones, and the **eye hole** is the **obturator foramen**. The **middle of the mask** is the joint called the **pubic symphysis**.

Location

Pelvic region.

Study Tips

Palpate (*feel by touch*):

- The curve along the superior aspect of your hip is the **iliac crest**.

- Follow the iliac crest posteriorly with your fingers until you feel a bump. This is the **posterior superior iliac spine** (PSIS).

- Place your hand on your buttock and push inward. The bony bump you feel is the **ischial tuberosity**.

Good landmarks on a coxal bone: **acetabulum** and the **obturator foramen**. The acetabulum marks the lateral aspect, and the obturator foramen marks the anterior aspect of the bone.

Key to Illustration		
Hip Bone (B)	4. Sacroiliac joint	11. Posterior inferior iliac spine (PIIS)
B1 Ilium	5. Iliac crest	12. Posterior superior iliac spine (PSIS)
B2 Ischium	6. Pubic crest	
B3 Pubis	7. Ischial tuberosity	13. Anterior superior iliac spine (ASIS)
1. Pubic symphysis	8. Lesser sciatic notch	
2. Obturator foramen	9. Ischial spine	14. Anterior inferior iliac spine (AIIS)
3. Acetabulum	10. Greater sciatic notch	

Note: The sacrum and coccyx are faded out because they are not part of the appendicular skeleton.

Sacrum

Coccyx

Male Pelvis
(anterior view)

The two pubis bones linked together look like a mask.

Acetabulum means "cup of vinegar."

Right Hip Bone
(lateral view)

1. _____

2. _____

3. _____

4. _____

5. _____

6. _____

7. _____

8. _____

9. _____

10. _____

11. _____

12. _____

13. _____

14. _____

Hip Bone (B)

B1 _____

B2 _____

B3 _____

Description

The **femur** is the only bone in the femoral region. It is the longest and strongest bone in the body and is commonly called the thigh bone. The **head** of the femur articulates with the acetabulum of the coxal bone to form the ball-and-socket joint in the hip. At the distal end, the patella (kneecap) covers the **patellar surface**. The **medial** and **lateral condyles** articulate with the proximal portion of the tibia to form the knee joint.

The **patella** is loosely held in place by ligaments.

Location

Femoral region (between hip and knee).

Study Tips

- Note that the **head** of the femur always points *medially*. This helps to distinguish a *left* femur from a *right* femur. It also helps to identify the medial condyle, because it is on the same side as the head of the humerus.

- To identify the posterior view: The **linea aspera** and **intercondylar groove** appear only in the posterior view.

- To identify the anterior from posterior in the patella: The anterior surface is rough and the posterior surface is much more smooth.

- **LAP** = Linea Aspera is Posterior

- For distinguishing the apex from the base of the patella, remember that apex is a general term meaning the *pointed tip* of a structure.

Palpate (*feel by touch*):

- You can easily feel your patella.

Key to Illustration		
Patella	**Femur**	
1. Base	3. Head	9. Intercondylar groove
2. Apex	4. Neck	10. Medial condyle
	5. Greater trochanter	11. Medial epicondyle
	6. Lesser trochanter	12. Lateral condyle
	7. Linea aspera	13. Lateral epicondyle
	8. Patellar surface	

Right Patella
(anterior view)

Right Femur
(anterior view)

Right Femur
(posterior view)

Patella

1. _____

2. _____

Femur

3. _____

4. _____

5. _____

6. _____

7. _____

8. _____

9. _____

10. _____

11. _____

12. _____

13. _____

Description

The **tibia** and the **fibula** are the two bones in the crural region between the knee and ankle. The **tibia** is commonly called the shinbone and is the larger of the two bones. At its **proximal** surface it articulates with the **distal** end of the femur to form the knee joint. At its distal end, the **inferior articular surface** articulates with the talus in the foot to form the ankle joint. The **anterior crest** is a ridge that runs along the shaft of the bone on the anterior surface. The **medial malleolus** is a large process that stabilizes the ankle joint.

The **fibula** is the smaller bone. The **head** of the fibula articulates with the proximal end of the tibia. The **lateral malleolus** articulates with the distal end of the tibia and with the talus in the ankle.

Location

Crural region (between knee and ankle).

Study Tips

- To distinguish between the **tibia** and the **fibula**: The tibia is the larger of the two bones, so use the phrase "Tough Tibia, Fragile Fibula."

- To distinguish between the two different ends of the fibula: The **lateral malleolus** is more tapered and triangular in shape and is located near the ankle; the **head** of the **fibula** is more rounded in shape and is located near the knee.

Palpate (*feel by touch*):

- Touch the large bump on the medial side of your ankle. This is the **medial malleolus** of the tibia.

- Now touch the bony bump on the lateral side of your ankle. This is the **lateral malleolus** of the **fibula**.

- Feel your patella (*kneecap*), and slide your hand straight down toward your ankle. The first small bump you feel below the knee is the **tibial tuberosity** of the tibia. The soft spot between your knee and the tibial tuberosity is the **patellar ligament**.

- Feel the long ridge of bone beginning below the **tibial tuberosity** and running down toward the ankle. This is the **anterior crest** of the tibia.

Key to Illustration		
Fibula	**Tibia**	
1. Head	4. Lateral condyle	7. Anterior crest
2. Lateral malleolus	5. Medial condyle	8. Medial malleolus
3. Inferior articular surface	6. Tibial tuberosity	9. Inferior articular surface

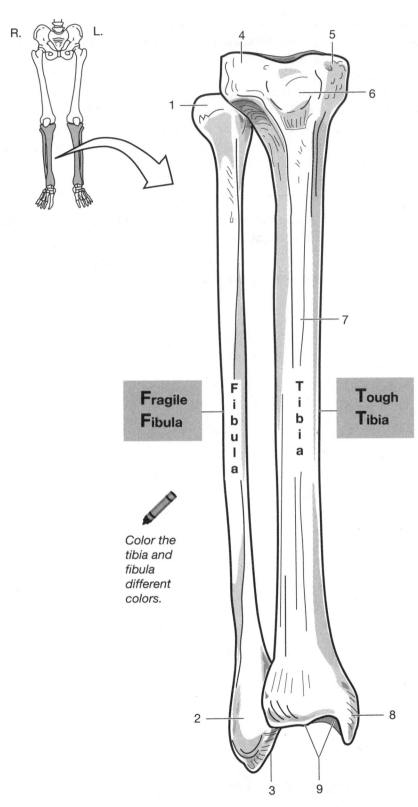

R.   L.

Fragile Fibula

Tough Tibia

Color the tibia and fibula different colors.

Right Fibula and Tibia
(anterior view)

Fibula

1. _____

2. _____

3. _____

Tibia

4. _____

5. _____

6. _____

7. _____

8. _____

9. _____

Description

Each foot contains a total of 26 bones and is divided into three groups of bones: **tarsals** (7), **metatarsals** (5), and **phalanges** (14). The tarsals are the ankle bones. The middle of the foot contains the metatarsals, and the phalanges are located in the toes or digits.

The most difficult part of the foot to learn is the *tarsals*. These seven bones include the **talus, calcaneus, navicular, cuboid,** and three **cuneiform** bones. The **calcaneus** (*heel bone*) is the largest bone in this group. The **talus** is curved on its superior aspect to articulate with the tibia. Positioned between the calcaneus and the metatarsals is the **cuboid** bone. The **navicular** lies between the talus and cuneiforms. The **cuneiform** bones are the smallest bones in this group and are identified by their position within the foot (*medial, intermediate,* and *lateral*).

Study Tips

- The **talus bone** articulates with the tibia. It is also the tallest bone when the foot is viewed laterally. You can easily recall this because the words *talus, tibia,* and *tallest* all begin with the letter *t*.

- The **great toe** has only two phalanges instead of three. It has a proximal phalanx and a distal phalanx but lacks a middle phalanx. This is the same pattern as the thumb of the hand.

Key to Illustration

Tarsals		Phalanges
1. Calcaneus	4. Medial cuneiform	8. Proximal phalanx
2. Talus	5. Intermediate cuneiform	9. Middle phalanx
3. Navicular	6. Lateral cuneiform	10. Distal phalanx
	7. Cuboid	

Color the different **tarsals** different colors.

Schematic of the Foot

Tarsals (*color*)

Metatarsals (*gray*)

Phalanges (*white*)

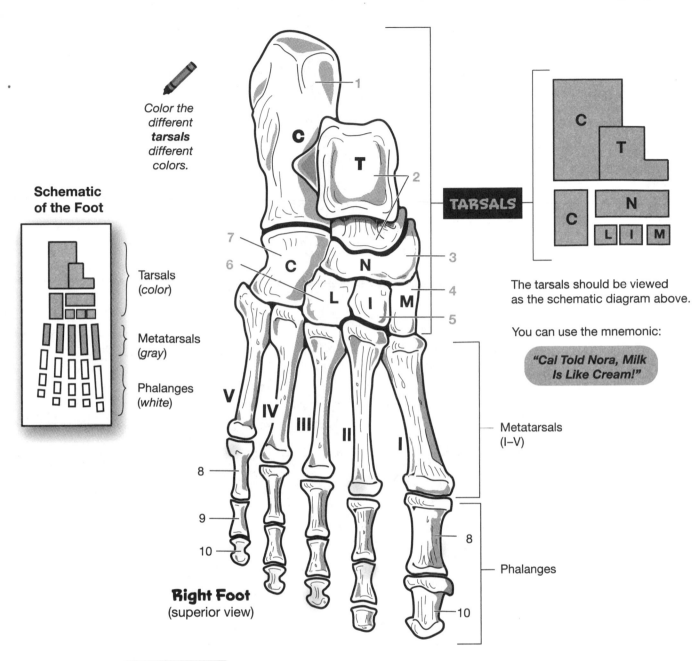

Right Foot
(superior view)

Metatarsals (I–V)

Phalanges

TARSALS

The tarsals should be viewed as the schematic diagram above.

You can use the mnemonic:

"Cal Told Nora, Milk Is Like Cream!"

TARSALS

1. _____

2. _____

3. _____

4. _____

5. _____

6. _____

7. _____

Phalanges

8. _____

9. _____

10. _____

Notes

JOINTS

JOINTS

Description

Joints (*articulations*) are formed when two or more bones meet. Common examples that quickly come to mind are freely moving joints like the knee and elbow joints. Sutures in the skull are also joints but are often overlooked because they have no movement associated with them. Joints are classified according to either their structure or their function (degree of movement).

The structural classification system has three categories:

Joint Example	Structural Classification	Functional Classification
Sutures in the skull	• **Fibrous joints**—only **fibrous connective tissue** anchors the joint together	• **Synarthrotic joints**—*no* movement
Pubic symphysis (in pelvis)	• **Cartilaginous joints**—only **cartilage** anchors the joint together	• **Amphiarthrotic joints**—*some* movement
Knee joint	• **Synovial joints**—many different **tissues/structures** are present, such as articular cartilage, joint capsule, synovial fluid, and ligaments	• **Diarthrotic joints**—*free* movement

Here are the key structures in any synovial joint:

- **Articular cartilage**—smooth layer of hyaline cartilage that covers the ends of long bones

- **Joint cavity**—potential space filled with synovial fluid

- **Joint capsule**—double-layered structure that surrounds the joint

    —outer layer = **fibrous capsule** (*dense irregular connective tissue*)

    —inner layer = **synovial membrane** (*loose connective tissue*)

- **Synovial fluid**—viscous, oily substance secreted by cells in the synovial membrane

- **Ligaments**—bands of fibrous connective tissue that connect one bone to another

Key to Illustration

1. Periosteum
2. Medullary cavity
3. Yellow marrow
4. Ligament
5. Fibrous capsule
6. Synovial membrane
7. Joint capsule
8. Joint cavity (filled with synovial fluid)
9. Articular cartilage

Note: *Structure is cut.*

Color the different joint structures different colors.

Typical Synovial Joint

1. _____

2. _____

3. _____

4. _____

5. _____

6. _____

7. _____

8. _____

9. _____

Description

The shoulder (*glenohumeral*) joint, like the hip joint, is classified as a ball-and-socket joint. This means the ball-like structure of one bone fits into the cup-like structure of a different bone. In this case, the **head of the humerus** ("ball") joins with the **glenoid cavity** ("socket") of the scapula. Notice in the illustration on the facing page that the smooth articular cartilage, the structure covering the head of the humerus and lining the glenoid cavity, reduces friction in the joint. Joints sacrifice stability for mobility and vice versa. The flatter head of the humerus connecting to the shallow depression of the glenoid cavity gives this joint a very wide range of motion but also makes it less stable and more likely for a dislocation to occur.

The following five major ligaments serve to stabilize the shoulder joint:

- **Coracoclavicular ligament**—connects the coracoid process of the scapula to the clavicle.

- **Acromioclavicular ligament**—connects the acromion of the scapula to the clavicle.

- **Coracoacromial ligament**—connects the coracoid process of the scapula to the acromion of the scapula.

- **Coracohumeral ligament**—connects the coracoid process of the scapula to the greater tubercle of the humerus.

- **Glenohumeral ligament**—three bands of fibrous tissue that connect the outer rim of the glenoid cavity to the lesser tubercle and neck of the humerus and are located anteriorly.

Common shoulder joint injuries include rotator cuff injuries, labral tears, dislocations, and bursitis. The rotator cuff refers to a group of four muscles surrounding the shoulder joint. Sprains or strains in any of these muscles or their associated tendons are examples of rotator cuff injuries. A labral tear is a break in the labrum, a piece of fibrocartilage along the rim of the acetabulum, which is often caused by injury, abnormal shape of the joint, or excessive repetitive motions. A dislocation is often caused by a direct injury such as a fall or an auto accident that results in the head of the humerus being moved out of the acetabulum. Bursitis refers to inflammation of a small fluid-filled sac called a bursa (sing.; plural *bursae*), which reduces friction in the joint.

Functions

The shoulder joint has a very wide range of motion including flexion, extension, abduction, adduction, circumduction, and rotation.

Study Tips

To identify the five different ligaments, first identify the following key structures in the shoulder joint: head of the humerus, glenoid cavity, acromion, coracoid process, and clavicle. Second, recognize the naming system. The ligaments are named after the structures to which they anchor, so if a ligament connects the acromion to the clavicle it is called the acromioclavicular ligament. Once you know the naming system, the ligaments actually name themselves!

Key to Illustration

Bony and Cartilaginous Structures

1. Intertubercular groove
2. Greater tubercle
3. Lesser tubercle
4. Articular cartilage on head of humerus
5. Acromion
6. Coracoid process
7. Articular cartilage in glenoid cavity

Ligaments and Bursae

8. Glenohumeral ligament
9. Coracohumeral ligament
10. Coracoacromial ligament
11. Acromioclavicular ligament
12. Coracoclavicular ligament
13. Tendon of the biceps brachii
14. Bursae

The illustrations below increase in structural complexity from the bones of the shoulder joints, to the ligaments, to the bursae.

Clavicle

5

6

4

3

2

7

Scapula

Humerus

1

Right Shoulder Joint: Bones
(anterior view)

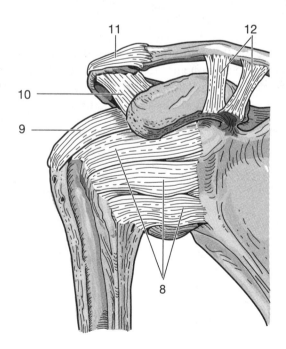

11

12

10

9

8

Right Shoulder Joint: Ligaments
(anterior view)

Color the different ligaments different colors.

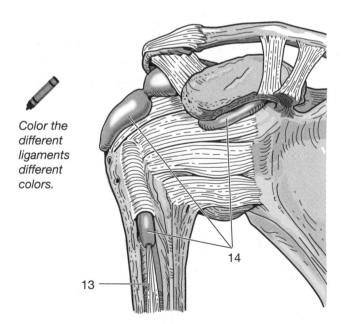

14

13

Right Shoulder Joint: Ligaments and Bursae
(anterior view)

Description

The hip (*coxal*) joint, like the shoulder joint, is classified as a ball-and-socket joint. This means the ball-like structure of one bone fits into the cup-like structure of a different bone. In this case, the **head of the femur** ("ball") joins with the **acetabulum** ("socket") of the hip bone. Joints sacrifice mobility for stability and vice versa. The large, knob-like head of the femur connecting to the deep depression of the acetabulum gives this joint greater stability but also makes its range of motion less wide than that of the shoulder joint.

The following three major ligaments serve to stabilize the hip joint:

- **Iliofemoral ligament**—strongest of the three, divided into two parts, connects the ilium to the femur and is located on the superior and anterior parts of the joint.

- **Ischiofemoral ligament**—connects the ischium to the femur and is located on the posterior portion of the joint.

- **Pubofemoral ligament**—connects the pubis to the femur and is located on the inferior portion of the joint.

Common hip joint injuries include fractures, labral tears, bursitis, and osteoarthritis. Fractures are common in elderly women with osteoporosis. The term "broken hip" is a misnomer because it is the neck of the femur that typically fractures rather than any of the bones of the hip. This is because the neck of the femur is a stress point in the joint. A labral tear is a break in the labrum, a piece of fibrocartilage along the rim of the acetabulum, which is often caused by injury, abnormal shape of the joint, or excessive repetitive motions. Bursitis refers to inflammation of a small fluid-filled sac called a bursa (sing.; plural *bursae*), which reduces friction in the joint. Osteoarthritis is caused by normal wear and tear due to aging leading to pain and inflammation. When this condition is severe a person may need to have hip replacement surgery.

Function

The hip joint has a wide range of motion including flexion, extension, abduction, adduction, circumduction, and rotation

Study Tips

To identify the three different ligaments, first locate the three different bones of the hip: ilium, ischium, and pubis. Second, recognize the naming system. The ligaments are named after the structures to which they anchor, so if a ligament connects the ilium to the femur it is called the iliofemoral ligament. Once you know the naming system, the ligaments actually name themselves!

Key to Illustration

1. Iliofemoral ligament	2. Pubofemoral ligament	3. Ischiofemoral ligament

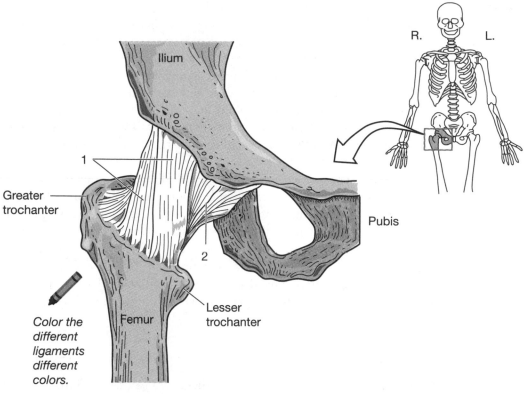

Ilium

1

Greater
trochanter

2

Femur

Lesser
trochanter

Pubis

R.   L.

Color the
different
ligaments
different
colors.

Right Hip Joint
(anterior view)

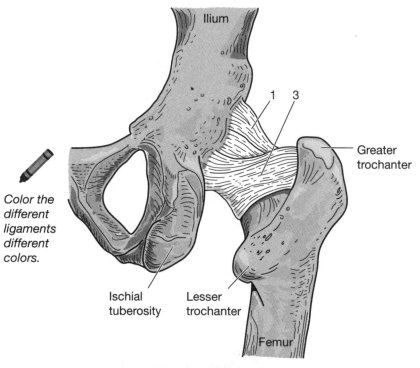

Ilium

1   3

Greater
trochanter

Color the
different
ligaments
different
colors.

Ischial
tuberosity

Lesser
trochanter

Femur

Right Hip Joint
(posterior view)

Description

The **elbow joint** is a hinge joint where the distal end of the humerus articulates with the proximal ends of both the radius and the ulna. When you touch your elbow you are feeling the **olecranon process** of the ulna, which articulates with a depression in the humerus called the olecranon fossa. There are three major ligaments that stabilize the elbow joint and hold it together—the **annular ligament, radial collateral ligament**, and **ulnar collateral ligament**. The **annular ligament** wraps around the head of the radius and attaches it to the ulna. The **radial collateral ligament** is best seen in the lateral view and attaches to the lateral epicondyle of the humerus and the annular ligament. The medial view of the elbow reveals the **ulnar collateral ligament**. It's composed of several bands of fibrous tissue running in different directions and attaches to the medial epicondyle of the humerus, the coronoid process of the ulna, and the olecranon process of the ulna.

There are several common injuries to the elbow joint. In young children under age 5, a radial head dislocation can occur called *nursemaid's elbow* where the head of the radius is moved out of its normal position in the joint. This is often caused by a parent who quickly grabs a child's hand and lifts her by one arm over a barrier. In adults, repetitive motions such as twisting motions of the wrist—like a construction worker turning a screwdriver or an office worker using a keyboard and mouse—can lead to a condition called *tennis elbow*. The problem here is small tears in the tendons that anchor arm muscles to the elbow. A typical symptom is pain at the lateral epicondyle of the humerus. Its name comes from the fact that tennis players with an improper backhand often suffer from this injury.

Study Tips

- *Annular* means "ring-shaped," which accurately describes the shape of the **annular ligament**.

- *Collateral* indicates the position of these two ligaments. Both are *laterally* within the elbow joint. They are also named after the bones to which they attach. If you identify the radius and ulna first, then the radial collateral ligament and the ulnar collateral ligament actually name themselves.

Key to Illustration		
1. Radial collateral ligament	**2.** Annular ligament	**3.** Ulnar collateral ligament

Color the different bones different colors.

Color the different ligaments different colors.

Medial epicondyle

Humerus

Lateral epicondyle

Olecranon process

Ulna

Radius

Schematic of the Bones of the Right Elbow Joint
(posterior view)

Radial collateral ligament

Ulnar collateral ligament

Annular ligament

Schematic of the Ligaments of the Right Elbow Joint
(posterior view)

Humerus

Lateral epicondyle

Articular capsule

Olecranon process

2

1

Radius

Ulna

Ligaments of the Elbow Joint
(lateral view)

Humerus

Color the different ligaments different colors.

Articular capsule

2

Medial epicondyle

Radius

3

Ulna

Ligaments of the Elbow Joint
(medial view)

Olecranon process

Description

The **knee joint** is a complex hinge joint where the distal end of the **femur** articulates with the proximal end of the **tibia**. The **patella** (*kneecap*) is part of this joint because it articulates with the patellar surface of the femur. The **quadriceps femoris tendon** covers the front of the patella and continues inferiorly to become the **patellar ligament**. Let's examine the various ligaments and other connective tissues that hold the knee joint together and stabilize it. On the posterior surface of the knee are two popliteal ligaments—the **oblique popliteal liagment** anchors the femur to the tibia, and the **arcuate popliteal ligament** anchors the femur to the fibula. On either side of the knee are two collateral ligaments—the **tibial** (*medial*) **collateral ligament** connects the femur to the tibia, and the **fibular** (*lateral*) **collateral ligament** connects the femur to the fibula. In the middle of the knee are two cruciate ligaments that connect the tibia to the femur—the **anterior cruciate ligament** (**ACL**) and the **posterior cruciate ligament** (**PCL**). Note that the term "cruciate" means cross-shaped, which helps you visualize that they cross over each other. Also in the middle of the knee are two pads of fibrocartilage called **menisci** (sing. *meniscus*)—the **lateral meniscus** and the **medial meniscus**. They act as cushions and provide lateral stability for the joint.

Knee injuries are quite common. A strain or tear in the **ACL** or **tibial collateral liagment** are two of the more common sports injuries in athletes playing strenuous sports like football, basketball, and soccer. A direct blow to the knee can cause a dislocation of the patella, which is very painful.

Study Tips

Many of the important structures in the knee joint are in pairs, so learning them as pairs is an organized way to approach it:

- two collateral ligaments
- two menisci
- two cruciate ligaments

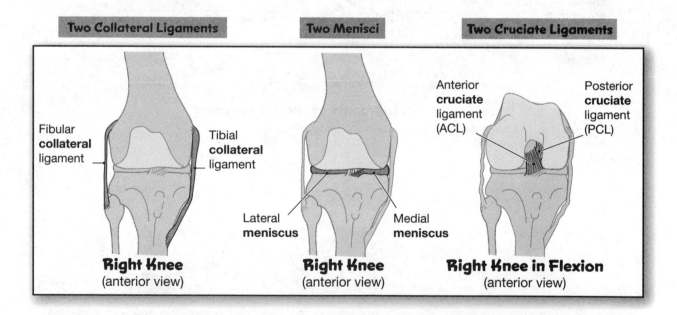

| Two Collateral Ligaments | Two Menisci | Two Cruciate Ligaments |

Right Knee (anterior view)

Right Knee (anterior view)

Right Knee in Flexion (anterior view)

Key to Illustration

1. Tibial (*medial*) collateral ligament
2. Fibular (*lateral*) collateral ligament
3. Oblique popliteal ligament
4. Arcuate popliteal ligament
5. Lateral meniscus

6. Medial meniscus
7. Anterior cruciate ligament (ACL)
8. Posterior cruciate ligament (PCL)
9. Patellar ligament
10. Quadriceps femoris tendon

Color the different bones different colors.

Schematic of the Bones of the Right Knee Joint
(anterior view)

Femur
Patella
Fibula
Tibia

Color the popliteal ligaments.

Femur
Muscle (cut)
Bursa
1
3
2
4
Tibia
Fibula

Right Knee
(posterior view)

Femur
2
5
8
7
Tibia
1
6
9 *(folded back)*
Patella
Fibula
10 *(cut)*

Right Knee
(deep anterior view)

Color all the ligaments and menisci different colors.

Femur
8
7
2
1
6
5
Tibia
Fibula

Right Knee
(deep posterior view)

Notes

MUSCULAR SYSTEM

Basic Structure of Skeletal Muscle

Description   A whole skeletal muscle is packaged like a series of tubes within other tubes. A muscle is first divided into bundles of long tubes called **fascicles.** Each fascicle is a bundle of skeletal muscle cells or fibers. Each skeletal muscle cell is a bundle of **myofibrils.** Each myofibril is composed primarily of myofilaments made of two key protein filaments—myosin and actin. These proteins are part of a repeated unit called a **sarcomere** (see pp. 214–215), the structural and functional unit for muscle contraction. The ends of a sarcomere are defined by the **Z-lines,** made of protein.

All of these different bundles of tubelike structures are held together with connective tissue. The **epimysium** (*epi* = upon, *mys* = muscle) is a tough, fibrous, connective tissue that completely surrounds the outside of a whole muscle. Within the whole muscle, the **perimysium** (*peri* = around, *mys* = muscle) fills the space around and between fascicles. Surrounding each skeletal muscle cell is another connective tissue called the **endomysium** (*endo* = within, *mys* = muscle) that mainly serves to bind one skeletal muscle cell to another.

Analogies   Three analogies are given for structures at the level of the sarcomere.

1. An **actin,** or thin, filament compares with a **double-stranded chain of pearls.** Each **pearl** is equivalent to **one molecule of actin.** (Note that this analogy does not include the troponin and tropomyosin proteins.)

2. The myosin filament has **myosin heads** (cross bridges) branching off of it, which later attach to actin during muscle contraction. From the lateral view, these heads appear angled like the **tail feathers in an arrow.**

3. The **myosin heads** attach to actin and pull on it with a regular movement. To visualize this movement, imagine the **heads moving** like a **boat rower's oars.** Unlike the oar movements, however, the heads do not all move at the same time.

Location   All skeletal muscles in the human body (more than 600 in all).

Function   Contraction for the purpose of movement.

Key to Illustration		
1. Whole muscle (biceps brachii from the illustration)	4. Myofibril	8. Myosin heads
	5. Epimysium	9. Myosin (*thick*) filaments
2. Fascicle	6. Perimysium	10. Actin (*thin*) filaments
3. Skeletal muscle cell (*fiber*)	7. Endomysium	11. Z-line

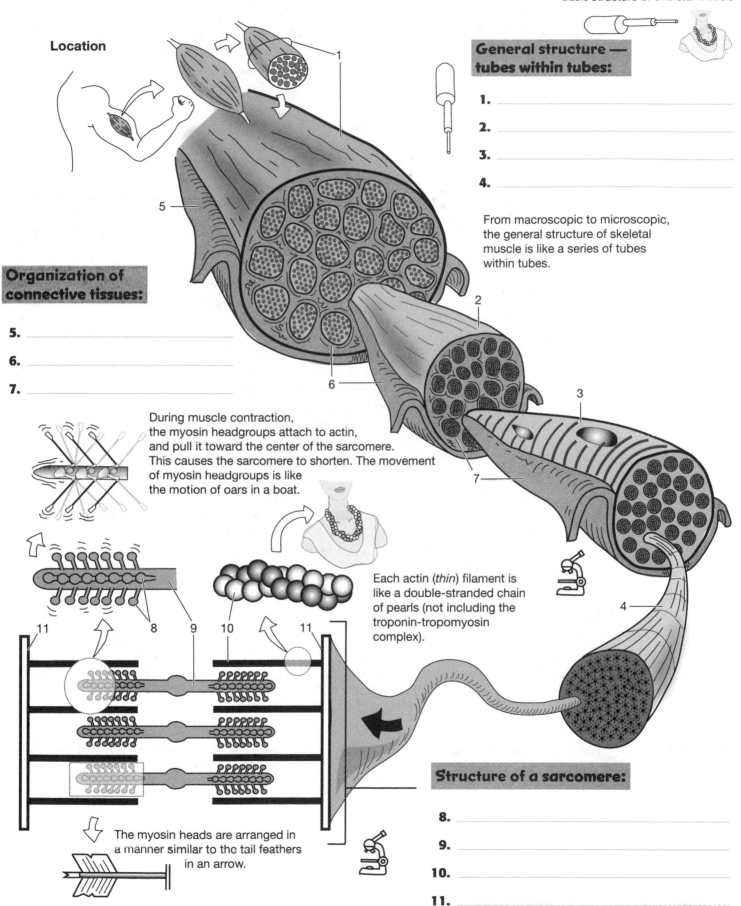

Location

General structure — tubes within tubes:

1. _____
2. _____
3. _____
4. _____

From macroscopic to microscopic, the general structure of skeletal muscle is like a series of tubes within tubes.

Organization of connective tissues:

5. _____
6. _____
7. _____

During muscle contraction, the myosin headgroups attach to actin, and pull it toward the center of the sarcomere. This causes the sarcomere to shorten. The movement of myosin headgroups is like the motion of oars in a boat.

Each actin (*thin*) filament is like a double-stranded chain of pearls (not including the troponin-tropomyosin complex).

The myosin heads are arranged in a manner similar to the tail feathers in an arrow.

Structure of a sarcomere:

8. _____
9. _____
10. _____
11. _____

Description   Muscle cells contract, but they are only able to do this because they contain many contractile proteins called **actin** and **myosin**. In skeletal muscle cells, these proteins are arranged in repeating structural and functional units called **sarcomeres** (see p. 210). In order to understand the process of muscle contraction, you must first master the structural complexity of these two proteins.

Actin   **Actin** is one of the key proteins required for muscle contraction and is nicknamed the "thin" filament because it has a much smaller diameter than myosin. One **molecule of actin** is a globular protein like a single pearl and contains a binding site for myosin where the myosin head attaches. These actin molecules are chemically bonded together to form a long strand. When two strands are wound together like a double-stranded chain of pearls, it is called an **actin filament**.

   The actin filament has two other regulatory proteins that are associated with it, namely, **tropomyosin** (**TM**) and **troponin** (**TP**). Imagine weaving two ribbons through the double-stranded chain of pearls from the previous paragraph, one along each strand. The ribbon is the protein TM. Connected to each TM at regular intervals is the other protein called TP that is composed of three subunits, so it looks like a molecular snowman. Together, TP and TM constitute a regulatory unit called the **troponin-tropomyosin complex** (*TP-TM complex*) that regulates whether or not the myosin heads bind to actin. The function of the TP is to bind calcium. During muscle contraction, calcium is released within the muscle cell and it binds to TP. This induces a shape change in the protein, which, in turn, causes the TP-TM complex to slide and shift to an open state, exposing the binding sites for the myosin heads to attach to actin. During muscle relaxation, calcium is removed from TP, so the TP-TM complex slides back to its closed position and the myosin heads are no longer able to bind to actin. It's important to note that during muscle contraction, the actin filaments move toward the center of the sarcomere (see p. 214).

Myosin   **Myosin** is the other key protein required for muscle contraction. It is nicknamed the "thick" filament because it has a diameter that is roughly double that of actin. One **molecule of myosin** consists of two subunits twisted together, and it has two parts: (1) *myosin heads*, and a (2) *myosin tail*. Each molecule looks like two cotton swabs twisted together with their ends cut off. Imagine you took about 300 of these cut cotton swabs, bent their ends, and combined them together in a cylindrical bundle. You now have a **myosin filament**. The bent, cottony tips would be sticking out in all directions. That's what the protein structure is really like. Each myosin head contains two binding sites: one for actin and another for ATP (see p. 216). During muscle contraction, ATP binds to the ATP binding site. Using free energy from ATP hydrolysis, the myosin head attaches to the actin and pulls on it, sliding the actin filament past myosin during muscle contraction (see p. 214). While the myosin heads do move during muscle contraction, it is important to note that the myosin filaments do not move. Only the actin filaments move during muscle contraction.

Analogy   For actin, each **actin molecule** is like a **pearl** and each **actin filament** is like a **double-stranded chain of pearls**. For myosin, each **myosin molecule** is like a **double-headed cotton swab** and the arrangement of the **myosin heads** in a filament is like the **tail feathers in an arrow**.

Study Tips   Notice that the terms **molecule** and **filament** are applied in the same way to both actin and myosin. That makes studying easier. In both cases, the molecule is the most basic structural unit and many molecules bonded together form a long chain called a filament.

ACTIN and MYOSIN FILAMENTS

ACTIN

One molecule of actin

Color the actin molecule.

Portion of one actin filament

Color the actin filament the same color as the actin molecule.

Each actin filament is like a double-stranded chain of pearls.

Portion of one actin filament + TP-TM complex

Ca⁺⁺

Tropomyosin (TM)

Troponin (TP)

Color the troponin (TP).

MYOSIN

A myosin molecule is like two cotton swabs twisted together with their ends cut off.

The myosin heads are arranged in a manner similar to the tail feathers in an arrow.

myosin tail

myosin heads

One molecule of myosin

One myosin filament

Color the myosin molecule.

Color the myosin filament the same color as the myosin molecule.

Portion of one myosin filament

Description

In the **sliding filament mechanism** of muscle contraction, the stimulus for skeletal muscle contraction is always a nerve impulse—no nerve impulse, no contraction. The **sarcomere** is the functional unit of muscle contraction, and its ends are marked by proteins called **Z-lines**. In the middle are the two proteins needed for muscle contraction: (1) **actin** (*thin*) **filaments**, and (2) **myosin** (*thick*) **filaments**. The myosin filaments have oar-like structures called **myosin heads** that stick out. Their function is to attach to and pull on actin.

During contraction, all the myosin heads pull on actin together, causing actin to slide past myosin as the actin is pulled toward the center of the sarcomere. The end result is that the sarcomere shortens. This shortening can be visualized most easily by noting that the Z-lines move toward each other. Also, note that neither the length of the myosin nor the actin changes during contraction. The thing that does change is the amount of *overlap* between the myosin and actin, which increases.

As goes the sarcomere, so goes the skeletal muscle cells and the whole muscle. When the sarcomeres contract, this causes the skeletal muscle cells to shorten, which shortens the whole muscle.

Relaxation occurs when nerve impulses stop. Then the linkages between the myosin heads and the actin filaments are broken, allowing the actin to slide past the myosin. This lengthening of the sarcomere continues until it has returned to its original position. These cycles of contraction and relaxation are repeated as needed.

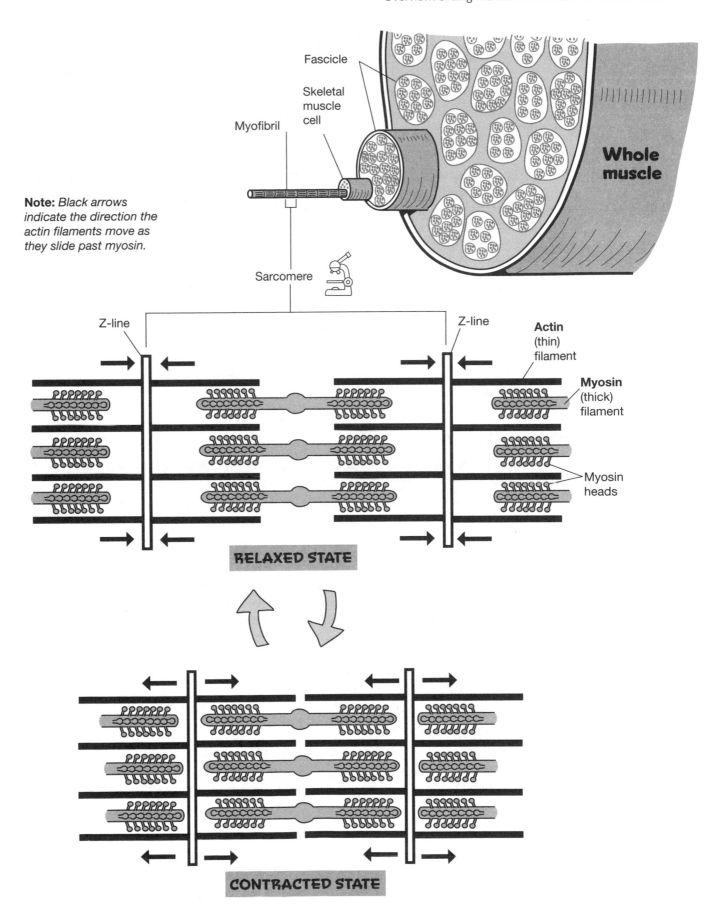

Fascicle

Skeletal muscle cell

Myofibril

Note: *Black arrows indicate the direction the actin filaments move as they slide past myosin.*

Sarcomere

Whole muscle

Z-line

Z-line

Actin (thin) filament

Myosin (thick) filament

Myosin heads

RELAXED STATE

CONTRACTED STATE

Description

All muscles can perform only one major function—contraction. At the microscopic level, it is the sarcomeres that shorten or contract. The process involves actin filaments sliding past myosin in a four-step contraction cycle: (1) **attachment**, (2) **pulling**, (3) **detachment**, and (4) **reactivation**.

A step-by-step summary of the contraction cycle is as follows:

(1) Attachment: Myosin head attaches to actin.

- The regulatory proteins **troponin** and **tropomyosin** already have shifted to expose the **myosin binding sites**.
- The products of **ATP hydrolysis**, **ADP** and **P**, are still bound to the myosin head from the previous cycle.
- The myosin head attaches to the actin molecule (*peg in a hole analogy*).

(2) Pulling: Myosin head pulls on actin.

- The myosin head bends inward, which releases **ADP** and **P**.
- The myosin molecule bends, thereby pulling the actin filament toward the center of the sarcomere.
- This pulling action is called a *power stroke*.

(3) Detachment: Myosin head releases from actin.

- As a new **ATP** molecule enters the cycle and binds to a myosin head, it forces the head to detach from actin.
- The link between myosin and actin is broken.

(4) Reactivation: Myosin head reenergizes itself for another cycle.

- ATP hydrolysis occurs, producing a net release of free energy that activates the myosin head by moving it to a high energy position.
- The myosin head now is ready to attach to actin, and the contraction cycle is repeated.

Analogies

- **Peg in a hole analogy**

  The peg fits in the hole. In this case, the actin binding site on the myosin head is like the peg, and the myosin binding site on the actin molecule is like the hole. In reality, it is a chemical bond rather than a peg and hole connection that allows the myosin head to bind to actin. But this analogy helps visualize the links formed between myosin and actin.

- **Cotton swab analogy**

  Each myosin molecule is like a double-headed cotton swab (*with one end cut off*) that is able to bend like a hinge in two places.

- **Rower analogy**

  As the myosin heads move through their four-step cycle, they somewhat resemble the movement of oars in a boat. However, they do not move in a synchronized rhythm.

LOCATION

Sarcomere

Ca+2 Troponin

Tropomyosin

Myosin binding site

ADP P

① **Attachment**

COTTON SWAB ANALOGY

A myosin molecule is like two cotton swabs twisted together whose ends have been cut off and it is able to bend like a hinge in two places

Hinge 1

Hinge 2

(ATP hydrolysis occurs)

ADP P

④ **Reactivation**

P

ADP

② **Pulling**

ATP

(New ATP molecule enters the cycle)

Color the ATP, ADP, and P the same color.

Actin binding site

ATP

③ **Detachment**

ROWER ANALOGY

During muscle contraction the myosin headgroups attach to actin and pull them toward the center of the sarcomere. This causes the sarcomere to shorten. This movement of the myosin headgroups is rather like the motion of oars in a boat.

MUSCULAR SYSTEM: General Structure and Function

Description   ATP is the common currency that fuels cellular activities and the primary energy molecule used by all cells. Muscle contraction demands enormous amounts of ATP—even those that last only a few seconds. For example, a contracting muscle cell uses about 2 million molecules of ATP every second! Wow! Without a constant supply of ATP, muscle contraction ceases. This begs the question: Where does all of this ATP come from?

Here is a summary of the four major sources of energy for muscle contraction:

1 ATP hydrolysis (see p. 74): Muscle cells store a local supply of ATP, but it does not provide much energy.

- Time allowed for muscle contraction: about 5 seconds.

2 Phosphorylation from creatine phosphate: Whereas ATP is the *primary* energy molecule used by cells, **creatine phosphate (CP)** is the *secondary* energy molecule found *only* in muscle cells. Creating it is a two-part process. First, the small chemical **creatine (C)** is produced by various organs, including the liver, and is delivered to muscle cells. Then, within the muscle cells, a phosphate group is added to C with the help of an enzyme.

The function of CP is to transfer its phosphate group directly onto ADP to create more ATP. In a sense, CP "recharges" ADP to make more ATP. One CP yields one ATP. During the relaxed state, muscle cells make excess ATP, which quickly undergoes ATP hydrolysis, supplying a phosphate group to transform C into CP.

In the contracting state, the levels of ADP are rising within the muscle cells, so CP transfers its phosphate group onto the excess ADP to form more ATP, which can be used to power more muscle contractions. But this process doesn't last long.

- Time allowed for contraction: about 15 seconds.

3 Fermentaton (see p. 478): Fermentation is the production of two **lactate** molecules from the breakdown of a single **glucose** molecule when no oxygen is present in the cell. This process occurs in the cytosol and yields a net gain of **two ATP. Glucose** is provided by either the blood or the breakdown of local **glycogen** within the muscle tissue. In **glycolysis**, a glucose molecule normally is converted into pyruvate. When oxygen is present, pyruvate enters the mitochondria to begin aerobic respiration. But under conditions of no oxygen, the pyruvate is converted into lactate instead.

Lactate is a waste product that accumulates in muscle cells, then diffuses out and is transported into liver cells. Although two ATP is not much for each cycle of fermentation, it doubles the time of contraction compared with the previous two mechanisms.

- Time allowed for muscle contraction: about 30 seconds.

4 Aerobic respiration (see p. 468): Aerobic respiration is the essential mechanism used to deliver a constant supply of ATP for muscle contraction. Without it, we couldn't sustain muscle contraction for long periods of time. In aerobic respiration, oxygen is required to produce more ATP through the **citric acid cycle** and **electron transport systems** inside the **mitochondrion**. The **pyruvate** produced in glycolysis enters the citric acid cycle in the mitochondrion instead of getting converted into lactate. The glucose needed for glycolysis comes from the same sources it did in fermentation.

In addition to glucose, energy sources include **fatty acids** from adipose tissue and **amino acids** from the breakdown of proteins. Instead of producing only two ATP (as in fermentation), the new total is about **36 ATP** for every glucose molecule broken down. As an example, if you plan to run a marathon, you will be relying on aerobic respiration to fuel all those muscle contractions for running!

- Time allowed for muscle contraction: hours.

ECF

Sarcolemma (*plasma membrane of skeletal muscle cell*)

ICF

① **ATP hydrolysis**

ATP or ATP → ADP + P + Free energy (*for muscle contraction*)

② **Phosphorylation from creatine phosphate (CP)**

ADP

excess ATP

CP

relaxed state

C

ADP

contracting state

ATP or ATP

Color the ATP, ADP, and CP molecules.

③ **Fermentation**

GLYCOLYSIS

Glycogen (*stored in muscle*)

Glucose

Pyruvate

2 ATP

Lactate

GLYCOLYSIS

④ **Aerobic respiration**

Glucose

Pyruvate

Glycogen (*stored in muscle*)

Amino acids

Fatty acids

Mitochondrion

CITRIC ACID CYCLE

ELECTRON TRANSPORT SYSTEM (ETS)

36 ATP + CO_2 + H_2O

KEY

C = Creatine

CP = Creatine phosphate

ATP = Adenosine triphosphate

ADP = Adenosine diphosphate

ECF = Extracellular fluid

ICF = Intracellular fluid

Key to Illustration

Facial Muscles (F)

F1 Frontalis

F2 Temporalis

F3 Platysma

F4 Orbicularis oculi

F5 Masseter

F6 Orbicularis oris

F7 Occipitalis

Neck Muscles (N)

N1 Sternohyoid

N2 Sternocleidomastoid

Shoulder Muscles (S)

S1 Trapezius

S2 Deltoid

S3 Infraspinatus

S4 Teres minor

S5 Teres major

Arm Muscles (AR)

AR1 Triceps brachii

AR2 Biceps brachii

AR3 Brachialis

Forearm Muscles (FO)

FO1 Pronator teres

FO2 Brachioradialis

FO3 Flexor carpi radialis

FO4 Palmaris longus

FO5 Extensor carpi radialis longus

FO6 Flexor carpi ulnaris

FO7 Extensor digitorum

FO8 Extensor carpi ulnaris

Thorax Muscles (T)

T1 Pectoralis major

T2 Serratus anterior

Abdominal Muscles (AB)

AB1 Rectus abdominis

AB2 External oblique

AB3 Internal oblique

AB4 Transverse abdominis

Hip Muscles (H)

H1 Gluteus medius

H2 Gluteus maximus

Back (B)

B1 Latissimus dorsi

Pelvis/Thigh (TH)

TH1 Tensor fasciae latae

TH2 Iliopsoas

TH3 Sartorius

TH4 Pectineus

TH5 Adductor longus

TH6 Gracilis

TH7 Rectus femoris

TH8 Vastus lateralis

TH9 Vastus medialis

TH10 Adductor magnus

TH11 Biceps femoris

TH12 Semitendinosus

TH13 Semimembranosus

Leg (L)

L1 Fibularis (*peroneus*) longus

L2 Tibialis anterior

L3 Extensor digitorum longus

L4 Gastrocnemius

L5 Soleus

- Facial Muscles (F)
- Neck Muscles (N)
- Shoulder Muscles (S)
- Arm Muscles (AR)
- Forearm Muscles (FO)

F1
F2
F4
F5
F6
F3
N1
N2
S1
Clavicle
S2
T1
T2
AR1
AR2
AR3
AB1
AB2
FO1
FO2
AB3
FO3
AB4
FO4
TH1
TH3
TH2
TH4
TH5
TH6
TH7
TH8
TH9
L1
L2
L4
L3
L5

F7
N2
S1
S2
S3
S4
S5
AR1
B1
AR3
FO2
Iliac crest
FO5
FO6
FO7
H1
FO8
H2
TH10
TH11
TH12
TH13
L4
L5
L5
Calcaneal tendon

- Thorax Muscles (T)
- Back Muscles (B)
- Abdominal Muscles (AB)
- Hip Muscles (H)
- Pelvis/Thigh Muscles (TH)
- Leg Muscles (L)

Anterior View

Posterior View

Color the different muscles different colors.

Description

Muscle Name	Action
1. Frontalis	Wrinkles skin of forehead, elevates eyebrows; draws scalp anteriorly
2. Orbicularis oculi	Depresses upper eyelid, elevates lower eyelid, tightens skin around eyes
3. Levator labii superioris	Elevates upper lip
4. Zygomaticus major	Retracts and elevates corner of mouth
5. Zygomaticus minor	Retracts and elevates upper lip
6. Risorius	Draws corner of mouth to the side
7. Platysma	Depresses mandible, pulls lower lip back and down (as in pouting)
8. Orbicularis oris	Closes, protrudes, and purses lips (kissing muscle)
9. Depressor labii inferioris	Depresses lower lip
10. Mentalis	Elevates and protrudes lower lip
11. Depressor anguli oris	Depresses corner of mouth
12. Buccinator	Compresses cheek inward (as in whistling)
13. Levator labii superioris alaeque nasi	Elevates upper lip; opens nostril (as in snarling)
14. Nasalis	Compresses bridge, depresses tip of nose; elevates corners of nostrils
15. Masseter	Elevates mandible and closes jaw
16. Temporalis	Elevates and retracts mandible
17. Corrugator supercilii	Pulls skin inferiorly and anteriorly; wrinkles brow

Color the muscles of the face.

Note: Platysma cut away at mandible.

1. _____
2. _____
3. _____
4. _____
5. _____
6. _____
7. _____
8. _____
9. _____

10. _____
11. _____
12. _____
13. _____
14. _____
15. _____
16. _____
17. _____

MUSCULAR SYSTEM: Skeletal Muscles

Superficial and Deep Muscles of the Neck, Shoulder, Thorax, and Abdomen

Description

Muscle Name	Action
Neck (N)	
N1. Platysma	Depresses mandible; pulls lower lip back and down (as in pouting)
N2. Sternocleidomastoid	Simultaneous contraction of both muscle heads: flexes neck. Individual action of each muscle head: rotates head to shoulder on opposite side
Shoulder (S)	
S1. Trapezius	Extends neck; retracts scapula
S2. Deltoid	Flexes, extends, abducts; medially and laterally rotates arm
Thorax (T)	
T1. Pectoralis major	Primary muscle of arm flexion; adducts and medially rotates arm; with arm fixed, pulls chest forward (*as in forced inspiration*)
T2. Pectoralis minor	Depresses and protracts scapula, elevates ribs
T3. Serratus anterior	Abducts and stabilizes scapula
Abdomen (A)	
A1. External oblique	Compresses anterior abdominal wall; flexes trunk; rotates trunk; depresses lower ribs
A2. Internal oblique	Compresses anterior abdominal wall; flexes trunk; rotates trunk; depresses lower ribs
A3. Transverse abdominis	Compresses anterior abdominal contents
A4. Rectus abdominis	Compresses anterior abdominal wall; flexes trunk

Analogies

- In the anterior, superficial view, the anterior portion of the **serratus anterior** looks like the blade of a **serrated knife**.

- The three abdominal muscles on the lateral surface of the abdomen—*external oblique, internal oblique,* and *transverse abdominis*—are layered on top of each other like a ham sandwich. The **internal oblique** is the **ham** (*in the middle*), and the **external oblique** and **transverse abdominis** are the layers of **bread**.

Study Tips

- Mnemonic for the four abdominal muscles: *"Really? Everything is terrible?"*

Really?	*Rectus abdominis*
Everything	*External oblique*
Is	*Internal oblique*
Terrible?	*Transverse abdominis*

- *Rectus* means *straight*—this indicates the muscle fiber direction.

- *Transverse* (transverse abdominis) indicates that muscle fibers run in line with the transverse plane line.

- Place your hands in your front pockets. Your fingers follow the fiber direction of the external oblique muscle. Now, cross your arms with palms on shoulders; your fingers indicate the fiber direction of the internal obliques. (Note: This also works for the external and internal intercostal muscles located between the ribs.)

Pectoralis major is a fan-shaped muscle.

Clavicle

Humerus

S1

N1

N2

S1

S2

T1

T2

Sternum

Triceps brachii

Biceps brachii

T3

Superficially, the anterior portion of the **serratus anterior** looks like the blade of a *serrated* knife.

T3

A4

A4

Linea alba

A2

A1

A4

Iliac crest

A3

Tensor fasciae latae

A4

Sartorius

A1

A2

A3

Iliopsoas

Adductor longus

Pectineus

The lateral abdominal muscles (A1–A3) are layered on top of each other like a sandwich.

Color the muscles in each regional area—neck, shoulder, thorax, abdomen— with different colors.

Description

Muscle Name	Action
1. Coracobrachialis	Flexion and abduction of arm/shoulder
2a. Biceps brachii, short head	Flexes and supinates forearm (*turns the corkscrew and pulls out the cork*)
2b. Biceps brachii, long head	Flexes and supinates forearm
3a. Triceps brachii, long head	Extends forearm; stabilizes shoulder joint
3b. Triceps brachii, medial head	Extends forearm
3c. Triceps brachii, lateral head	Extends forearm
4. Brachialis	Flexes forearm
5. Pronator teres	Flexes forearm; pronates forearm
6. Brachioradialis	Flexes forearm
7. Flexor carpi radialis	Flexes wrist; abducts hand
8. Palmaris longus	Weak wrist flexor
9. Flexor carpi ulnaris	Flexes and adducts hand
10. Anconeus	Adducts ulna during forearm rotation; weak forearm extensor
11. Extensor carpi radialis longus	Extends wrist; abducts hand
12. Flexor carpi ulnaris	Flexes and abducts hand
13. Extensor carpi ulnaris	Extends and adducts hand
14. Extensor digitorum	Extends hand, extends digits 2–5

Study Tips

- Most muscles that act as **flexors** are best seen in the anterior view of the upper limb, and muscles that act as **extensors** are best seen on the posterior view of the upper limb.

- In the anterior view, use the **brachioradialis** as a landmark in the forearm. It is the widest muscle on the lateral surface and inserts on the styloid process of the radius. Medially from the brachioradialis, use the mnemonic "*Foolish People Follow*" for the following muscles: *F*lexor carpi radialis, *P*almaris longus, and *F*lexor carpi ulnaris.

- In the posterior view, use the **extensor digitorum** muscle as a landmark for the forearm. To correctly identify it, find the four tendons anchoring to the phalanges in all the fingers except the thumb. All these tendons are associated with this muscle. Then, use the mnemonic "*Eat Food, Except Broccoli*" to learn the adjacent muscles in the forearm. Laterally from the extensor digitorum is the first part of the phrase, "*Eat Food,*" for the *E*xtensor carpi ulnaris and *F*lexor carpi ulnaris. Medially from the extensor digitorum is the last part of the phrase, "*Except Broccoli,*" for the *E*xtensor carpi radialis longus and the *B*rachioradialis.

- The **PALM**aris *longus* muscle anchors directly into the middle of the PALM.

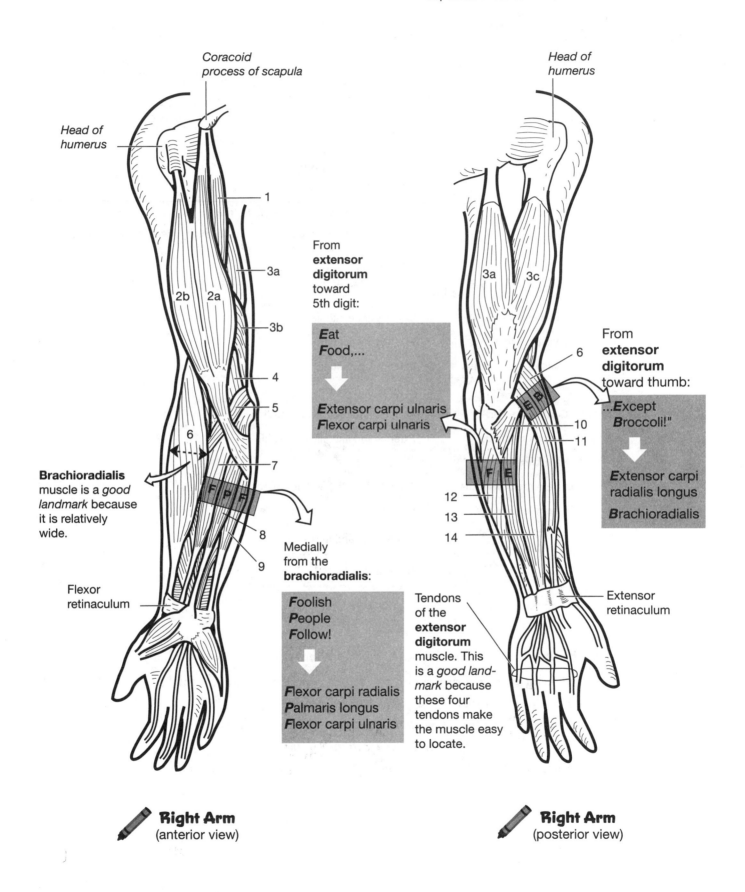

Coracoid process of scapula

Head of humerus

Head of humerus

From **extensor digitorum** toward 5th digit:

*E*at
*F*ood,...

*E*xtensor carpi ulnaris
*F*lexor carpi ulnaris

From **extensor digitorum** toward thumb:

...*E*xcept *B*roccoli!"

*E*xtensor carpi radialis longus
*B*rachioradialis

Brachioradialis muscle is a *good landmark* because it is relatively wide.

Flexor retinaculum

Medially from the **brachioradialis**:

*F*oolish
*P*eople
*F*ollow!

*F*lexor carpi radialis
*P*almaris longus
*F*lexor carpi ulnaris

Tendons of the **extensor digitorum** muscle. This is a *good landmark* because these four tendons make the muscle easy to locate.

Extensor retinaculum

Right Arm (anterior view)

Right Arm (posterior view)

Muscles of the Thigh: Anterior and Posterior Views

Description

Muscle Name	Action
1. Iliopsoas muscle	
a. Psoas major	Flexes thigh or flexes trunk on thigh (*as in during a bow*); also effects lateral flexion of vertebral column
b. Iliacus	Flexes thigh or flexes trunk on thigh (*as in during a bow*)
2. Tensor fasciae latae	Abducts, flexes, and medially rotates thigh
3. Sartorius	Flexes and laterally rotates thigh; flexes knee
4. Pectineus	Adducts, flexes, and medially rotates thigh
5. Adductor longus	Adducts, flexes, and laterally rotates thigh
6. Adductor magnus	Anterior part flexes and medially rotates thigh; posterior part extends and laterally rotates thigh
7. Gracilis	Adducts hip and flexes leg
8. Quadriceps femoris muscle	
a. Rectus femoris	Extends knee and flexes thigh at hip
b. Vastus lateralis	Extends leg at knee
c. Vastus intermedius	Extends leg at knee
d. Vastus medialis	Extends leg at knee
9. Gluteus medius	Abducts and medially rotates hip
10. Gluteus maximus	Extends and laterally rotates hip
11. Hamstrings	
a. Semimembranosus	Extends thigh; flexes knee; medially rotates leg
b. Semitendinosus	Extends hip and flexes knee
c. Biceps femoris, long head	Extends hip and flexes knee
d. Biceps femoris, short head	Flexes knee

Study Tips

- The **sartorius** muscle looks like a sash and is the *longest* muscle in the body.

- The **quadriceps femoris** is a group of four muscles on the anterior thigh:
  Rectus femoris (*name indicates location—femoral region*)
  Vastus **lateralis** (*name indicates location—on **lateral** aspect of thigh*)
  Vastus **medialis** (*name indicates location—on **medial** aspect of thigh*)
  Vastus **intermedius** (*located deep and **intermediate** to vastus lateralis and vastus medialis*)

- The **hamstrings** are a group of three muscles of the posterior thigh:
  Biceps femoris (*is located **by** itself on the lateral aspect of the thigh*)
  Semi**m**embranosus (*is the **most medial** of all the hamstring muscles*)
  Semi**tendin**osus (*is the hamstring muscle with the longest **tendon***)

- The **tensor fasciae latae** is often mispronounced to sound like the coffee drink—*latte*—but it is actually pronounced LAY-TEE.

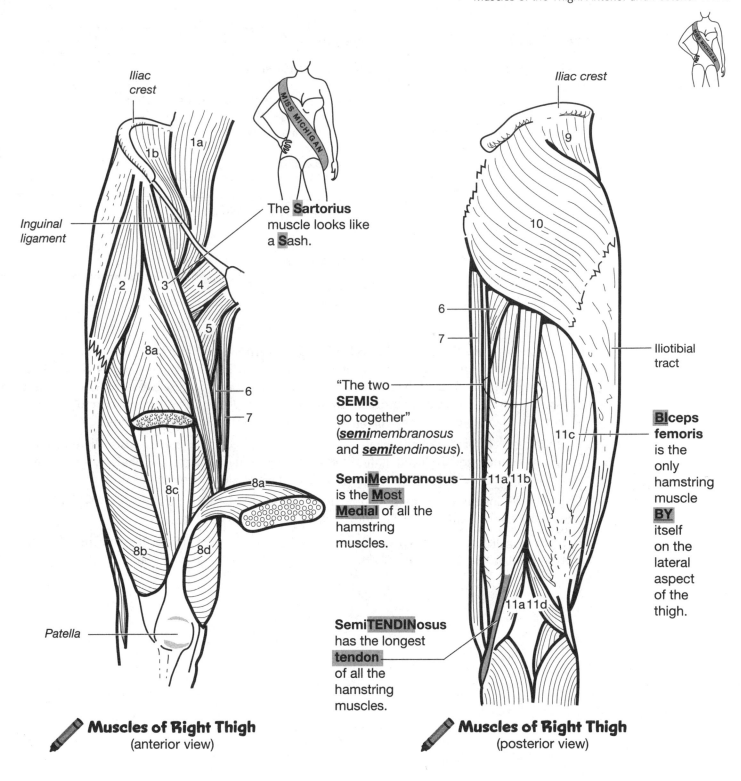

Iliac crest

Inguinal ligament

The **Sartorius** muscle looks like a **S**ash.

Iliac crest

Iliotibial tract

"The two **SEMIS** go together" (*semi*membranosus and *semi*tendinosus).

SemiM**embranosus** is the **M**ost **M**edial of all the hamstring muscles.

SemiTENDIN**osus** has the longest **tendon** of all the hamstring muscles.

BIceps **femoris** is the only hamstring muscle **BY** itself on the lateral aspect of the thigh.

Patella

Muscles of Right Thigh
(anterior view)

Muscles of Right Thigh
(posterior view)

Superficial Muscles That Move the Ankle, Foot, and Toes: Anterior and Posterior Views

Description

Muscle Name	Actions
1. Tibialis anterior	Dorsiflexes and inverts foot
2. Extensor digitorum longus	Dorsiflexes, everts foot, and extends digits 2–5
3. Fibularis (*peroneus*) longus	Plantar flexes and everts foot
4. Soleus	Plantar flexes foot
5a. Gastrocnemius, medial head	Flexes knee and plantar flexes foot
5b. Gastrocnemius, lateral head	
6. Extensor hallicus longus	Extends great toe; dorsiflexes ankle; everts foot
7. Plantaris	Plantar flexion of foot

Study Tips

- In the anterior view, use the **tibia** as a landmark to learn muscles in a sequence either medially or laterally from the tibia. Use the following mnemonic: *T*ake *E*than *F*ishing: *S*ounds *G*ood! The first part of the phrase, *T*ake *E*than *F*ishing, gives the sequence of muscles laterally from the tibia (*T*ibialis anterior, *E*xtensor digitorum longus, *F*ibularis longus). The second part of the phrase, *S*ounds *G*ood, gives the sequence medially from the tibia (*S*oleus, *G*astrocnemius).

- To double-check that you have identified the muscles correctly, use the **extensor digitorum longus** muscle as a landmark. To locate it, find the four tendons anchoring to the phalanges in all the toes except the great toe. Follow these tendons up into this muscle to correctly identify it.

- The **fibularis longus** has a long tendon that inserts into the fifth metatarsal. This tendon *loops around the lateral malleolus* of the fibula, which makes it easy to locate.

- The **soleus** (*soleus* = fish) is so named because it looks like a flat fish. It is located *deep* to the gastrocnemius, as a flat fish would rest *deep* on the bottom of a body of water.

Superficial Muscles That Move the Ankle, Foot, and Toes: Anterior and Posterior Views

Laterally from the tibia:

Take	**T**ibialis anterior
Ethan	**E**xtensor digitorum longus
Fishing ...	**F**ibularis longus

Patella

Patellar tendon

Tibial tuberosity

Tibia
(good landmark)

Medially from tibia:

... Sounds
Good!

Soleus
Gastrocnemius

Superior
extensor
retinaculum

*Lateral
malleolus
of fibula*

Inferior
extensor
retinaculum

Right Leg
(anterior view)

*Medial
malleolus
of tibia*

Tendons of the
**extensor
digitorum longus**
muscle. This is a *good
landmark muscle* because
these four tendons on its
insertion point make it
easy to locate.

Calcaneal
tendon

Calcaneous

Right Leg
(posterior view)

*(Gastrocnemius—
cut and removed)*

The **soleus**
muscle is
named
because it
looks
like a
sole
flatfish.

Filet of soleus?

231

MUSCULAR SYSTEM: Neck to Gluteal Region

Superficial and Deep Muscles of the Neck, Shoulder, Back, and Gluteal Region

Description

Muscle Name	Action
1. Sternocleidomastoid	Simultaneous contraction of both muscle heads: flexes neck forward. Individual action of each muscle head: rotates head to shoulder on opposite side
2. Trapezius	Elevates, retracts, depresses, or rotates scapula upward; elevates
3. Deltoid	Flexes, extends, abducts; medial and laterally rotates arm
4. Infraspinatus*	Abducts and laterally rotates arm
5. Teres minor*	Adducts, extends, and laterally rotates arm
6. Teres major	Extends, medially rotates, and adducts arm
7. Latissimus dorsi	Extends, adducts, and medially rotates arm
8. External oblique	Compresses anterior abdominal wall; flexes trunk; rotates trunk; depresses lower ribs
9. Gluteus medius	Abducts and medially rotates hip
10. Gluteus maximus	Extends and laterally rotates hip
11. Semispinalis capitis	Extends head and rotates to opposite side
12. Splenius capitis	Extends and hyperextends the head
13. Levator scapulae	Elevates scapula and draws it medially; with scapula fixed, flexes neck laterally to same side
14. Rhomboid minor	Retracts, adducts, and stabilizes the scapula
15. Rhomboid major	Adducts, retracts, elevates, and rotates scapula; stabilizes scapula
16. Supraspinatus*	Abducts arm; stabilizes shoulder joint
17. Serratus anterior	Abducts and stabilizes scapula
18. Serratus posterior inferior	Depresses last four ribs
19. Internal oblique	Compresses anterior abdominal wall; flexes trunk; rotates trunk; depresses lower ribs

* Rotator cuff muscles: **supraspinatus**, **infraspinatus**, **teres minor**, and **subscapularis**. The only one of these muscles not shown in either the table or the illustration is the **subscapularis**.

Study Tip

- The rotator cuff = the "SITS" muscle group.

  S Supraspinatus

  I Infraspinatus

  T Teres MINOR

  S Subscapularis

A pro baseball pitcher injured his rotator cuff, so now he **SITS** down in the **MINOR** leagues (*minor* indicates teres *minor* instead of teres *major*).

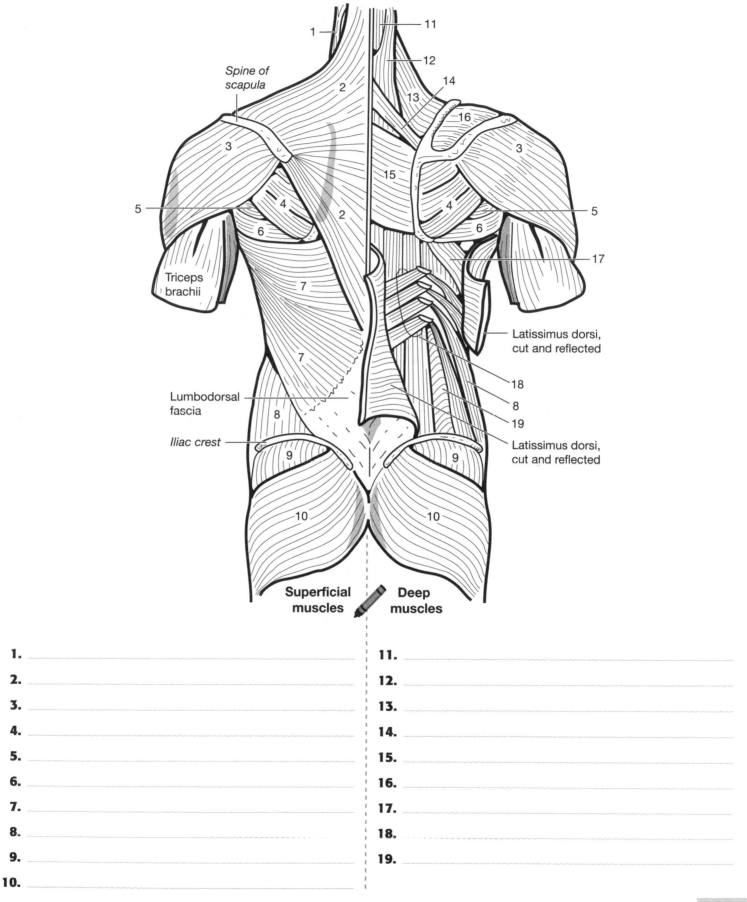

Spine of scapula

Triceps brachii

Lumbodorsal fascia

Iliac crest

Latissimus dorsi, cut and reflected

Latissimus dorsi, cut and reflected

Superficial muscles   **Deep muscles**

1. _____
2. _____
3. _____
4. _____
5. _____
6. _____
7. _____
8. _____
9. _____
10. _____

11. _____
12. _____
13. _____
14. _____
15. _____
16. _____
17. _____
18. _____
19. _____

Notes

NERVOUS SYSTEM

Description This module provides an overview of the nervous system. Along with the endocrine system, the **nervous system** is one of the important regulators of the body. The nervous system helps maintain our homeostasis. For example, by sensing changes in blood pressure or body temperature, it can trigger the appropriate responses to keep our vital functions running smoothly. Without the nervous system, we literally would be disconnected from our inner and outer worlds. Imagine not being able to sense anything—no sights, no sounds, no smells! What if all 600+ muscles in your body were paralyzed? They would be unable to move because you would have nothing to stimulate them to contract. This will give you an appreciation of the importance of the nervous system.

To illustrate further, consider how much nervous system activity is involved in the seemingly simple act of getting your morning cup of coffee. All the sights, sounds, smells, and tastes of the coffee must be detected by various **sensory receptors**. These receptors are connected to **sensory nerves** that carry the information to the **brain** like data along a computer cable. All this information must be routed to the correct interpretation center in the cerebrum of the brain. When you decide to take that first sip, you need to contract your biceps brachii to bring the cup to your mouth. This is motor output that begins as a nerve impulse in the brain, runs down the spinal cord, and is carried by a **motor nerve** to the muscle. In summary, without the nervous system, no sensory input would go to the brain and no motor output to muscles or glands.

Organization The nervous system can be divided into two major divisions: the **central nervous system** (CNS) and the **peripheral nervous system** (PNS). Here is a brief summary of each:

- Peripheral nervous system (PNS): consists of all the **sensory nerves** and **motor nerves** that travel through the body like electrical wiring in a house; more broadly, any nerve tissue outside the CNS.

- Central nervous system (CNS): consists of the **brain** and **spinal cord**.

- **THOUGHTS, MEMORIES:** Your **brain** can remember how coffee looks, smells, and tastes. In addition, it can make associations with other memories, such as that trip you took where you had your best cup of coffee ever!

Mmmm ... coffee ... the elixir of life!

- **SIGHT:** The visual image of the coffee cup must be detected by neurons in the retinas of the eyes, causing nerve impulses to travel along the optic nerves to the back of the brain, where the image is interpreted.

- **SMELL:** The aroma of the coffee is due to aromatic molecules that travel through the air, into the nasal cavity, and are then detected by **sensory receptors** in the olfactory organ. This triggers nerve impulses to be generated and sent along a **sensory nerve** to be interpreted in the brain as a coffee smell.

Spinal cord

Motor nerve

- **TASTE:** The taste of coffee is detected by special receptors within the taste buds of the tongue. This information is carried along sensory nerves and interpreted in the brain.

Aromatic molecules

Note:
Gray arrows show direction of sensory input; colored arrows show direction of motor output.

Biceps brachii

Sensory nerve

- **MOVEMENT:** The conscious movement of all skeletal muscles in the body requires stimulation from motor nerves. For example, the action of lifting the coffee cup to your lips requires contraction of the biceps brachii muscle. This begins as a conscious thought in the cerebrum. Next, a nerve impulse travels down the spinal cord and out through a motor nerve, stimulating the muscle to contract.

- **TOUCH:** Sensing the coffee cup's smooth handle requires stimulation of tactile receptors in the skin. They cause a nerve impulse to be sent along sensory nerves, up the **spinal cord** and to the **brain**, where the sensation is interpreted.

Color the sensory and motor nerves different colors.

Description

There are many different structural types of neurons. For the sake of comparison, we will examine only three different types: **unipolar**, **bipolar**, and **multipolar** neurons. They are named after the number of long cellular processes (*dendrites, dendritic process, axon*) that branch off the cell body. A *uni*polar neuron has *one* very short process off the cell body that connects to a long, single axon with dendrites at one end and synaptic knobs on the other. A *bi*polar neuron has *two* separate processes—one long dendritic process that connects to the dendrites and one long axon that connects to the synaptic knobs. A *multi*polar neuron has *many* dendrites branching off the cell body. It also has a single axon that terminates in the synaptic knobs.

Every neuron has three basic parts: a **dendrite**(s), a **cell body**, and an **axon.** A neuron may have more than one dendrite but only one axon. At the end of the axon are the **synaptic knobs.** Here there is a small space called a **synapse** that connects the neuron to a muscle or a gland or another neuron.

Information flows through a neuron in a one-way direction. The starting point is always the dendrites—sensory receptors that receive various types of stimuli. With a strong enough stimulus—a threshold stimulus—a potential is generated that passes from the dendrites to the cell body. From here, an action potential is generated that travels along the axon to the end of the synaptic knobs. The synaptic knob produces and releases a chemical called a **neurotransmitter.** This chemical messenger diffuses across the synapse and binds to a receptor in the muscle cell, glandular cell, or neuron to which it is connected (see pp. 252–253). After the neurotransmitter binds to its receptor, it induces a response. For example, it may stimulate muscle tissue to contract, or cause cells in a gland to release a hormone, or stimulate a neuron to fire and generate a nervous impulse.

Location

- **Unipolar neurons**—*ex.*: sensory neurons of the peripheral nervous system. These neurons are very common.

- **Bipolar neurons**—*ex.*: photoreceptors in the retina of the eye. These neurons are rare.

- **Multipolar neurons**—*ex.*: most common type of neuron in the brain and spinal cord; motor neurons in the peripheral nervous system.

Function

All neurons conduct nerve impulses.

Study Tip

- The term *synaptic knobs* has many other names so don't let this confuse you. These include *synaptic terminals*, *axon terminals*, *synaptic end bulbs*, and *presynaptic terminals*. Phew! I think it's time to pick one and eliminate the others, don't you?

Key to Illustration		
1. Unipolar neuron	**D**	= Dendrite
2. Bipolar neuron	**C**	= Cell body
3. Multipolar neuron	**A**	= Axon
	SK	= Synaptic knob

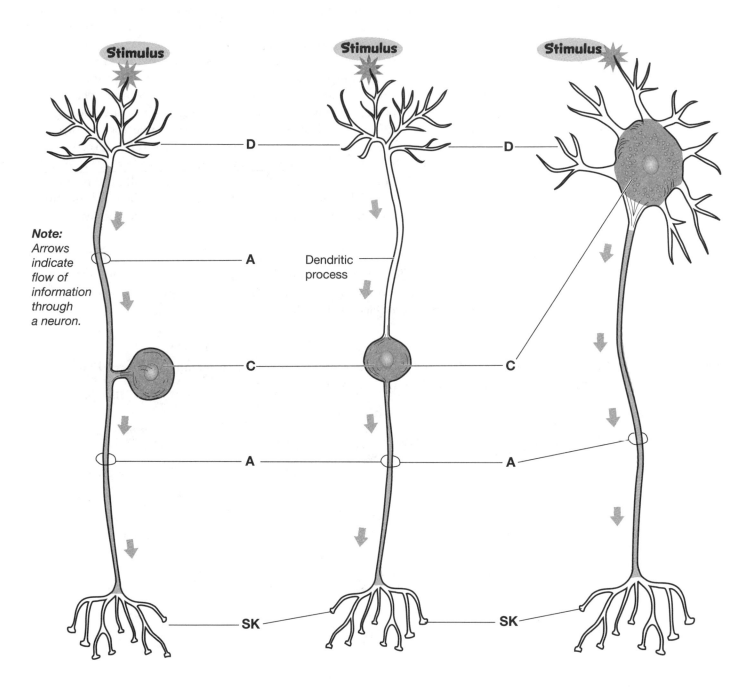

Note: Arrows indicate flow of information through a neuron.

Dendritic process

1. _____

2. _____

3. _____

 Color all the dendrites one color and the synaptic knobs a different color.

D = _____

C = _____

A = _____

SK = _____

Description

A multipolar neuron is one type of **neuron**. It has many processes called **dendrites** that respond to stimuli. The **soma** (*cell body*) is the regional area that includes the nucleus, cytoplasm, and various organelles. The rough endoplasmic reticulum is found in large clusters called **Nissl bodies**. The cell body tapers off into a funnel-shaped structure called the **axon hillock** that becomes the **axon**. Nervous impulses are conducted along the length of the axon. The axon ends in the **synaptic knob**.

Axons can be either **myelinated** or **unmyelinated**. In the process of **myelination**, the axon is wrapped in a sheath of lipid and protein that insulates the nervous impulse and speeds impulse conduction (see pp. 250–251). This process begins in the fetus and continues into late adolescence. In the central nervous system (CNS) cells called **oligodendrocytes** are responsible for the myelination process. In the peripheral nervous system (PNS), **neurolemmocytes** (*Schwann cells*) perform this task by wrapping themselves around the axon many times. These layers of plasma (*cell*) membrane constitute the **myelin sheath**. During this process, the nucleus and other organelles are pushed to the outer surface of the neurolemmocyte (*Schwann cell*). This outer layer is called the **neurilemma**. Segments of unwrapped axon between neurolemmocytes (*Schwann cells*) are called **nodes of Ranvier**.

Analogy

Each neuron has only one **axon**. The axon is like an **electrical cord**. The axon conducts a nervous impulse like the copper wires in the cord conduct electricity. The **myelin sheath** serves to insulate the axon like the **plastic casing** around the electrical cord.

Location

Multipolar neurons are the most common type of neuron in the brain and spinal cord but also occur as motor neurons in the peripheral nervous system.

Function

All neurons conduct nerve impulses.

Key to Illustration

1. Dendrite
2. Cell body (*soma*)
3. Nucleus of neuron
4. Nissl bodies
5. Axon hillock
6. Axon
7. Neurolemmocyte (*Schwann cell*)
8. Nodes of Ranvier
9. Nucleus of neurolemmocyte (*Schwann cell*)
10. Neurilemma
11. Endoneurium
12. Synaptic knob

MYELINATION

Multipolar Neuron in Peripheral Nervous System (PNS)

Note: Arrows indicate direction of nerve impulse.

Myelin sheath

Note: Arrows indicate direction of flow of electricity.

• The **axon** conducts an **action potential** like the copper wires in an electrical cord conduct an electric current.

• The **myelin sheath** is like the plastic casing around an electrical cord. It insulates the electrical impulse.

Description

Even though there are billions of neurons in the nervous system, they all function in the same basic manner. Generally, neurons respond to a **stimulus** that leads to a predictable **response**. For a neuron, the type of stimulus varies but the response is the same—secreting **neurotransmitters** from the end of the neuron. Notice in the illustration on the facing page that the stimulus occurs at a structure called the **dendrite** at the top of the neuron while the response occurs at a structure called the **synaptic knob** at the end of the neuron. As an overview to neuron function, there are four key physiology concepts that need to be understood: (1) **membrane potential**, (2) **graded potential**, (3) **action potential**, and (4) **synaptic transmission**. Let's look at each of these concepts.

A **membrane** (*resting*) **potential** (see p. 88) is a separation of positive (+) and negative (−) charges across a plasma membrane that results in an electrical potential or a voltage. This potential energy is in the form of an ion gradient that is required in order to generate either a **graded potential** or an **action potential**. The membrane potential applies to the plasma membrane surrounding all parts of the neuron. Nevertheless, as shown in the illustration, it is typically discussed with the **dendrite** because it is impossible to stimulate a neuron if it does not already have a membrane potential. As an analogy, a membrane potential is like a battery because it has a positive (+) end and a negative (−) end. When placed in a flashlight, the potential energy in the battery is converted into kinetic energy, a flow of electricity that lights the bulb. In a neuron, this potential energy can be converted into kinetic energy to generate a nerve impulse.

A **graded potential** is a local change in membrane potential that spreads across a small region of the neuron and then stops. The term *graded* refers to the fact that they vary in size and are proportional to the strength of the stimulus. As shown in the illustration, the graded potential occurs at the **dendrite** or the **cell body**. As an analogy, a graded potential is like a **small, local ripple effect** that spreads for a short distance and then fades out. Graded potentials can only send cellular signals over short distances, but typically with no effect.

An **action potential** (see p. 246) is essentially a nerve impulse that travels along the **axon** where there is a change in membrane potential due to the opening and closing of ion channels in rapid succession. During this process, sodium ions (Na^-) enter the axon, and potassium ions (K^+) exit it. To generate an action potential, a **threshold stimulus** of a minimum intensity is required. Anything less than this minimum results in no action potential while anything at or above the threshold always results in an action potential. This is called the *all-or-none* principle. Think of an action potential like a **wave of electric current** rapidly moving along the axon in a one-way direction. Action potentials allow neurons to send cellular communication signals over long distances.

Synaptic transmission (see p. 252) is the process of a neuron secreting a chemical messenger called a neurotransmitter to signal an adjacent cell (such as another neuron, muscle cell, or gland cell) to respond. The signal may either stimulate or inhibit that cell. As shown in the illustration, synaptic transmission occurs at the **synaptic knob** at the end of the neuron. After a neurotransmitter is released from the neuron, it crosses a fluid-filled space, docks with a receptor in the adjacent cell, and induces a response. This is analogous to a **speedboat crossing a channel** and pulling into its dock.

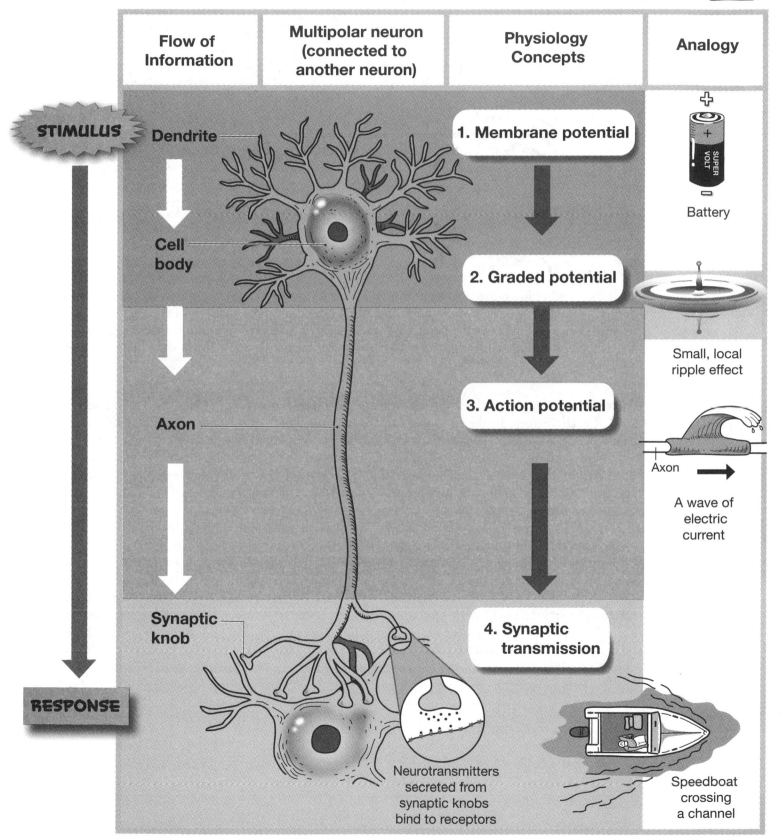

Flow of Information	Multipolar neuron (connected to another neuron)	Physiology Concepts	Analogy
STIMULUS			
Dendrite		1. Membrane potential	Battery
Cell body		2. Graded potential	Small, local ripple effect
Axon		3. Action potential	Axon — A wave of electric current
Synaptic knob		4. Synaptic transmission	
RESPONSE	Neurotransmitters secreted from synaptic knobs bind to receptors		Speedboat crossing a channel

Description

Your body has more than 600 skeletal muscles, which constitute most of your body mass. More than 45 miles of nerves are "wired" through your body like electrical wiring in a house. Each nerve is a collection of **neurons** (*nerve cells*), connective tissue, and blood vessels. Motor neurons connect to muscle tissue to stimulate it to contract.

Neurons are the fundamental cells in nervous tissue. Though they are of various types, they all share certain features. Surrounding the nucleus of every neuron is a region called the **cell body**, where most of the organelles are found. This is the metabolic center of the cell. Branching out from the cell, like tentacles from an octopus, are two types of processes—**dendrites** or **axons**. As a rule, each neuron has only one axon per cell but may have one or more dendrites.

The neuron in the illustration is called a multipolar (*multi* = many) neuron because it has *many* dendrites branching out of the cell body. The dendrites act as the sensory receptors of the cell, and the axon conducts an action potential (*nerve impulse*) along the length of the cell like copper wires in an electrical cord. The **synaptic knobs** are doorknob-like structures located at the end of the axon that contain chemical messengers called **neurotransmitters**. Any time a neuron is stimulated, it conducts a nerve impulse along its axon and usually responds by releasing neurotransmitters.

Nerve impulses flow through a neuron in a one-way direction:

dendrite $\longrightarrow$ cell body $\longrightarrow$ axon $\longrightarrow$ synaptic knobs

First, a stimulus must arrive at the dendrite. The type of stimulus needed to cause a response depends on the type of neuron. For example, sensory neurons have dendrites in the skin that respond to various stimuli such as temperature, touch, and pressure. When dendrites detect a stimulus, it triggers a signal to be sent to the cell body, then along the axon, and finally to the synaptic knobs.

Neurons Connect to . . .

Neurons can connect to three different things: (1) **other neurons**, (2) **glands**, and (3) **muscles**. Let's look at each, in turn. In the nervous system, neurons connect to other neurons like paper dolls in a chain. When this occurs, the orientation of one cell to another always follows a predictable "head to tail"–type pattern. More specifically, the synaptic knobs ("tail") of the first cell always connect to the dendrites ("head") of the second cell. The neurotransmitters move between the two cells in a specialized junction called the **synapse**. For example, your brain is composed of billions of neurons that connect to each other.

Sometimes neurons connect to an endocrine gland to stimulate the release of other chemical messengers called **hormones**. For example, neurons are wired to your adrenal glands, located on top of your kidneys. As part of a normal response called the *fight-or-flight* response, neurotransmitters stimulate hormone-producing cells within the gland to secrete epinephrine, or *adrenaline*. It travels through the blood to specific target organs in the body. As one example, epinephrine targets the heart, resulting in an increase in heart rate, which may be needed to help you flee a dangerous situation.

Last, neurons that connect to muscle tissue (*skeletal, cardiac,* and *smooth*) are called motor neurons. Acetylcholine (ACh) is the name of the neurotransmitter released by motor neurons that stimulates all the skeletal muscles in your body to contract. As ACh is released, it crosses a synapse and docks with receptors in the skeletal muscle cells to induce a response—contraction!

NEURON: FUNCTIONAL OVERVIEW

White arrows indicate direction of flow of electricity.

The **axon** conducts an **action potential** like the copper wires in an electrical cord conduct an electric current.

Synaptic knobs

Neurotransmitters

Nucleus

Cell body

Axon

Dendrite

R

Neuron releases neurotransmitters

S

Neuron stimulated at a dendrite

ONE WAY *Information flows through a neuron in a one-way direction—from __dendrite__ to __cell body__ to __axon__ to __synaptic knobs__.*

KEY

S Stimulus

R Response

NEURONS CONNECT TO ...

1. Other neurons

Synapse

Neurotransmitters

Color the neurons.

Neurotransmitters

2. Glands

Hormone

Blood

To target cells

3. Muscles

Muscle contracts!

Skeletal muscle

Synapse

Description

Action potentials (*nerve impulses*) travel at great speeds along the long, slender **axons** of neurons. Like waves of electric current, they always move in a one-way direction from the cell body toward the synaptic knobs. Their purpose is to send a signal to other neurons, muscles, or glands. The axon is like a long cylinder of plasma membrane that separates the **intracellular fluid (ICF)** from the **extracellular fluid (ECF)**.

The chemical composition of these two fluids differs in several ways. For example, the ECF has a greater concentration of **sodium ions (Na^+)**, and the ICF has a greater concentration of **potassium ions (K^+)**. Within the axon's plasma membrane are voltage-gated (V-G) ion channels. These proteins may be in one of two states—*open* or *closed*. They are triggered to open and close by a rapid change in membrane potential (voltage). Let's examine two different V-G channels: **Na^+ channels** and **K^+ channels**. Think of them like floodgates in a dam. When opened, they allow many ions to pass through the plasma membrane. V-G **Na^+ channels** only allow **Na^+** to diffuse into the axon, and V-G **K^+ channels** only allow **K^+** to diffuse out of the axon.

Conducting a nerve impulse involves three key steps: (1) *membrane potential*, (2) *depolarization*, and (3) *repolarization*. Let's look at each step.

① **Membrane** *(resting)* **potential:** *the state of the neuron when it is NOT conducting a nerve impulse.*

Membrane potentials are also called **resting potentials** (see p. 88). When a neuron is **not** conducting a nerve impulse, it is in this state. All living plasma membranes are *polarized*. Just as a battery has a positive (+) end and a negative (−) end, the same is true for a plasma membrane. The outside surface of the membrane is positive (+), and the inner surface is negative (−). Just as a battery stores a small voltage, the membrane also stores a voltage, normally about −70 mV. This can be measured with a voltmeter that has one electrode in the ECF and the other in the ICF. This voltage is a type of potential energy that can be used to conduct a nerve impulse. This potential energy comes from two ion gradients: (1) the **Na^+ gradient** and the (2) **K^+ gradient**. These gradients are maintained by a protein pump not shown in the illustration called the **Na^+-K^+ pump** (see p. 90).

Action potential: *the state of the axon when it is conducting a nerve impulse.*

② **Depolarization**

In order to conduct a nerve impulse, a stimulus strong enough to open a V-G NA^+ channel—called the **threshold**—must arrive at the neuron. The type of stimulus is unique to the type of neuron. For example, tactile receptors in the skin would be responsive only to touch. After this threshold is reached, it leads to the opening of V-G **Na^+ channels** in the first segment of the axon. This causes a sudden influx of **Na^+** resulting in a *depolarization*. This rapid shift of positive charges inside the axon reverses the normal polarity of the plasma membrane. In other words, the inside becomes more positive with respect to the outside. This sudden change in potential triggers the next segment of the axon to open its V-G **Na^+ channels**, and a chain reaction ensues. Like a tidal wave, depolarization moves forward along each segment of the axon.

③ **Repolarization**

In the wake of the depolarization "tidal wave," a *repolarization* occurs. The change in potential from the depolarization triggers V-G **K^+ channels** to open, and **K^+** diffuses out of the axon. This rapid outflux of positive charge helps flip the polarity almost back to what it used to be in the resting potential.

After repolarization, more **K^+** diffuses out of the axon through other special membrane proteins called **leakage channels** to re-establish the original resting potential. These proteins are passive channels that do not open and close. After a short resting period called a **refractory period**, the neuron is once again ready to conduct another nerve impulse. Lastly, the excess Na^+ in the ICF is pumped out of the axon by the Na^+-K^+ pump (see p. 90).

The cycle then repeats itself: *resting potential, depolarization, repolarization.* And on and on it goes!

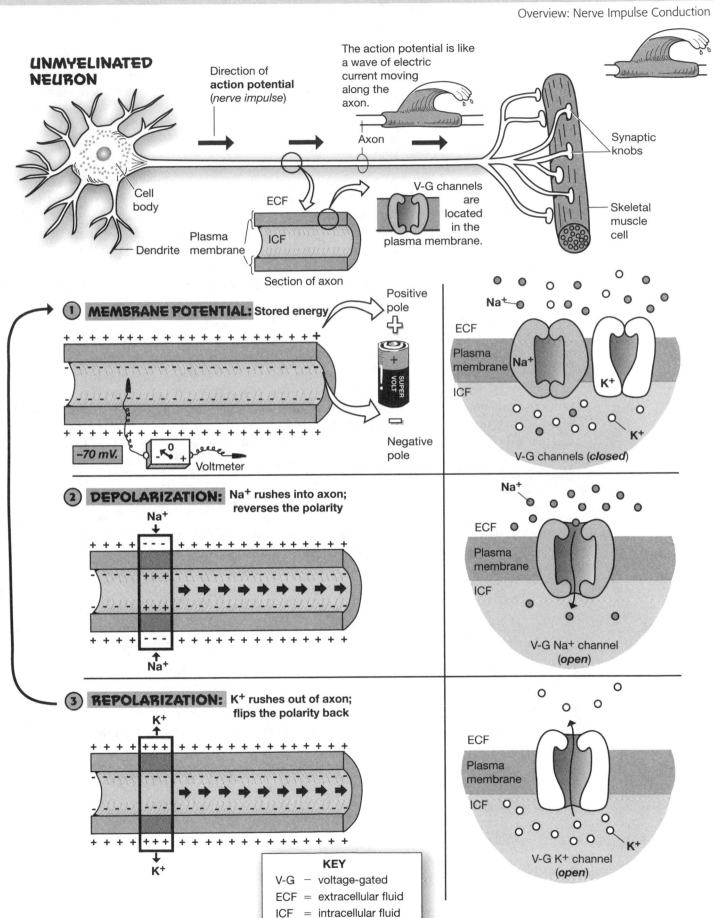

UNMYELINATED NEURON

Direction of **action potential** (*nerve impulse*)

The action potential is like a wave of electric current moving along the axon.

Axon

Synaptic knobs

Skeletal muscle cell

Cell body

Dendrite

Plasma membrane

ECF

ICF

Section of axon

V-G channels are located in the plasma membrane.

① MEMBRANE POTENTIAL: Stored energy

Positive pole

Negative pole

−70 mV.

Voltmeter

SUPER VOLT

ECF

Plasma membrane

ICF

Na⁺

K⁺

V-G channels (*closed*)

② DEPOLARIZATION: Na⁺ rushes into axon; reverses the polarity

Na⁺

Na⁺

ECF

Plasma membrane

ICF

V-G Na⁺ channel (*open*)

③ REPOLARIZATION: K⁺ rushes out of axon; flips the polarity back

K⁺

K⁺

ECF

Plasma membrane

ICF

K⁺

V-G K⁺ channel (*open*)

KEY

V-G = voltage-gated

ECF = extracellular fluid

ICF = intracellular fluid

Description

Like an electric current travels through the copper wires in an electrical cord, nerve impulses travel along axons. The two general types of impulse conduction are: (1) continuous conduction, and (2) saltatory conduction (see p. 250). This module explains **continuous conduction**, which has the following key features:

- Occurs only in **unmyelinated neurons**.
- Is slower than saltatory conduction (about 2 m/sec).

Speed is determined mainly by myelination and diameter of the **axon**. Some axons are thick, and others are thin. The general rule is: "Thicker is quicker." Therefore, thin, unmyelinated axons are much slower than thick myelinated axons. For example, in one second, impulses in thick myelinated axons can travel longer than the length of a football field (100 yards), whereas those in thin unmyelinated would not have covered enough ground to make a first down (10 yards). That's a big difference in speed.

Unmyelinated neurons lack a myelin sheath around their axons. The **extracellular fluid (ECF)** comes into direct contact with the axon. This allows for easy exchange of ions across the plasma membrane of the axon.

Impulse conduction involves a repeated cycle of depolarization and repolarization all along the length of the axon (see p. 246). This is caused by the rapid opening and closing of two different voltage-gated (V-G) channels: (1) **Na^+ channels** and (2) **K^+ channels**. Like any gate, these channels can be in two states—either *open* or *closed*. They are triggered to open and close by a rapid change in membrane potential (voltage). First, a **threshold** stimulus triggers V-G Na^+ channels to open, and sodium ions diffuse into the **intracellular fluid (ICF)** within the axon.

Then V-G K^+ channels open, and K^+ ions diffuse out of the axon and into the **extracellular fluid (ECF)**. This flux of ions across the membrane creates an action potential. With less surface area on the axon to depolarize, the impulse travels faster.

Analogy

Continuous conduction is like a domino effect in which each adjacent segment of the axon must be stimulated one after the other. The dominos are like the voltage-gated channels that must open and close in rapid succession.

Sequence of Events

The Na^+ channels are shown in the facing illustration. **Note:** K^+ channels are not shown in the illustration.

1. V-G **Na^+ channels** are stimulated to open, and Na^+ ions diffuse into the axon, resulting in a depolarization. This is followed by V-G **K^+ channels** opening to cause a repolarization.

2. In the adjacent part of the axon, V-G **Na^+ channels** again are stimulated to open, and Na^+ ions diffuse into the axon, resulting in another depolarization. This is followed by V-G **K^+ channels** opening to cause another repolarization.

3. This process of stimulating each adjacent segment of the axon continues along the entire length of the axon.

UNMYELINATED NEURON

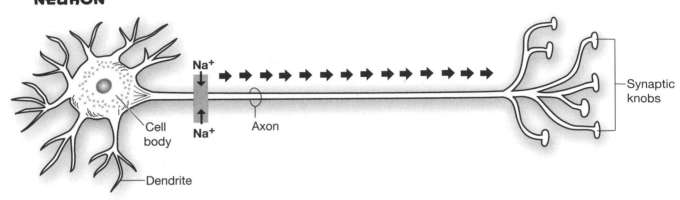

Cell body

Dendrite

Na$^+$

Na$^+$

Axon

Synaptic knobs

Na$^+$

Plasma membrane

ECF

Voltage-gated Na$^+$ channel (*open*)

ICF

DOMINO EFFECT:
Continuous conduction in an unmyelinated neuron is like a domino effect, in which each segment of the axon must be stimulated, one after the other.

SEQUENCE OF EVENTS

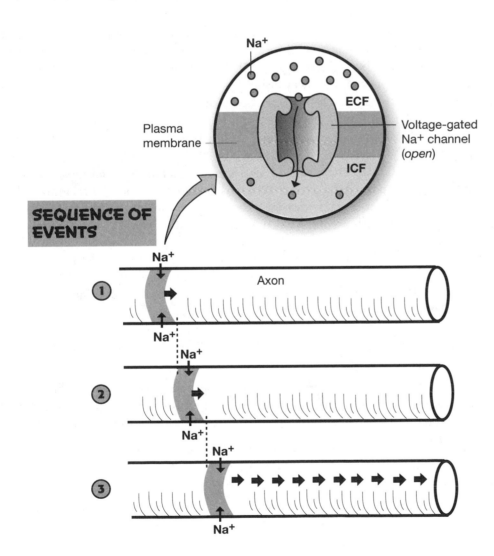

① Na$^+$

Axon

Na$^+$

② Na$^+$

Na$^+$

③ Na$^+$

Na$^+$

Description

The two general types of nerve impulse conduction are: (1) continuous conduction (see p. 248) and (2) saltatory conduction. This module explains the latter, **saltatory** (*saltare* = leaping) **conduction**. Saltatory conduction has the following key features:

- Occurs only in **myelinated neurons**.
- Is faster than continuous conduction (about 120 m/sec).

Speed of impulse conduction is mainly determined by myelination and diameter of the **axon**. Some axons are thick and others are thin. The general rule is: "Thicker is quicker." Therefore, thick myelinated axons are much faster than thin unmyelinated axons. For example, in one second, impulses in thick myelinated axons can travel longer than the length of a football field, whereas those in thin unmyelinated axons would not have covered enough ground to make a first down. That's a huge difference in speed.

Myelinated neurons in the peripheral nervous system have their axons covered by **neurolemmocytes** (*Schwann cells*). During development of the nervous system, these cells wrap themselves around the axon, forming layers of plasma membrane collectively referred to as the **myelin sheath**. This covering serves as insulation like the plastic coating around the copper wires in an electrical cord. Short segments of exposed axon between the neurolemmocytes are called **nodes of Ranvier**.

Impulse conduction involves a repeated cycle of depolarization and repolarization all along the length of the axon (see p. 246). This is caused by the rapid opening and closing of two different voltage-gated (V-G) channels: (1) Na^+ **channels** and (2) K^+ **channels**. Like any gate, these channels can be in two states—either *open* or *closed*. They are triggered to open and close by a rapid change in membrane potential (voltage). First, V-G Na^+ channels are stimulated to open to allow sodium ions to diffuse into the **intracellular fluid** (ICF) within the axon. Then V-G K^+ channels open, and K^+ ions diffuse out of the axon and into the **extracellular fluid** (ECF). This flux of ions across the membrane creates an action potential.

In myelinated neurons, V-G Na^+ channels and V-G K^+ channels are located mostly at nodes of Ranvier. Why? This is the only part of the axon where ions can cross the plasma membrane because ions within the ECF come in direct contact with the axon. Because ions can't cross the barrier of the myelin sheath, voltage-gated channels are not needed in segments of the axon covered by myelin.

Analogy

Saltatory conduction is like skipping a stone across the surface of the water, as the impulse seems to "jump" from node to node. Strictly speaking, the impulse does not actually jump. But this image is useful to illustrate that saltatory conduction is up to 60 times faster than continuous conduction, which is more like a domino effect.

Sequence of Events

The Na^+ channels are shown in the facing illustration. (**Note:** V-G K^+ channels are not shown in the illustration.)

1. V-G Na^+ **channels** at the **first node** are stimulated to open, and sodium ions diffuse into the axon, causing depolarization. The action potential generated is shuttled through the ICF of the axon to the next node of Ranvier.

2. The action potential triggers the V-G Na^+ **channels** at the **second node** to open, and sodium ions diffuse into the axon, causing another depolarization. The action potential generated is again shuttled through the axon to the next node of Ranvier.

3. The action potential triggers the V-G Na^+ **channels** at the **third node** to open, and sodium ions diffuse into the axon, causing another **depolarization**. The action potential generated is again shuttled through the axon to the next node of Ranvier. This process repeats itself until the action potential is conducted along the entire length of the axon.

Study Tip

Remember the alliteration: <u>S</u>altatory is <u>S</u>peedy like <u>S</u>kipping a <u>S</u>tone.

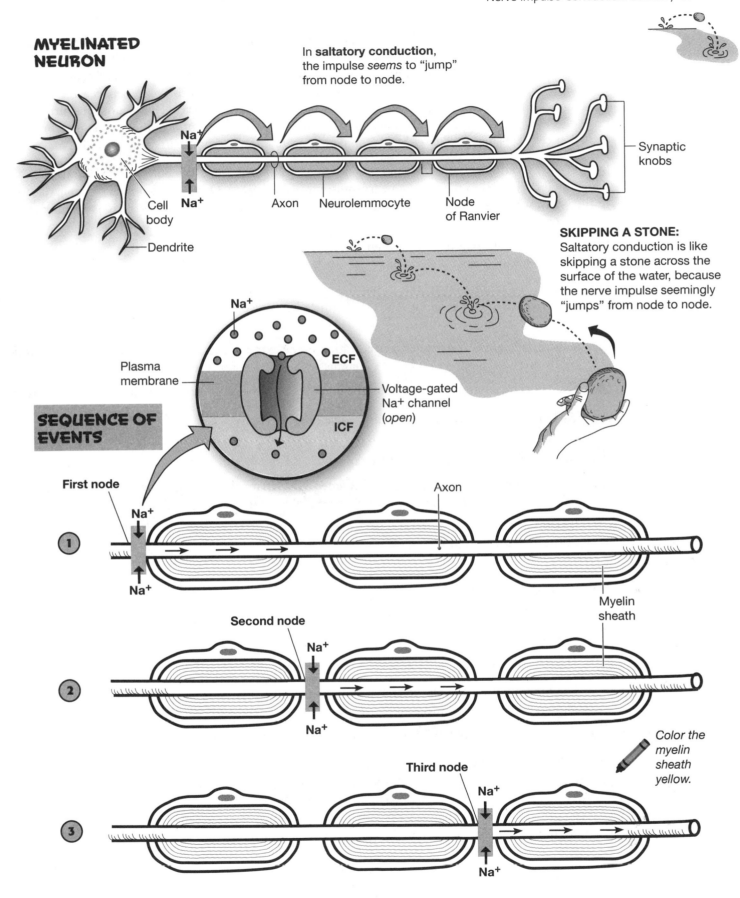

MYELINATED NEURON

In **saltatory conduction**, the impulse *seems* to "jump" from node to node.

Synaptic knobs

Cell body

Axon

Neurolemmocyte

Node of Ranvier

Dendrite

Na⁺

Na⁺

SKIPPING A STONE:
Saltatory conduction is like skipping a stone across the surface of the water, because the nerve impulse seemingly "jumps" from node to node.

Na⁺

Plasma membrane

ECF

Voltage-gated Na⁺ channel (*open*)

ICF

SEQUENCE OF EVENTS

First node

Na⁺

Na⁺

Axon

Myelin sheath

1

Second node

Na⁺

Na⁺

2

Color the myelin sheath yellow.

Third node

Na⁺

Na⁺

3

Description

Synaptic transmission is the method by which neurons communicate with other cells. The illustration shows the synaptic knobs of a motor neuron connecting to a skeletal muscle cell in the neuromuscular junction (NMJ), whereas the synaptic knobs of neuron #1 connect to the dendrites of neuron #2 in a neuro-neuro junction (NNJ). The synaptic cleft is the narrow, fluid-filled space between the synaptic knob and the muscle cell/nerve cell. In both cases, a chemical messenger called a neurotransmitter must cross this gap, like a ferryboat crossing a channel, and dock at the receptor in the plasma membrane of the other cell. This may lead to either a stimulatory or inhibitory response in the other cell.

Steps 1–4: *Secretion of neurotransmitters*. These steps are the same for both an NMJ and an NNJ. The only difference is in the type of neurotransmitter(s) involved. Only acetylcholine (ACh) is released in the NMJ, whereas NNJs have dozens of different possible neurotransmitters. Here is a summary of this mechanism:

1 Neurotransmitters are synthesized in synaptic knobs and stored within vesicles at high concentrations.

2 When a nerve impulse travels along the axon to the synaptic knobs, it triggers the opening of voltage-gated (V-G) calcium (Ca^{++}) channels because of a change in voltage. Note that these channels normally are *closed*.

3 Because calcium is at a higher concentration in the extracellular fluid, it diffuses into the synaptic knobs through the open V-G Ca^{++} channels, where it helps trigger exocytosis (see p. 86).

4 The neurotransmitter is released into the fluid-filled synapse, or synaptic cleft, and diffuses across it.

Steps 5–7: *Inducing a response in either muscle or nerve tissue*.
For the NMJ:

5 ACh binds to the ACh-gated Na^+ channel (*closed state*).

6 The binding of ACh induces the ACh-gated Na^+ channel to *open*, allowing only sodium ions (Na^+) to diffuse into the muscle cell resulting in depolarization.

7 If the polarization is past threshold, an action potential is generated in the muscle cell that may lead to a muscle contraction.

When nerve impulses from the motor neuron cease, ACh in the synapse is quickly broken down, and muscle action potentials stop.

For the NNJ:

5 The neurotransmitter (NT) binds to NT-gated Na^+ channel (*closed state*).

6 The binding of the neurotransmitter induces the NT-gated Na^+ channel to *open*, allowing only sodium ions (Na^+) to diffuse into the neuron, resulting in depolarization.

7 If the depolarizing potential reaches threshold, it can trigger an action potential (*nerve impulse*).

Alternatively, the neurotransmitter may bind to a *different* NT-gated channel that is specific for either potassium (K^+) or chloride (Cl^-). The illustration shows a NT-gated K^+-gated channel as an example. When this channel opens, K^+ diffuses out of neuron #2. This loss of positive charge leads to hyperpolarization—an increase in the value of the membrane potential (see p. 88). This inhibitory pathway prevents an action potential from being induced in neuron #2.

When nerve impulses from neuron #1 cease, the neurotransmitter in the synapse has two fates: (1) like ACh, it is broken down, or (2) it is reused by being transported back into the synaptic knob or to a neighboring neuroglial cell.

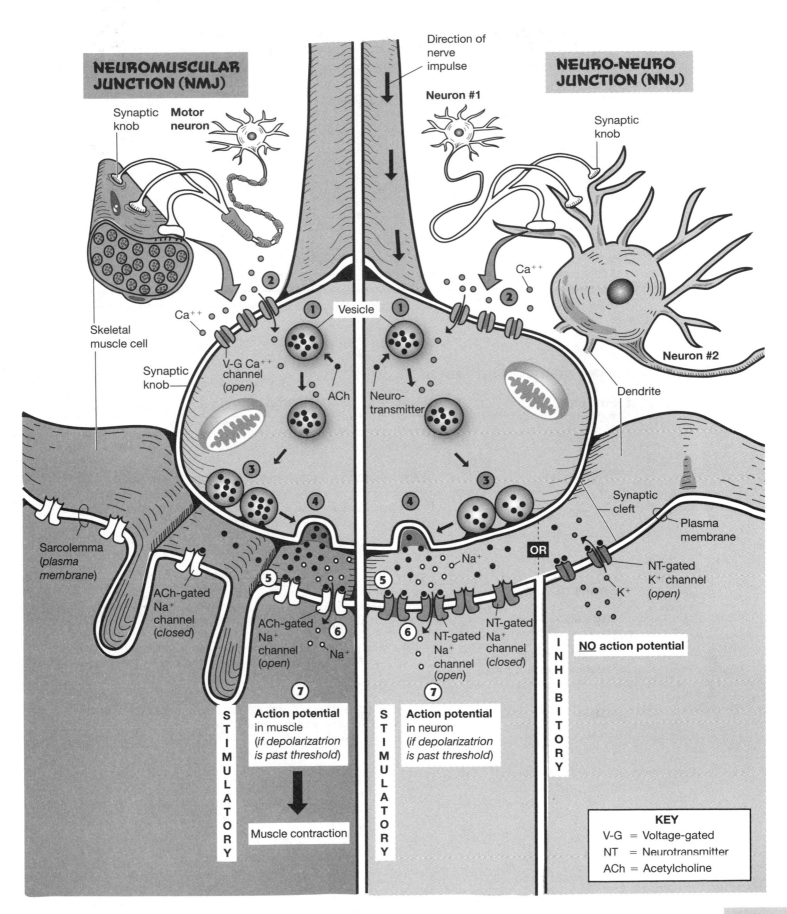

NEUROMUSCULAR JUNCTION (NMJ)

Synaptic knob

Motor neuron

Skeletal muscle cell

Synaptic knob

Ca^{++}

V-G Ca^{++} channel (*open*)

ACh

Sarcolemma (*plasma membrane*)

ACh-gated Na$^+$ channel (*closed*)

ACh-gated Na$^+$ channel (*open*)

Na$^+$

S T I M U L A T O R Y

Action potential in muscle (*if depolarizatrion is past threshold*)

Muscle contraction

Direction of nerve impulse

Neuron #1

Vesicle

Neuro-transmitter

Na$^+$

NT-gated Na$^+$ channel (*open*)

S T I M U L A T O R Y

Action potential in neuron (*if depolarizatrion is past threshold*)

NEURO-NEURO JUNCTION (NNJ)

Synaptic knob

Ca^{++}

Neuron #2

Dendrite

Synaptic cleft

Plasma membrane

NT-gated K$^+$ channel (*open*)

K$^+$

NT-gated Na$^+$ channel (*closed*)

OR

I N H I B I T O R Y

NO action potential

KEY

V-G	=	Voltage-gated
NT	=	Neurotransmitter
ACh	=	Acetylcholine

Description

Peripheral **nerves** spread throughout the body like electrical wiring through a house. They are a collection of bundles of microscopic axons. Examples include the sciatic and femoral nerves in the thigh and the brachial nerve in the arm. Nerves are visible to the naked eye and structured as a series of tubes within tubes. Each tubelike structure is wrapped in a protective connective tissue. Each of the labeled structures in the illustration is explained in the tables below.

Tubes within Tubes

Structure	Description
1. Nerve	long, macroscopic, cable-like structure containing bundles of axons
2. Fascicle	a single bundle of axons
3. Myelinated axon	an axon wrapped in a protective myelin sheath
4. Axon	long, thin part of a neuron that carries nerve impulses along its length

Connective Tissue Organization

Structure	Description
5. Epineurium	the thick layer of dense irregular connective tissue that wraps around the outside of a nerve
6. Perineurium	the cellular connective tissue layer that wraps around each fascicle
7. Endoneurium	the thin layer of areolar connective tissue that wraps around each axon and binds one myelinated axon to another within a fascicle

Neuronal Structures

Structure	Description
8. Neurolemmocyte (*Schwann cell*)	a cell that wraps itself around the axon of a neuron numerous times during development of the nervous system; the end result is that a myelin sheath is created
9. Myelin sheath	the layers of the neurolemmocyte's cell membrane that form a protective, insulating coating of lipoprotein around the axon

Location

Sciatic nerve

1

5

General structure—tubes within tubes:

1. _____

2. _____

3. _____

4. _____

Blood vessels

From macroscopic to microscopic, the general structure of a nerve is like a series of tubes within tubes.

(Compare with skeletal muscle structure, see p. 211)

2

3
4

6

(See neuron structure, p. 240)

8

8

8

Organization of connective tissues:

5. _____

6. _____

7. _____

7

9

4

Color the neurolemmocytes.

Neuronal structures:

8. _____

9. _____

Description

The **brachial plexus** (*brachial* = pertaining to the upper limb between shoulder and elbow, *plexus* = network) is a complex neural network that begins with the roots of spinal nerves C5-T1 and ends with the branches that are the start of the peripheral nerves. Moving out from the spinal cord, it is composed of the following sequence of structures: *roots*, some of which join to become *trunks*, that then divide to become *divisions*, that then join to become *cords*, which, finally, become *branches*. Let's examine each of these structures:

- **Roots**—the anterior branches of the spinal nerves. There are five roots: *C5, C6, C7, C8*, and *T1*.

- **Trunks**—bundles of axons. There are three trunks: *superior, middle*, and *inferior*.

- **Divisions**—trunks divide into the two pathways they follow, *anterior* and *posterior*.

- **Cords**—there are three cords: *lateral, posterior*, and *medial*. They are named by their positions relative to the axillary artery—a major artery that supplies blood to the upper limbs.

- **Branches**—the branches (*terminal branches*) are the peripheral nerves. There are five major nerves shown in both the schematic below and the illustration on the facing page: *musculocutaneous nerve, median nerve, axillary nerve, radial nerve*, and *ulnar nerve*. These are the nerves that innervate the shoulder and the upper limbs.

Study Tips

To recall the correct sequence for parts of the brachial plexus from *roots* to *branches,* use the mnemonic: *R*acing *T*ortoise *D*efeats *C*ocky *B*unny

Schematic of a Brachial Plexus

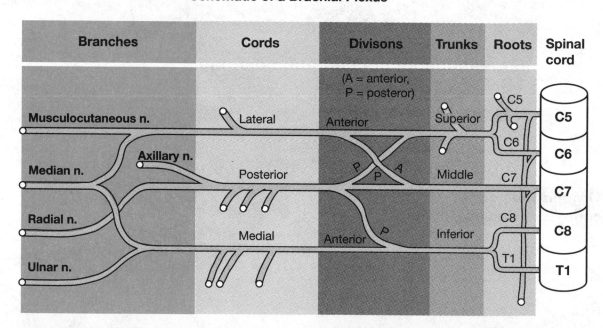

Branches	Cords	Divisons	Trunks	Roots	Spinal cord

(A = anterior, P = posteror)

Musculocutaneous n. — Lateral — Anterior — Superior — C5 — C5

Axillary n. — C6

Median n. — Posterior — P P A — Middle — C7 — C6

Radial n. — C8 — C7

Medial — Anterior P — Inferior — C8

Ulnar n. — T1 — C8

T1

Key to Illustration

1. Axillary nerve
2. Musculocutaneous nerve
3. Median nerve
4. Ulnar nerve

5. Radial nerve
  a. Deep branch—radial nerve
  b. Superficial branch—radial nerve
6. Palmar digital nerves

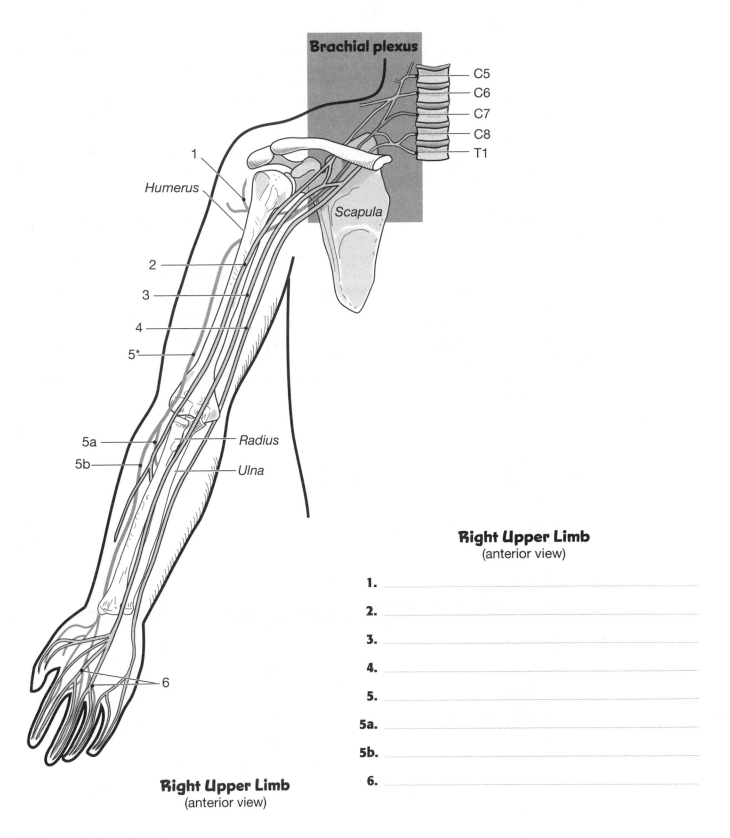

Brachial plexus

C5
C6
C7
C8
T1

1
Humerus
2
3
4
5*
5a — Radius
5b — Ulna
6

Scapula

Right Upper Limb
(anterior view)

Right Upper Limb
(anterior view)

1. _____
2. _____
3. _____
4. _____
5. _____
5a. _____
5b. _____
6. _____

* This nerve is shown in color
  simply to make it easier to
  follow its pathway.

Description

All the nerves that innervate the pelvic region and the lower limbs originate from two neural networks: the **lumbar plexus** and the **sacral plexus**. Due to the overlap between them, they are often referred to collectively as the **lumbosacral plexus**. Let's examine each in more detail:

The **lumbar plexus** (*lumbar* = lower back, *plexus* = network) begins with the anterior branches of spinal nerves L1-L4, and it also receives a contribution from T12. The neural network here is less complicated than the brachial plexus. It simply passes anteriorly and posteriorly to become the major peripheral nerves that supply the abdominal wall muscles, external genitals, and parts of the lower limbs. The largest nerve in this network is the **femoral nerve**, which passes anteriorly through the thigh.

The **sacral plexus** (*sacral* = relating to the sacrum, *plexus* = network) begins with the anterior branches of spinal nerves L4-S4. Like the lumbar plexus, it also simply passes anteriorly and posteriorly to become the major peripheral nerves that supply the buttocks, perineum, and lower limbs. The largest nerve in this network, the **sciatic nerve**, is actually two nerves combined. It passes posteriorly through the thigh, and as it approaches the knee, it divides into the *tibial nerve* and the *common fibular nerve*. It is important to note that it is the longest and widest nerve in the body. Associated with this nerve is a medical condition called **sciatica**, which refers to pain in the leg due to damage to the sciatic nerve. The pain—tingling, numbness, and weakness—typically runs from the lower back down the leg. While it most often occurs in only one leg at a time, it can occur in both. The most common cause is a herniated disc, but it has many other possible causes such as narrowing of the spinal canal or a bone spur.

Key to Illustration

1. Iliohypogastric nerve	11. Superficial fibular nerve
2. Ilioinguinal nerve	12. Deep fibular nerve
3. Genitofemoral nerve	13. Dorsal digital nerves
4. Lateral femoral cutaneous nerve	14. Superior gluteal nerve
5. Obturator nerve	15. Inferior gluteal nerve
6. Femoral nerve	16. Pudendal nerve
7. Posterior femoral cutaneous nerve *(cut)*	17. Posterior femoral cutaneous nerve
8. Sciatic nerve	18. Sciatic nerve
9. Saphenous nerve	19. Tibial nerve
10. Common fibular nerve	20. Common fibular nerve
	21. Sural nerve

T12
L1
L2
L3
L4
L5

Lumbar plexus

Lumbo-sacral plexus

S1
S2
S3
S4

Sacral plexus

1
2
3
4
5

Inguinal ligament

6
7
8

Femur

9

Tibia

10
11

Fibula

12

13

Right lower limb
(anterior view)

Right Lower Limb
(anterior view)

1. _____
2. _____
3. _____
4. _____
5. _____
6. _____
7. _____
8. _____
9. _____
10. _____
11. _____
12. _____
13. _____

Right Lower Limb
(posterior view)

14. _____
15. _____
16. _____
17. _____
18. _____
19. _____
20. _____
21. _____

14
15
16
17
18
19
20
21

Right lower limb
(posterior view)

Description

This module gives an overview of how sensory receptors detect a stimulus, convert it into an electrical impulse, and send it along a **sensory neuron** to be interpreted in the brain. A **sensory receptor** is a specialized structure to detect a specific stimulus. Many are simply the dendrites of sensory neurons encapsulated by some other tissue. They are found in locations such as the skin, joints, tendons, and blood vessel walls. Each type of receptor responds to a single type of stimulus—touch, temperature, pressure, pain. Receptors can be organized into the following major categories:

- **Mechanoreceptors:** detect stimuli that compress, bend, or stretch cells; allow for sensations such as touch, pressure, and vibration. *Ex.:* Lamellated corpuscles: mostly detect deep pressure; locations include lower dermis of the skin, joints, tendons, muscles, periosteum, and pancreas.

- **Chemoreceptors:** detect changes in chemical concentrations (H^+, O_2, CO_2) in a solution

- **Nociceptors:** detect pain resulting from tissue damage

- **Photoreceptors:** sensitive to light that strikes the retina of the eye

- **Thermoreceptors:** detect changes in temperature

- **Osmoreceptors:** detect changes in osmotic pressure of body fluids

Sensory transduction refers to the process of converting the stimulus energy into a nerve impulse. This conversion is essential, allowing all the sensory information to travel through the nervous system along sensory nerves to eventually be interpreted in the **brain**. Without it, no sensations would be perceived.

Let's use the example of a **lamellated corpuscle** as our sensory receptor. It primarily detects pressure but also stretching and vibration. Its structure consists of a single dendrite surrounded by an oblong-shaped **capsule** of up to 60 layers of fibrous connective tissue, giving it a sliced-onion appearance. Between the layers is a viscous material. In the capsule's core is a fluid-filled space that surrounds the dendrite.

Imagine pressing your finger against a flat surface and feeling the pressure that results. The illustration shows the process of sensory transduction. Here is a summary of each step:

① Stimulus: pressure

The normal membrane potential is symbolized by the " + " signs outside the membrane and the " − " signs inside the membrane (see p. 88). The act of pressing serves as a stimulus to compress the concentric layers within the capsule. This, in turn, deforms the plasma membrane of the dendrite.

② Graded potential generated

In response to this membrane deformation, the membrane becomes more permeable to Na+ (sodium ions). This triggers protein channels to open, and Na+ rapidly diffuses into the dendrite, thereby depolarizing the membrane. The small current generated is called a **graded potential**.

③ Action potential generated

A graded potential has different orders of magnitude. If it reaches a threshold or minimum strength, it can induce an **action potential**, or nerve impulse, in the first **node of Ranvier**. Because this sensory neuron is myelinated, it results in saltatory conduction (see p. 250). This is where the nerve impulse is shuttled from the first node to the second, then third, and so on. Finally, the nerve impulse travels from the sensory nerve and up the **spinal cord**, where it is interpreted in the cerebral cortex.

That's pressure!

Brain

Spinal cord

Sensory neuron

Lamellated corpuscle

Capsule

Color the capsule.

Axon

Myelin sheath

Node of Ranvier

KEY

S Stimulus

R Response

(1) **Stimulus: Pressure**

S PRESS!

PRESS!

+ + + + + + + + + + + + + + + +
− − − − − − − − − − − − − − − −
+ + + + + + + + + + + + + + + +

(2) **Graded potential generated**

R

Na⁺ Na⁺ Na⁺ Na⁺

Na⁺ Na⁺ Na⁺ Na⁺

(3) **Action potential generated**

NERVE IMPULSE!

Na⁺

Na⁺

Node of Ranvier

Description

The **spinal cord** is a long, slender structure that links the body with the brain. Most of the cord is protected by the bony vertebrae because it runs through the vertebral canal of the vertebral column. Three layers of protective membranes called **meninges** surround the spinal cord and brain. From outermost to innermost, these are as follows:

- **Dura mater:** thickest and strongest; contains fibrous connective tissue
- **Arachnoid:** thin layer made of simple squamous epithelia; lacks blood vessels
- **Pia mater:** tightly adheres to the spinal cord and follows every surface feature; supplies many blood vessels directly to the spinal cord

Below the arachnoid is a potential space called the **subarachnoid space** that is filled with **cerebrospinal fluid.** This serves as a cushion to protect the spinal cord and functions as a medium through which to deliver nutrients and remove wastes. Extending laterally off the spinal cord are 31 pairs of **spinal nerves.** Each short spinal nerve immediately branches into different parts, such as the smaller dorsal ramus, the larger ventral ramus, and the rami communicantes. Each spinal nerve, along with these branches, is *mixed*, meaning it carries both sensory and motor axons. These become the various peripheral nerves that spread throughout the body. The spinal nerves and their associated structures are:

- **Dorsal root:** contains only *sensory* axons
- **Dorsal root ganglion:** contains the neuron cell bodies of sensory neurons
- **Ventral root:** contains only motor axons
- **Dorsal ramus:** branch of each spinal nerve that innervates certain parts of the skin and muscles of the neck and back
- **Ventral ramus:** branch of each spinal nerve that innervates the ventral and lateral body surfaces, structures in the body wall, and the limbs
- **Rami communicantes:** branches of each spinal nerve that contain axons related to the autonomic nervous system (ANS); consist of a pair of rami—one white ramus and one gray ramus

The spinal cord contains areas of **gray matter** and **white matter.** The white matter is located in the outer portion of the spinal cord and consists of myelinated axons that run along its length. It is divided into the following three regional areas called **funiculi** (sing. *funiculus*): a **posterior funiculus, lateral funiculi,** and an **anterior funiculus.** The **white commissure** is a narrow band of white matter that connects the anterior funiculi together.

The **gray matter** is located in the inner portion of the spinal cord and includes short neurons called interneurons along with cell bodies, dendrites, and axon terminals of other neurons. The three regional areas of gray matter are referred to as horns: **posterior horn, lateral horn, anterior horn.** In the center of the spinal cord is a small passageway called the **central canal** that contains cerebrospinal fluid. The horizontal band of gray matter that surrounds the central canal is called the **gray commissure.**

Analogy

The gray matter in the middle of the spinal cord looks like a butterfly. Note that the exact shape of the gray matter changes from one segment of the spinal cord to another, so it does not always look exactly like a butterfly.

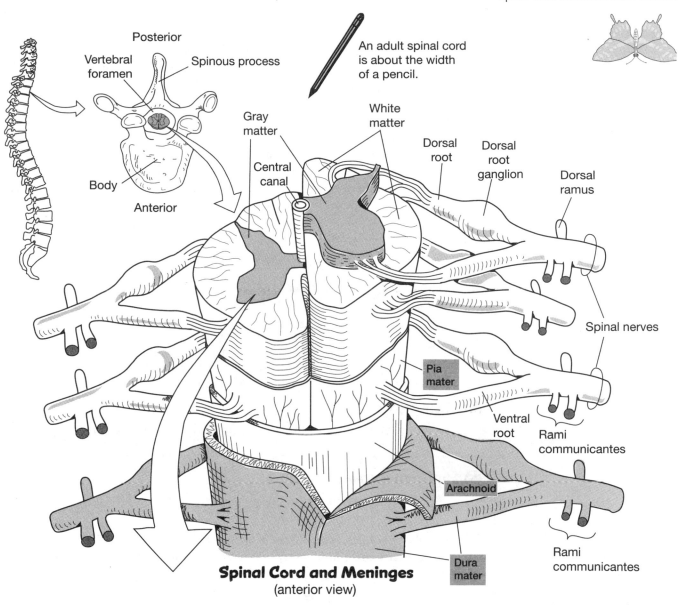

Posterior

Vertebral foramen

Spinous process

Body

Anterior

An adult spinal cord is about the width of a pencil.

Gray matter

White matter

Central canal

Dorsal root

Dorsal root ganglion

Dorsal ramus

Spinal nerves

Pia mater

Ventral root

Rami communicantes

Arachnoid

Dura mater

Rami communicantes

Spinal Cord and Meninges
(anterior view)

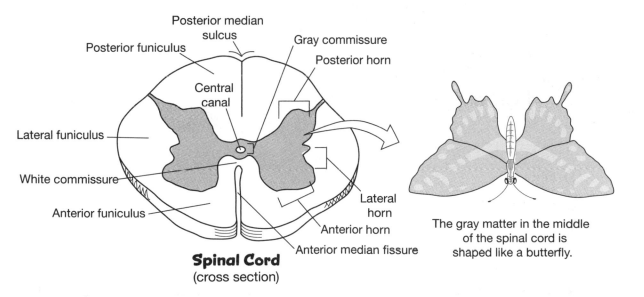

Posterior median sulcus

Posterior funiculus

Gray commissure

Posterior horn

Central canal

Lateral funiculus

White commissure

Anterior funiculus

Lateral horn

Anterior horn

Anterior median fissure

Spinal Cord
(cross section)

The gray matter in the middle of the spinal cord is shaped like a butterfly.

Description

A **reflex** is a rapid, involuntary response to a stimulus. A reflex arc refers to the neural pathway involved in a single reflex. The two general reflex categories are somatic reflexes and autonomic reflexes. Only **somatic** (*body*) **reflexes**—which result in the contraction of skeletal muscles—will be illustrated in this module. An example is the withdrawal reflex. If the skin on your finger is poked by a thumbtack, this stimulus is detected by pain receptors that trigger nerve impulses to contract flexor muscles to allow you to pull your hand away. During a physical exam, a physician may strike your patellar tendon with a percussion hammer. This stimulates your quadriceps femoris muscle to contract, resulting in extension of the knee joint.

The autonomic reflexes are so named because they are part of the autonomic nervous system (ANS). These reflex arcs stimulate responses in smooth muscle, cardiac muscle, and glands that allow for control of functions such as digestion, urination, and heart rate.

Somatic Reflex Arc

The parts of a typical somatic reflex arc are:

Input

① Receptors

—located at the end of sensory neurons

—respond to a specific stimulus (*ex.:* touch, pressure, or pain stimuli).

—different for different stimuli

In the illustration on the facing page, a thumbtack is shown penetrating the skin; this would be detected by pain receptors in the skin.

② Sensory Neuron

—conducts impulses from the receptor to the spinal cord (or brain)

Processing

③ Integrating Center

—regions of gray matter in the spinal cord or brain that act as the integrating center; simple reflexes pass only through the spinal cord

A polysynaptic reflex is shown on the facing page; it always includes at least one short **interneuron** in the gray matter of the spinal cord that transmits nerve impulses from the sensory neuron to the motor neuron

—monosynaptic reflexes lack the interneuron and simply directly connect the sensory neuron to the motor neuron

Output

④ Motor Neuron

—conducts impulses from the integration center to the effector

⑤ Effector

—the last component in a somatic reflex arc; skeletal muscle

Analogy

① Input

② Processing

③ Output

This is like typing the letter k in a word processor on your laptop. Pressing on the keyboard is the input. This sends a message to the microchip—the brain of the computer or processing center. Then the microchip sends a signal for the final output—the appearance of the letter k on the screen.

SOMATIC REFLEX ARC

S OUCH!

1 Receptor

2 Sensory neuron

3 Integrating center

Interneuron

Gray matter

Skin

4 Motor neuron

Spinal cord

5 Effector

R

Skeletal muscle contracts

Color the sensory neuron, the interneuron, and the motor neuron different colors.

KEY

S Stimulus

R Response

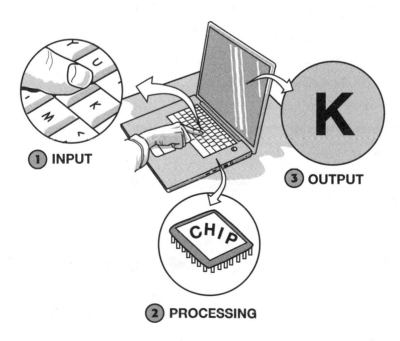

1 INPUT

2 PROCESSING

3 OUTPUT

K

CHIP

Description   The **meninges** (plural; *meninx* = singular) are the three protective membranes that surround both the brain and the spinal cord. The **cranial meninges** cover the brain (shown in the illustration on the facing page) while the **spinal meninges** (see p. 262) cover the spinal cord. They are continuous, similar in structure, and maintain the same layering pattern from outermost to innermost.

    The table below describes the three different meninges as listed from outermost to innermost.

Meninges	Key Points
Dura mater	• Outermost meninx or membrane • Term means "tough mother" (dura = tough, mater = mother) because it is thick, fibrous, and strong • Composed of dense irregular connective tissue • Vascular—contains blood vessels • Two layers of dura mater for the cranial meninges: (1) an outer periosteal layer, and (2) an inner meningeal layer; this permits the formation of either sinuses or partitions; when separated, the layers can form a space like the superior sagittal sinus shown on the facing page where venous blood is collected from the brain; when they stick together they form a partition like the falx cerebri that runs through the longitudinal fissure, between the cerebral hemispheres • One layer of dura mater for the spinal meninges
Arachnoid mater	• Middle meninx or membrane • Term means "spider-like mother" (arachnoid = spider-like, mater = mother) due to the spider-like arrangement of collagen fibers that connects the thin arachnoid layer to the underlying pia mater • Composed of simple squamous epithelia • Avascular—does not contains blood vessels • Above it is a very narrow space called the subdural space that may contain some interstitial fluid • Below it is a larger space called the subarachnoid space that contains cerebrospinal fluid (CSF) • Arachnoid granulations (arachnoid villi) are pouch-like extensions of the arachnoid mater that contain CSF and allow it to be reabsorbed into the venous blood in the superior sagittal sinus
Pia mater	• Innermost meninx or membrane • Term means "delicate mother" (pia = delicate, mater = mother) because it is thin and transparent • Composed of collagen fibers and elastic fibers • Vascular—contains blood vessels • Tightly adheres to the outer surface of every contour of both the brain and the spinal cord

Study Tips   To recall the layering of the meninges from outermost to innermost, use the mnemonic—*Don't Anger Pablo!*

Key to Illustration		
1. Skin of scalp	**4a.** Dura mater - periosteal layer	**6.** Pia mater
2. Periosteum	**4b.** Dura mater - meningeal layer	**7.** Falx cerebri
3. Bone of skull	**5.** Arachnoid mater	

Cranial Meninges

Meninges

Color the meninges (#4 - #6) different colors.

Cerebral cortex

Right cerebral hemisphere

Left cerebral hemisphere

Blood vessel

Arachnoid granulation

Superior sagittal sinus (filled with blood)

Subarachnoid space (filled with CSF)

Subdural space (filled with interstital fluid)

Description

Let's consider the process of creating, circulating, and reabsorbing **cerebrospinal fluid** (CSF). First, what is CSF and why do we need it? CSF is a solution filtered from the blood plasma that contains mostly water and dissolved substances such as sodium ions (Na^+), chloride ions (Cl^-), oxygen, glucose, and other nutrients. It is produced by a structure in the brain called a **choroid plexus** that contains permeable blood capillaries. The total amount of CSF in the central nervous system at any given moment is roughly 150 mL. On a daily basis, the average adult produces about 500 mL of CSF. The rate of CSF production is roughly the same as the rate of reabsorption. If this were not the case, the pressure around the brain would increase and possibly damage the delicate neurons in the brain. After being produced, the CSF circulates around the brain and spinal cord before being returned back into the blood. It has three important functions:

- **Protective cushion**—it gives the brain buoyancy and provides a fluid cushion that protects it against injury.

- **Nutrient delivery system**—it delivers oxygen, glucose, and other nutrients to the billions of neurons in the brain.

- **Waste removal system**—it removes waste products such as carbon dioxide (CO_2) by transporting them into the blood.

To understand the process of creating, circulating and reabsorbing CSF, refer to the numbers on the illustration on the facing page and follow the process step by step:

1. **Creation:** CSF is produced by the two **choroid plexuses**—one in each lateral ventricle of each cerebral hemisphere. CSF then fills the lateral ventricles, third ventricle, and associated spaces.

2. **Creation:** CSF is also produced by the **choroid plexus** in the fourth ventricle and it fills that space. A small amount of it moves down the narrow central canal of the spinal cord while most of it flows through the subarachnoid space.

3. **Circulation:** CSF flows through the subarachnoid space and circulates around the brain and spinal cord like radiator coolant around a car engine. Note that the arrows in the illustration show the direction that CSF moves.

4. **Reabsorption:** The CSF flows into and through the **arachnoid granulations** *(arachnoid villi)* from the subarachnoid space. It is reabsorbed back into the venous blood in the superior sagittal sinus.

Analogy

CSF circulates around the brain and spinal cord like radiator coolant circulates around a car engine.

Study tips

Focus your attention on two spaces in the illustration on the facing page: (1) **spaces indicated by the letter 'c'** and the (2) **superior sagittal sinus indicated by letter 'b.'** The spaces with the letter 'c' include the subarachnoid space, the ventricles, and the central canal of the spinal cord. These spaces are where the CSF is circulating and you are instructed to color the CSF yellow because it is filtered blood plasma and plasma is typically pale yellow in color. Although CSF is normally colorless, this helps you make a connection to its origin. You are also instructed to color the venous blood in the superior sagittal sinus blue because it is deoxygenated blood. In the ongoing cycle of creation, circulation, and reabsorption of CSF, the important concept to understand is that this is a *cycle* where the CSF returns to where it came from. It was created by filtering it *from* the blood and then it is reabsorbed by returning it back *to* the blood.

Creation, Circulation, and Reabsorption of Cerebrospinal Fluid (CSF)

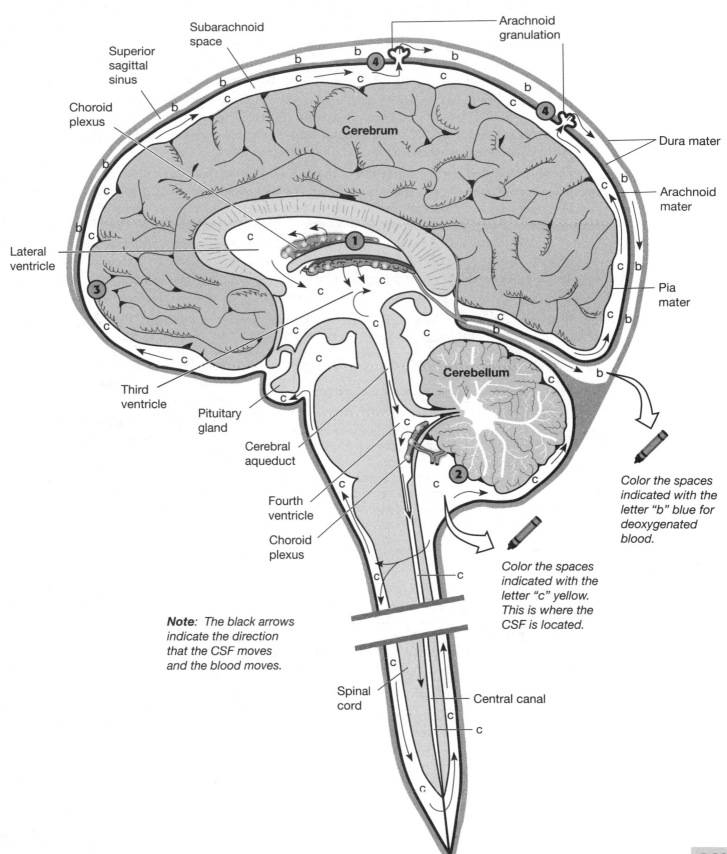

Arachnoid granulation

Subarachnoid space

Superior sagittal sinus

Choroid plexus

Cerebrum

Dura mater

Arachnoid mater

Lateral ventricle

Pia mater

Third ventricle

Pituitary gland

Cerebellum

Cerebral aqueduct

Fourth ventricle

Choroid plexus

Color the spaces indicated with the letter "b" blue for deoxygenated blood.

Color the spaces indicated with the letter "c" yellow. This is where the CSF is located.

Note: The black arrows indicate the direction that the CSF moves and the blood moves.

Spinal cord

Central canal

269

Description

The three *largest regions* of the brain are the **brainstem, cerebellum,** and **cerebrum.**

1. The **brainstem** is located at the base of the brain and contains regulatory centers to control things we take for granted, such as respiration, cardiovascular activities, and digestion.

2. The **cerebellum** is located posterior to the brainstem and inferior to the cerebrum. It is divided into two left or right halves, or **hemispheres,** and is extensively folded to increase surface area. Its general function is to work with the cerebrum to coordinate skeletal muscle movements, and it also allows the body to maintain proper balance and posture.

3. The **cerebrum** is the largest part of the brain and contains billions of neurons. Like the cerebellum, it is divided into two **hemispheres.** The deep division between these two hemispheres is called the **longitudinal fissure.** The term **fissure** indicates a deep groove or depression that separates major sections of the brain.

The surface of the cerebrum is not smooth but is folded into many little hills and gulleys. Each hill is called a **convolution** (or *gyrus*), and each valley is a shallow groove called a **sulcus.**

The cerebrum is the part of the brain associated with higher brain functions including planning, reasoning, analyzing, and storing/accessing memories. Ironically, without it, you would not be able to read and learn about the brain as you are doing now. It also perceives and interprets sensory information and coordinates various motor functions such as those involved in speech. The cerebrum is divided into four major lobes named after the bones that cover them: *frontal, parietal, temporal,* and *occipital.*

Brainstem

The **brainstem** consists of three parts: *medulla oblongata, pons,* and *midbrain.* The table below gives a description and general function of each part.

Brainstem Region	Description	General Functions
Medulla oblongata	Between spinal cord and pons	Respiratory control center; cardiovascular control center
Pons	Between medulla and midbrain; bulges out as widest region in brainstem	Controls respiration along with medulla; relays information from cerebrum to cerebellum
Midbrain	Between diencephalon and pons; includes corpora quadrigemina (*sensory relay station*) and cerebral aqueduct (*connects third and fourth ventricles; contains cerebrospinal fluid*)	Visual and auditory reflex centers; provides pathway between brainstem and cerebrum

Diencephalon

The **diencephalon** is located above the brainstem and contains three parts: *epithalamus, thalamus,* and *hypothalamus.* The table below gives a description and general function of each part.

Brainstem Region	Description	General Functions
Epithalamus	Roof of third ventricle; includes pineal gland; choroid plexus (forms cerebro-spinal fluid)	Pineal gland makes hormone melatonin, which regulates day-night cycles.
Thalamus	Two egg-shaped bodies that surround the third ventricle	Relays sensory information to cerebral cortex; relays information for motor activities; information filter
Hypothalamus	Forms floor of third ventricle; between thalamus and chiasm	Controls autonomic centers for heart rate, blood pressure, respiration, digestion, hunger center, thirst center, regulation of body temperature, production of emotions

Brain: Largest Regions, Brainstem, and Diencephalon

Convolution (*gyrus*)

Sulcus

Surface features of the cerebrum

BRAIN (superior view)

Frontal lobe

(*Anterior*)

Frontal lobe

LEFT HEMISPHERE

RIGHT HEMISPHERE

Longitudinal fissure

Central sulcus

Parietal lobe

Occipital lobe

(*Posterior*)

BRAIN (lateral view)

Lateral sulcus   Cerebrum   Central sulcus

Frontal lobe

Parietal lobe

Occipital lobe

Temporal lobe

Brainstem

Cerebellum

Color the different lobes of the cerebrum different colors.

BRAIN (midsagittal view)

Corpus callosum

Third ventricle

Cerebrum

Pineal gland (*body*)

Corpora quadrigemina

Cerebral aqueduct

Arbor vitae

Cerebellum

Fourth ventricle

Central canal

Spinal cord

DIENCEPHALON

T

H

E

KEY

E = epithalamus
T = thalamus
H = hypothalamus

STUDY TIP

To recall the parts of the diencephalon, use the mnemonic:

"**E**xpect **T**otal **H**armony!" ➡ **E**pithalamus **T**halamus **H**ypothalamus

BRAINSTEM

M

P

M

Spinal cord

STUDY TIP

To recall the parts of the brainstem, use the mnemonic:

"**M**ake **P**eace **M**onday!" ➡ **M**idbrain **P**ons **M**edulla oblongata

271

Description

Though small in size, the **hypothalamus** in the **brain** controls many vital body functions. In general, it controls autonomic nervous system (**ANS**) centers for heart rate, blood pressure, respiration, and digestion. It also serves as our hunger center and thirst center, regulator of body temperature, and producer of emotions. It even plays a role in memory. As part of the diencephalon, it is located between the **thalamus** and the **optic chiasm**. The **anterior commissure** and **mamillary body** form the borders on either side. The **infundibulum** is a stalk-like structure that connects the hypothalamus to the **pituitary gland**. Neurons in the hypothalamus are wired directly into the **posterior pituitary** gland to create a vital link between the nervous system and the endocrine system.

The hypothalamus consists of many different nuclei—clusters of cell bodies. These nuclei correlate to the specific functions listed below.

Functions

The hypothalamus has five major functions:

1. Controls the autonomic nervous system (**ANS**) and the **endocrine system**.

   For the **ANS**, which is divided into two divisions—sympathetic and parasympathetic:
   - The **anterior nucleus** controls the parasympathetic division.
   - The **dorsomedial nucleus** controls the sympathetic division.

   For the **endocrine system**:
   - Notice the colored lines in the illustration linking the paraventricular nucleus and the supraoptic nucleus to the posterior pituitary gland.
   - The **paraventricular nucleus** releases **OT** (*oxytocin*), the hormone that stimulates uterine contractions during labor and delivery.
   - The **supraoptic nucleus** releases **ADH** (*antidiuretic hormone*), important for maintaining water balance.

2. Regulates emotional behavior.
   - The **mammillary body** contains relay stations for the sense of smell. In addition, it links the hypothalamus to the limbic system—our "emotional brain" (see p. 282). The limbic system helps us regulate basic emotions such as rage, fear, happiness, and sadness.

3. Regulates body temperature.
   - The **preoptic area** works like a thermostat to maintain our normal body temperature of 98.6°F by regulating shivering and sweating mechanisms.

4. Regulates food and water intake.
   - The **anterior nucleus** is the thirst center, checking dissolved solutes in the blood. As solute levels increase, fluid intake is stimulated.
   - The **ventromedial nucleus** is the hunger center, checking levels of glucose and other nutrients in the blood. As these nutrient levels decrease, food intake is stimulated.

5. Regulates sleep/wake cycles.
   - These cycles are called *circadian rhythms*.
   - The **suprachiasmatic nucleus** is the control center.

HYPOTHALAMUS FUNCTIONS

1. Regulates the ANS and the endocrine system

ANS
- Anterior nucleus
- Dorsomedial nucleus

Endocrine system
- Paraventricular nucleus (OT)
- Supraoptic nucleus (ADH)

2. Regulates emotional behavior
- Mammillary body

98.6°F

3. Regulates body temperature
- Preoptic area

4. Regulates water and food intake
- Anterior nucleus—thirst center
- Ventromedial nucleus—satiety center

5. Regulates sleep/wake cycles
- Suprachiasmatic nucleus

Color each of the different nuclei a different color.

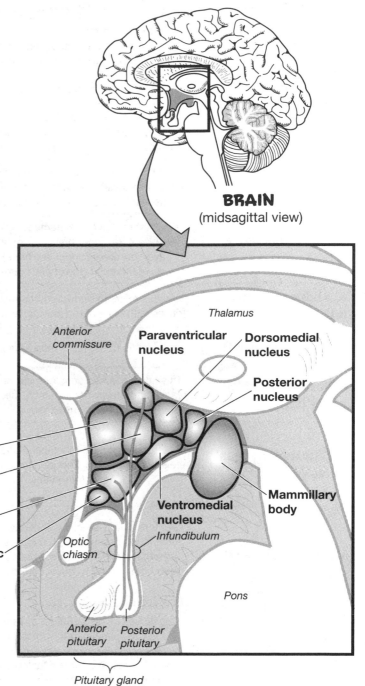

BRAIN
(midsagittal view)

Thalamus

Anterior commissure

Paraventricular nucleus

Dorsomedial nucleus

Posterior nucleus

Preoptic area

Anterior nucleus

Supraoptic nucleus

Suprachiasmatic nucleus

Optic chiasm

Infundibulum

Ventromedial nucleus

Mammillary body

Pons

Anterior pituitary

Posterior pituitary

Pituitary gland

HYPOTHALAMUS
(midsagittal view)

Description   The table below gives an overview of the twelve pairs of cranial nerves. Please note that many of the cranial nerves have other functions not listed in this table, so it is not intended to be a comprehensive list.

Cranial Nerve Name (Roman numeral)	Comments	Primary Function	Primary Function Specifics
1. Olfactory nerve (I)	*Olfactory* refers to sense of smell. Two key structures are associated with this nerve: the olfactory tract and the olfactory bulb.	Sensory	Smell
2. Optic nerve (II)	*Optic* refers to vision.	Sensory	Vision
3. Oculomotor nerve (III)	*Oculomotor* refers to muscles that move the eyeballs. It connects to four different eyeball muscles and one eyelid muscle.	Motor	Movement of eyeballs and upper eyelids
4. Trochlear nerve (IV)	*Trochlea* means "pulley-like" and refers to a structure associated with the superior oblique muscles that the trochlear nerve controls.	Motor	Movement of eyeballs
5. Trigeminal nerve (V)	Largest cranial nerve, named *trige*minal because it has *three* branches: ophthalmic (eye), maxillary (upper jaw bone), and mandibular (lower jaw bone).	Mixed	Sensory: Transmits impulses for touch, temperature, and pain Motor: Chewing
6. Abducens nerve (VI)	*Abducens* refers to *abduction*, the movement it induces in the lateral rectus muscles it controls.	Motor	Movement of eyeballs
7. Facial nerve (VII)	*Facial* refers to the facial muscles it controls.	Mixed	Sensory: Proprioception and taste Motor: Facial expression
8. Vestibulocochlear nerve (VIII)	This nerve has two branches, the *vestibular* branch and the *cochlear* branch, that connect to the inner ear.	Sensory	Transmits impulses for balance and equilibrium (vestibular branch) and for hearing (cochlear branch)
9. Glossopharyngeal nerve (IX)	*Glosso-* refers to the tongue and *pharyngeal* refers to the pharynx. This identifies where the nerve terminates.	Mixed	Sensory: Taste and general sensory information from posterior tongue; monitoring of both blood pressure and dissolved blood gases (O_2 and CO_2) Motor: Elevates pharynx during swallowing
10. Vagus nerve (X)	The term vagus means wandering and it is most fitting because it is the longest cranial nerve that wanders from the brain stem and then branches off to many different organs in the body.	Mixed	Sensory: General sensory information from pharynx; monitoring of both blood pressure and dissolved blood gases (O_2 and CO_2) Motor: Smooth muscle contraction or relaxation in organs of the digestive tract; decreasing heart rate; stimulation of digestive secretions
11. Accessory nerve (XI)	This nerve can be identified by its position; it runs parallel to the spinal cord.	Motor	Swallowing movements; head and shoulder movements
12. Hypoglossal nerve (XII)	The term hypoglossal means "below the tongue" and refers to where the nerve terminates.	Motor	Tongue movements during speech and swallowing

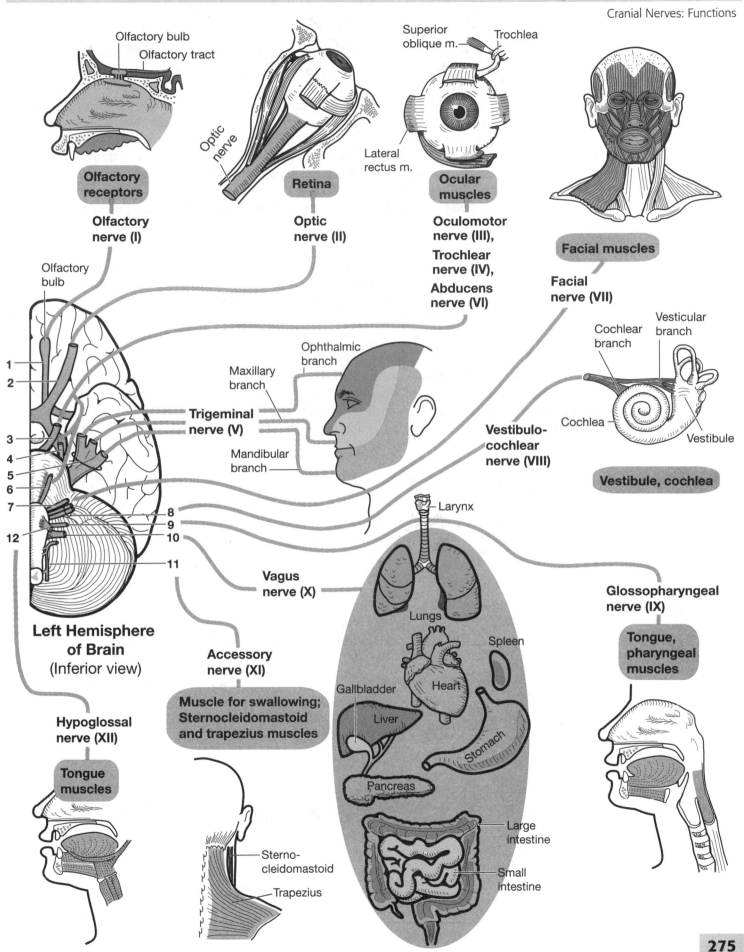

Olfactory bulb
Olfactory tract

Optic nerve

Superior oblique m.
Trochlea

Lateral rectus m.

Olfactory receptors

Olfactory nerve (I)

Retina

Optic nerve (II)

Ocular muscles

Oculomotor nerve (III), Trochlear nerve (IV), Abducens nerve (VI)

Facial muscles

Facial nerve (VII)

Cochlear branch
Vesticular branch

Cochlea
Vestibule

Olfactory bulb

Ophthalmic branch
Maxillary branch

Trigeminal nerve (V)

Mandibular branch

Vestibulo-cochlear nerve (VIII)

Vestibule, cochlea

1
2

3
4
5
6
7
8
9
10
11
12

Larynx

Left Hemisphere of Brain (Inferior view)

Accessory nerve (XI)

Vagus nerve (X)

Lungs

Spleen

Gallbladder
Heart
Liver
Stomach
Pancreas

Glossopharyngeal nerve (IX)

Tongue, pharyngeal muscles

Hypoglossal nerve (XII)

Muscle for swallowing; Sternocleidomastoid and trapezius muscles

Tongue muscles

Sterno-cleidomastoid
Trapezius

Large intestine
Small intestine

Description

The heart contains ventricles that fill with blood, whereas the brain contains **ventricles** that are constantly filled with cerebrospinal fluid. In total, the brain has four ventricles inside it: *lateral ventricle* (of left hemisphere), *lateral ventricle* (of right hemisphere), *third ventricle*, and *fourth ventricle*. This entire network is referred to as the **ventricular system** in the brain. The lateral ventricles are the largest of the four and do not directly connect to each other because they are separated by a thin partition called the **septum pellucidum**. Both do connect to the third ventricle in the region of the diencephalon via small passageways called **interventricular foramina**. The third ventricle is connected to the fourth ventricle via a passageway called the **cerebral aqueduct** *(aqueduct of Sylvius)*. The fourth ventricle is located in the pons (of the brainstem) and the cerebellum. It communicates via a very narrow passageway called the **central canal** that runs through the middle of the spinal cord.

Analogy

To visualize the relative positions of the ventricles, compare the whole **ventricular system** with the **hollow head of a ram**. The **fourth ventricle** is like the **neck of the ram**, the **third ventricle** is like the head, and the **lateral ventricles** are like the **two horns**. The ram's horns also follow the same general shape of the paired lateral ventricles.

Study Tip

The first and second ventricles are not numbered because they are the lateral ventricles. If you think of the two lateral ventricles as *first ventricle* and *second ventricle,* the numbering makes sense in relation to the **third ventricle** and **fourth ventricle**. Ah, the goofy things that anatomists do! As the saying goes, "You are not a good anatomist unless you know 87 different names for the same structure."

Key to Illustration	
1. Lateral ventricles	2. Third ventricle
1a. Anterior horns of lateral ventricles	3. Cerebral aqueduct (*aqueduct of Sylvius*)
1b. Posterior horns of lateral ventricles	4. Fourth ventricle
1c. Inferior horns of lateral ventricles	5. Central canal

Anterior view

The ventricular system in the brain is like the neck, head, and horns of a ram if they were hollow.

Lateral view

Color the ventricles and write labels on the illustrations.

Anterior view

Lateral view

This module will describe some of the selected functional areas of the cerebral cortex. These areas have been divided into three general groups: **sensory areas**, **motor areas**, and **association areas**. Note that the words *cortex* and *area* are often used interchangeably.

SENSORY AREAS   Control regions where sensations are perceived

1. **Primary somatic sensory cortex**	This important region is shown in dark gray behind the central sulcus. General sensory input (e.g., touch, temperature, pressure, and pain) from all parts of the body is perceived here.
2. **Gustatory cortex**	Located in the parietal lobe; taste sensations are perceived here, such as the flavors of the ice cream shown in the icon.
3. **Auditory cortex**	Located in the temporal lobe; auditory stimuli are processed by the brain here.
4. **Visual cortex**	Located in the occipital lobe; visual images are perceived here (like the star shown in the icon).

MOTOR AREAS   Control centers for conscious muscle movements

1. **Primary motor cortex**	This important area is shown in color in front of the central sulcus. It controls voluntary muscle movements throughout the body, including those of the hands and feet, arms and legs, face and tongue.
2. **Premotor cortex**	This area serves as the "choreographer" for the primary motor cortex. It decides which muscle groups will be used and how they will be used prior to stimulating the primary motor cortex.
3. **Motor speech area** (*Broca's area*)	This area controls and coordinates the muscles involved in normal, fluent speech. Damage to this area can result in strained speech with disconnected words. Located in only one hemisphere, usually left.
4. **Frontal eye field**	This area controls muscle movements of the eye, such as those needed to read this page.

ASSOCIATION AREAS   Control regions—near sensory areas—involved in recognizing and analyzing incoming information

1. **Prefrontal area**	This area is most highly developed in humans and other primates. It regulates emotional behavior and mood and also is involved in planning, learning, reasoning, motivation, personality, and intellect.
2. **Somatic sensory association area**	This area allows you to *predict* that sandpaper is rough, for example, even without looking at it. It also stores memories about previous sensory experiences, so you can determine when blindfolded, for example, that the object placed in your hand was a pair of scissors.
3. **Sensory speech area** (*Wernicke's area*)	This area seems to be an important part of language development—processing words we hear being spoken. It also appears important for children when they are sounding out new words. Damage to this area may result in deficiencies in recognizing written and spoken words.
4. **Auditory association area**	This area allows you to comprehend, interpret, analyze, and question what you are hearing. For example, it enables you to recognize a familiar song or disregard noise.
5. **Visual association area**	This allows you to associate the perceived image of the star with the letters "S-T-A-R". You connect the word "star" with the image of a star.

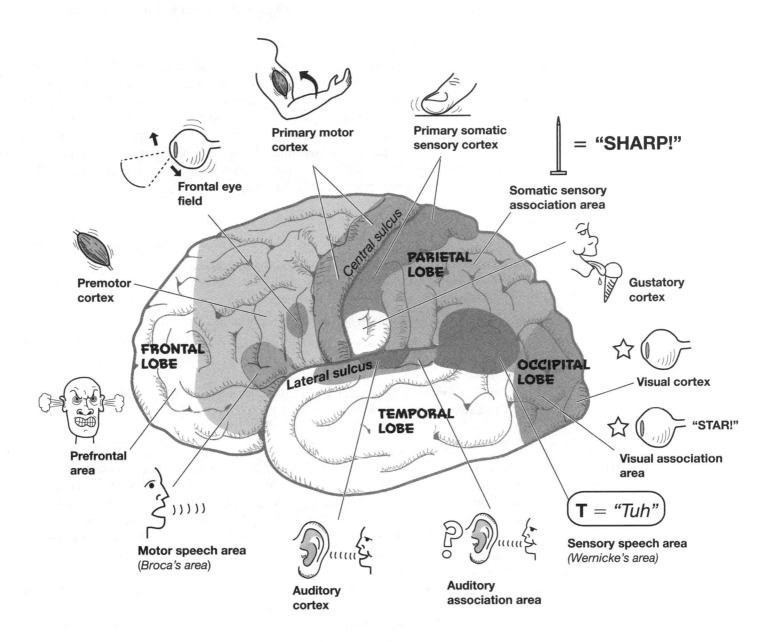

Primary motor cortex

Primary somatic sensory cortex

= "SHARP!"

Frontal eye field

Somatic sensory association area

Premotor cortex

Central sulcus

PARIETAL LOBE

Gustatory cortex

FRONTAL LOBE

Lateral sulcus

OCCIPITAL LOBE

Visual cortex

TEMPORAL LOBE

☆ "STAR!"

Visual association area

Prefrontal area

T = "Tuh"

Motor speech area (*Broca's area*)

Sensory speech area (*Wernicke's area*)

Auditory cortex

Auditory association area

Left Cerebral Hemisphere
(lateral view)

Description

This module explains electroencephalogram (EEG), sleep/wakefulness, and reticular formation.

EEG

An **electroencephalogram** (**EEG**) is to the nervous system what an **ECG** (or **EKG**) is to the cardiovascular system (see p. 348). By detecting the electrical activity of the billions of neurons in the brain, an EEG measures normal brain function. **Electrodes** are placed around the scalp to detect the small voltage generated from these nerve impulses. Wires connect the electrodes to an EEG machine, where the voltage is amplified so it can be seen as characteristic waves. Each person has her own unique brain waves, displayed as a graph of **voltage** (millivolts or mV) over **time**. The height of the wave is the **amplitude** (in mV), and the frequency is measured in **hertz** (Hz), or cycles per second. The following is a summary of the four common types of brain waves:

Sleep/Wakefulness

- **Alpha** waves (8–13 Hz) occur when a person is calm and awake with eyes closed. This is like the restful state of the brain.

- **Beta** waves (14–30 Hz) occur when we are mentally alert, such as balancing your checkbook or studying a map to find your way to your destination.

- **Theta** waves (4–7 Hz) are common in children but abnormal in wakeful adults.

- **Delta** waves (< 4 Hz) occur normally in wakeful infants and in adults during deep sleep or while under anesthesia. In wakeful adults, they indicate brain damage.

EEGs also can be used to detect abnormalities in the brain, such as epilepsy, brain tumors, and traumatic injuries. The absence of any brain waves is the legal definition of what is called *brain dead*.

Reticular Formation

The brain never really "turns off." Neurons are sending impulses to each other constantly whether a person is awake or asleep, but the brain needs to have methods to control these states of consciousness. Sleep states are regulated through the hypothalamus. Wakeful states are controlled by the **reticular formation**—a cluster of neurons centrally located and extending like columns through the brainstem—*midbrain*, **pons**, and **medulla oblongata**. One subset of these neurons—the **reticular activating system** (**RAS**) acts like an alarm system that keeps the brain alert. The RAS sends stimulatory impulses through nerve pathways to other key parts of the brain, such as the cerebral cortex, thalamus, hypothalamus, cerebellum, and spinal cord. Sensory input is delivered to the RAS so it can filter out everything that is not important. Otherwise, we would go insane from sensory overload. Only the most important sensory input is routed to the cerebral cortex so we can be aware of it. For example, you probably are unaware of the sensation of your silky, long-sleeved shirt against your arm, but you would notice if someone suddenly were to pull on it.

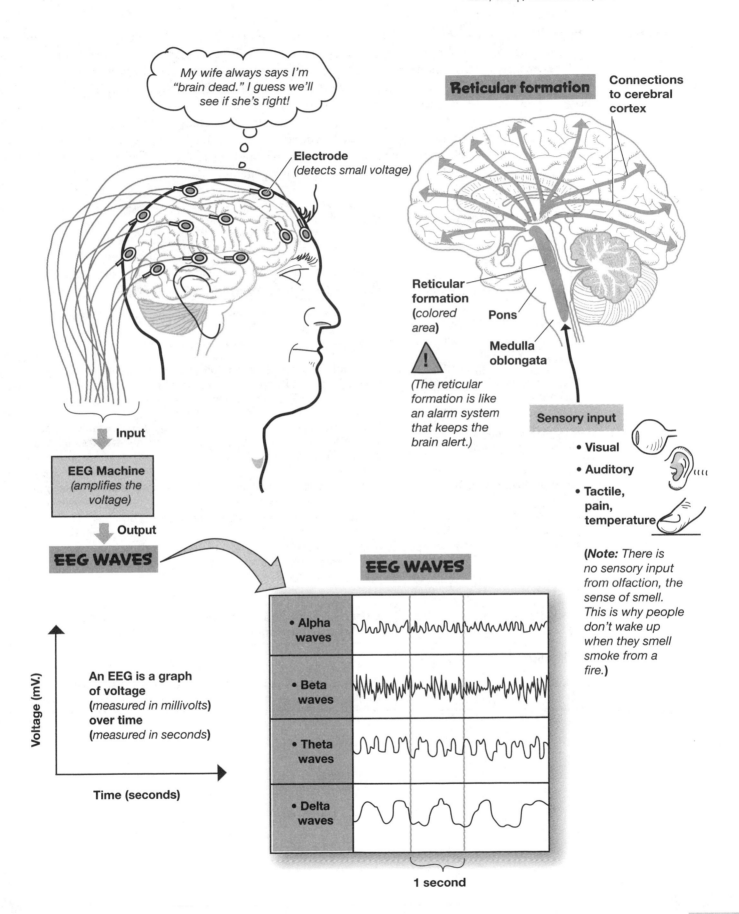

Description

The limbic (*limbus* = border) **system** is called our "emotional brain" because it processes our basic emotions such as fear, happiness, and sadness. To achieve this, it must integrate different brain centers, such as sensory systems and conscious control centers. Physically, it is a collection of structures found mostly in the cerebrum and diencephalon (thalamus, hypothalamus, epithalamus). It is so named because these structures form a border around the diencephalon.

Scientists still do not universally agree on the structures that compose the limbic system. The table below lists the most commonly recognized structures, with similar structures grouped together. The term *nuclei* refers to clusters of cell bodies in the CNS, and *tracts* are bundles of axons in the CNS. The structures in the table are illustrated on the facing page.

Limbic System Structure	Description
Cerebral Cortex Structures	
Cingulate (*cingulum* = girdle, to surround) **gyrus**	A mass of cerebral cortex located above the corpus collosum; its name describes the fact that it surrounds the diencephalon
Parahippocampal gyrus	A mass of cerebral cortex in the temporal lobe; works with the hippocampus to store memories
Nuclei **Hippocampus** (*hippocampos* = seahorse)	Anatomists thought this nucleus looked like a seahorse; located above the parahippocampal gyrus; connects to the diencephalon through the fornix; important in learning and storing long-term memories
Amygdaloid (*amygda* = almond shaped) **body**	Nucleus located in front of and connected to the hippocampus; directly involved in emotions such as fear; can store memories and correlate them with specific emotions
Anterior thalamic nucleus	Located in the thalamus; relays visceral sensations to the cingulate gyrus
Septal nucleus	Nucleus located near the anterior commissure
Mammillary (*mammilla* = nipple) **body**	Nipple-like structures in the floor of the hypothalamus that serve as relay stations for reflexes related to sense of smell
Tracts **Fornix** (arch)	A tract of white matter that connects the hippocampus to the hypothalamus; a portion of it is shaped like an arch
Other Structures **Olfactory bulbs**	All serve to interpret various odors in the brain; odors are linked to specific memories, so certain odors can trigger powerful emotions

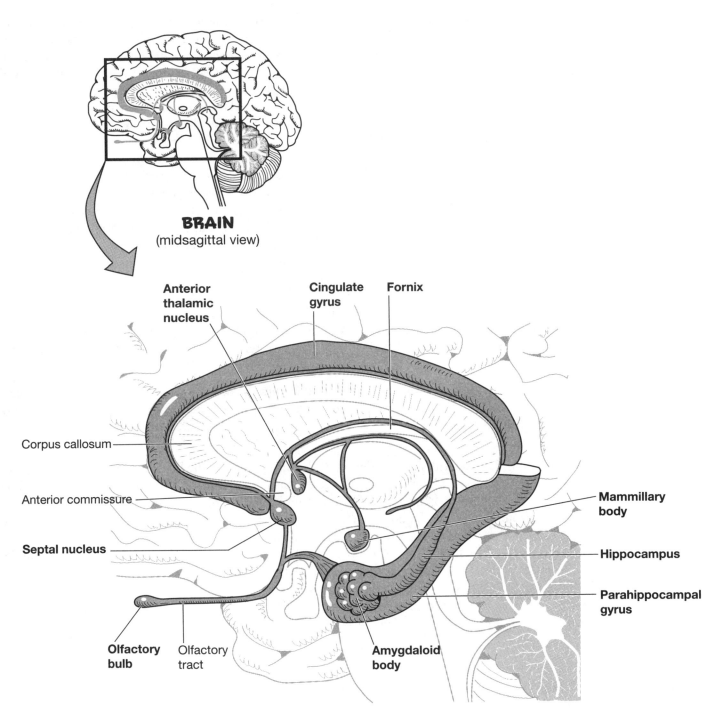

BRAIN
(midsagittal view)

Anterior
thalamic
nucleus

Cingulate
gyrus

Fornix

Corpus callosum

Anterior commissure

Septal nucleus

Mammillary
body

Hippocampus

Parahippocampal
gyrus

Olfactory
bulb

Olfactory
tract

Amygdaloid
body

Midsagittal View

Note: *Major structures in the limbic system are shown in color.*

Description

This module explains the concept of lateralization between the left and right cerebral hemispheres. Looking at the illustration of the **brain**, the left and right cerebral hemispheres appear very similar. In fact, they are anatomically similar, and the two hemispheres work together for many functions. This is evidenced by the **corpus callosum**—a thick band of axons connecting the left and right hemispheres. Each hemisphere also has functional specialization. There is lateralization in which certain functions are found in only one of the hemispheres. For example, in most people, the Broca's area for speech production is found only in the left hemisphere. We can make generalizations about the functional differences between the two hemispheres that apply to most people. Consider your left hemisphere your "analytical" hemisphere and your right your "creative" hemisphere.

The illustration lists the functional differences. Your left brain excels at language and logic. It deals with information in an organized, logical way as a scientist would. It helps you work with mathematical equations, write, and follow directions step by step. In contrast, your right hemisphere excels at musical and artistic abilities. It helps you understand shape and pattern relationships useful for facial recognition and drawing. It also is the seat of insight and inspiration.

But these generalizations are not set in stone. Here are some variables that are exceptions to the rule:

- **Individual differences.** Some individuals have one or more control centers in the hemisphere opposite from the one where it normally is found.

- **Gender differences**. Lateralization is greater in males than females. In typical females, a portion of the corpus callosum is thicker, indicating greater hemispheric integration. This means that both hemispheres work together more frequently.

- **Age differences.** Children can "rewire" their brains more easily than adults. For example, if part of the brain is damaged or surgically removed in a child, the opposite hemisphere can take over and compensate.

One hemisphere doesn't actually dominate the other. Even so, the hemisphere that controls spoken and written language is designated as the *categorical* (or "dominant") hemisphere. As mentioned previously, this is the left hemisphere for most people and correlates to handedness. Because nerves cross over from one side of the body to the opposite side in the brain, motor activity on the right side of the body is controlled by the left hemisphere, and vice versa.

The same usually holds true for handedness. About 91 percent of the population is right-handed, and in most of these people the left hemisphere is the categorical one. Interestingly, the situation is a bit different for "lefties." In the majority of them, the left hemisphere is still their categorical one. In only about 15 percent is the right hemisphere categorical. In summary, although the two hemispheres work together all the time, they also specialize in specific functions.

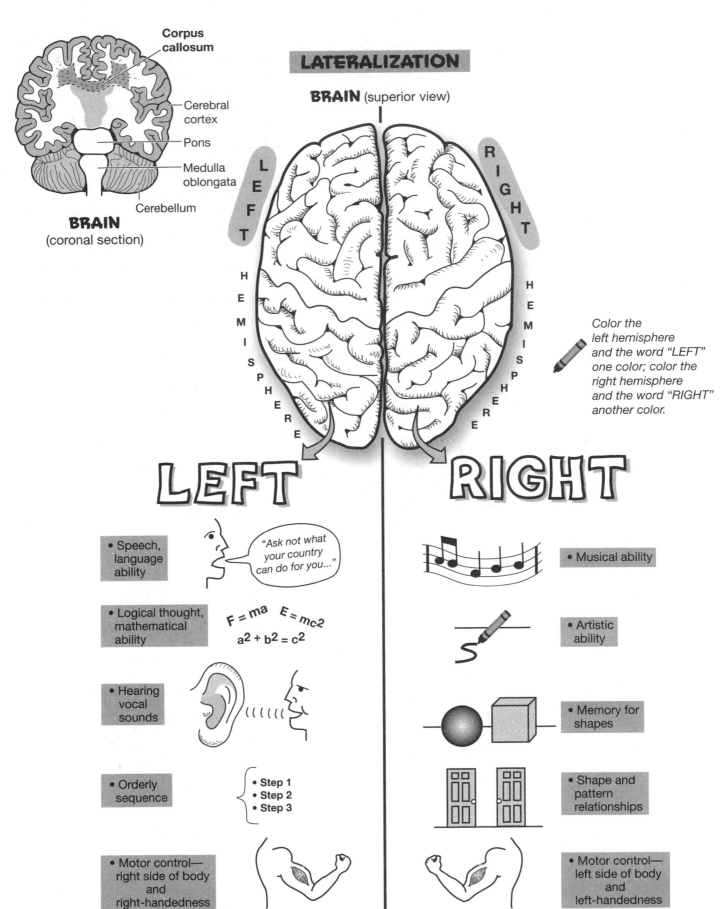

Corpus callosum

Cerebral cortex

Pons

Medulla oblongata

Cerebellum

BRAIN
(coronal section)

LATERALIZATION

BRAIN (superior view)

LEFT HEMISPHERE

RIGHT HEMISPHERE

Color the left hemisphere and the word "LEFT" one color; color the right hemisphere and the word "RIGHT" another color.

LEFT

- Speech, language ability

"Ask not what your country can do for you..."

- Logical thought, mathematical ability

$F = ma$ $E = mc^2$
$a^2 + b^2 = c^2$

- Hearing vocal sounds

- Orderly sequence

  - Step 1
  - Step 2
  - Step 3

- Motor control—right side of body and right-handedness

RIGHT

- Musical ability

- Artistic ability

- Memory for shapes

- Shape and pattern relationships

- Motor control—left side of body and left-handedness

Description

Though the **brain** makes up only 2 percent of body mass, it uses about 20 percent of the body's glucose and oxygen. The brain never really shuts off; it's even active during sleep. This is because the billions of **neurons** in the brain are constantly sending nerve impulses along their axons, which requires large amounts of **ATP**. This process occurs in mitochondria within the **cell body** of the neuron. These hungry cells require a constant supply of **glucose** and **oxygen** to produce this ATP through **aerobic respiration**. For each glucose molecule broken down, up to 36 ATP can be produced.

Unlike skeletal muscle cells, which contain a stored form of glucose called **glycogen**, the brain has no glycogen stores. Skeletal muscle also contains a protein called **myoglobin** that temporarily binds and stores large amounts of oxygen.

Unfortunately, the brain has no myoglobin deposits. To compensate, the brain has elaborate networks of blood vessels to directly deliver all the glucose and oxygen it needs through the blood. The **carotid arteries** transport oxygenated blood to the brain. But to protect vital neurons, a brain barrier system restricts what specific substances can be transported from the blood to the brain tissues. This prevents toxins that may enter the bloodstream from damaging neurons. Fortunately, this system is permeable to glucose, oxygen, carbon dioxide, sodium, potassium, and chloride. As a result, it doesn't present a problem for delivering nutrients to the neurons or eliminating wastes from them.

Cardiac Arrest

If a person has cardiac arrest, blood stops flowing to the brain. Of all the different cells in the body, neurons in the brain are most likely to die if blood flow is not restored. This is why **CPR** (cardiopulmonary resuscitation) is performed. Without it, neurons begin to die in about 3 to 4 minutes. In fact, this occurs any time blood flow to the brain is interrupted.

Brain Metabolism and Brain Function

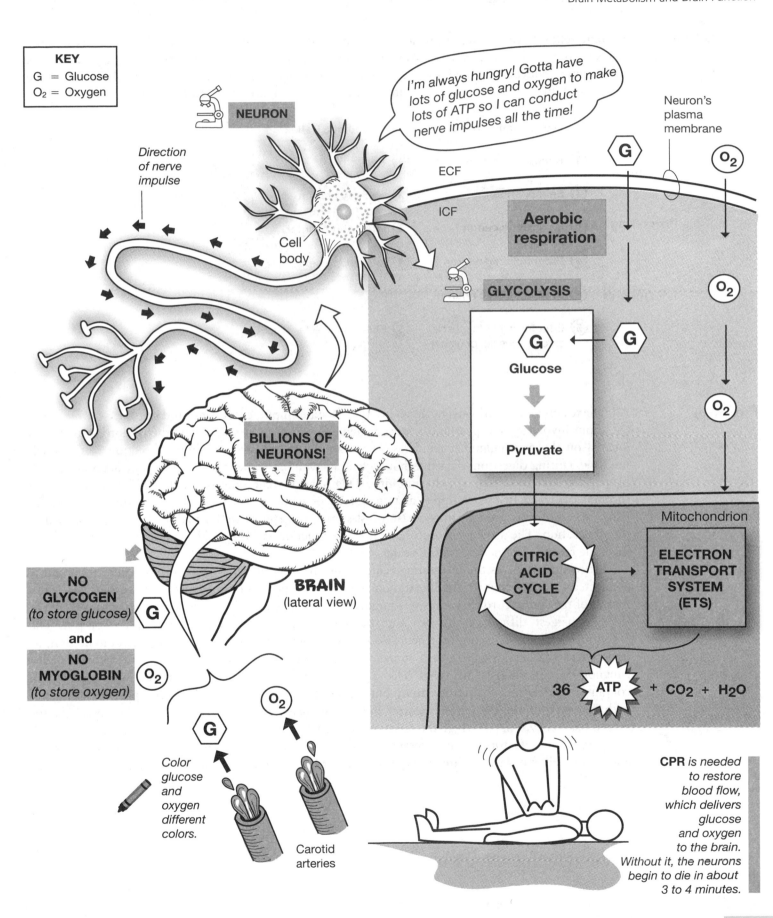

KEY
G = Glucose
O₂ = Oxygen

NEURON

I'm always hungry! Gotta have lots of glucose and oxygen to make lots of ATP so I can conduct nerve impulses all the time!

Neuron's plasma membrane

Direction of nerve impulse

Cell body

ECF

ICF

Aerobic respiration

GLYCOLYSIS

G

Glucose

⬇
⬇

Pyruvate

BILLIONS OF NEURONS!

Mitochondrion

CITRIC ACID CYCLE

ELECTRON TRANSPORT SYSTEM (ETS)

BRAIN
(lateral view)

NO GLYCOGEN *(to store glucose)*

G

and

NO MYOGLOBIN *(to store oxygen)*

O₂

36   ATP   + CO₂ + H₂O

G

O₂

Color glucose and oxygen different colors.

Carotid arteries

CPR *is needed to restore blood flow, which delivers glucose and oxygen to the brain. Without it, the neurons begin to die in about 3 to 4 minutes.*

Description

Somatic reflexes were covered previously (see p. 264). This module compares and contrasts somatic reflexes and autonomic reflexes. Somatic reflexes stimulate the contraction of skeletal muscle. **Autonomic reflexes** are used by the autonomic nervous system (ANS) to control many of the involuntary functions we take for granted, such as digestion, urination, and heart rate.

	Autonomic Reflex	Somatic Reflex
Input	① Receptors	① Receptors
	② Sensory neuron	② Sensory neuron
Processing	③ Integrating center	③ Integrating center
Output	④ Preganglionic neuron	④ Integrating center
	⑤ Postganglionic neuron	
	⑥ Effector (smooth muscle, cardiac muscle, or gland)	⑤ Effector (skeletal muscle)

Pathway

All reflexes involve a pathway called an arc that includes input, processing, and output.

Input involves a receptor and a **sensory neuron**. In an autonomic reflex, the receptor typically is located on the organ/gland it regulates. For example, the illustration shows a section of the small intestine. During digestion, the smooth muscle in its wall has to be stimulated to contract and thereby propel digested food through the intestine. This has to occur only when digested food is moving through the tract. In this example, a stretch receptor is located in the wall of the small intestine to detect when food is present. In contrast, most somatic receptors are located in the skin, as shown in the illustration. The sensory neuron has the same function in both reflexes: It conducts the impulse from the receptor to the integrating center.

Processing involves the **integrating center**—the spinal cord and/or the brain—where the input is connected to the output. In the illustration both reflex types contain a short **interneuron** in the gray matter of the spinal cord that transmits impulses from the sensory input to the motor output.

The biggest difference between somatic and autonomic reflexes is seen in the output. The autonomic reflex uses two motor neurons—a **preganglionic neuron** and a **postganglionic neuron**, as opposed to the single neuron in a somatic reflex. This introduces an additional structure called a **ganglion**—a collection of neuronal cell bodies outside the CNS. Notice that the autonomic reflex contains a ganglion in its motor output, but the somatic reflex does not. Now compare that ganglion with the one found in the sensory neuron. Both contain cell bodies, but only the ganglion in the motor output has a new synapse where the preganglionic neuron meets the postganglionic neuron.

Finally, the **effector** is also different. In somatic reflexes skeletal muscle is the only effector. An autonomic reflex has three different types of effectors: smooth muscle, cardiac muscle, or glands.

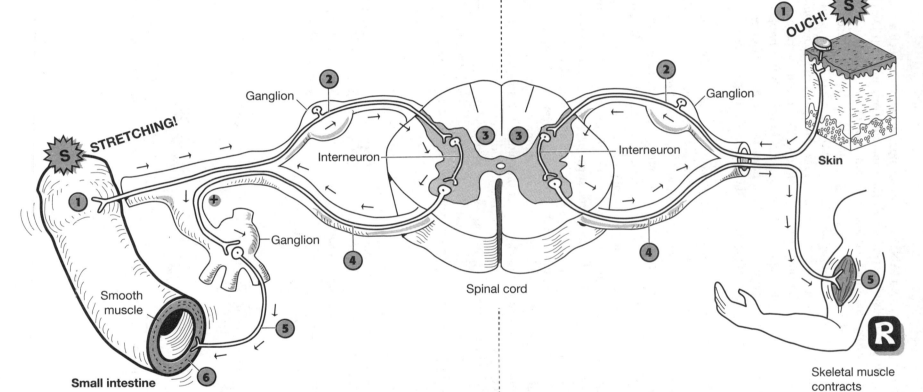

AUTONOMIC REFLEX

SOMATIC REFLEX

Ganglion

②

STRETCHING!

Ⓢ

Interneuron

①

④

+

Ganglion

Smooth muscle

⑤

Small intestine

⑥

③ ③

Spinal cord

②

Ganglion

Interneuron

④

OUCH!

①

Ⓢ

Skin

⑤

Ⓡ

Skeletal muscle contracts

Ⓡ Smooth muscle contracts

Color the same parts of the reflex arc the same colors for both types of reflexes.

KEY

 Stimulus

 Response

NERVOUS SYSTEM: Autonomic Nervous System (ANS)

Autonomic Reflexes Compared with Somatic Reflexes

Description

The **autonomic nervous system (ANS)** controls involuntary activities such as digestion and heart rate. It is subdivided into two divisions: **sympathetic (SD)** and **parasympathetic (PD)**. These two divisions work together to regulate normal functions and maintain **homeostasis**. Each division has its own groups of nerves that connect to the major visceral organs in the body. But there are some exceptions. For example, the adrenal gland is controlled only by the SD because the PD has no nerves connected to it.

This module gives an overview of the SD, also referred to as the *thoracolumbar division*, as motor output arises from these sections of the spinal cord. Because it prepares the body for emergency situations, it is nicknamed the "fight-or-flight" system. Consider what happens when you get nervous before a test you are not prepared to take. Your heart rate increases, your pupils dilate, your breathing rate increases, and your palms perspire. All these are a direct result of your sympathetic division in action.

The SD has a more complex anatomical structure than the PD. The illustration on the facing page shows only the motor output to the effector. The nerves involved are like the electrical wiring in a house. They consist of a **preganglionic neuron** linked to a **postganglionic neuron**. Preganglionic neurons arise from the T1–L2 segments of the spinal cord. Notice that the preganglionic axons are quite short, and the postganglionic axons are relatively long. Recall that ganglia (sing. *ganglion*) are clusters of neuron cell bodies. The **paravertebral ganglia**—two columns of ganglia like strings of beads that run along either side of the spinal cord—are unique to the SD. Included in this chain are the cervical ganglia, which consist of three parts: **superior**, **middle**, and **inferior**. Postganglionic neurons are wired from these hubs to target organs in the head and thoracic cavity.

Three other ganglia, all located near the abdominal aorta and classified as the **prevertebral ganglia**, are: (1) **celiac**, (2) **superior mesenteric**, and (3) **inferior mesenteric**. Postganglionic neurons arising from these hubs connect to all the target organs and glands below the diaphragm. The ANS has three possible effectors: smooth muscle, cardiac muscle, and glands.

To clarify the illustration, when you see the postganglionic neurons connecting to organs, they actually are connecting to muscle in the walls of these target organs. Notice that many of the postganglionic neurons pass through **plexuses** (sing. *plexus*), or neural networks, to link with their target organs.

Analogy

Regulating the visceral organs is like driving a car. Sometimes you have to step on the gas, and other times you have to step on the brake. Most of the time, the SD is like stepping on the gas. For example, sympathetic impulses sent along the accelerator nerve to the heart increase the heart rate. Unfortunately, it does not work this way with all organs/glands. There are exceptions to this rule.

Study Tips

Sympathetic = Stressful. This simple alliteration helps you distinguish the function of the SD from the PD.

Summary

ITEM	DESCRIPTION
Nickname	*Fight-or-flight* system
Location of preganglionic neuron cell bodies	T1–L2 regions of spinal cord (*thoracolumbar*)
Length of preganglionic axon	Short
Length of postganglionic axon	Long
Location of ganglia	Near the spinal cord
Neurotransmitters	Preganglionic neurons secrete acetylcholine (ACh). Postganglionic neurons secrete norepinephrine (NE).

Motor
Output
to
Effector

Brain

Sublingual
salivary gland

Submandibular
salivary gland

Parotid
salivary
gland

Heart

Plexus

KEY

Preganglionic neurons

Postganglionic neurons

Lungs

Adrenal
gland

Cervical
ganglia

Superior

Middle

Inferior

T1

T5

Liver

Splanchnic n.

Plexus

Spleen

Pancreas

Kidney

T8   T8

Celiac
ganglion

Plexus

Stomach

Postganglionic
neurons to
blood vessels,
sweat glands,
arrector pili
muscles, and
adipose tissue

Splanchnic n.

L2

Superior
mesenteric
ganglion

Large
intestine

Small
intestine

Inferior
mesenteric
ganglion

Plexus

Rectum

Paravertebral
ganglion

Paravertebral
ganglion

*Highlight the different
pathways to
different effectors.*

Ductus
deferens

Seminal
vesicle

Uterus

Ovary

Prostate
gland

Urinary
bladder

Testis

Description

The autonomic nervous system (ANS) controls involuntary activities such as digestion and heart rate. It is subdivided into two divisions: **sympathetic** (**SD**) and **parasympathetic** (**PD**). These two divisions work together to regulate normal functions and maintain **homeostasis**. Each division has its own groups of nerves that connect to the major visceral organs in the body. But there are some exceptions. For example, the lacrimal gland is controlled only by the PD, because the SD has no nerves that connect to it.

This module gives an overview of the PD—also referred to as the *craniosacral division* because motor output arises from some *cranial* nerves and from the *sacral* region of the spinal cord. Because the PD is most active when the body is at rest, it is nicknamed the "resting and digesting" system. Consider what happens when you lie down on the couch after eating a heavy meal. Your digestive function increases, pupils constrict, heart rate decreases, breathing rate decreases, and blood pressure decreases. All of these illustrate your parasympathetic division in action.

The PD has a simpler anatomical structure than the SD. The illustration on the facing page shows only the motor output to the effector. The nerves involved are like the electrical wiring in a house and consist of a **preganglionic neuron** linked to a **postganglionic neuron**. Preganglionic neurons arise from the following cranial nerves (CN): **III** (oculomotor), **VII** (facial), **IX** (glossopharyngeal), and **X** (vagus). In the sacral region of the spinal cord, preganglionic neurons arise from **S2–S4** to become the **pelvic nerves**, which connect to target organs in the pelvic region.

As a general rule, the preganglionic axons tend to be quite long, and the postganglionic axons are relatively **short**. Recall that ganglia (sing. *ganglion*) are clusters of neuron cell bodies. Four important ganglia are located in the head, where preganglionic neurons link to postganglionic neurons: (1) **ciliary**, (2) **pterygopalatine**, (3) **submandibular**, and (4) **otic**. These major hubs connect to target organs in the head, such as the eye, lacrimal gland, and salivary glands. The ANS has three possible effectors: *smooth muscle*, *cardiac muscle*, and *glands*. To clarify the illustration, when you see postganglionic neurons connecting to organs, they actually are connecting to muscle tissue in the walls of these organs.

The most important nerve in the PD is the **vagus nerve** (CN X) because it accounts for as much as 80% of all the preganglionic axons in the PD. It innervates all the thoracic organs and most of the abdominal organs. Notice that it runs through **plexuses** (sing. *plexus*) or neural networks to link with all its target organs.

Analogy

Regulating the visceral organs is like driving a car. Sometimes you have to step on the gas, and other times you have to step on the brake. Most of the time, the PD is like stepping on the brake. For example, parasympathetic impulses sent along the vagus nerve to the heart decrease the heart rate. Unfortunately, it does not work this way with all organs/glands. There are exceptions to this rule.

Study Tip

Parasympathetic = Peaceful. This simple alliteration helps you distinguish the function of the SD from the PD.

Summary

Item	Description
Nickname	*Resting and digesting* system
Location of preganglionic neuron cell bodies	Brainstem and S2–S4 regions of spinal cord (*craniosacral*)
Length of preganglionic axon	Long
Length of postganglionic axon	Short
Location of ganglia	Usually near or within the wall of the target organ
Neurotransmitters	Preganglionic neurons secrete acetylcholine (ACh). Postganglionic neurons secrete acetylcholine (ACh).

Parasympathetic Division (PD) of the ANS

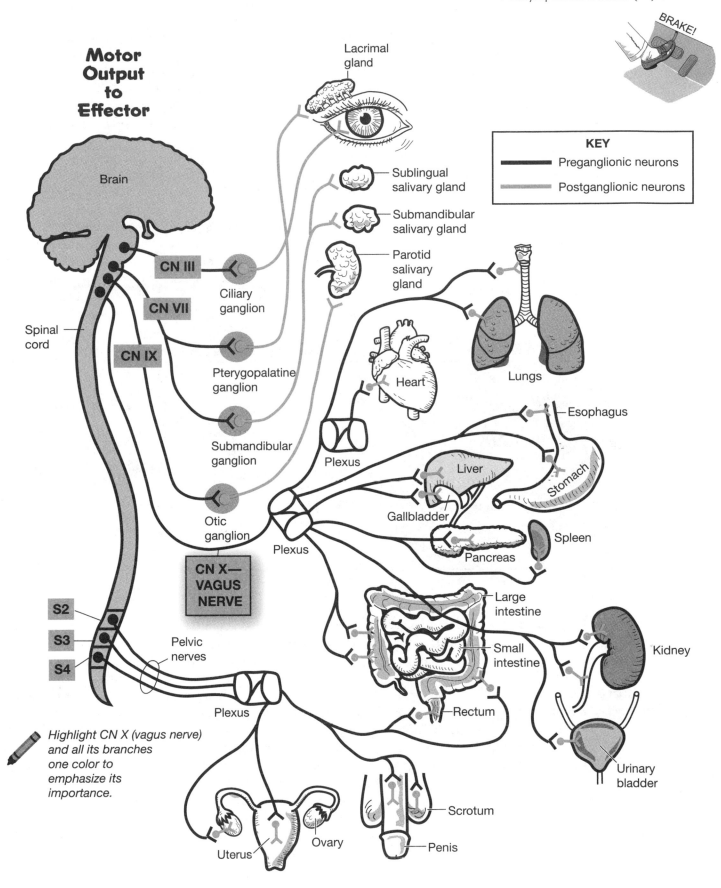

Motor Output to Effector

Brain

Spinal cord

CN III

CN VII

CN IX

Ciliary ganglion

Pterygopalatine ganglion

Submandibular ganglion

Otic ganglion

CN X— VAGUS NERVE

S2

S3

S4

Pelvic nerves

Plexus

BRAKE!

Lacrimal gland

Sublingual salivary gland

Submandibular salivary gland

Parotid salivary gland

KEY

━━━ Preganglionic neurons

━━━ Postganglionic neurons

Heart

Plexus

Lungs

Esophagus

Liver

Stomach

Gallbladder

Spleen

Pancreas

Plexus

Large intestine

Small intestine

Kidney

Rectum

Urinary bladder

Scrotum

Uterus

Ovary

Penis

Highlight CN X (vagus nerve) and all its branches one color to emphasize its importance.

Description

The autonomic nervous system (ANS) uses reflexes to regulate many of the unconscious activities we take for granted, such as digestive activities, heart rate, and pupillary responses. These responses are induced by specific neurotransmitters combining with specific receptors. The three major neurotransmitters secreted in the ANS are **acetylcholine (ACh)**, **norepinephrine (NE)**, and **epinephrine (EPI)**. Acetylcholine binds to either **muscarinic (M)** or **nicotinic (N)** receptors, and NE and EPI bind to either alpha or beta receptors. There are two types of **alpha (μ) receptors** (*alpha 1* and *alpha 2*) and three kinds of **beta (β) receptors** (*beta 1*, *beta 2*, and *beta 3*).

When a neurotransmitter binds to a receptor, it induces a response in its **target cells**. The response is determined by which specific neurotransmitter combines with which specific receptor. For example, when **NE** binds to an alpha 1 receptor, it typically elicits a different response than when NE binds to a beta 1 receptor. Although there are exceptions to the rule, we can generalize about the different responses induced by various neurotransmitter/receptor combinations. When ACh binds to muscarinic receptors, for example, it sometimes results in an excitatory response. Other times, it results in an inhibitory response. And when ACh binds to nicotinic receptors, it usually produces an excitatory response. When NE (or EPI) binds to an alpha receptor, it usually induces an excitatory response, such as constriction of smooth muscle within blood vessels walls. When NE binds to a beta receptor, it often induces an inhibitory response, such as dilation of the bronchioles in the lungs.

ACh Receptors

NE and EPI Receptors

*Color the **neurotransmitters** and their **receptors** different colors.*

Cholinergic and Adrenergic Neurons

In the ANS, nerve fibers that secrete *only* ACh are called **cholinergic neurons**, and nerve fibers that secrete *only* NE or EPI are called **adrenergic neurons**. In the parasympathetic division, all the **preganglionic** neurons and **postganglionic** neurons are cholinergic. The ACh released from the preganglionic neurons binds to **nicotinic neurons** in the dendrites of the postganglionic neurons. This stimulates the **postganglionic neuron**, in turn, to release ACh, which binds to **muscarinic receptors** in the **effector** (*smooth muscle, cardiac muscle,* or a *gland*).

In the sympathetic division, the preganglionic neurons also are cholinergic, so they release ACh into the **synaptic cleft**. The ACh then binds to the nicotinic receptors in dendrites of the postganglionic neurons. Because all the postganglionic neurons are adrenergic, they release NE, which binds to alpha or beta receptors in the effector. In one special case that does not involve a postganglionic neuron, the sympathetic division also directly innervates hormone-producing cells in the inner medulla of the adrenal gland. The preganglionic neurons secrete ACh, which binds to the nicotinic receptors in the hormone-producing cells. This stimulates the secretion of both NE and EPI into the blood. These hormones travel through the bloodstream to their target cells, which contain either alpha or beta receptors.

PARASYMPATHETIC Division

Preganglionic neurons   Ganglion   Postganglionic neurons   Effector *(cardiac muscle, smooth muscle, or a gland)*

Acetylcholine (ACh)   Acetylcholine (ACh)

Color the cholinergic neurons, adrenergic neurons, and neurotransmitters different colors.

SYMPATHETIC Division

Preganglionic neurons   Ganglion   Postganglionic neurons   Effector *(cardiac muscle, smooth muscle, or a gland)*

Acetylcholine (ACh)

Epinephrine (EPI), Norepinephrine (NE)

or

Adrenal gland   Hormone-producing cell   Blood vessel

α or β

KEY

C = Cholinergic neuron	M = Muscarinic receptor	∝ = Alpha receptor	
A = Adrenergic neuron	N = Nicotinic receptor	β = Beta receptor	
S Stimulus	R Response		

Notes

ENDOCRINE SYSTEM

ENDOCRINE SYSTEM

Overview: Hormone Function and Glands

Description   Along with the nervous system, the endocrine system is one of the great regulators of activities in the human body. Its major purpose is to maintain **homeostasis** and to regulate various processes such as growth and human development. It consists of many different glands containing specific cells that synthesize and release chemical messengers called **hormones**. These hormones enter the bloodstream, where they travel to a specific cell called a **target cell**. This target cell has a receptor for that specific hormone. Once the hormone binds to the receptor, it induces a response in that cell. The general concept of how the endocrine system functions is illustrated below:

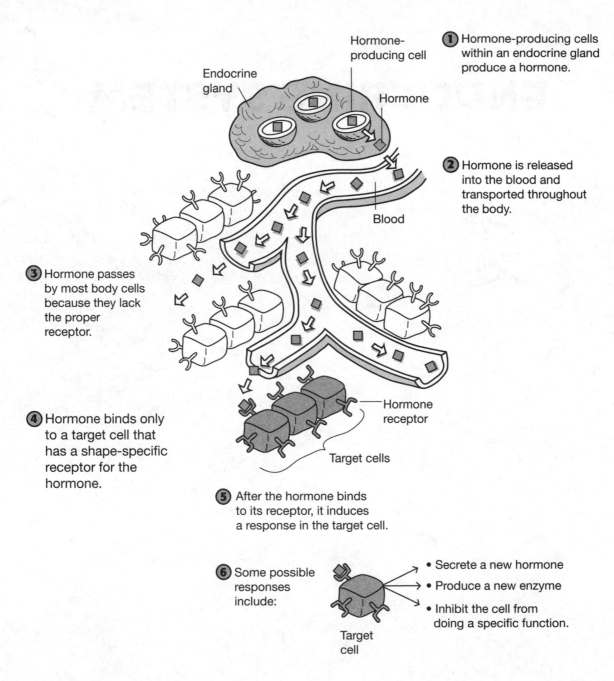

1 Hormone-producing cells within an endocrine gland produce a hormone.

2 Hormone is released into the blood and transported throughout the body.

3 Hormone passes by most body cells because they lack the proper receptor.

4 Hormone binds only to a target cell that has a shape-specific receptor for the hormone.

5 After the hormone binds to its receptor, it induces a response in the target cell.

6 Some possible responses include:
- Secrete a new hormone
- Produce a new enzyme
- Inhibit the cell from doing a specific function.

Endocrine gland

Hormone-producing cell

Hormone

Blood

Hormone receptor

Target cells

Target cell

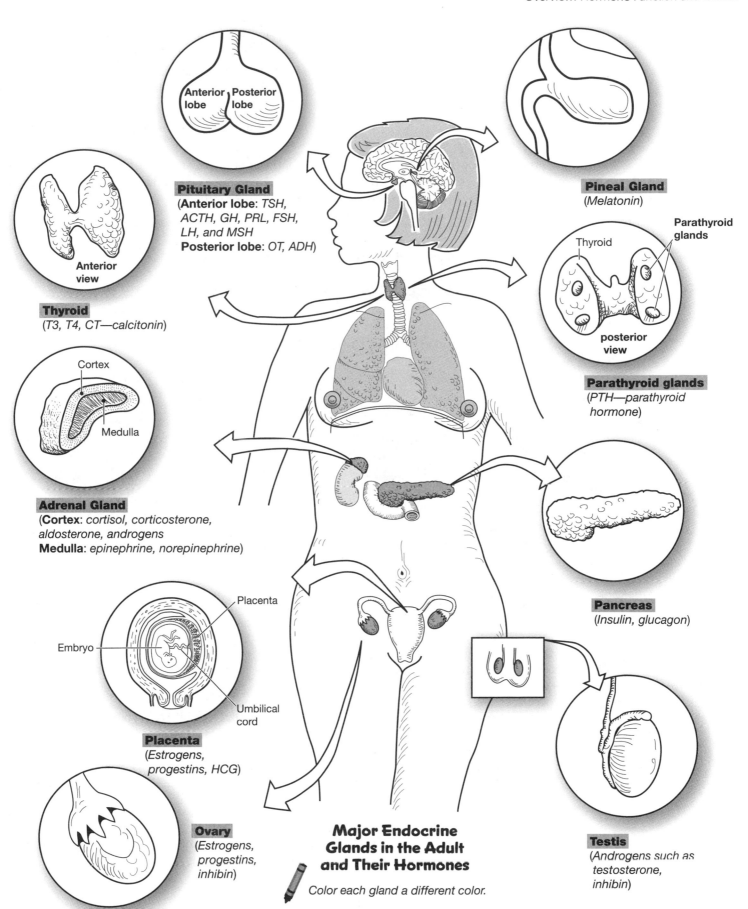

Pituitary Gland
(**Anterior lobe:** *TSH, ACTH, GH, PRL, FSH, LH, and MSH*
Posterior lobe: *OT, ADH*)

Anterior lobe | Posterior lobe

Pineal Gland
(*Melatonin*)

Thyroid
(*T3, T4, CT—calcitonin*)

Anterior view

Cortex

Medulla

Adrenal Gland
(**Cortex:** *cortisol, corticosterone, aldosterone, androgens*
Medulla: *epinephrine, norepinephrine*)

Thyroid

Parathyroid glands

posterior view

Parathyroid glands
(*PTH—parathyroid hormone*)

Placenta

Embryo

Umbilical cord

Placenta
(*Estrogens, progestins, HCG*)

Pancreas
(*Insulin, glucagon*)

Ovary
(*Estrogens, progestins, inhibin*)

Major Endocrine Glands in the Adult and Their Hormones

Color each gland a different color.

Testis
(*Androgens such as testosterone, inhibin*)

Description

Steroid hormones are lipid-soluble and synthesized from **cholesterol**. Because they must be transported through the watery blood, they need special transporters called **protein carriers**. These carriers also prevent hormones from being degraded by enzymes in the blood. Once the hormones arrive at their target cells, they enter the nucleus to stimulate DNA to begin protein synthesis. The effects of any steroid hormone have the following key features:

- **Slower action:** They produce a response in hours or days after the initial binding of hormone with its receptor—slower than nonsteroid hormone.

- **No amplification:** The effects produced are proportional to the amount of hormone secreted; a greater amount of hormone is needed to induce a response compared with a nonsteroid, in which the effects of a hormone are amplified by a chain reaction.

Steroid

Examples of steroid hormones are given in the table below.

Steroid Hormones	Site of Secretion
● Aldosterone ● Cortisol ● Androgens (e.g., testosterone)	Adrenal glands (cortex)
● Testosterone	Testes
● Estrogens ● Progesterone	Ovaries
● Calcitriol	Kidneys

Sequence

The following sequence of events is a simplified mechanism describing how steroid hormones induce a response in their target cells, illustrated on the facing page. Note that thyroid hormones (T_3 and T_4) use a similar mechanism of action.

(1) The steroid hormone travels through the blood with the help of a protein carrier. After being released from this carrier, the hormone diffuses through the plasma membrane into the **cytoplasm** and into the **nucleus** through the **nuclear pore**.

(2) Once inside the nucleus, the hormone binds with its **hormone receptor** to form a **hormone-receptor complex**.

(3) The hormone-receptor complex binds to a **shape-specific receptor on DNA** like a key in a lock.

(4) This stimulates DNA to begin the process of protein synthesis. First, DNA makes a copy of itself in the form of a mobile, single-stranded molecule called **mRNA** (messenger RNA). The smaller mRNA is able to leave the nucleus through the nuclear pore, and it moves into the cytoplasm.

(5) The mRNA molecule binds to the **ribosome**, the protein factory of the cell.

(6) The message in mRNA is read by the ribosome, and it links free **amino acids** to form a new protein.

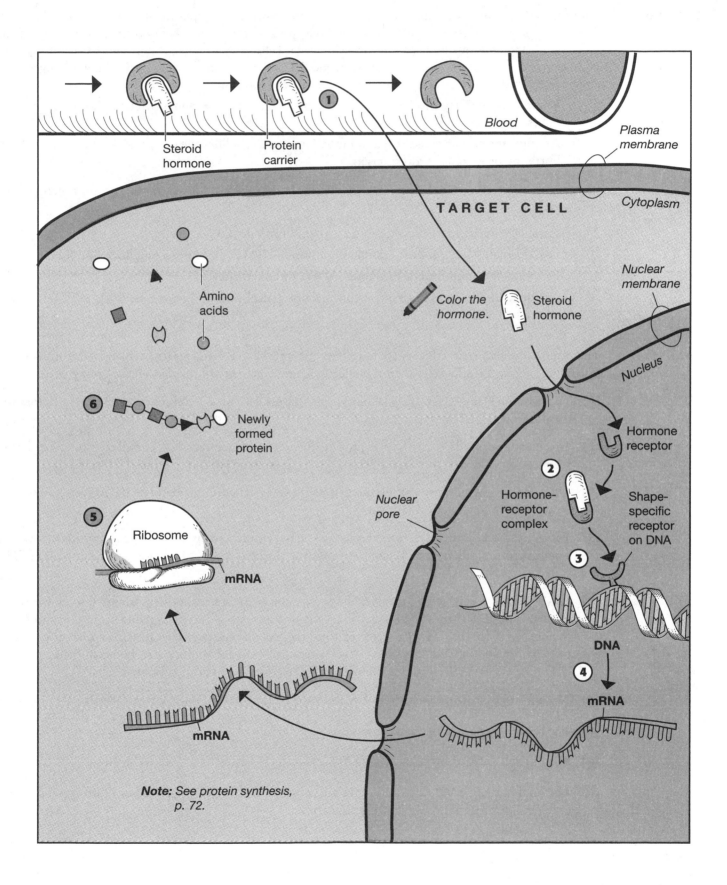

Blood

Steroid hormone

Protein carrier

(1)

Plasma membrane

Cytoplasm

TARGET CELL

Nuclear membrane

Amino acids

Color the hormone.

Steroid hormone

Nucleus

Hormone receptor

Newly formed protein

(6)

Hormone-receptor complex

(2)

Shape-specific receptor on DNA

(3)

(5)

Ribosome

Nuclear pore

mRNA

DNA

(4)

mRNA

mRNA

Note: See protein synthesis, p. 72.

Description Nonsteroid hormones are water soluble, so they are easily transported through the blood. Many are protected and transported by protein carriers that prevent them from being degraded by enzymes in the blood. Once the hormones arrive at their **target cells**, they first bind at receptors located at the plasma membrane. This triggers a cascade inside the cell that eventually leads to cellular changes. The effects of a nonsteroid hormone have the following key features:

- **Rapid action:** They produce a response within *seconds* or *minutes* after the hormone initially binds to its receptor.

- **Amplification:** The chain reaction in the target cell amplifies the effects of the hormone; a little hormone produces a big response.

Nonsteroid hormones can be subdivided into different chemical categories. The four major subgroups are:

- Proteins—*ex.:* **insulin, glucagon,** and **growth hormone (GH)**
- Glycoproteins—*ex.:* **follicle-stimulating hormone (FSH)**, and **luteinizing hormone (LH)**
- Peptides—*ex.:* **antidiuretic hormone (ADH)** and **oxytocin (OT)**
- Amino acid derivatives—*ex.:* **epinephrine, norepinephrine,** and **thyroxine (T_4)**

Mechanism The following sequence of events describes one of the mechanisms by which *some* nonsteroid hormones induce a response in their target cells. These events are illustrated on the facing page.

1. The **hormone** travels through the blood to its target cell, which contains membrane **receptors** for the hormone.

2. The hormone binds with the receptor to form a **hormone-receptor complex**. This complex is referred to as the *first messenger* because it begins a chain reaction within the target cell.

3. Many hormone-receptor complexes trigger proteins called **G proteins** to be converted from an inactive form to an activated form.

4. The **activated G protein,** in turn, triggers an enzyme called **adenylate cyclase** to become activated. The function of adenylate cyclase is to catalyze the conversion of **adenosine triphosphate (ATP)** into **cyclic AMP (cAMP)**.

5. **cAMP** acts as an enzyme to catalyze the conversion of an **inactive protein kinase** into an **activated protein kinase.** The function of any kinase is to transfer a phosphate group from one substance to another. Note that cAMP is referred to as the *second messenger* because it's a critical player in this chain reaction, and a threshold level is required to finally induce a response in the target cell. Also note that cAMP is not the *only* second messenger used within all target cells.

6. The **activated protein kinase** transfers a phosphate group from ATP onto a protein.

7. The result is a **phosphorylated protein.**

8. The phosphorylated proteins eventually produce cellular changes within the target cell.

9. The final fate of **cAMP** is that it is either deactivated within the target cell, or it diffuses out of the cell. This ensures that the chain reaction stops.

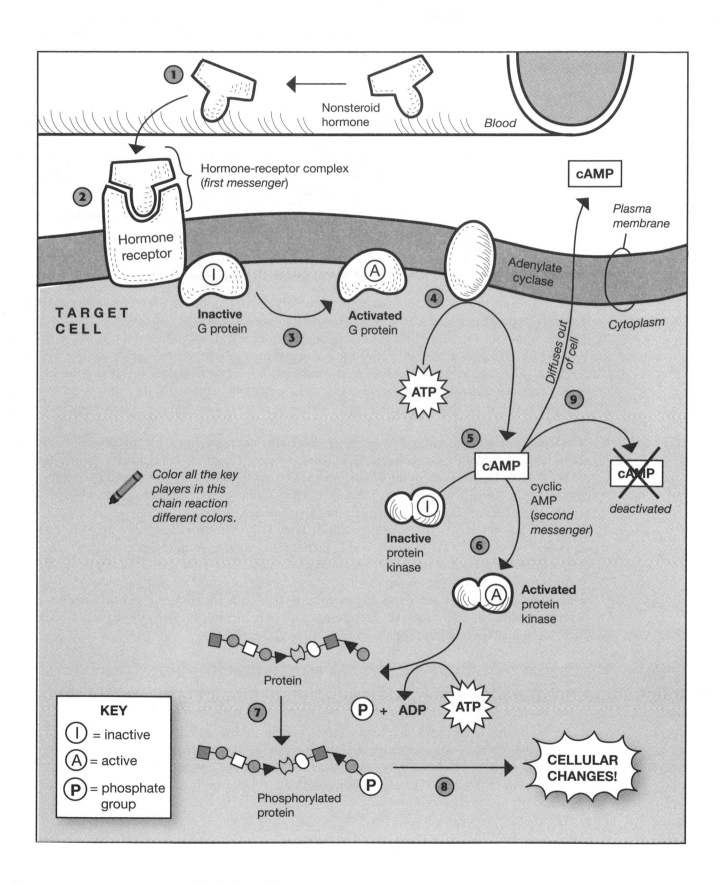

Description

Regulation of **blood glucose** levels is controlled by two hormones: **insulin** and **glucagon**. These two hormones are antagonists because they have opposite functions. Insulin causes blood glucose levels to decrease, and glucagon causes blood glucose levels to increase. Both hormones are produced by cells within the **pancreas**.

Mechanism

Let's examine the mechanism of action. The normal level of blood glucose is about 70–100 mg/dL and has to be maintained. Glucose levels typically rise after a meal rich in carbohydrates, such as baked potatoes or corn flakes. These high glucose levels are detected by insulin-secreting cells within the pancreas. This triggers these cells to secrete more insulin—the "feeding hormone"—into the blood. As insulin travels through the blood, it induces two major responses:

1. Stimulates the storage of excess glucose in the form of **glycogen** within the liver and some other tissues.

2. Stimulates the uptake of glucose by most body cells. Insulin binds to receptors in the plasma membrane, which opens channels for glucose to pass through.

The glucose then is metabolized to produce large amounts of the energy molecule ATP. Insulin secretion stops when blood glucose levels fall back to normal levels.

If you haven't eaten anything for a long time, your blood glucose levels fall below normal. This is detected by glucagon-secreting cells in the pancreas. In response to this stimulus, these cells secrete glucagon—the "starvation hormone"—into the blood. Glucagon targets the liver and some other tissues, such as skeletal muscle, where glycogen is stored, and stimulates the conversion of glycogen into **glucose**. The result is that the blood glucose level rises. Glucagon secretion stops when blood glucose rises to normal levels.

Diabetes mellitus is caused by the body's inability to produce or use insulin, which results in abnormally high blood glucose levels. It is one of the most common endocrine disorders and is one of the leading causes of death in the United States. One method of diagnosis is detecting high levels of glucose in the urine. If left untreated, it can result in serious health problems, such as cardiovascular disease, vision problems, kidney damage, and nerve damage.

There are two types of diabetes mellitus, **type 1** (**insulin-dependent**) and **type 2** (**non-insulin-dependent**). Type 1 is a result of the body's inability to produce insulin, and there is no cure. It often occurs in children so it is referred to as *juvenile diabetes*. This is the less common of the two types. Treatment requires self-monitoring of blood glucose levels and daily insulin injections or the use of an insulin pump. Type 2 is far more common than Type 1. Sufferers are typically obese adults who produce insulin, but their body cells become resistant to it. Treatment involves behavior modification in the form of diet, exercise, and weight loss.

Study Tips

- A simple association to distinguish insulin from glucagon is to remember that insulin is nicknamed the "feeding hormone" and glucagon is nicknamed the "starvation hormone."

- The pancreas is included in the endocrine system and digestive system because it is both an endocrine gland and an exocrine gland. Only about 1 percent of its cells serve the endocrine function to secrete hormones like insulin and glucagon. The remaining 99 percent of its cells serve the exocrine function to secrete digestive enzymes for the purpose of chemically digesting lipids, carbohydrates, and proteins.

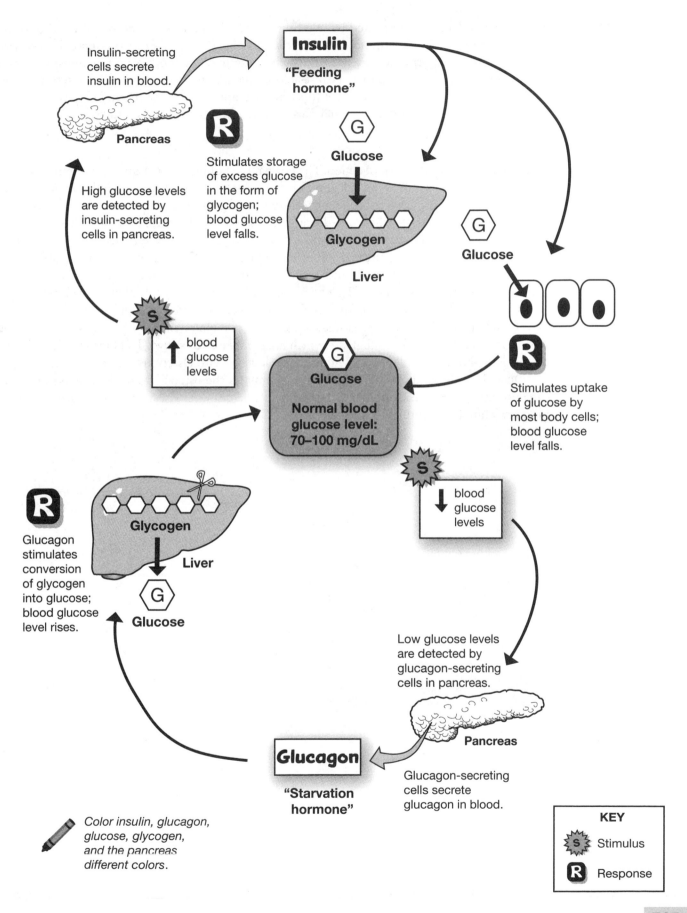

Insulin-secreting cells secrete insulin in blood.

Pancreas

High glucose levels are detected by insulin-secreting cells in pancreas.

Insulin

"Feeding hormone"

R Stimulates storage of excess glucose in the form of glycogen; blood glucose level falls.

G **Glucose**

Glycogen

Liver

G **Glucose**

R Stimulates uptake of glucose by most body cells; blood glucose level falls.

S ↑ blood glucose levels

G **Glucose**

Normal blood glucose level: 70–100 mg/dL

S ↓ blood glucose levels

R Glucagon stimulates conversion of glycogen into glucose; blood glucose level rises.

Glycogen

Liver

G **Glucose**

Low glucose levels are detected by glucagon-secreting cells in pancreas.

Pancreas

Glucagon

"Starvation hormone"

Glucagon-secreting cells secrete glucagon in blood.

Color insulin, glucagon, glucose, glycogen, and the pancreas different colors.

KEY

S Stimulus

R Response

Description

Blood calcium levels are regulated by two hormones: **calcitonin (CT)** and **parathyroid hormone (PTH)** or parathormone. CT is produced by the **thyroid gland**, and PTH is produced by the **parathyroid glands**. These two hormones are **antagonists** because they have opposite functions: CT causes blood calcium levels to decrease, and PTH causes blood calcium levels to increase. It's interesting to note that PTH is essential for life, but CT is not (only in pregnant women). Without PTH, calcium levels would fall so low that it would impair normal muscle contraction and blood clotting.

Mechanism

Let's examine the mechanism of action. The normal level of blood calcium is in the range of 9–11 mg/dL, which has to be maintained. As calcium-rich foods such as milk, broccoli, and tofu are ingested, blood calcium levels rise. Calcitonin-producing cells within the thyroid gland sense this, and they secrete CT into the blood. CT targets the bone-degrading cells within the skeletal system called **osteoclasts**. The plasma membranes in osteoclasts contain receptors for CT. Once CT binds to this receptor, it inhibits these cells from doing their normal job of degrading bone. CT does not affect **osteoblasts**—the bone-forming cells—they carry on with their normal function of depositing calcium salts in bone. This causes the blood calcium levels to fall. CT stops being produced when blood calcium levels return to normal.

If you haven't eaten any calcium-rich foods or taken any calcium supplements for a long time, your blood calcium levels fall. This is detected by the two pairs of parathyroid glands located on the posterior surface of the thyroid gland, which are stimulated to secrete PTH into the blood. PTH targets **osteoclasts**—bone-degrading cells within the skeletal system. Osteoclasts contain receptors for PTH.

Once PTH docks at its PTH receptor, it stimulates the osteoclasts to decompose bone and release calcium into the blood. Consequently, the blood calcium level rises. PTH stops being produced when blood calcium levels return to normal.

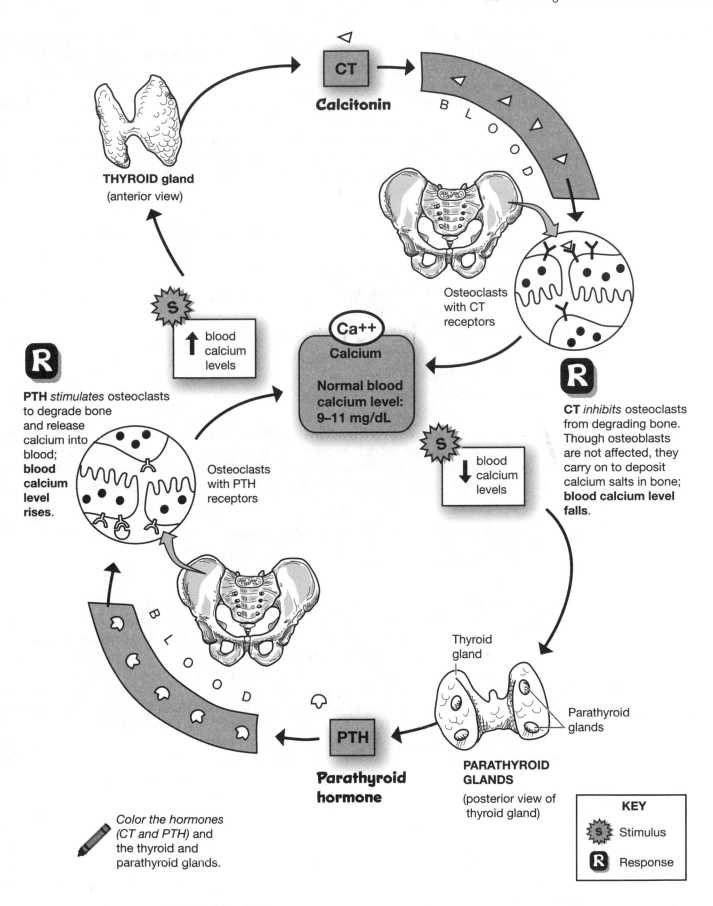

CT
Calcitonin

B L O O D

THYROID gland
(anterior view)

Osteoclasts
with CT
receptors

S ↑ blood calcium levels

Ca++
Calcium

**Normal blood
calcium level:
9–11 mg/dL**

R

CT *inhibits* osteoclasts
from degrading bone.
Though osteoblasts
are not affected, they
carry on to deposit
calcium salts in bone;
**blood calcium level
falls**.

R

PTH *stimulates* osteoclasts
to degrade bone
and release
calcium into
blood;
**blood
calcium
level
rises**.

Osteoclasts
with PTH
receptors

S ↓ blood calcium levels

B L O O D

Thyroid
gland

Parathyroid
glands

PTH

**Parathyroid
hormone**

**PARATHYROID
GLANDS**

(posterior view of
thyroid gland)

*Color the hormones
(CT and PTH) and
the thyroid and
parathyroid glands.*

KEY

S Stimulus

R Response

Description

The **pituitary gland** (*hypophysis*) is a small but powerful gland located at the base of the brain. It consists of two distinct lobes: the **anterior lobe** (*adenohypophysis*) and the **posterior lobe** (*neurohypophysis*). Because each lobe functions differently and secretes different hormones, each is presented in a separate module. This one covers the **anterior pituitary**.

Surrounding the pituitary is a protective pocket of bone called the **sella turcica** of the **sphenoid bone**. The pituitary is nicknamed the "master gland" because it makes many different hormones that control the other endocrine glands in the body. But the **hypothalamus** in the brain actually controls the pituitary gland with specific bundles of neurons. This provides a *vital link* between the nervous system and the endocrine system. The pituitary is connected to the hypothalamus by a slender stem-like structure called the **infundibulum** (*pituitary stalk*).

Hormones Produced

Five different types of **hormone-producing cells** within the anterior pituitary directly secrete the following seven major hormones:

- Adrenocorticotropic hormone (**ACTH**)
- Follicle-stimulating hormone (**FSH**)
- Growth hormone (**GH**)
- Luteinizing hormone (**LH**)
- Prolactin (**PRL**)
- Thyroid-stimulating hormone (**TSH**)
- Melanocyte-stimulating hormone (**MSH**)

Sequence

The anterior pituitary relies on a specialized group of blood vessels collectively called a portal system. The body has two important portal systems: the hepatic portal system (*in the liver*) and the **hypophyseal portal system** (*in the anterior pituitary*). Portal systems link two capillary networks to more efficiently deliver chemical messengers to a specific location before they are released to the general circulation:

Superior hypophyseal artery ⟶ Primary capillary network ⟶ Hypophyseal portal veins ⟶ Secondary capillary network ⟶ Anterior hypophyseal veins

Mechanism

The anterior pituitary contains a mass of various **hormone-producing cells** (**HPCs**) controlled by a group of neurons in the **ventral hypothalamus**. These neurons can either stimulate or inhibit hormone production by secreting either **releasing factors** (**RFs**) or inhibiting factors, respectively. Let's examine the stimulation process. **RFs** stimulate the **HPCs** to secrete hormones. Different RFs are needed to make the seven major hormones produced by the anterior pituitary. These RFs are produced in the neuron's **cell body** and released from the **axon terminal**. After a nerve impulse triggers them to release, they diffuse rapidly into the **primary capillary network** of the portal system, and then the hypophyseal portal veins quickly shuttle them into the secondary capillary network surrounding the HPCs. As the RFs diffuse out of the blood, they immediately bind to shape-specific receptors in the HPC's plasma membrane, which induces the production of specific hormones. These hormones then enter the secondary capillary network and exit the anterior pituitary via the anterior hypophyseal vein. As the hormones travel through the blood, they eventually bind to receptors in their target cells and induce a response.

For example, a thyroid-stimulating releasing factor is needed to stimulate the secretion of thyroid-stimulating hormone (TSH). After TSH leaves the anterior lobe via the **anterior hypophyseal vein,** it travels through the blood to all the different organs but binds only at those targets that have receptors for TSH. As its name implies, TSH targets the thyroid gland to induce the secretion of thyroid hormones such as triiodothyronine (T_3) and thyroxine (T_4).

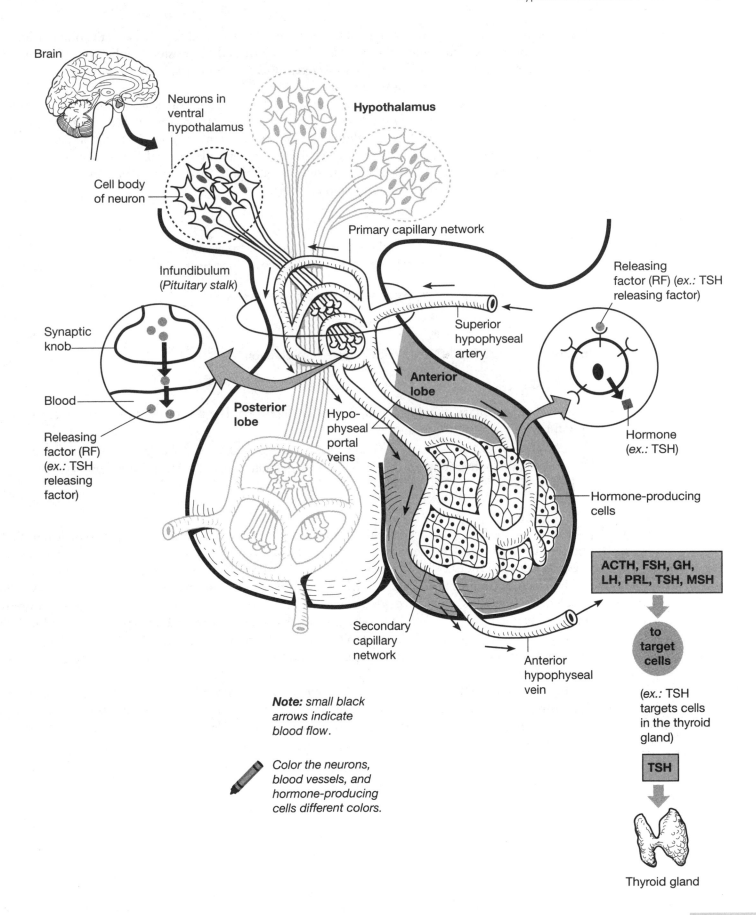

Brain

Neurons in ventral hypothalamus

Hypothalamus

Cell body of neuron

Primary capillary network

Infundibulum (*Pituitary stalk*)

Releasing factor (RF) (*ex.*: TSH releasing factor)

Synaptic knob

Superior hypophyseal artery

Anterior lobe

Blood

Posterior lobe

Hypo-physeal portal veins

Releasing factor (RF) (*ex.*: TSH releasing factor)

Hormone (*ex.*: TSH)

Hormone-producing cells

ACTH, FSH, GH, LH, PRL, TSH, MSH

Secondary capillary network

Anterior hypophyseal vein

to target cells

(*ex.*: TSH targets cells in the thyroid gland)

Note: small black arrows indicate blood flow.

Color the neurons, blood vessels, and hormone-producing cells different colors.

TSH

Thyroid gland

Description

The **pituitary gland** (*hypophysis*) is a small but powerful gland located at the base of the brain. It consists of two distinct lobes: the **anterior lobe** (*adenohypophysis*) and the **posterior lobe** (*neurohypophysis*). Because each lobe functions differently and secretes different hormones, each is presented in a separate module. This one highlights the **posterior pituitary**.

Surrounding the pituitary is a protective pocket of bone called the **sella turcica** of the **sphenoid bone**. The pituitary is nicknamed the "master gland" because it makes many different hormones that control the other endocrine glands in the body. But the **hypothalamus** in the brain actually controls the pituitary gland with specific bundles of neurons. This provides a *vital link* between the nervous system and the endocrine system. The pituitary is connected to the hypothalamus by a slender, stem-like structure called the **infundibulum** (*pituitary stalk*).

Hormones Produced

The cells in the posterior pituitary do *not* produce any hormones. Instead, neurons originating in the hypothalamus that extend into the posterior pituitary produce, store, and release the following two hormones:

- **Antidiuretic hormone (ADH)**
- **Oxytocin (OT)**

In short, both ADH and OT are released by **neurosecretion**.

Sequence

The posterior pituitary contains a simple **capillary network**. The following flowchart indicates blood flow through the posterior pituitary:

Inferior hypophyseal a. $\longrightarrow$ capillary network $\longrightarrow$ Posterior hypophyseal v.

Mechanism

Two different clusters of neurons in the hypothalamus produce hormones: the **paraventricular nuclei** and the **supraoptic nuclei**. The former produces OT, and the latter produces ADH. Each neuron in the cluster is a long, slender cell that has a kind of head and a tail, the **cell body** and the **axon terminal**.

Hormones are produced in the cell body and then stored in membrane-bound chambers called vesicles. These vesicles travel through the neurons and into the axon terminals located in the posterior pituitary. Nerve impulses trigger the release of these hormones from the axon terminals and into the blood. ADH and OT diffuse into the capillary network and exit the pituitary gland via the posterior hypophyseal vein. As they travel through the blood, they bind to receptors in their target cells and induce a response.

For example, ADH targets cells in the kidney to reabsorb more water into the blood, thereby decreasing urine output. OT has multiple targets, one of which is the smooth muscle in the wall of the uterus. During childbirth, OT is released to trigger the muscular contractions needed for labor and delivery.

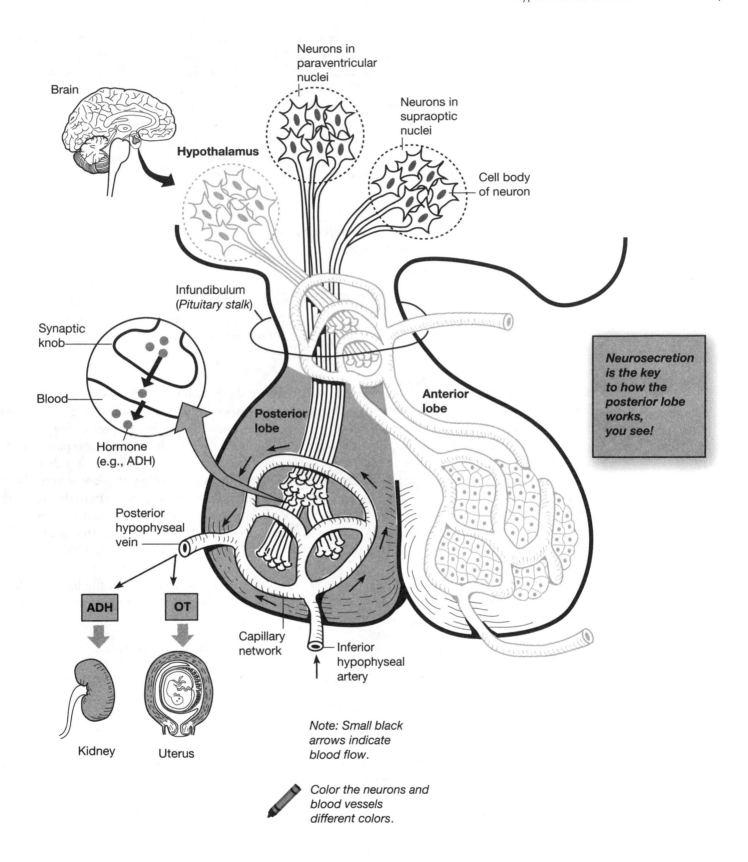

Brain

Neurons in paraventricular nuclei

Neurons in supraoptic nuclei

Hypothalamus

Cell body of neuron

Infundibulum (*Pituitary stalk*)

Synaptic knob

Blood

Hormone (e.g., ADH)

Posterior lobe

Anterior lobe

Neurosecretion is the key to how the posterior lobe works, you see!

Posterior hypophyseal vein

ADH

OT

Capillary network

Inferior hypophyseal artery

Kidney

Uterus

Note: Small black arrows indicate blood flow.

Color the neurons and blood vessels different colors.

Description The highly vascular **adrenal** (*suprarenal*) **glands** are small structures located on top of both kidneys. They are divided into two major regions: the *outer* **adrenal cortex** and the *inner* **adrenal medulla**. The adrenal cortex constitutes most of the mass of the adrenal glands. Each of these regions contains different types of cells that produce different hormones. This module focuses only on the adrenal medulla. **Catecholamines** are a group of chemicals derived from the amino acid **tyrosine**. Catecholamine-producing cells (*chromaffin cells*), or **CPCs**, are located within the medulla and are directly stimulated by nerves from the **sympathetic division** of the autonomic nervous system (see p. 290). Because the sympathetic division helps the body respond to stressful situations on its own, the secretion of catecholamines simply boosts this response.

Hormones Produced The catecholamine-producing cells in the adrenal medulla secrete the following two hormones:

- **Norepinephrine (NE)** (*noradrenaline*)
- **Epinephrine** (*adrenaline*)

As you can see from the illustration, they have similar chemical structures. In fact, the body is able to convert norepinephrine into epinephrine with the help of a converting enzyme. About 80 percent of the CPCs secrete epinephrine, and 20 percent secrete norepinephrine.

Mechanism When faced with an emergency or stressful situation, the body reacts in multiple ways collectively referred to as the **fight-or-flight** response. Most of these responses are the result of the sympathetic nervous system at work, but the release of catecholamines intensifies this response. For example, imagine that you discover a fire raging through your home. Your body needs to prepare you for physical activity so you can flee the situation and survive. As you recognize the fire as a danger, this conscious thought sends a nerve impulse from your **cerebral cortex** to your **hypothalamus** in the brain. Then an impulse is stimulated along nerve fibers in the brain that connect the hypothalamus to the spinal cord. The nerve impulse carried along the last nerve connects the spinal cord directly to the CPCs in your adrenal medulla. These sympathetic nerves release the neurotransmitter **acetylcholine (ACh)**, and it binds to ACh receptors in the CPCs. After binding, it stimulates the CPCs to secrete norepinephrine and epinephrine into the blood.

Like all hormones, these catecholamines travel through the blood like cars on the highway. Their final destination—**target cells**—contain receptors for catecholamines. The binding of the hormone to its receptor in the various target cells produces the following responses:

1. Heart rate and force of contraction increase (this speeds blood circulation throughout the body)

2. Blood vessels feeding kidneys and digestive tract constrict (this reduces blood flow to those tissues)

3. Bronchioles in lungs dilate (this moves air in and out of lungs faster)

4. Liver converts glycogen to glucose (this increases blood glucose levels)

All these responses are for the purpose of helping you flee the burning house. To run away from the fire, more blood will have to flow to your skeletal muscles to deliver more glucose and oxygen to the muscle cells. These nutrients will provide energy for muscle contraction as you flee. The following responses also occur (not included in the illustration):

- Metabolic rate increases.
- Urine output decreases.
- Blood is redirected from the digestive system toward the brain, skeletal muscles, and heart.

Medulla

- Catecholamine-producing cells in the **adrenal medulla** secrete **catecholamines** into the blood.

Adrenal gland

Kidney

Norepinephrine (NE)

Epinephrine

CATECHOLAMINES

FIRE!

Cerebral cortex

Hypo-thalamus

Pituitary gland

Color the neurons, NE, and epinephrine different colors.

Acetylcholine

Synaptic knob

Catecholamine-producing cell (CPC)

Medulla

NE and epinephrine

R

1. Heart rate and force of contraction **increase**.

2. Blood vessels feeding kidneys and digestive tract **constrict**.

3. Bronchioles in lungs **dilate**.

4. Liver converts glycogen to glucose; **blood glucose rises**.

Glycogen

Glucose

KEY

- **S** Stimulus
- **R** Response
- **+** Stimulate

ENDOCRINE SYSTEM

The Adrenal Cortex: Aldosterone

Description
The highly vascularized **adrenal** (*suprarenal*) **glands** are small structures located on top of both kidneys. They are divided into two major regions: the *outer* **adrenal cortex** and the *inner* **adrenal medulla**. The adrenal cortex constitutes most of the mass of the adrenal glands. Each of these regions contains different types of cells that produce different hormones. This module focuses on one aspect of the adrenal cortex, namely, its role in producing the hormone **aldosterone**.

Hormones Produced

The **adrenal cortex** is divided into three different zones, each with its own hormone-producing cells. The first zone in the outer cortex is called the **Zona glomerulosa**. Various cells in this region secrete a group of hormones called **mineralocorticoids**. They are so named because they affect mineral homeostasis (such as Na^+ and K^+) and come from the cortex of the adrenal gland. Of this group, aldosterone is the most potent and physiologically significant, so we will focus on its mechanism of action.

Mechanism
Aldosterone controls the levels of Na^+ (sodium) and K^+ (potassium) ions in extracellular fluids (such as the blood). The net result of its action is to reabsorb Na^+ ions into the blood and simultaneously excrete K^+ ions into the urine. Because "water follows the ions," as Na^+ is reabsorbed, water also is reabsorbed.

1 As shown in the illustration, two types of stimuli stimulate aldosterone release:

1. a decrease in Na^+ levels in the plasma or an increase in K^+ levels, and
2. a decrease in **blood volume** and/or **blood pressure**.

The kidneys detect these stimuli and cause a chemical chain reaction referred to as the **renin-angiotensin-aldosterone (RAA) mechanism**.

This is the most important pathway to trigger aldosterone release, the details of which are discussed in another module (see p. 512). In short, certain cells in the nephrons of the kidney are stimulated to release an enzyme called renin into the blood. This eventually leads to the activation of a hormone called **angiotensin II,** which stimulates secretory cells in the adrenal cortex to release aldosterone into the blood. The primary targets for aldosterone are the renal tubules in the **nephrons** of the kidneys. The response is that protein transporters reabsorb sodium and excrete potassium. Because water follows the sodium ions, it is also reabsorbed. This leads to an increase in blood volume and a corresponding increase in blood pressure. As blood pressure returns to normal, it shuts off the RAA mechanism.

2 Changes in plasma levels of either Na^+ or K^+ also can directly stimulate the secretory cells in the adrenal cortex to release aldosterone. Of these two ions, the secretory cells are *more sensitive* to changes in K^+.

Adrenal cortex

- Aldosterone-producing cells in the **adrenal cortex** secrete **aldosterone** into the blood

Adrenal gland

Kidney

Aldosterone

Major Pathways for Release of Aldosterone

KEY

- S Stimulus
- R Response

S ↓ Na⁺ in blood and/or ↑ K⁺ in blood

① ②

S ↓ blood volume and/or blood pressure

Kidney

Renin-angiotensin-aldosterone (RAA) mechanism

(see p. 512)

Adrenal gland

Adrenal cortex

Aldosterone (targets kidney tubules)

Color *aldosterone*.

R

Kidney

↑ reabsorption of Na⁺ and water,

↑ K⁺ excretion

↑ blood volume and ↑ blood pressure

ENDOCRINE SYSTEM

The Adrenal Cortex: Cortisol and Stress

Description

The **adrenal** (*suprarenal*) **glands**, located on top of both kidneys, are divided into two major regions: the *outer* **adrenal cortex** and the *inner* **adrenal medulla**. Each of these regions contains different types of cells that produce different hormones. This module focuses on the role of the adrenal cortex, which produces the hormone **cortisol**.

Hormones Produced

The **adrenal cortex** is divided into three different zones, each with its own hormone-producing cells. The second zone in the middle of the cortex is called the **Zona fasciculata**. Various cells in this region secrete a group of hormones called **glucocorticoids**—so named because they affect glucose homeostasis and come from the cortex of the adrenal gland. This group consists of the following three hormones: cortisol, corticosterone, and cortisone. Of these three, cortisol is the most abundant and potent. Because it accounts for nearly all the hormonal activity here, we will examine its mechanism of action.

Mechanism

We can't live without glucocorticoids, because they help us maintain normal blood glucose levels, normal blood volume, and resistance to stress—to name just a few functions. First let's examine how cortisol is secreted, then the responses it produces. Cortisol follows a predictable daily cycle in which its levels *peak after we awake* and *dip to their lowest point after we fall asleep*. This cycle is regulated by a negative feedback loop. As blood levels of glucocorticoids fall, this stimulates neurons in the **hypothalamus** to secrete **CRH** (corticotropin-releasing hormone) into the blood. CRH targets cells in the **anterior pituitary** and stimulates them to secrete **ACTH** (adrenocorticotropic hormone). As ACTH travels through the blood, it targets cells in the adrenal cortex to secrete cortisol. As cortisol travels to its target cells in the body, it induces the following responses:

1. **Protein breakdown:** Proteins are broken down into **amino acids** (mostly in skeletal muscle). These amino acids enter the blood and travel to body cells, where they can be used to either make new proteins or be used to produce ATP.

2. **Gluconeogenesis:** This is a process that occurs in liver cells, where a noncarbohydrate such as an amino acid or lactic acid is converted into glucose. The glucose then is used by body cells to produce ATP.

3. **Lipolysis:** This is the process of taking triglycerides stored in adipose tissue and converting them into fatty acids and glycerol. Cells can then use the glycerol and fatty acids to produce ATP.

4. **Stress resistance:** The glucose produced by liver cells in gluconeogenesis allows body cells to produce ATP. This supply of ATP enables the body to battle many different stresses.

Stress also triggers the release of cortisol. Chronic stress can lead to elevated levels of cortisol, and this can affect your health negatively. This leads to depression of the immune system. As a result, wounds may take longer to heal, and you may be more susceptible to contracting a virus such as the common cold.

Adrenal cortex

- Cortisone-producing cells in the **adrenal cortex** secrete **cortisol** into the blood.

Adrenal gland

Kidney

Cortisol

The **STRESS** is killing me!

Cerebral cortex

Hypothalamus

CRH

Pituitary gland

Posterior lobe

Anterior lobe

ACTH

Color ACTH and cortisol different colors.

Adrenal cortex

Cortisol

R

1. **Protein breakdown**

Amino acid

2. **Gluconeo-genesis;** blood glucose rises

G Glucose

3. **Lipolysis**

4. **Stress resistance**

STR   ESS

KEY

S   Stimulus

R   Response

317

Description

The highly vascularized **adrenal** (*suprarenal*) **glands** are small structures located on top of both kidneys. They are divided into two major regions: the *outer* **adrenal cortex** and the *inner* **adrenal medulla**. The adrenal cortex constitutes most of the mass of the adrenal glands. Each of these regions contains different types of cells that produce different hormones. This module focuses on one aspect of the adrenal cortex—its role in producing the masculinizing hormones called **androgens** (*andros* = male human being).

Hormones Produced

● The **adrenal cortex** is divided into three different zones, each with its own hormone-producing cells. The innermost zone is called the **Zona reticularis**. Various cells in this region secrete a group of male sex steroids called androgens. The most potent type of androgen is **testosterone**. Like all steroidal hormones, it is derived from **cholesterol**.

Mechanism

Testosterone is produced by the testes in the male at high levels during puberty and adulthood. The adrenal cortexes of both sexes produce small amounts of an androgen called **dehydroepiandrosterone** (**DHEA**). This is converted into **testosterone** by other tissue cells. At the onset of puberty, testosterone triggers similar responses in both sexes, such as an increase in underarm hair and pubic hair. But in the male, the amount of testosterone produced by the adrenal glands is insignificant compared with the quantity produced by the testes. So let's focus on the responses testosterone produces in the female throughout her lifetime:

① **Develops underarm and pubic hair**—occurs before puberty

② **Triggers the growth spurt**—also occurs before puberty

③ **Increases the sex drive**—occurs in the adult female

④ **Converts testosterone into estrogens**—postmenopausal women convert testosterone into estrogens (female sex hormones).

Details of the mechanism by which the adrenal glands produce androgens is not well understood. Production seems to be stimulated by **adrenocorticotropic hormone** (**ACTH**) secreted by the **anterior pituitary**.

Adrenal cortex

- Androgen-producing cells in the **adrenal cortex** secrete an androgen that is converted into **testosterone** by other tissue cells.

Adrenal gland

Kidney

OH

Testosterone

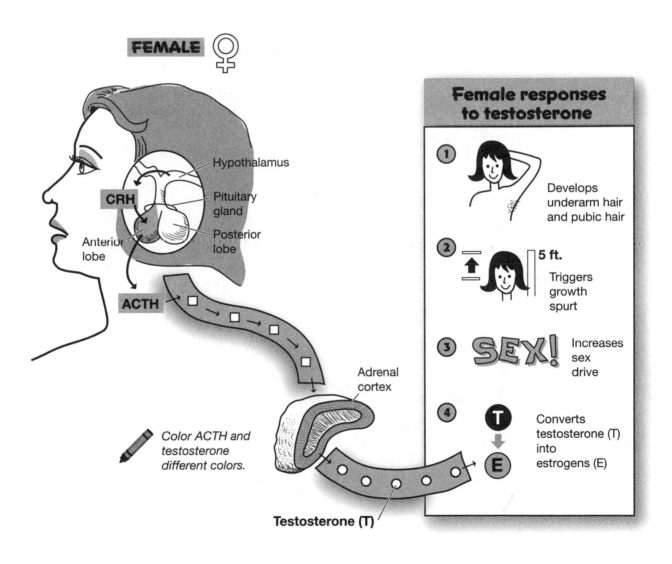

FEMALE

Hypothalamus

CRH

Pituitary gland

Anterior lobe

Posterior lobe

ACTH

Adrenal cortex

Color ACTH and testosterone different colors.

Testosterone (T)

Female responses to testosterone

1. Develops underarm hair and pubic hair

2. 5 ft. Triggers growth spurt

3. **SEX!** Increases sex drive

4. **T** → **E** Converts testosterone (T) into estrogens (E)

Notes

SPECIAL SENSES

Description

The eyeball rests in the orbit in the skull and is surrounded by fatty tissue. Six major muscles control the movement of the eyeball:

Muscle Name	Action	Cranial Nerve Innervation
Superior rectus	Eyeball moves up	Oculomotor n. (III)
Inferior rectus	Eyeball moves down	Oculomotor n. (III)
Medial rectus	Eyeball moves medially	Oculomotor n. (III)
Lateral rectus	Eyeball moves laterally	Abducens n. (VI)
Superior oblique	Eyeball rotates; moves down and laterally	Trochlear n. (IV)
Inferior oblique	Eyeball rotates; moves up and laterally	Oculomotor n. (III)

These muscles are innervated by three different cranial nerves: oculomotor (III), trochlear (IV), and abducens (VI) nerves. In total, there are 12 pairs of cranial nerves, and they are referred to with Roman numerals. In this case, we only need to know the Roman numerals for number 3 (III), number 4 (IV), and number 6 (VI). The lacrimal gland produces tears, which are spread across the surface of the eye during blinking. The tears accumulate in a pooling area called the **medial canthus**. Then they enter openings called the **superior lacrimal punctum** and the **inferior lacrimal punctum** that lead to passageways called the **superior lacrimal caniculus** and the **inferior lacrimal caniculus**. From here they travel to the **lacrimal sac** and down the **nasolacrimal duct** and are drained into the nose. This explains why your nose runs when you are crying heavily.

Analogy

- The rectus muscles are positioned like points on a compass (north, south, east, and west).

Study Tips

- The rectus muscle name tells you which direction it moves the eyeball. For example, the *superior* rectus moves the eyeball *superiorly* or upward.

- Rectus means *straight*. This tells you that the muscle fibers run in *straight* direction.

- Oblique means *diagonal*. This tells you that the muscle fibers run in a *diagonal* orientation.

- The Lacrimal gland is located on the Lateral side of the eyeball. Use it as a landmark.

Key to Illustration

Lacrimal Apparatus
1. Superior lacrimal punctum (*opening*)
2. Superior lacrimal caniculus
3. Lacrimal sac
4. Inferior lacrimal punctum (*opening*)
5. Inferior lacrimal caniculus
6. Nasolacrimal duct
7. Opening of nasolacrimal duct

Eyeball Muscles
8. Superior oblique muscle
9. Superior rectus muscle
10. Lateral rectus muscle
11. Inferior oblique muscle
12. Inferior rectus muscle
13. Medial rectus muscle

The rectus muscles are positioned like points on a compass (north, south, east, and west).

Lacrimal gland

Medial canthus

Inferior nasal concha

Trochlea

Lacrimal gland

Levator palpebrae superioris muscle (cut)

Note: *Arrows indicate the direction the eye will move when a muscle is contracted.*

Lacrimal Apparatus

1. _____
2. _____
3. _____
4. _____
5. _____
6. _____
7. _____

Eyeball
(anterior view)

8. _____
9. _____
10. _____
11. _____
12. _____
13. _____

Optic nerve

Trochlea

Color different muscles different colors.

Eyeball in Orbit
(lateral view)

Description

The eye is divided into three tunics or sheaths: fibrous tunic, vascular tunic, and neural tunic. The **fibrous tunic** is a thick, tough layer of connective tissue that consists of the **sclera** (*white of the eyeball*) on its posterior portion, which becomes the transparent **cornea** on its anterior portion. The cornea is cup-shaped and filled with a liquid called **aqueous humor**. The **vascular tunic** consists of the vascular, dark-brown **choroid** coat on its posterior portion, which becomes the **ciliary body**, and the **iris** (*colored part of the eye*) on its anterior portion.

The lens of the eye is held in place by **suspensory ligaments** that anchor it to the ciliary body. When the smooth muscle in the ciliary body contracts, it can change the shape of the lens, allowing the eye to focus on near versus distant objects. The iris covers the front of the lens. In the center of the iris is a hole called the **pupil** that allows light to enter the eye. Dilation and constriction of the pupil controls the amount of light that enters the eye. The **neural tunic** is the **retina** and covers only the posterior portion of the eye. This thin layer contains photoreceptors called **cones** and **rods**. Rods help us see in low-light situations, and cones help us see color and give us sharper, clearer images.

Filling the **posterior cavity** of the eye is a jellylike substance called **vitreous humor**. It maintains the normal shape of the eyeball. On the back of the retina is a small disk, called the **macula lutea**, that contains cones but no rods. At the center of this structure is a shallow depression called the **fovea centralis,** which has the highest concentration of cones and gives us our sharpest vision. In contrast, the one region of the retina that lacks photoreceptors is called the **optic disc** (*blind spot*). No images form here. It is located where the optic nerve leaves the eye.

Analogy

The **macula lutea** is like a **target**, and the **fovea centralis** is like the **bull's-eye**.

Study Tip

- To recall the order of the layers in the back of the eye, use the mnemonic: *S*cared *C*ats *R*un. This indicates the layers from outermost to innermost, or *S*clera, *C*horoid, *R*etina.

- Cones are for Color vision.

Key to Illustration

1. Iris	**9.** Ciliary body
2. Cornea	**10.** Posterior cavity (filled with vitreous humor)
3. Edge of pupil	**11.** Central vein
4. Lens	**12.** Central artery
5. Anterior chamber (filled with aqueous humor)	**13.** Optic nerve
6. Posterior chamber (filled with aqueous humor)	**14.** Fovea centralis
7. Anterior cavity	**15.** Macula lutea
8. Suspensory ligaments	**16.** Optic disc (*blind spot*)

To recall the layers on the posterior portion of the eyeball:

| **S**cared **C**ats **R**un! | = | **S**clera **C**horoid **R**etina |

Color all the different layers different colors.

Sclera   Choroid   Retina

The macula lutea is like the target and the fovea centralis is the bull's-eye.

Blood vessels branching off central a. and central v.

1. _____
2. _____
3. _____
4. _____
5. _____
6. _____
7. _____
8. _____
9. _____
10. _____
11. _____
12. _____
13. _____
14. _____
15. _____
16. _____

Description

The ear is divided into three regional areas: external ear, middle ear, and inner ear. The **external ear** extends from the ear itself to the tympanic membrane. The **auricle** is made of elastic cartilage and directs sound waves into the **external auditory meatus**. At the end of this passageway is the delicate **tympanic membrane**, or eardrum. This cone-shaped connective tissue vibrates when sound waves strike it.

The **middle ear** is an air-filled space that extends medial to the tympanic membrane and up to the inner ear. It contains three ear **ossicles**: the **malleus, incus,** and **stapes**. These bones act as a lever system both to transmit and to amplify sound waves from the tympanic membrane to the inner ear. The **auditory tube**, or **Eustachian tube**, connects the middle ear to the nasopharynx. Because this tube is short and horizontal in children, it is easier for bacteria to enter the middle ear, causing middle ear infections. The function of this tube is to equalize pressure on both sides of the tympanic membrane.

The **inner ear** is a bony complex of fluid-filled chambers that contain receptors for both hearing and equilibrium. The receptors for equilibrium are located in the **semicircular canals**, and those for hearing are located in the **cochlea**. As the ear ossicles vibrate, they transfer vibrations to the **oval window** of the inner ear. This sends shock waves through fluid-filled chambers within the cochlea that eventually dissipate out through the **round window** to the air in the **tympanic cavity**. These vibrations stimulate hair cells in the **spiral organ** (*organ of Corti*).

Each hair cell is linked to a nerve fiber. The hair cell transforms this mechanical, vibrational force into an electrical stimulus that carries nerve impulses along the cochlear nerve branch to the temporal lobe in the cerebrum, where it is interpreted as sound.

Analogy

The **tympanic membrane** looks like the **cones** on a stereo speaker. The **malleus** looks like a **hammer**, the **incus** looks like an **anvil**, and the **stapes** looks like a **stirrup**. The **cochlea** of the inner ear looks like a snail shell.

Key to Illustration

1. Auricle	19. Posterior semicircular canal
2. Lobule	20. Stapes in oval window
3. External auditory meatus	21. Round window
4. Tympanic membrane	22. Spiral organ (*organ of Corti*)
5. Tympanic cavity	23. Cochlear duct
6. Malleus	24. Saccule
7. Incus	25. Utricle
8. Stapes	26. Vestibular duct
9. Auditory (*Eustachian*) tube	27. Cochlear duct
10. Semicircular canals	28. Tympanic duct
11. Vestibule	29. Tectorial membrane
12. Vestibular branch of auditory nerve	30. Stereocilia
13. Facial nerve	31. Outer hair cells
14. Cochlear branch of auditory nerve	32. Supporting cells
15. Cochlea	33. Basilar membrane
16. Anterior semicircular canal	34. Inner hair cell
17. Semicircular ducts	35. Nerve fibers of cochlear nerve
18. Lateral semicircular canal	

Malleus looks like a hammer.

Incus looks like an anvil.

The tympanic membrane looks like the cones on a stereo speaker and vibrates like it, too.

Stapes looks like a stirrup.

Middle Ear

Elastic cartilage

Inner ear

External ear

Middle ear

Cochlea looks like a snail shell.

Inner Ear
(internal view)

Nerve fibers

Spiral organ

Cochlea
(cross section)

SPECIAL SENSES: Taste

Description

The surface of the tongue is covered with many small epithelial projections called **papillae**. There are three different types of papillae: **filiform**, **fungiform**, and **circumvallate**. The filiform papillae are located on the tip of the tongue; the fungiform papillae are located posterior to the filiform; and the **circumvallate papillae** are found in a V-shaped strip along the posterior margin of the tongue.

Taste buds are located along the sides of the papillae. Each type of papilla contains a different number of taste buds. In total, the average adult has about 10,000 taste buds, but this number decreases with age. A taste bud is composed of two different types of cells—**gustatory** (*taste*) **cells** and **supporting cells**.

The gustatory cells are modified neurons that have **microvilli** (*taste hairs*) that protrude onto the surface of a papilla. Chemicals in food bind at receptors in these microvilli, which triggers a nervous impulse in the gustatory cells. This follows a nerve pathway to the gustatory cortex in the cerebrum of the brain, where the taste is interpreted.

Analogy

The **fungiform** (*fungus*, mushroom) **papilla** is dome-shaped like a **mushroom cap**, and the **filiform** (*filum*, thread) **papilla** looks like a **flame**.

Key to Illustration

1. Epiglottis	5. Lingual tonsil	9. Taste bud
2. Palatopharyngeal arch	6. Circumvallate papilla	10. Gustatory (*taste*) cells
3. Palatine tonsil	7. Fungiform papilla	11. Supporting cells
4. Palatoglossal arch	8. Filiform papilla	12. Microvilli (*taste hairs*)

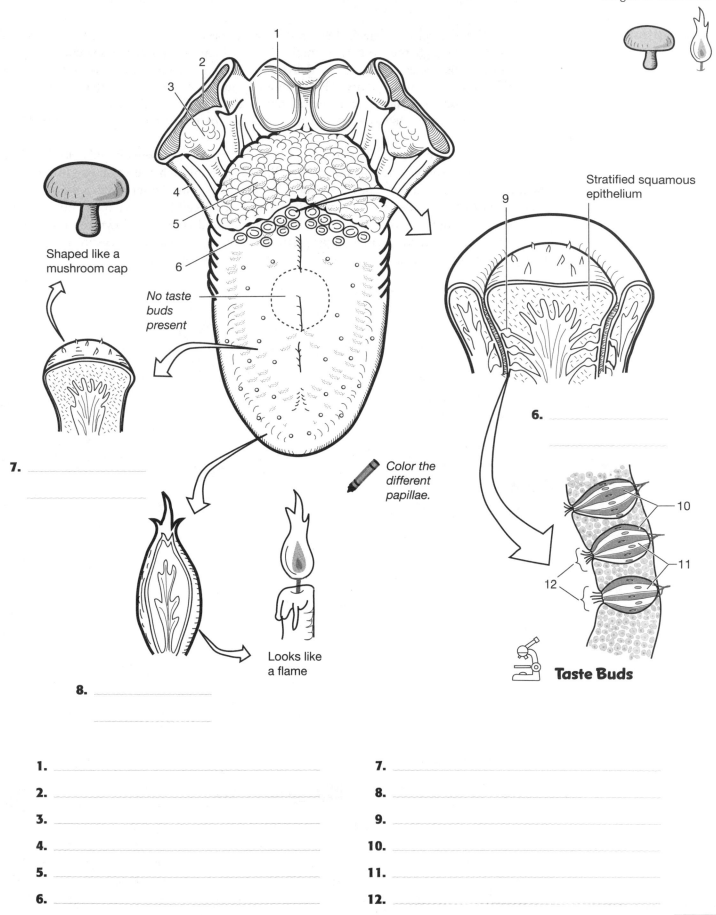

Shaped like a
mushroom cap

Stratified squamous
epithelium

No taste
buds
present

Color the
different
papillae.

Looks like
a flame

Taste Buds

6. _____

7. _____

8. _____

1. _____   7. _____

2. _____   8. _____

3. _____   9. _____

4. _____   10. _____

5. _____   11. _____

6. _____   12. _____

Description

Olfaction refers to the sense of smell. Let's follow what occurs when you smell a cup of coffee. As you inhale air through the nostrils it enters the nasal cavity, and the nasal conchae produce turbulent airflow. This disperses the air and delivers aromatic molecules to the two **olfactory organs** located on the roof of the nasal cavity. These small organs are coated with a thick layer of mucus produced by the **olfactory glands**. The aromatic molecules diffuse through this mucus and bind to a receptor in the numerous **olfactory cilia**. These cilia are extensions of modified neurons called **olfactory receptor cells**.

Once the aromatic molecule binds to the receptor, it triggers a nervous impulse in the olfactory receptor cell. As this impulse travels along the cell, it passes through the **olfactory foramen** in the **cribriform plate** of the ethmoid. Then it reaches the **olfactory bulb**, the terminal portion of the first cranial nerve. Within the olfactory bulb, the olfactory receptor cell forms synaptic connections with other neurons. The impulse is transferred to these neurons and continues down the **olfactory tract**. Finally, the impulses are carried to the appropriate olfactory interpretation areas in the brain. These include regions in both the frontal and the temporal lobes of the cerebrum.

Key to Illustration

1. Olfactory bulb	4. Supporting cells	8. Lamina propria
2. Olfactory tract	5. Olfactory foramen	9. Olfactory epithelium
3. Olfactory receptor cells (neurons)	6. Cribriform plate of ethmoid	10. Mucous layer
	7. Olfactory gland	11. Olfactory cilia

Neuron

Smells like
a French roast

Frontal
sinus

Opening to
nasal cavity

Aromatic
molecules

Upper lip

Palatine process
of maxilla

Pituitary
gland

Sphenoidal
sinus

Opening to
auditory tube

Note:
Large arrows
indicate
direction
of nervous
impulse.

Note: Small arrows
indicate direction
of air flow.

1. _____
2. _____
3. _____
4. _____
5. _____
6. _____
7. _____
8. _____
9. _____
10. _____
11. _____

BLOOD

BLOOD

Cell Types

Description   Blood is a specialized type of connective tissue because it contains cells, fibers, and a liquid ground substance. It is composed of two major parts: **plasma** and **formed elements**. The **plasma** is a straw-colored fluid that contains mostly water, proteins, and other solutes. The **formed elements** consist of the following blood cells and cell fragments scattered in the plasma: **erythrocytes** (*red blood cells*), **leukocytes** (*white blood cells*), and **thrombocytes** (*platelets*).

Formed Elements of Blood

Cell	Description	Function
Erythrocytes (*red blood cells*)	Constitute 99.9 percent of all blood cells; biconcave disks; no nucleus in mature cell; filled with the protein hemoglobin	• Transport O_2 from lungs to body cells • Transport CO_2 from blood to lungs
Leukocytes (*white blood cells*)	Less than 0.1 percent of all blood cells; five different types; **some have visible granules in the cytoplasm, others do not**; nucleus present in all types	• Fight against pathogens, such as bacteria and viruses
Thrombocytes (*platelets*)	Less than 0.1 percent of formed elements; cell fragments; no nucleus; contain enzymes	• Involved in blood clotting

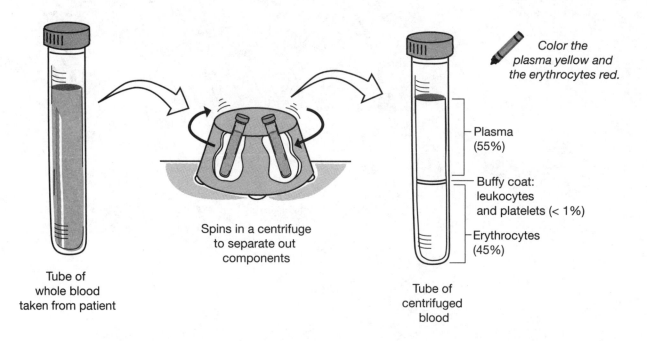

Color the plasma yellow and the erythrocytes red.

Tube of whole blood taken from patient

Spins in a centrifuge to separate out components

Plasma (55%)

Buffy coat: leukocytes and platelets (< 1%)

Erythrocytes (45%)

Tube of centrifuged blood

Key to Illustration		
1. Erythrocytes (*red blood cells*)	**Leukocytes** (*white blood cells*) 2. Monocyte 3. Lymphocyte 4. Neutrophil 5. Eosinophil 6. Basophil	**7. Thrombocytes** (*platelets*)

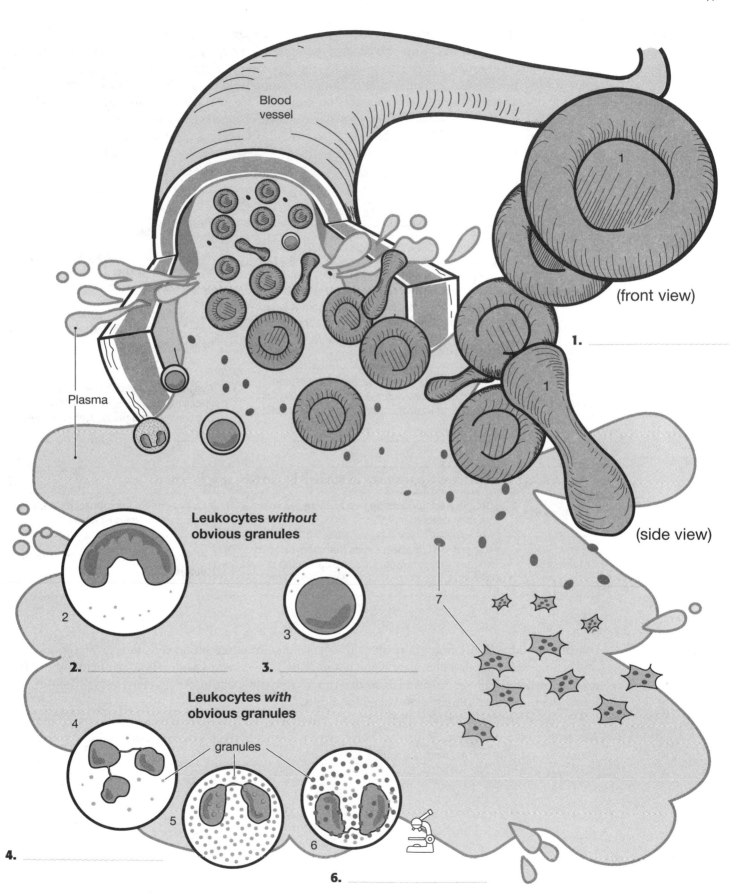

Blood vessel

(front view)

Plasma

(side view)

1. _____

Leukocytes *without* obvious granules

2. _____   **3.** _____

Leukocytes *with* obvious granules

granules

4. _____

6. _____

5. _____

Description

For the purpose of distinguishing one cell type from another in lab, it is handy to divide **leukocytes** (*white blood cells* or *WBCs*) into two groups: **granulocytes** (*contain granules*) and "**agranulocytes**" (*no granules*). These granules are secretory vesicles and lysosomes located in the cytoplasm. Granulocytes contain obvious, visible granules and include **neutrophils**, **eosinophils**, and **basophils**. The term *agranulocytes* is confusing because it means "no granules" when, in fact, granules are present. They are simply smaller and more difficult to see. The "agranulocytes" include **lymphocytes** and **monocytes**.

The general function of all leukocytes is to defend the body against various pathogens, such as bacteria and viruses.

Features and Functions of WBCs

Granulocyte	Features/Appearance in Stained Blood Smear	Functions
Neutrophil	Nucleus has three to five lobes; contains least amount of visible granules; named after the fact that granules are *neutral*, does not stain well	Phagocytic cell; engulfs bacteria and debris in tissues
Eosinophil	Nucleus usually has two lobes; numerous large granules that stain brightly; named after dye used to stain granules red, *eosin* dye; other dyes stain granules different colors	Phagocytic cell; fights parasitic infections; engulfs anything labeled with antibodies; reduces inflammation
Basophil	Nucleus usually masked by deep purple/blue granules; contains many variably sized granules; named after the *basic* stain used to stain granules, hematoxylin	Assists in damaged tissue repair by releasing histamine (vasodilator) and heparin (anticoagulant) from granules

"Agranulocyte"	Features/Appearance in Stained Blood Smear	Functions
Lymphocyte	Round nucleus takes up nearly entire cell volume; slightly larger than RBC	Part of immune response; defends against pathogens or toxins
Monocyte	Largest WBC; nucleus varies from horseshoe to kidney shape; nucleus takes up about half of cell volume	Phagocytic cell; engulfs pathogens and debris in tissues

Study Tip

A **white blood cell differential** measures the percentage of white blood cells in a given blood sample and is a normal part of a complete blood count (CBC). It is used to assess if the white blood cell levels are abnormally high or low and help determine a possible cause. As a reference, it's important to know the normal percentages of white blood cells.

To rank the white blood cells from most common to least common, use the following mnemonic: *N*ever *L*et *M*onkeys *E*at *B*ananas. This gives you the correct order: *N*eutrophils, *L*ymphocytes, *M*onocytes, *E*osinophils, and *B*asophils.

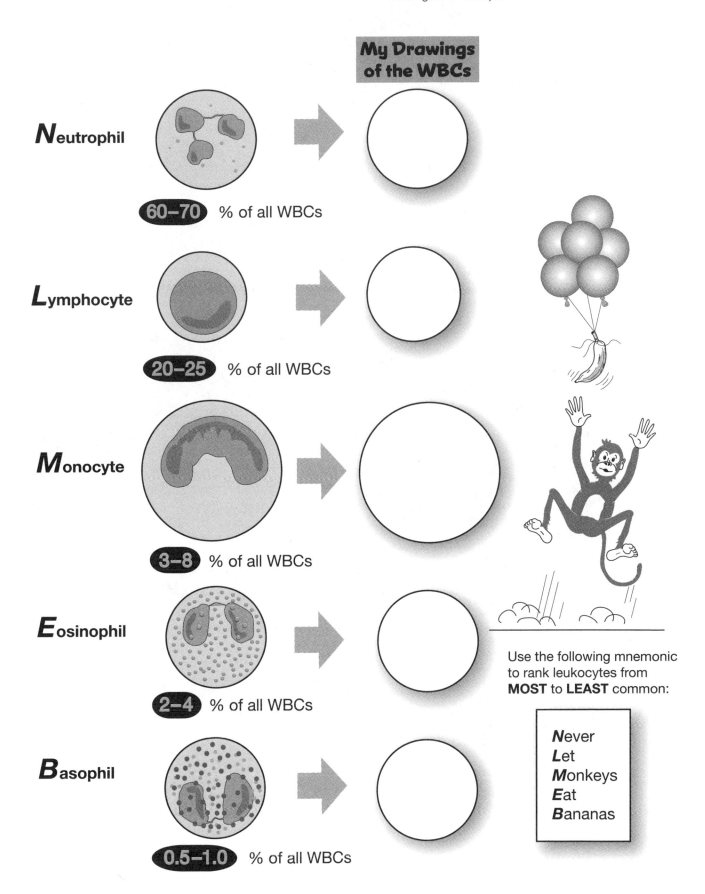

My Drawings of the WBCs

Neutrophil

60–70 % of all WBCs

Lymphocyte

20–25 % of all WBCs

Monocyte

3–8 % of all WBCs

Eosinophil

2–4 % of all WBCs

Basophil

0.5–1.0 % of all WBCs

Use the following mnemonic to rank leukocytes from **MOST** to **LEAST** common:

Never
Let
Monkeys
Eat
Bananas

Notes

CARDIOVASCULAR SYSTEM

CARDIOVASCULAR SYSTEM

Overview

Description

The **cardiovascular system** consists of the **heart** and all the **blood vessels**. Functionally, the heart is like a double pump with each pump connected to the other through a long series of blood vessels. It consists of two receiving chambers called **atria** and two pumping chambers called **ventricles**. The left side of the heart always pumps oxygenated blood, and the right side always pumps deoxygenated blood.

The illustration on the facing page shows blood flow through the heart, through the **pulmonary circuit**, and through the **systemic circuit**. The pulmonary circuit refers to all the blood vessels that take deoxygenated blood from the right ventricle of the heart to the lungs, and then return oxygenated blood to the left atrium. After this oxygenated blood is pumped from the left atrium to the left ventricle, it is pumped out to the rest of the body. The blood vessels that transport this oxygenated blood to the body are part of the systemic circuit.

All gas exchange occurs within **capillaries**. Capillaries are microscopic blood vessels only one cell layer thick. Their wall is made of **simple squamous epithelium**. These flat cells easily permit the diffusion of gases, such as **oxygen** (O_2) and **carbon dioxide** (CO_2). Oxygen diffuses out of the blood and into body cells to be used in the process of cellular respiration. Carbon dioxide is a normal by-product of cellular respiration and gradually builds up within body cells. Carbon dioxide diffuses from the body cells into the capillary.

Beginning in the right atrium, this is a flowchart for the blood flow:

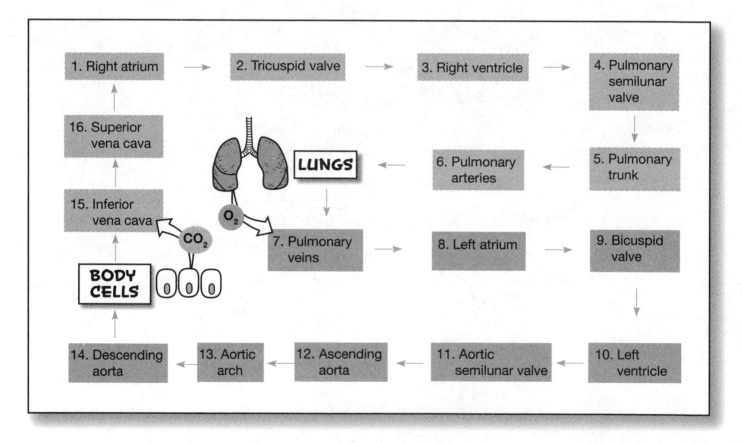

Study Tip

In this pathway that begins and ends with the right atrium, here is a handy tip to recall which atrioventricular valve—bicuspid or tricuspid—comes first, and which comes second, in the pathway.

You ride your **tricycle** (**tricuspid**) before your **bicycle** (**bicuspid**).

Systemic circuit

Capillaries in head, neck, upper limbs

CO_2

O_2

Body cells

Capillaries in right lung

Pulmonary circuit

Capillaries in left lung

Pulmonary circuit

CO_2

O_2

CO_2

O_2

6   13   6

16   7   7

12   5   7   7

8

1   11

4   9

2   10

3

Right lung

Left lung

15

Labeling

Label all the structures (1–16) from the flowchart on the facing page.

Color all blood vessels and heart chambers either red for oxygenated blood or blue for deoxygenated blood.

Body cells

O_2

14

Capillaries in chest, abdomen, lower limbs

CO_2

KEY

Oxygenated blood

Deoxygenated blood

Systemic circuit

Description

Coronary circulation refers to the blood supply to the heart. The **coronary arteries** supply oxygenated blood to the heart, and the **cardiac veins** carry deoxygenated blood back to the heart. The following flowchart summarizes coronary circulation through the blood vessels.

Base of aorta → left and right coronary arteries → branches of coronary arteries

(circumflex a., anterior interventricular a., marginal a., posterior interventricular a.) →

coronary capillaries → cardiac veins → coronary sinus → right atrium

When **coronary arteries** become blocked, the blood supply to the heart is reduced. This deprives cardiac muscle cells of oxygen. If this blockage persists over many years, it may lead to a **myocardial infarction** (*heart attack*).

Analogy

A **sulcus** is like a **shallow groove** or **gulley**.

Study Tip

To distinguish the anterior from the posterior view of the heart, use the **coronary sinus** as a landmark for the posterior view. It is often difficult to see on a dissected specimen because it is normally covered with a horizontal band of fatty tissue. Good landmarks for the anterior view include the **pulmonary trunk, anterior interventricular artery, circumflex artery**, and **ascending aorta**.

Key to Illustration

Blood vessels (B)

B1 Superior vena cava
B2 Inferior vena cava
B3 Ascending aorta
B4 Aortic arch
B5 Descending aorta
B6 Brachiocephalic trunk
B7 L. common carotid a.
B8 L. subclavian a.
B9 Pulmonary trunk
B10 Pulmonary arteries
B11 Pulmonary veins
B12 R. coronary a. (in coronary sulcus)
B13 L. coronary a.

B14 Circumflex a.
B15 Anterior interventricular a. (in anterior interventricular sulcus)
B16 Marginal a.
B17 Anterior cardiac v.
B18 Great cardiac v.
B19 Small cardiac v.
B20 Coronary sinus
B21 Posterior v. of l. ventricle
B22 Middle cardiac v.
B23 Posterior interventricular a. (in posterior interventricular sulcus)

Structures (S)

S1 Ligamentum arteriosum
S2 Apex (tip) of heart
S3 Auricle of r. atrium
S4 Auricle of l. atrium

Chambers (C)

C1 R. atrium
C2 L. atrium
C3 R. ventricle
C4 L. ventricle

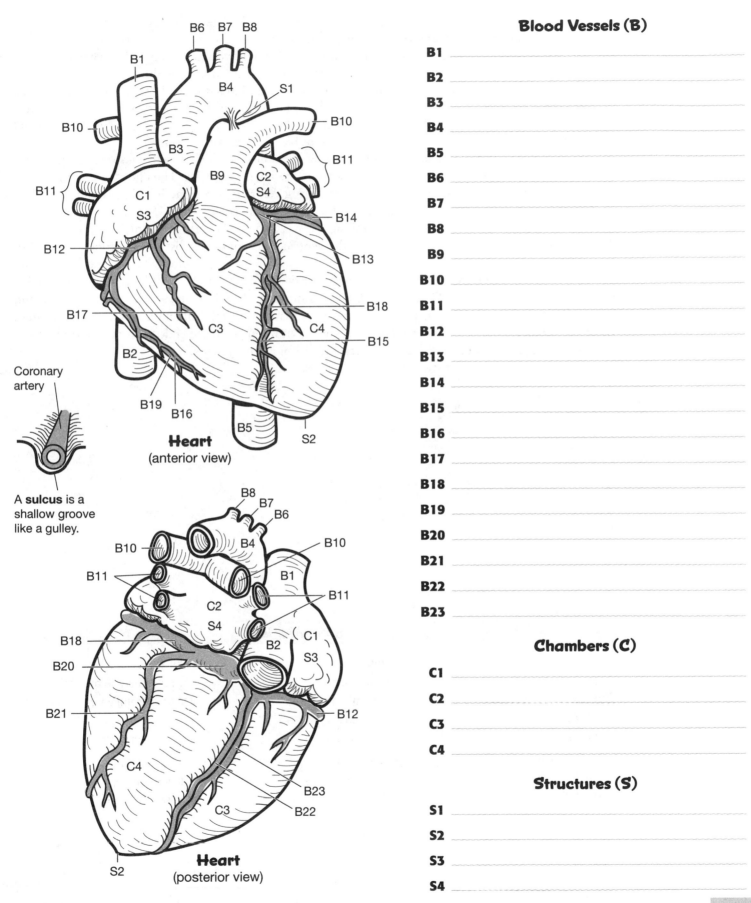

Coronary artery

A **sulcus** is a shallow groove like a gulley.

Heart
(anterior view)

Heart
(posterior view)

Blood Vessels (B)

B1 _____
B2 _____
B3 _____
B4 _____
B5 _____
B6 _____
B7 _____
B8 _____
B9 _____
B10 _____
B11 _____
B12 _____
B13 _____
B14 _____
B15 _____
B16 _____
B17 _____
B18 _____
B19 _____
B20 _____
B21 _____
B22 _____
B23 _____

Chambers (C)

C1 _____
C2 _____
C3 _____
C4 _____

Structures (S)

S1 _____
S2 _____
S3 _____
S4 _____

Description

The heart is divided into left and right halves and has four chambers, two atria and two ventricles. The **atria** are the first chambers to receive blood from the body. They fill with blood, contract, and transfer blood to the pumping chambers, or **ventricles**. The **right ventricle** pumps deoxygenated blood to the lungs, and the **left ventricle** pumps oxygenated blood to the rest of the body. The heart has two different types of valves: **atrioventricular (AV) valves** and **semilunar valves**. The AV valves are located between the atria and the ventricles. The one on the right side of the heart has three valve flaps, so it is called the **tricuspid valve**, and the one on the left side has two valve flaps, so it is called the **bicuspid (mitral) valve**. These valves permit a one-way flow of blood from atria to ventricles.

Long, fibrous, cord-like structures called **chordae tendineae** anchor the valve flaps to the **papillary muscles**—long, cone-shaped, muscular extensions of the inner ventricles. The chordae tendineae and papillary muscles help keep the AV valves closed during ventricular contraction. The semilunar valves are located at the base of each major artery that leaves each ventricle.

On the right side is the pulmonary semilunar valve, and on the left is the aortic semilunar valve. These valves prevent backflow of blood into the ventricles. From outermost to innermost, the wall of the heart is made of three layers: epicardium, myocardium, and endocardium. The **epicardium** (*visceral pericardium*) is made of fibrous connective tissue and is the innermost layer of the pericardial sac that surrounds the heart. The **myocardium** is composed of multiple layers of cardiac muscle and many blood vessels and nerves. The **endocardium** is made of simple squamous epithelium and lines the inside of all the heart chambers and valves. It is continuous with the endothelium of blood vessels that enter and exit the heart, such as the aorta and the pulmonary veins.

Key to Illustration

Blood Vessels (B)
- B1. Superior vena cava
- B2. Inferior vena cava
- B3. Ascending aorta
- B4. Aortic arch
- B5. Descending aorta
- B6. Brachiocephalic trunk
- B7. L. common carotid a.
- B8. L. subclavian a.
- B9. Pulmonary trunk
- B10. Pulmonary arteries
- B11. Pulmonary veins

Chambers (C)
- C1. R. atrium
- C2. L. atrium
- C3. R. ventricle
- C4. L. ventricle

Structures (S)
- S1. Ligamentum arteriosum
- S2. Interventricular septum
- S3. Chordae tendineae
- S4. Papillary muscle
- S5. Apex

Valves (V)
- V1. Pulmonary semilunar
- V2. Aortic semilunar
- V3. Tricuspid leaflet
- V4. Bicuspid leaflet

Wall Layers (W)
- W1. Epicardium
- W2. Myocardium
- W3. Endocardium

Each **semilunar valve flap** opens like a **kangaroo pouch** to catch blood and prevent it from falling back into the ventricle.

The **semilunar valves** resemble a modified **peace sign** when closed.

To distinguish between the left and right ventricles, note that the wall of the **left ventricle** is thicker than the wall of the right ventricle.

The atrioventricular valves and associated structures resemble a parachute. The **valve flaps** are the **parachute,** the **parachute cords** are the **chordae tendineae,** and the **paratrooper** is the **papillary muscle.**

KEY for COLORED ARROWS

◼	Oxygenated blood
◼	Deoxygenated blood

345

Description

The **heart** has its own internal regulation system to achieve two important functions: (1) triggering the heartbeat, and (2) coordinating the timing between contraction of the atria and contraction of the ventricles. This control system is referred to as the **intrinsic conduction system**. Without this control system to ensure that the heart chambers completely fill with blood before contracting, the heart would be a very inefficient pump.

The intrinsic conduction system consists of the following six structures:

1. Sinoatrial (SA) node
2. Internodal pathway
3. Atrioventricular node
4. Atrioventricular (AV) bundle
5. Bundle branches (right and left)
6. Purkinje fibers

This system is a network of interconnecting cardiac muscle cells that spread through the atria and the ventricles. Notice that this system can be structurally divided into two main parts: the **nodes** (SA and AV nodes) and the **pathway** (internodal pathway, AV bundle, bundle branches, and Purkinje fibers). Let's examine each, in turn.

The nodes consist of clusters of specialized cardiac muscle cells that contain very few contractile proteins—myosin and actin—found in normal cardiac muscle cells. These cells are the body's only autorhythmic muscle cells. Instead of contracting, they serve as a kind of "spark plug," or stimulus, to establish the heartbeat. The **SA node** is located within the wall of the right atrium and is referred to as the *primary pacemaker*. The **AV node** is located in the septum between the two atria and is referred to as the *secondary pacemaker*.

A long network of specialized cardiac muscle cells—the pathway—ensures that the heart chambers are stimulated to contract in a coordinated manner. Like smooth muscle cells, cardiac muscle cells can stimulate adjacent cells. Think of the pathway like a series of dominoes. Knocking over the first domino will cause all the others to fall over. Similarly, because cardiac muscle cells interdigitate with each other like pieces of a jigsaw puzzle, stimulating the first cell will quickly stimulate all the others in the pathway.

The **internodal pathway** radiates from the SA node and extends to the left and right atria and AV node. The **AV bundle** is a relatively short segment of cells that extends from the AV node and penetrates into the top of the interventricular septum. It's the only electrical connection between the atria and the ventricles. From here, it splits into two pathways—the right and left bundle branches, which extend through the interventricular septum toward the apex of the heart. The bundle branches extend into the walls of their respective ventricles and into the papillary muscles to become the Purkinje fibers.

Impulse Pathway

The flow of the impulse for contraction always moves in the following sequence:

SA node ⟶ internodal pathway ⟶ AV node ⟶ AV bundle ⟶
bundle branches ⟶ Purkinje fibers

Let's examine this sequence with respect to the coordination of the filling and contracting of the atria and the ventricles. First, the autorhythmic cells within the SA node trigger the impulse to spread to the left and right atria through the internodal pathway. Simultaneously, the atria fill with blood and expand. The impulse causes the atria to contract, which forces blood into the ventricles. A delay in ventricular contraction is needed to allow the ventricles to fill with blood. This delay comes in the form of the time it takes to stimulate the AV node and send the impulse down the AV bundle and bundle branches. By the time the impulse has spread to the Purkinje fibers, the ventricles have finished filling with blood, and the ventricles are stimulated to contract.

Intrinsic Conduction System

Specialized cardiac muscle cells in the SA node are like the spark plug that establishes the heartbeat.

The intrinsic conduction system is an **ELECTRICAL EVENT** that leads to the contracting of the heart chambers, a mechanical event

Impulse Pathway

1 _____
↓
2 _____
↓
3 _____
↓
4 _____
↓
5 _____
↓
6 _____

1 Sinoatrial (SA) node

2 Internodal pathway

3 Atrioventricular (AV) node

4 Atrioventricular (AV) bundle

5 Bundle branches (right and left)

6 Purkinje fibers

Papillary muscle

Note: Colored arrows indicate direction of impulse.

Apex

*The **intrinsic conduction system** is shown in white. Color it all one color.*

347

Description
An electrocardiogram (ECG or EKG) is a graph of the heart's electrical activity as expressed in **millivolts (mV)** over time. The instrument used to obtain an ECG is called an **electrocardiograph**. A person is connected to the electrocardiograph with electrodes placed on the arms and legs (*limb leads*) and along the chest (*chest leads*). Each of the limb leads gives a slightly different picture of the heart's electrical activity. An ECG is used to detect if the electrical conduction pathway within the heart is normal and if any damage has been done to the heart.

In a typical lead II recording, three different waves appear: **P, QRS complex,** and **T.** Each wave represents an electrical event called a *depolarization* or a *repolarization*. These electrical events stimulate cardiac muscle within the heart wall to either contract or relax. Consequently, these events lead to the contraction and relaxation of the heart chambers—atria and ventricles.

- **P wave:** *atrial depolarization*—at the end of the P wave, both atria have depolarized, which causes the atria to contract.

- **QRS complex:** *ventricular depolarization*—at the end of the QRS complex, both ventricles have depolarized, which causes the ventricles to contract. **Note:** Atrial repolarization also occurs during this period, but it is masked by the ventricular depolarization.

- **T wave:** *ventricular repolarization*—at the end of the T wave, both ventricles have repolarized, which causes the ventricles to relax.

Reading an ECG is an art that requires a lot of expertise. Even so, we can explain the general way in which they are interpreted. Two types of variations may signal abnormalities:

1. Variation in *wave height* (they may be *elevated* or *depressed*)

2. Variation in normal *time intervals*

Let's consider changes in wave height. For example, if a P wave is elevated, it may indicate atrial enlargement. If the QRS complex is elevated, it may indicate ventricular enlargement. A tall and pointed T wave may indicate myocardial ischemia.

The following three **normal time intervals** are examined on any ECG:

- **P–R interval** = time from beginning of P wave to start of QRS complex; 0.2 sec.

- **S–T segment** = time from end of S wave to beginning of T wave; 0.1 sec.

- **Q–T segment** = time from beginning of QRS complex to end of T wave; 0.4 sec.

Any time interval longer or shorter than normal also may indicate an abnormality. For example, a P–Q interval much longer than normal indicates a blockage in the normal conduction pathway. This could be caused by scar tissue resulting from a childhood infection of rheumatic fever. Another example is a longer than normal Q–T segment, which may indicate myocardial damage.

Normal Electrocardiogram or ECG (Lead II)

QRS Complex

R

Electrocardiograph

S–T segment

P wave

T wave

Q

P–R interval

S

Q–T interval

Millivolts (mV)

Time (seconds)

Color the P wave, QRS complex, and T wave different colors.

Description

The heart is essentially two pumps that work together as one synchronized unit. The left side pumps only oxygenated blood, and the right side pumps only deoxygenated blood. The **cardiac cycle** refers to all the pumping actions that occur within the heart during one entire heartbeat. It consists of both the atria and the ventricles filling with blood and then contracting. It begins with contraction of the atria and ends with refilling of the atria. On average, this continuous cycle takes about 800 msec. to complete in an adult.

The atria and the ventricles have repeated patterns of contraction (systole) and relaxation (diastole). The atria function as the receiving chambers of blood from the body. After they fill with blood, they contract and force blood downward into the true pumps—the more powerful ventricles. Because these chambers have the more difficult task of pumping the blood to the body, they have a thicker layer of cardiac muscle in their walls *and a larger volume*. Recall that the heart has two pairs of valves: atrioventricular (AV) valves and semilunar valves. The AV valves function as one-way valves to permit blood to flow from the atria to the ventricles. The closing of the semilunar valves prevents blood from flowing back into the ventricles after being ejected.

The electrocardiogram (ECG) is related to the cardiac cycle. An ECG measures electrical changes in the heart muscle that induce the contractions of the chambers. **These contractions lead to pressure changes within the chambers, which, in turn, induce the heart valves to either open or close.**

Steps

To understand the sequence of events, the cardiac cycle has been divided into five steps, as follows:

(1) Atrial contraction
- **ECG connection:** from P wave to Q wave
- **AV valves** *open*
- **Semilunar valves** *closed*
- **Description:** The left and right atria contract simultaneously, causing the atrial pressure to increase. This forces blood through the AV valves and into the ventricles. The ventricles are relaxed and filling with blood. The semilunar valves are closed because the pressure in the ventricles is too low to force them open.

(2) Isovolumetric ventricular contraction
- **ECG connection:** begins with R wave
- **AV valves** *closed*
- **Semilunar valves** *closed*
- **Description:** The ventricles begin contracting in this phase, causing the ventricular pressure to increase. Higher ventricular pressure relative to the atria closes the AV valves. The volume of blood in the ventricles remains constant (isovolumetric). All heart valves are closed.

(3) Ventricular ejection
- **ECG connection:** from S wave to T wave
- **AV valves** *closed*
- **Semilunar valves** *open*
- **Description:** The ventricles continue contracting during this phase. Like wringing out a wet rag, the ventricles wring the blood out, beginning at the apex and moving upward. This allows for the maximum volume of blood to be ejected—about 70 mL per ventricle. The high ventricular pressure forces the semilunar valves to open, and blood is ejected into the pulmonary arteries and the aorta.

(4) Isovolumetric ventricular relaxation
- **ECG connection:** begins at end of T wave
- **AV valves** *closed*
- **Semilunar valves** *closed*
- **Description:** The ventricles are relaxing, and the heart valves are all closed. The volume of blood in the ventricles remains constant (isovolumetric). Because ventricular pressures are higher than atrial pressures, no blood flows into the ventricles. Ventricular pressure is quickly decreasing during this phase.

(5) Passive ventricular filling
- **ECG connection:** after T wave to next P wave
- **AV valves** *open*
- **Semilunar valves** *closed*
- **Description:** As blood flows into the atria, the atrial pressure increases until it exceeds the ventricular pressure. This forces the AV valves open, and blood fills the ventricles. This is the main way the ventricles are filled. At the end of this phase, the ventricles will be about 70–80% filled.

Cardiac Cycle

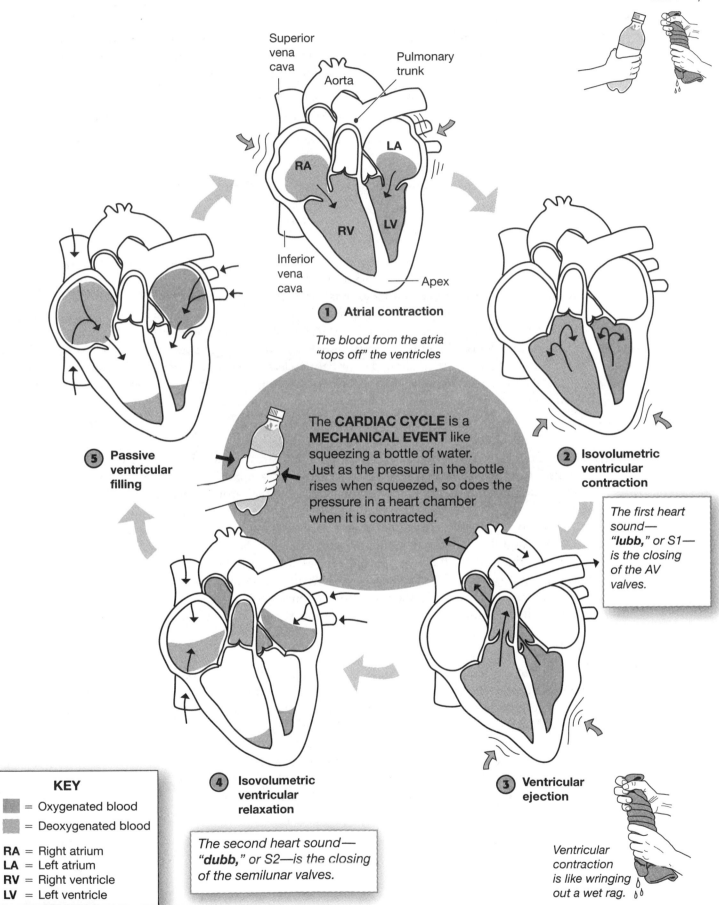

Superior vena cava

Aorta

Pulmonary trunk

LA

RA

LV

RV

Inferior vena cava

Apex

① Atrial contraction

The blood from the atria "tops off" the ventricles

The **CARDIAC CYCLE** is a **MECHANICAL EVENT** like squeezing a bottle of water. Just as the pressure in the bottle rises when squeezed, so does the pressure in a heart chamber when it is contracted.

② Isovolumetric ventricular contraction

*The first heart sound— "**lubb**," or S1— is the closing of the AV valves.*

⑤ Passive ventricular filling

④ Isovolumetric ventricular relaxation

③ Ventricular ejection

Ventricular contraction is like wringing out a wet rag.

*The second heart sound— "**dubb**," or S2—is the closing of the semilunar valves.*

KEY

	= Oxygenated blood
	= Deoxygenated blood

RA = Right atrium
LA = Left atrium
RV = Right ventricle
LV = Left ventricle

Description

In the central nervous system (CNS), the **cardiovascular (CV) center** in the **medulla oblongata** is the command-and-control center for regulating heart function. It uses reflex pathways to control heart rate. The three major peripheral sensory receptors that provide input to the CV center are:

1. **Proprioceptors**—measure tension changes in muscles and joints

2. **Chemoreceptors**—detect changes in blood acidity by sensing changes in CO_2 and H^+ levels

3. **Baroreceptors**—located in the carotid sinus, aortic arch, and other arteries; detect changes in blood pressure

The CV center sends its motor output to the heart through two different nerves:

1. **Vagus nerve**—to decrease heart rate

2. **Cardiac accelerator nerve**—to increase heart rate

The internal pacemakers within the heart—the **SA node** and **AV node**—help set its normal rate and rhythm. But the heart has to be able to respond to various stimuli so it can increase or decrease heart rate as needed. For example, during exercise, the heart rate increases, which, in turn, elevates cardiac output, thereby providing more oxygen and glucose to skeletal muscle tissue. As you exercise, proprioceptors detect increased tension in skeletal muscles, and chemoreceptors detect increased levels of CO_2 in the blood. This triggers nerve impulses to be sent along sensory neurons to the CV center. Motor output then is carried along the cardiac accelerator nerve to stimulate the heart rate to increase.

Now let's use the variable of blood pressure changes to follow the reflex pathway shown in the illustration for both the parasympathetic and sympathetic divisions.

Parasympathetic Control

The **vagus nerve** (cranial nerve X) is part of the parasympathetic division of the ANS and sends impulses to the heart to decrease the heart rate. It is a mixed nerve—meaning that it contains both sensory and motor neurons. Baroreceptors, located in the carotid sinus, aortic arch, and other arteries, detect changes in blood pressure.

As blood pressure increases above normal levels, impulses are sent along sensory neurons to the CV center and then carried along motor neurons within the vagus nerve to the heart. The vagus nerve innervates the heart at the SA node and AV node. It functions to inhibit the heart, thereby reducing the heart rate and strength of contractions. This, in turn, decreases cardiac output, which lowers blood pressure back to normal. At the same time, impulses also are sent out along vasomotor nerves (not shown in illustration) that stimulate vasodilation, thereby further lowering the blood pressure.

Sympathetic Control

The **cardiac accelerator nerve** is part of the sympathetic division of the ANS and sends impulses to the heart to increase the heart rate. As blood pressure decreases below normal levels, this is detected by baroreceptors that send impulses along a neural network over to the CV center, then to the spinal cord and to the heart. The final nerve in this pathway—the **cardiac accelerator nerve**—innervates the heart at the SA node, AV node, and cardiac muscle in the ventricular wall (ventricular myocardium). The response is that both the heart rate and strength of contractions increase. This, in turn, increases cardiac output, which increases blood pressure. At the same time, impulses are sent out along vasomotor nerves (not shown in illustration) that stimulate vasoconstriction and thereby further increase blood pressure.

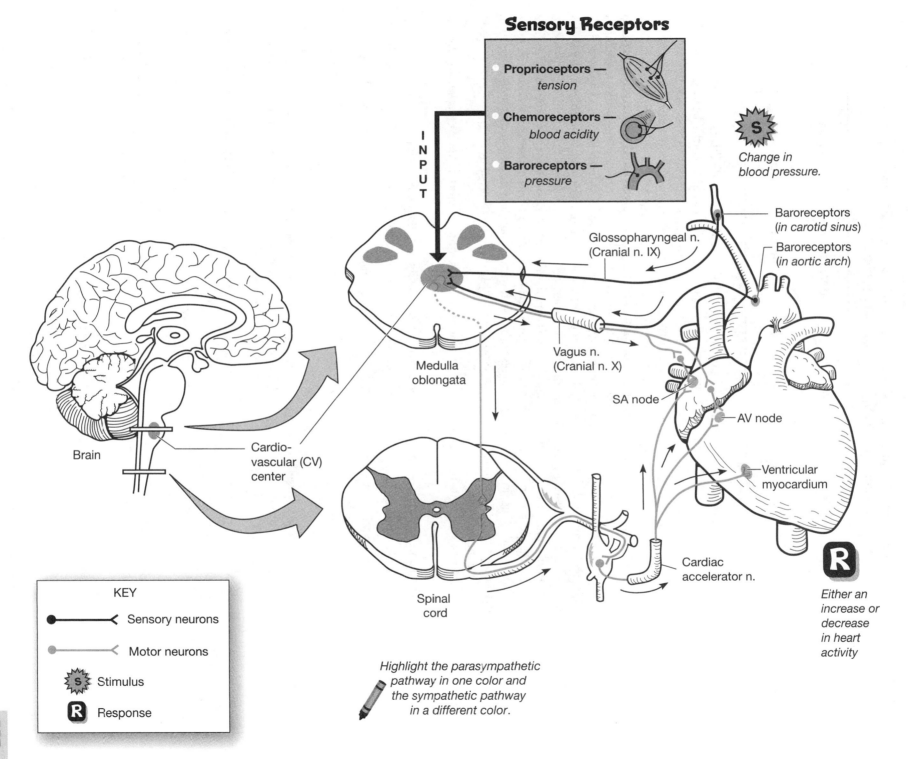

Sensory Receptors

- **Proprioceptors** — *tension*
- **Chemoreceptors** — *blood acidity*
- **Baroreceptors** — *pressure*

INPUT

S *Change in blood pressure.*

Baroreceptors (*in carotid sinus*)

Baroreceptors (*in aortic arch*)

Glossopharyngeal n. (Cranial n. IX)

Vagus n. (Cranial n. X)

SA node

AV node

Ventricular myocardium

Medulla oblongata

Brain

Cardio-vascular (CV) center

Spinal cord

Cardiac accelerator n.

R *Either an increase or decrease in heart activity*

KEY

● ───< Sensory neurons

● ───< Motor neurons

S Stimulus

R Response

Highlight the parasympathetic pathway in one color and the sympathetic pathway in a different color.

CARDIOVASCULAR SYSTEM: Blood Vessels—Structure and Function

Arteries, Arterioles, Capillaries, Venules, and Veins

Description

The body has five fundamental types of blood vessels: *arteries, arterioles, capillaries, venules,* and *veins*. All of them connect in the following predictable pattern:

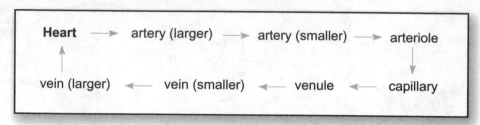

Arteries always carry blood away from the heart. They are thicker-walled than veins because the blood within them is at a higher pressure. All **veins** always carry blood back to the heart. Because the pressure within them is lower, they are thinner-walled. Larger veins contain valves—similar to one-way semilunar valves in the heart—that assist the low-pressure venous blood return to the heart. When a person is standing, gravity works against venous blood flow. Like a staircase breaks up the distance between the first and second floors into smaller steps, valves break up larger veins into smaller segments. Arteries and veins connect at the microscopic level by capillary networks. **Capillaries** are the smallest blood vessels in the body and are important functionally because gas and fluid exchange occurs here. Their entire wall is often a single cell layer in thickness. **Arterioles** and **venules** are microscopic vessels that feed and drain capillaries, respectively. Nearest the capillaries, they are structurally similar to a capillary except they have small amounts of smooth muscle around them. The smooth muscle around an arteriole can be stimulated to either constrict (*vasoconstriction*) or relax (*vasodilation*). This, in turn, causes changes in blood flow to a capillary and systemic blood pressure. Vasoconstriction of arterioles decreases blood flow and increases blood pressure, and vasodilation does just the opposite.

Arteries and veins have three major layers in their walls: *tunica externa, tunica media,* and *tunica interna*:

- **Tunica externa**—connective tissue layer made mostly of collagen fibers
- **Tunica media**—layers of smooth muscle and some elastic fibers
- **Tunica interna**—endothelial layer (simple squamous epithelium) with underlying loose connective tissue

Study Tips

- Don't confuse yourself by trying to distinguish between arteries and veins as to whether they carry oxygenated blood or deoxygenated blood. *This does not work* because some arteries/veins carry oxygenated blood and others carry deoxygenated blood.

- To recall the general function of arteries, use the phrase: *Arteries Away!* **Arteries** always carry blood *away* from the heart.

- To recall one structural difference between arteries and veins: *Veins have Valves* (arteries do not have valves).

- To distinguish an artery from a vein under the microscope, remember that **arteries** always have a **thicker tunica media**.

Key to Illustration		
1. Tunica interna (*tunica intima*)	**3.** Tunica externa (*tunica adventitia*)	**4.** Lumen of blood vessel
2. Tunica media		

CARDIOVASCULAR SYSTEM: Blood Vessels—Structure and Function

Arteries, Arterioles, Capillaries, Venules, and Veins

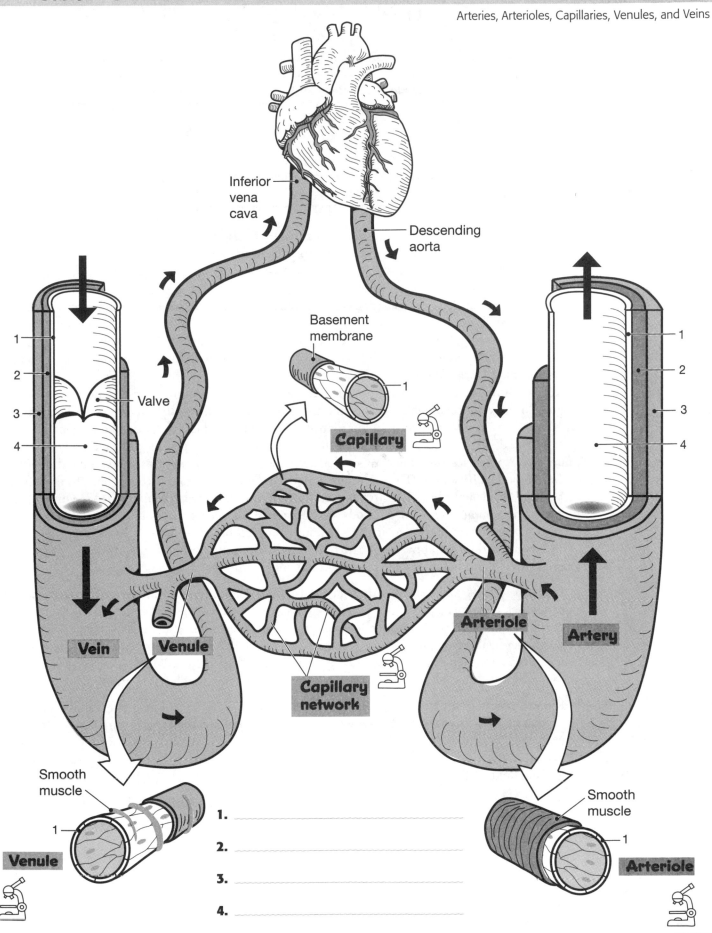

Inferior vena cava

Descending aorta

Basement membrane

Valve

Capillary

Vein

Venule

Capillary network

Arteriole

Artery

Smooth muscle

Venule

Smooth muscle

Arteriole

1. _____

2. _____

3. _____

4. _____

Description   The illustration gives an overview of the general pattern of circulation. Blood always follows a predictable circuit through blood vessels. There are five fundamental types of blood vessels in the body: *arteries, arterioles, capillaries, venules,* and *veins*. All of them connect in the following pattern:

The schematic illustration on the facing page shows three of these five: *arteries, veins,* and *capillaries*. **Arteries** always carry blood away from the heart. They are thicker-walled than veins because the blood within them is at a higher pressure. **As distance from the heart increases, blood pressure decreases.** All **veins** always carry blood back to the heart. Because the pressure within them is lower, they are thinner-walled. Arteries and veins connect together at the microscopic level by capillary networks. **Capillaries** are the smallest blood vessels in the body and are very important functionally because gas exchange and fluid exchange occurs here. Oxygen exits the blood to be used by body cells, and carbon dioxide enters the blood from cells. The liquid plasma is filtered out of the blood to become **interstitial fluid** (*tissue fluid*).

Let's follow the general pattern of circulation. Veins carrying low-pressure, deoxygenated blood drain into the **vena cava**, which drains into the heart's **right atrium** (**RA**). This receiving chamber fills with blood, contracts, and forces blood into the **right ventricle** (**RV**). All this deoxygenated blood is then pumped out of the right ventricle to go to the lungs to get oxygenated. In the lungs, the pulmonary capillaries function only for gas exchange, in which oxygen diffuses into the blood and carbon dioxide diffuses out. The oxygenated blood is then transported through veins to the **left atrium** (**LA**). The LA fills with blood, contracts, and forces blood into the **left ventricle** (**LV**). This oxygenated blood is then pumped out to the body via the **aorta**. The heart feeds its own cardiac muscle first through **coronary capillaries** so it can continue pumping blood every minute of every day. Arteries carry oxygenated blood above the heart to the capillaries in the brain, trunk, and upper limbs. Other arteries carry blood below the heart to the following major areas:

- **Digestive organs and spleen**—After gas and fluid exchange occurs at the **splenic** and **mesenteric capillaries**, deoxygenated blood is carried by veins to the **hepatic portal system** in the liver. Note that capillaries in this system are not for the typical purpose of gas and fluid exchange. Instead, these highly permeable capillaries are specialized for delivering nutrients absorbed by the digestive tract to liver cells before entering the general circulation. The liver cells serve as special processing centers that perform many functions. For example, they store glucose as glycogen and detoxify alcohols.

- **Kidney**—Another unique group of permeable capillaries is the **glomerular capillaries**. Like the capillaries in the hepatic portal system, these are also not for the typical purpose of gas and fluid exchange. Instead, they are specialized to filter only the blood plasma, place it in a separate tubular system, and process this liquid into urine. These capillaries lead into the **peritubular capillaries**, where gas and fluid exchange does occur.

- **Gonads**—In the male, gas and fluid exchange occurs at capillaries in the testes, whereas in the female, this occurs at capillaries in the ovaries.

- **Liver, lower limbs**

Brain

Capillary (c.)

Trunk, upper limbs

Lungs

Pulmonary c.

Vena Cava *(large veins)*

Aorta *(large artery)*

RA   LA

RV   LV

Color the uncolored portion of the illustration blue.

Spleen

Splenic c.

Coronary c.

Liver

Heart

Hepatic portal system ✱

Mesenteric c.

Digestive organs

Kidney

Peritubular c.   Glomerular c.✱

Gonads

Liver

Lower limbs

LOWER Presure Side

HIGHER Presure Side

✱ *Note: These are specialized capillaries.*

Description

Fluid flow through any tube—whether water through a garden hose or blood through a vessel—follows a pattern. To understand **blood flow**, you must think of the fluid not as a single unit but as moving in concentric layers. The two types of flow patterns are: (1) **laminar flow**, and (2) **turbulent flow**. As another clinical example, turbulent blood flow is heard moving through a defective heart valve as a "murmur."

Types of Blood Flow

(1) **Laminar** (*lamina* = layer) **flow** is the normal flow pattern in healthy blood vessels.

The concentric layers of fluid flow together, but not at the same rate. As shown in the illustration, the outer layers travel more slowly than the inner layers. Why? As the fluid in the outermost layer rubs against the vessel wall, it is hindered by the force of friction and moves the slowest. In contrast, the fluid in the center of the tube is interfered with the least, so it moves the fastest. Normal, healthy blood vessels are smooth on their inner surface. Like the smooth surface on the inside of a garden hose or a drainpipe, this allows the fluid to flow through most easily.

(2) **Turbulent flow** is the result of any disruption in the normal, laminar flow.

Some low levels of turbulent flow are normal as blood flows through all the vessels in the body, but high levels may indicate an abnormality. What causes it? Two things: (1) physical change in a vessel (*ex.:* constriction, sharp turn, or narrowing of a vessel) or (2) Disease state (*ex.:* atherosclerosis). Because large blood vessels connect to smaller ones, some narrowing is normal, but it is a gradual change that doesn't usually result in turbulent flow.

Let's consider an abnormality like the constriction shown in the illustration. As fluid strikes the wall of the constriction, its normal, linear path is disrupted. The result is the generation of swirling little currents—like backflows.

As a clinical example, the carotid arteries are the major vessels that deliver oxygenated blood to the brain. If a person has atherosclerosis, these arteries may become blocked because of plaque deposits. This can lead to a stroke or even death. Because turbulent flow produces its own unique sounds, a physician can detect this problem by placing a stethoscope on the patient's neck. This problem can be compounded because the swirling currents of turbulent flow may cause part of the plaque to break off—becoming a mobile fragment called an embolus. The embolus then may travel downstream and cause a blockage in some part of the brain.

Normal **blood flow** moves like water through a garden hose.

Faster   Slower

Blood vessel wall

① LAMINAR FLOW

Laminar is *Linear*: Each fluid layer flows in a straight line.

Color each concentric ring of fluid a different color.

Constriction

② TURBULENT FLOW

Buckle yourself in, and prepare for a bumpy ride!

CARDIOVASCULAR SYSTEM: Blood Vessels—Structure and Function

Description
The heart generates the blood pressure by ejecting blood into the arterial system. **Blood pressure (BP)** is a type of **hydrostatic pressure**—or force of a fluid against the wall of a tube. Arteries carry blood away from the heart, and veins carry it back to the heart. **As distance from the heart increases, blood pressure decreases.** Therefore, on average, arteries have much higher pressure than veins. As the ventricles contract and force blood out into the arteries, these vessels expand, and the pressure rises to a *maximum* pressure. As the ventricles relax, and no more blood is ejected into the arteries, the arteries recoil, and pressure falls to a *minimum* pressure. This constant cycle of rising and falling blood pressure is the source of our pulse.

When blood pressure is taken, it is measured in mm Hg and expressed numerically as the maximum pressure over the minimum pressure. This is referred to as the **systolic pressure** over the **diastolic pressure**. For example, a normal blood pressure reading might be 120/70, read as "120 over 70."

Blood pressure has to be homeostatically maintained at a normal level. If blood pressure exceeds normal levels, it can cause smaller vessels to rupture, leading to anything from a stroke even to blindness. If levels fall too low, this also can lead to numerous problems. For example, filtration is totally dependent on pressure. The kidneys constantly filter the blood to remove waste products, and the capillaries in all our organs filter the blood to form tissue fluid. No pressure, no filtration. The body has numerous physiological mechanisms to increase BP when it falls too low.

Analogy
Like water pressure in a garden hose results from the force of water against the inside of the hose, blood pressure results from the force of blood against the wall of the blood vessel.

Measuring BP
1. When a blood pressure cuff is wrapped around the upper arm, the intended purpose is to compress the **brachial artery**, which delivers oxygenated blood to the arm.

2. The valve is closed on the blood pressure instrument, then the bulb is squeezed until the cuff pressure exceeds the systolic pressure in the brachial artery. The result is that the artery closes so blood temporarily stops flowing through it.

3. As the valve is slowly opened, the pressure within the cuff decreases. Consequently, the brachial artery begins to open. Blood squirts through the constricted vessel, resulting in turbulent flow (see p. 358) that can be detected by a stethoscope. These characteristic "whooshing" sounds are collectively referred to as the **sounds of Korotkoff**, and the pressure at onset is the systolic pressure.

4. Turbulent blood flow continues until the brachial artery has expanded back to its normal size. The pressure reading at the last sound before the restoration of normal blood flow represents the diastolic pressure.

Blood pressure (BP) is a type of **hydrostatic pressure (HP)** or fluid force against the wall of a tube.

As **water pressure** in a garden hose results from the force of water against the inside of the hose, **blood pressure** results from the force of blood against the wall of a blood vessel.

The **heart** generates the blood pressure by ejecting blood into the arterial system.

Garden hose

HP

Blood vessel

BP

Measuring Blood Pressure

1

Brachial a.

Cuff

Pressure gauge

0

2

120

0

Bulb

Valve

3

Systolic pressure

120

0

Stethoscope

4

70

0

Diastolic pressure

$$\frac{120_{\text{mm Hg}}}{70_{\text{mm Hg}}} = \frac{\text{MAX}}{\text{MIN}}$$

Color the 120 and "MAX" the same color and the 70 and "MIN" a different color.

Description

Capillaries, the most microscopic blood vessels in the body, join **arterioles** to **venules**. Their diameter is so small that red blood cells must pass through single file. These vessels form an interconnecting network or mesh of vessels called capillary beds. All the various tissues in the body depend on these capillary beds to remove wastes and to deliver nutrients and fluid.

The tubelike structure of each capillary is simple. The wall of the vessel is made up of endothelial cells or simple squamous epithelial cells. Each of these cells is flat to allow for easy diffusion of solutes from blood to tissue cells and vice versa. Wrapped around the outside of these cells like a thin blanket are protein fibers called the basement membrane. Gaps between adjacent endothelial cells called intercellular clefts allow for passage of fluid and small solutes.

The three types of capillaries are: (1) continuous capillaries, (2) fenestrated capillaries, and (3) sinusoidal capillaries.

Type	Comments/Permeability	Location in Body
① **Continuous capillary**	• Most common • *Least permeable*	Skin, connective tissues, skeletal muscle, smooth muscle, and lungs
② **Fenestrated capillary**	• *Most permeable.* Endothelial cells contain small holes called fenestrations that increase permeability	Kidneys, small intestine, choroid plexuses in brain, and some endocrine glands
③ **Sinusoidal capillary**	• *Most permeable*—has the largest fenestrations and the largest intercellular clefts	Liver, red bone marrow, spleen, and some endocrine glands

Blood flow through a capillary bed is controlled by bands of smooth muscle called precapillary sphincters. When they are closed, blood bypasses the capillary bed by going through the **shunt**. When they are open, blood moves through the entire capillary bed (as illustrated). This allows the body to respond to different situations. For example, during exercise, the sphincters have to be opened in the capillaries in skeletal muscles to meet the muscle's increased oxygen demand and deal with increased production of carbon dioxide. During the digestion of a heavy meal, more blood has to be shunted to the capillaries in the digestive tract rather than to the skeletal muscles.

Solute Diffusion

Capillary beds are the sites of nutrient and waste exchange between the blood and body cells. Nutrients move from the blood into tissue cells, and wastes move from tissue cells into the blood. For example, the respiratory gases—oxygen and carbon dioxide—**diffuse** across the wall of the capillary. Oxygen is a vital nutrient that diffuses out of the blood and into tissue cells, and carbon dioxide is a waste product that diffuses from tissue cells into the blood.

Recall that diffusion depends on a gradient (see pp. 78–79). The rule for simple diffusion is that a solute moves from a region of higher solute concentration to a region of lower solute concentration. Oxygen is carried in the blood primarily by the protein hemoglobin. As cells consume oxygen to metabolize glucose in the cellular respiration process, this ensures that the gradient for oxygen is maintained. Carbon dioxide is a by-product of cellular respiration, so it maintains a gradient as it constantly accumulates within body cells.

ONE WAY →

Basement membrane

Intercellular cleft

Endothelial cell

Red blood cell in lumen

Color all the simple squamous epithelial cells the same color.

1 Continuous capillary

Intercellular cleft

Endothelial cell

Fenestrations

2 Fenestrated capillary

Intercellular cleft

Endothelial cell

Incomplete basement membrane

3 Sinusoidal capillary

Three types of capillaries

Capillary bed

Shunt

Arteriole end

Venule end

Precapillary sphincters (open)

Red blood cell

ONE WAY →

Blood always moves through a capillary bed in a one-way direction—from arteriole end to venule end.

KEY
- Oxygen
- Carbon dioxide

Arteriole end

Tissue cell

Venule end

Diffusion of oxygen and carbon dioxide across a capillary bed

Description

Gas exchange occurs across capillary beds. This is how oxygen is delivered to your cells and how carbon dioxide is removed from tissues. In addition to diffusion of gases, two processes are occurring simultaneously with diffusion: **filtration** and **reabsorption**. Let's examine each, in turn.

The filtration process is completely dependent on a force. No force, no filtration. In capillaries, this force is the blood pressure inside the capillaries and is called the **capillary hydrostatic pressure (CHP)**. It drives the filtration process and is measured in millimeters of mercury (Hg). Across a capillary bed, the CHP quickly drops as blood moves from the *higher* pressure **arteriole end** (35 mm Hg) to the *lower* pressure **venule end** (18 mm Hg). Working against the **CHP** is a counterforce called the **blood colloidal osmotic pressure (BCOP)**.

Focus on the last two words of this term: osmotic pressure. **Osmotic pressure** represents the force from water's tendency to move across a semipermeable membrane toward the solution with the greater concentration of nonpenetrating solutes (primarily proteins). The osmotic pressure is proportional to the solute concentration. The higher the solute concentration, the greater is the osmotic force. Because the concentration of nonpenetrating solutes in the blood is greater than the tissue fluid within the interstitial spaces, water's tendency is to move back into the blood.

Unlike the CHP, the BCOP remains constant across the capillary bed at 25 mm Hg. Why? Because the concentration of nonpenetrating solutes does not change from one end of the capillary bed to the other.

Formula

Filtration occurs *only* at the **arteriole end** of the capillary bed. Here is the formula to calculate the

net filtration pressure (NFP): $$\text{NFP} = \text{CHP} - \text{BCOP}$$

substituting the numbers: $$\text{NFP} = 35 \text{ mm Hg} - 25 \text{ mm Hg}$$
$$= +10 \text{ mm Hg}$$

This positive (+) number indicates a net movement of water and small solutes (such as sodium and chloride ions) out of the blood and into the interstitial spaces. This process of filtering the blood plasma is the only way the body has to create interstitial fluid. The interstitial fluid is then drained into the lymphatic system to become lymph (see pp. 398–399).

Reabsorption of water occurs *only* at the venule end. Let's calculate the NFP again:

$$\text{NFP} = 18 \text{ mm Hg} - 25 \text{ mm Hg}$$
$$= -7 \text{ mm Hg}$$

The negative (−) number indicates net movement of water back into the blood. In short, some water is always reabsorbed at the venule end of the capillary because of osmosis.

Analogy

The capillary hydrostatic pressure (CHP) is the blood pressure inside the capillary. The CHP is like the water pressure inside a garden hose that has been turned on. Now imagine that you punctured some tiny holes in the hose. The fluid squirting out of the hose is like the solution being filtered out of the blood to create the tissue fluid within the interstitial spaces.

Hole in hose

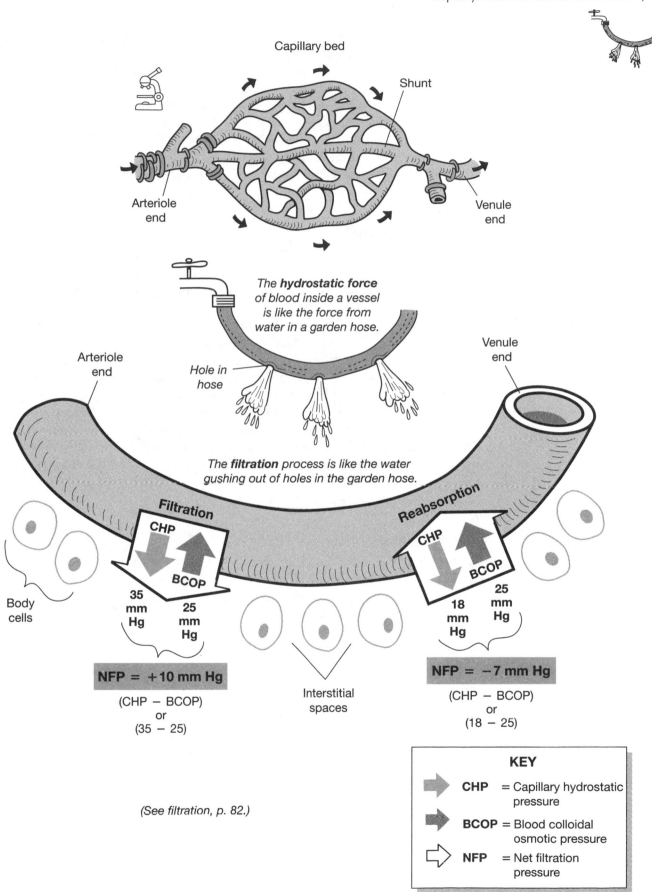

Capillary bed

Shunt

Arteriole
end

Venule
end

*The **hydrostatic force**
of blood inside a vessel
is like the force from
water in a garden hose.*

Arteriole
end

Venule
end

*Hole in
hose*

*The **filtration** process is like the water
gushing out of holes in the garden hose.*

Filtration

Reabsorption

CHP

BCOP

CHP

BCOP

Body
cells

35
mm
Hg

25
mm
Hg

18
mm
Hg

25
mm
Hg

NFP = +10 mm Hg

(CHP − BCOP)
or
(35 − 25)

Interstitial
spaces

NFP = −7 mm Hg

(CHP − BCOP)
or
(18 − 25)

(See filtration, p. 82.)

KEY	
CHP	= Capillary hydrostatic pressure
BCOP	= Blood colloidal osmotic pressure
NFP	= Net filtration pressure

Description

Venous return refers to the volume of blood returning to the heart through the veins. Veins are under much lower pressure than arteries, which makes it difficult to return venous blood to the heart. For example, the average pressure in the venous system is about 16 mm Hg, and the average pressure in the arterial system is about 60 mm Hg. Veins contain semilunar valves at regular intervals, which allow for blood flow in one direction and prevent backflow. Venous return is easier when a person is lying down because the force of gravity is not a factor. But when someone is standing up, consider how long a journey it is for the venous blood in your foot to return to the heart while moving against gravity. Although the heart's normal pumping action helps move the venous blood back to the heart, the body has two other mechanisms we will refer to collectively as venous "pumps": (1) **skeletal muscle "pump,"** and (2) **respiratory "pump."**

① Skeletal muscle "pump"

When you are walking, the normal muscle contractions in your legs greatly aid in venous return. They actually squeeze blood up through the peripheral veins toward the heart in an action called "milking." Here is how it works: The illustration shows a short segment of a vein with two semilunar valves. The **proximal valve** is nearer the heart, and the **distal valve** is farther from the heart. As the skeletal muscle around this vein contracts, the vessel is squeezed, thereby increasing the pressure inside the vein. This increased pressure forces the proximal valve to open, but the distal valve remains closed because of back pressure. The end result of this action is that some venous blood is moved up into the next section of the vein. By repeating this pattern, venous blood is given an extra push on its journey back to the heart.

If you have to stand for a long time, it is good to flex your gastrocnemius muscles (calf muscles) to prevent fainting. Fainting is caused by a decrease in venous return, which leads to decreased cardiac output, which causes a decrease in the blood supply to the brain. This means that less oxygen and glucose are being delivered to neurons in the brain, which causes fainting. By flexing your gastrocnemius muscles after standing still for a long time, you will put your skeletal muscle "pump" to work for you.

② Respiratory "pump"

Normal breathing facilitates venous return. The deeper the breathing, the greater is the assistance in this regard. As shown in the illustration, normal breathing involves a repeating cycle of contraction and relaxation of the **diaphragm**. When this muscle contracts for a normal inspiration, it moves downward and compresses the contents of the **abdominal cavity**, which increases the abdominal pressure. Simultaneously, the volume in the **thoracic cavity** increases, causing a decrease in pressure. This pressure gradient forces blood out of the various veins in the abdominal region and into the inferior vena cava (**IVC**) to allow blood to move back into the right atrium of the heart.

Venous "pumps"

The BIG SQUEEZE!

The venous "pumps" are external forces
that act on veins like squeezing
a hollow rubber tube
with your hands.

1 Skeletal muscle "pump"

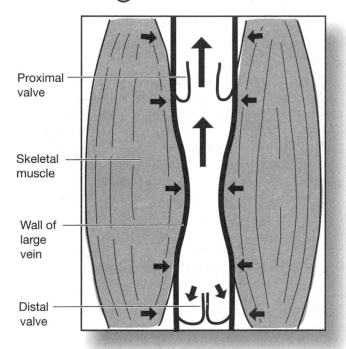

Proximal
valve

Skeletal
muscle

Wall of
large
vein

Distal
valve

*Color the blood
in the large
vein blue.*

2 Respiratory "pump"

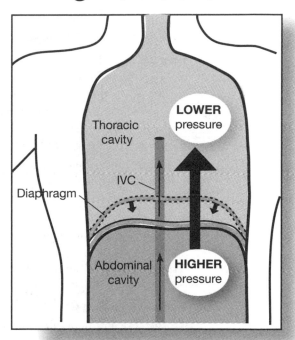

Thoracic
cavity

LOWER
pressure

IVC

Diaphragm

Abdominal
cavity

HIGHER
pressure

*Which way does a
pressure gradient go?
It always moves
air or fluids
from HIGH to LOW.*

*Color the ovals
for higher pressure
and lower pressure
different colors.*

Description   Cardiac output (CO) is the amount of blood pumped out of a ventricle in 1 minute and is expressed as milliliters of blood per minute. It is mathematically defined as the product of the **stroke volume** (**SV**) times the **heart rate** (**HR**). This is the mathematical equation: $CO = SV \times HR$

A normal stroke volume is the amount of blood pumped out of each ventricle in one beat—about 70 mL for an adult heart. A healthy adult male has a resting heart rate of about 70 beats per minute (bpm). By plugging these numbers into the formula given above, we can calculate a normal CO. As shown in the equation at the bottom of the illustration, it calculates to approximately 4,900 mL of blood per minute. If converted to liters, it is 4.9 L per minute.

The illustration shows this by assuming that we could collect blood continually (from only one ventricle) in a 6-liter glass beaker. At the end of 1 minute, it would have filled the beaker to 4.9 L of blood. The CO can be measured by taking a person's blood pressure and pulse, then doing some calculations.

For example, let's say your blood pressure was measured as 118/80 and pulse was 70 bpm. From this, calculate the pulse pressure (systolic pressure minus diastolic pressure). In this case, $118 - 80 = 36$. Then plug the numbers into this equation:

$$
\begin{aligned}
CO &= 2 \text{ mL} \times \text{pulse pressure} \times HR \\
&= 2 \times 36 \times 70 \\
&= 5{,}320 \text{ mL/min.}
\end{aligned}
$$

In general, factors that effect SV also affect HR. For example, *during exercise, both SV and HR increase.* Even so, let's consider the key factors that govern each of these two variables.

Factors affecting SV

- **Preload:** defined as the amount of tension in the ventricular cardiac muscle cells prior to contracting. The rule is that *the greater the tension on the cells, the more forcefully they contract.* This relationship is referred to as the **Frank-Starling Law.** For example, during exercise, more blood is returned to the heart through the venous return. As more blood enters the heart, it stretches cardiac muscles cells, resulting in a more forceful ejection of blood. For this reason, exercise increases SV.

- **Contractility:** defined as the degree to which cardiac muscle cells can shorten when stimulated by a specific chemical substance. For example, calcium ions have a positive influence on cardiac muscle contraction. Abnormally low levels of calcium result in an irregular heartbeat, decreasing the SV.

- **Afterload:** defined as the amount of force needed from ventricular cardiac muscle cells to eject blood from the ventricles and past the semilunar valves. Anything that impedes blood flow can increase afterload. For example, any blockage in the peripheral blood vessels (such as atherosclerosis) would restrict blood flow and increase the afterload. Hypertension (high blood pressure) also increases the afterload. As the afterload increases, the SV decreases.

Factors affecting HR

- **Age:** HR gradually decreases from childhood to adulthood. A newborn may have an HR of 120 bpm, whereas an adult male may have 70 bpm. In the elderly, the HR increases again relative to that of a young adult.

- **Sex:** On average, females have slightly higher resting HRs than males. The difference is about 5 bpm.

- **State of activity:** During certain phases of the sleep cycle, the HR decreases. But during exercise the HR temporarily increases.

- **Endurance training:** Marathon runners may have a resting HR of 50 bpm. This type of training increases heart size as well as SV. This allows for a normal CO with a lower HR.

- **Stress, anxiety:** Stress and anxiety increase HR.

The CO does not remain constant. It regularly rises and falls. Maintaining the CO in a normal range is primarily the job of the cardiovascular (CV) center in the medulla oblongata of the brain. Using reflex pathways in the autonomic nervous system, the CV center regulates the rhythm and force of the heart rate. Hormones also help to control cardiac activity. For example, epinephrine and norepinephrine are powerful cardiac stimulators. In short, the nervous system and hormonal regulators work together to regulate heart activity, thereby indirectly ensuring that a normal CO is achieved.

Cardiac output (CO) = stroke volume (SV) × heart rate (HR)

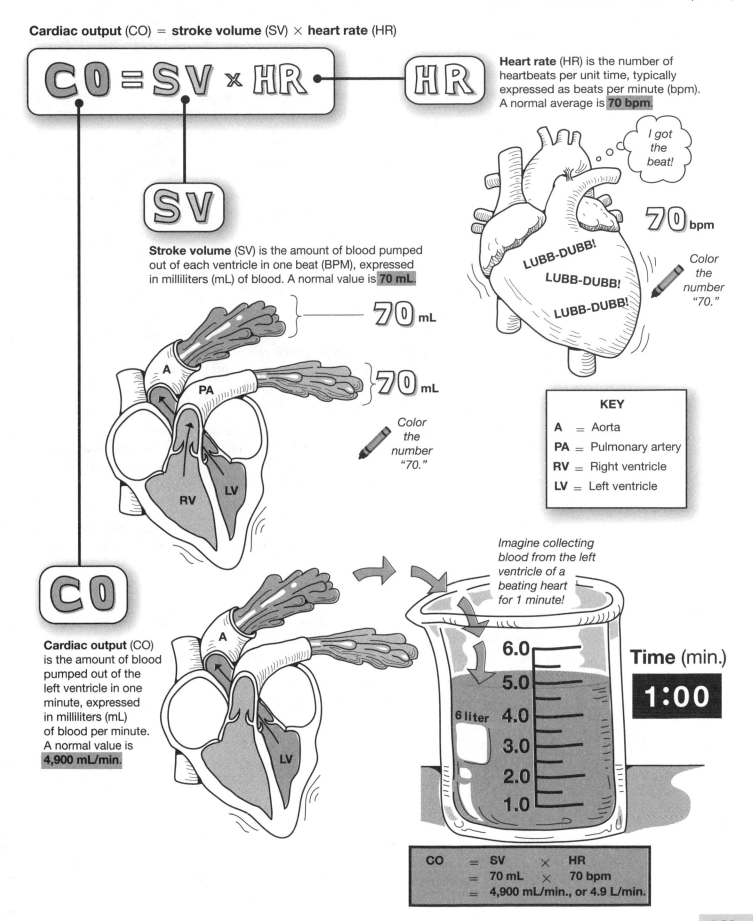

CO = SV × HR

HR

Heart rate (HR) is the number of heartbeats per unit time, typically expressed as beats per minute (bpm). A normal average is **70 bpm**.

I got the beat!

70 bpm

LUBB-DUBB!

LUBB-DUBB!

LUBB-DUBB!

Color the number "70."

SV

Stroke volume (SV) is the amount of blood pumped out of each ventricle in one beat (BPM), expressed in milliliters (mL) of blood. A normal value is **70 mL**.

70 mL

70 mL

Color the number "70."

A

PA

RV

LV

KEY		
A	=	Aorta
PA	=	Pulmonary artery
RV	=	Right ventricle
LV	=	Left ventricle

CO

Cardiac output (CO) is the amount of blood pumped out of the left ventricle in one minute, expressed in milliliters (mL) of blood per minute. A normal value is **4,900 mL/min.**

Imagine collecting blood from the left ventricle of a beating heart for 1 minute!

A

LV

6 liter

6.0
5.0
4.0
3.0
2.0
1.0

Time (min.)

1:00

CO	=	SV	×	HR
	=	70 mL	×	70 bpm
	=	4,900 mL/min., or 4.9 L/min.		

Description

The following four regulatory mechanisms are used to control **blood pressure** (**BP**): cardiovascular center, neural regulation, hormonal regulation, and autoregulation.

① **Neural regulation through the cardiovascular center**

Neural regulation uses reflex pathways to regulate normal blood pressure. As its name implies, the **cardiovascular center** (**CV**), in the **medulla oblongata**, is the nervous system's command-and-control center for regulating the heart rate and vasoconstriction. It receives peripheral sensory input from three major sources: **proprioceptors** (*detect changes in muscle tension*), **chemoreceptors** (*detect changes in blood acidity*), and **baroreceptors** (*detect changes in blood pressure*). The CV responds by sending the appropriate motor output to the heart and blood vessels. Depending on the stimulus, the **heart rate** (**HR**) either increases or decreases, and blood vessels are stimulated to constrict. The end result is that the BP either increases or decreases back to normal levels. For details of these reflex pathways, see p. 352.

② **Hormonal regulation**

Hormonal regulation involves the use of chemical messengers called **hormones** that bind to receptors at a target tissue in order to induce a response.

- Renin-angiotensin-aldosterone (**RAA**) system

  When blood pressure decreases, it stimulates the kidney to release the enzyme renin into the blood. This leads to the production of a powerful vasoconstrictor called angiotensin II. This, in turn, stimulates the adrenal cortex to produce the hormone aldosterone. The net result is an increase in blood pressure. For details of this mechanism, see p. 512.

- Epinephrine (**EPI**)/norepinephrine (**NE**)

  EPI and **NE** are produced by cells in the adrenal medulla in response to emergency or stressful situations. Two of the organs they target are the heart and blood vessels. The result is that heart rate increases and blood vessels constrict, respectively. Together, these two responses lead to an increase in blood pressure. For details of this mechanism, see p. 312.

- Antidiuretic hormone (**ADH**)

  Water loss from the blood, like what occurs during excessive sweating, stimulates the posterior lobe of the pituitary gland to produce **ADH**. This hormone targets the nephrons in the kidneys and causes more water to be reabsorbed into the blood, thus restoring normal blood pressure. For the details of this mechanism, see p. 506.

- Atrial natriuretic peptide (**ANP**)

  As blood volume increases, this stretches the atria in the heart, which stimulates cells in the atria to produce **ANP**. By increasing vasodilation, sodium ion excretion, and urine production and blocking the release of hormones like ADH and aldosterone, ANP decreases BP.

③ **Autoregulation**—the automatic control of blood flow to a tissue

These mechanisms are normally not as significant as the neural and hormonal mechanisms mentioned above in 1 and 2.

- **Physical changes**

  Changes in body temperature affect BP. For example, cold causes superficial blood vessels to constrict, which increases BP. In contrast, heat causes blood vessels to dilate, leading to a decrease in blood pressure.

- **Chemical changes**

  Different types of body cells, such as smooth muscles cells, endothelial cells, and macrophages, produce various chemicals that signal blood vessels to either dilate or constrict. For example, smooth cells produce lactic acid, which causes blood vessels to dilate. Other vasodilating chemicals include nitric oxide (NO), H^+, and K^+.

GOAL: Maintain Normal Blood Pressure

$$\frac{120 \text{ mm Hg}}{70 \text{ mm Hg}}$$

Color the 120 one color and the 70 a different color.

① Neural regulation through the cardiovascular (CV) center

- **Proprio-ceptors** *tension*
- **Chemo-receptors** *blood acidity*
- **Baro-receptors** *pressure*

S

INPUT

(see p. 352)

Cardio-vascular (CV) center

OUTPUT

R

⬇ **HR**   ⬆ **HR**   blood vessels **constrict**

② Hormonal regulation

- Renin-angiotensin-aldosterone (**RAA**) system
- Epinephrine (**EPI**) /norepinephrine (**NE**)
- Antidiuretic hormone (**ADH**)
- Atrial natriuretic peptide (**ANP**)

Kidney | Adrenal medulla | Pituitary gland (Posterior lobe) | Heart

RAA system | **EPI, NE** | **ADH** | **ANP**

⬆ **BP**   ⬆ **BP**   ⬆ **BP**   ⬇ **BP**

(see p. 494)

③ Autoregulation

R

Physical changes

⬇ **COLD** ➔ blood vessels **constrict**

S

⬆ **HOT** ➔ blood vessels **dilate**

R

Chemical signals

Smooth muscle cells ➔ Vasodilating chemicals ➔ blood vessels **dilate**

Description

What is the **hepatic portal system**? The term *hepatic* refers to the liver and the term *portal* refers to a *portal vessel* which is defined as a blood vessel that carries blood from one capillary network to another. All the blood vessels associated with this network constitute a *system*. In this case, the first capillary network is found in all the digestive organs and the second capillary network is located in the liver (the capillaries here are called sinusoids). The vessel that links them together is the **hepatic portal vein**. This wide vessel is labeled as #1 in the illustration on the facing page. Place your finger on it and notice that there are many vessels draining deoxygenated blood from the digestive organs and carrying it through the hepatic portal vein and into the liver.

What is the purpose of this system? The blood vessels in this system carry nutrient-rich blood to the liver so the nutrients can be processed in liver cells before entering the general circulation. Allow me to explain. The deoxygenated venous blood in this system is different from normal venous blood in that the level of nutrients such as glucose and amino acids is much higher. For example, after a meal, blood glucose levels rise and excess glucose needs to be delivered to liver cells where it is stored as glycogen. After all these nutrients have been processed in liver cells, then the deoxygenated blood is drained by the **heptic veins** (#2 in the illustration) which carry blood to the **inferior vena cava** (labeled in the illustration) where it enters the general circulation.

There are three major veins that drain the blood into the hepatic portal vein:

- **Superior mesenteric vein** (#10)
- **Inferior mesenteric vein** (#9)
- **Splenic vein** (#8)

Let's examine each of these veins by locating them on the illustration. Place your finger on the **superior mesenteric vein** (#10). Notice the blood vessels to the left that are draining blood from part of the large intestine. Also notice the blood vessels draining blood from the pancreas and the bottom of the stomach. Do you see the words *duodenum* and *ileum* in the illustration? These are two of the three parts of the small intestine and blood is being drained from both of them. The third part, the jejunum, is not shown in the illustration. Even so, you can see blood vessels branching off the right side of the superior mesenteric vein that have been cut. These are draining the rest of the small intestine that is not shown.

Now place your finger on the **inferior mesenteric vein** (#9). Notice that it drains blood from the rectum and the distal portion of the large intestine. As you follow it upwards, notice that it merges with the **splenic vein** (#8). The name *splenic* obviously informs us that it drains blood from the spleen, but if you look near its terminus, you will see that it also drains blood from part of the stomach. Lastly, if you look at the pancreas, you can see that it also drains blood from the pancreas via the **pancreatic veins** (#7).

Key to Illustration

1. Hepatic portal vein
2. Hepatic veins
3. Left gastric vein
4. Right gastric vein
5. Left gastroepiploic vein
6. Right gastroepiploic vein
7. Pancreatic vein
8. Splenic vein
9. Inferior mesenteric vein
10. Superior mesenteric vein

Inferior vena cava

Liver

Stomach

Spleen

Pancreas

Duodenum

Large intestine

Ileum

Large intestine

Rectum

1. _____
2. _____
3. _____
4. _____
5. _____

6. _____
7. _____
8. _____
9. _____
10. _____

Description

The **Fetal circulation** is different from adult blood circulation. This is to be expected if you give it some thought. Although the fetal heart is functioning and pumping the fetal blood, the fetal lungs are collapsed and filled with fluid, so they are not functioning for gas exchange. The kidneys are also not functioning to remove fetal waste products from the blood. None of the digestive organs are functioning because no food is being ingested, so nutrients cannot be obtained. What replaces these nonfunctional organs in the fetus? The placenta, the organ in the mother's uterus, is the site for exchanging blood gases (O_2 and CO_2), removing waste products, and obtaining nutrients. This is why it is so important for a pregnant woman to maintain a healthy diet. The fetus's lifeline is the umbilical cord, a bendable, cable-like structure that connects the fetus to the placenta.

Let's follow the pathway that blood flows in fetal circulation. Begin by placing your finger on structure number 4 on the illustration on the facing page. These are the pair of **umbilical arteries** that branch off the internal iliac arteries and carry the fetal blood to the placenta to become oxygenated. The newly oxygenated blood from the placenta flows through the single **umbilical vein** to the nonfunctional liver where it connects to the **ductus venosus** that drains into the **inferior vena cava**. The deoxygenated blood from the inferior vena mixes with the oxygenated blood from the ductus venosus to become mixed blood entering the heart. Blood then drains from both the **superior vena cava** and the **inferior vena cava** into the **right atrium** (**RA**). Blood in the RA follows two different paths. Some of the blood in the RA goes through an opening with a flap called the **foramen ovale** and enters the **left atrium** (**LA**). The remaining blood enters the right ventricle, and then flows into the pulmonary trunk. Interestingly enough, both pathways end up bypassing the nonfunctional lungs. Blood in the left atrium is pumped into the left ventricle, flows through the aortic arch, and moves into the systemic circuit. While a small amount of blood in the pulmonary trunk goes to the lungs, almost all of it is pumped through the **ductus arteriosus**—a short fetal blood vessel that connects the pulmonary trunk to the aorta and diverts the blood into the systemic circuit. As the mixed blood moves through the **common iliac artery** and into the **internal iliac artery**, it enters the **umbilical arteries** once again to repeat the cycle.

After birth, the umbilical cord is tied off and blood doesn't flow through it anymore. The following key changes occur:

- **Umbilical arteries** close and become fibrous cords at their distal ends.
- The **umbilical vein** closes to become the fibrous **ligamentum teres** (*round ligament*).
- **Ductus venosus** closes to become the fibrous **ligamentum venosum**.
- **Foramen ovale** closes and seals off to become a shallow depression in the interatrial septum called the **fossa ovalis**.
- **Ductus arteriosus**—closes to become the fibrous **ligamentum arteriosum**.

Key to Illustration		
1. Foramen ovale	**3.** Ductus venosus	**5.** Umbilical vein
2. Ductus arteriosus	**4.** Umbilical arteries	

Fetal Circulation

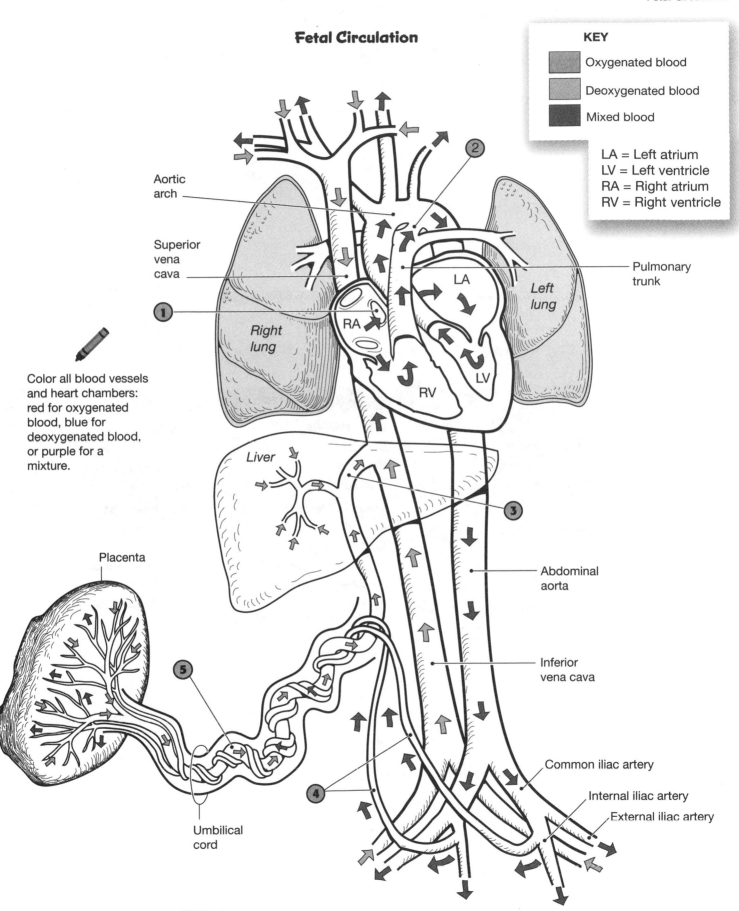

KEY

Oxygenated blood

Deoxygenated blood

Mixed blood

LA = Left atrium
LV = Left ventricle
RA = Right atrium
RV = Right ventricle

Aortic arch

Superior vena cava

Color all blood vessels and heart chambers: red for oxygenated blood, blue for deoxygenated blood, or purple for a mixture.

Right lung

Left lung

LA

RA

LV

RV

Pulmonary trunk

Liver

Abdominal aorta

Placenta

Inferior vena cava

Common iliac artery

Internal iliac artery

External iliac artery

Umbilical cord

Key to Illustration

Head/Neck (H)
H1. Internal carotid a.
H2. External carotid a.
H3. Common carotid a.
H4. Subclavian a.

Shoulder (S)
S1. Axillary a.

Thorax (T)
T1. Aortic arch
T2. Pulmonary trunk
T3. Pulmonary a.

Upper Limb (UL)
UL1. Brachial a.
UL2. Ulnar a.
UL3. Radial a.

Abdomen/Pelvis (A)
A1. Abdominal aorta
A2. Celiac trunk
A3. Superior mesenteric a.
A4. Renal a.
A5. Gonadal a.
(testicular a. in males, ovarian a. in females)

A6. Inferior mesenteric a.
A7. Common iliac a.
A8. Internal iliac a.
A9. External iliac a.

Lower Limb (LL)
LL1. Femoral a.
LL2. Popliteal a.
LL3. Anterior tibial a.
LL4. Fibular a.
LL5. Posterior tibial a.

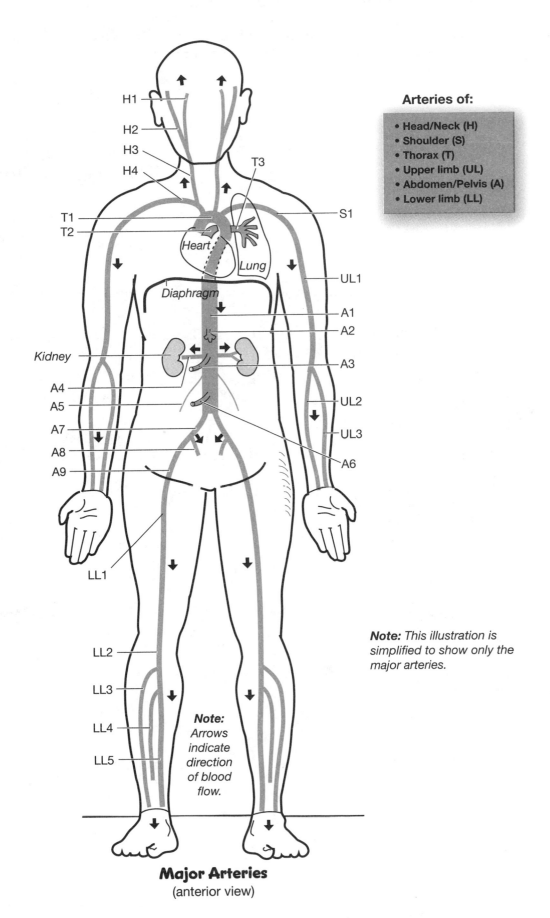

H1
H2
H3
H4
T3
T1
T2
Heart
Lung
S1
Diaphragm
UL1
A1
A2
Kidney
A3
A4
A5
UL2
A7
UL3
A8
A6
A9
LL1
LL2
LL3
LL4
LL5

Arteries of:

- **Head/Neck (H)**
- **Shoulder (S)**
- **Thorax (T)**
- **Upper limb (UL)**
- **Abdomen/Pelvis (A)**
- **Lower limb (LL)**

Note: This illustration is simplified to show only the major arteries.

Note: Arrows indicate direction of blood flow.

Major Arteries
(anterior view)

Description   The arterial system in the head and neck feeds oxygenated blood to the brain, face, eyes, neck, larynx, pharynx, esophagus, and other important structures.

Key to Illustration

Note: Colored arrows indicate direction of blood flow.

Lateral View

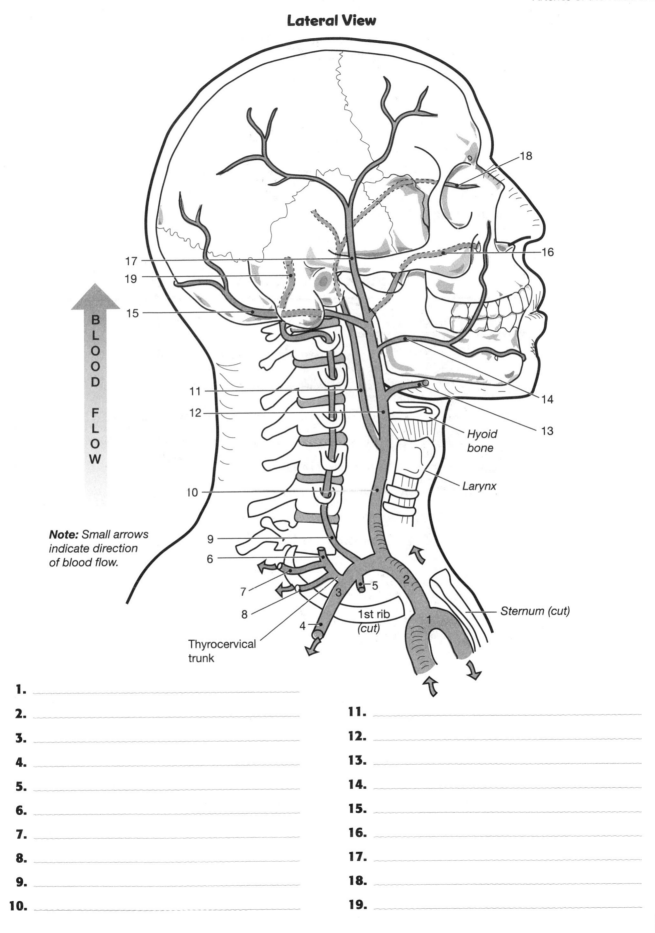

B
L
O
O
D

F
L
O
W

Note: Small arrows indicate direction of blood flow.

Hyoid bone

Larynx

Thyrocervical trunk

1st rib (cut)

Sternum (cut)

1. _____
2. _____
3. _____
4. _____
5. _____
6. _____
7. _____
8. _____
9. _____
10. _____

11. _____
12. _____
13. _____
14. _____
15. _____
16. _____
17. _____
18. _____
19. _____

Description   The arteries branching off the abdominal aorta feed oxygenated blood to the digestive organs, kidneys, spleen, gonads, urinary bladder, and other important structures in the abdominopelvic cavity and the lower limbs.

Study Tips
- Use diaphragm as a landmark
- Celiac trunk is first branch below the diaphragm
- Superior mesenteric is larger in diameter than the inferior mesenteric
- Renal and gonadal are paired
- Gonadal arteries are slender

Schematic of Major Branches

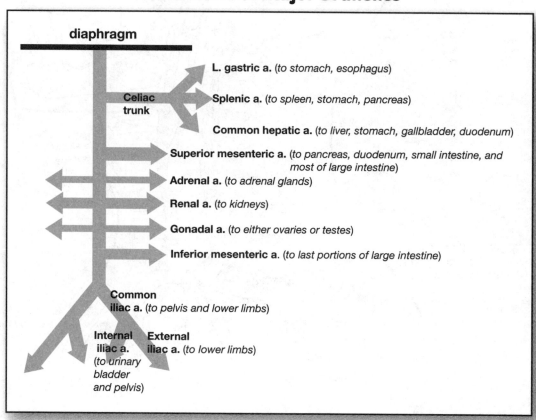

Key to Illustration		
1. Celiac truck	5. Superior mesenteric a.	9. Common iliac a.
2. Common hepatic a.	6. Renal a.	10. Internal iliac a.
3. Left gastric a.	7. Gonadal a.	11. External iliac a.
4. Splenic a.	8. Inferior mesenteric a.	

DIAPHRAGM

LIVER

SPLEEN

KIDNEY

KIDNEY

Adrenal a.

BLOOD FLOW

Schematic of Major Branches

diaphragm

Use the diaphragm as a landmark to locate specific branches (e.g., celiac trunk is first branch below the diaphragm).

The blood vessel pattern looks like a stick figure wearing boxing gloves (kidneys).

(Right)   (Left)

Anterior View

1. _____
2. _____
3. _____
4. _____
5. _____
6. _____

7. _____
8. _____
9. _____
10. _____
11. _____

Description   The arterial system in the upper limbs feeds oxygenated blood to all the tissues in the arms.

Key to Illustration

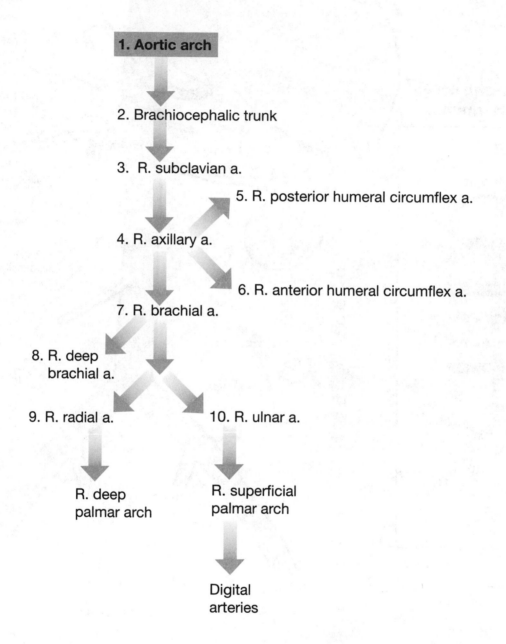

1. Aortic arch

2. Brachiocephalic trunk

3. R. subclavian a.

5. R. posterior humeral circumflex a.

4. R. axillary a.

6. R. anterior humeral circumflex a.

7. R. brachial a.

8. R. deep brachial a.

9. R. radial a.

10. R. ulnar a.

R. deep palmar arch

R. superficial palmar arch

Digital arteries

Note: Colored arrows indicate direction of blood flow.

Note: *Small arrows indicate direction of blood flow.*

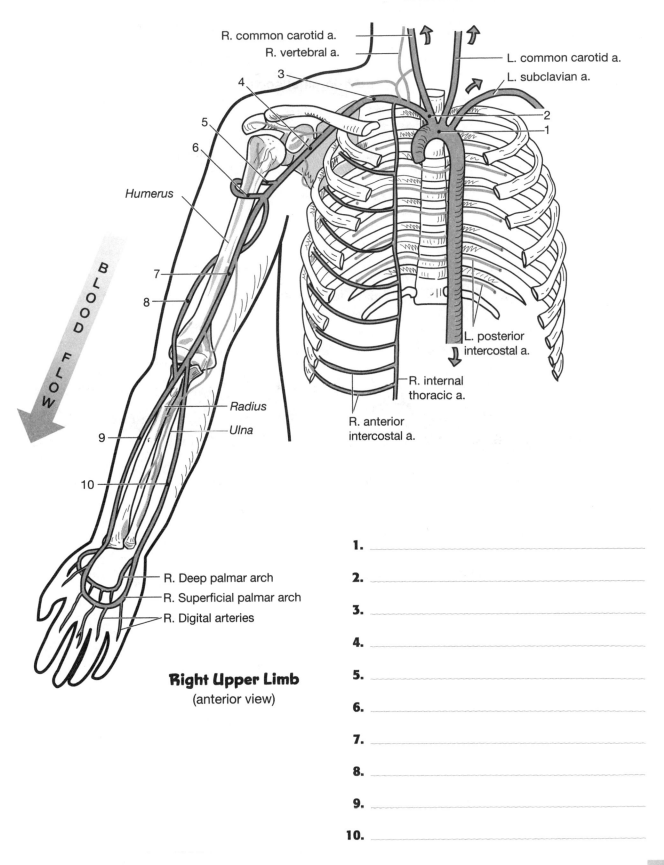

R. common carotid a.
R. vertebral a.
L. common carotid a.
L. subclavian a.

3
4
5
6

Humerus

7
8

2
1

L. posterior intercostal a.

R. internal thoracic a.

R. anterior intercostal a.

BLOOD FLOW

Radius

Ulna

9

10

R. Deep palmar arch
R. Superficial palmar arch
R. Digital arteries

Right Upper Limb
(anterior view)

1. _____
2. _____
3. _____
4. _____
5. _____
6. _____
7. _____
8. _____
9. _____
10. _____

Description   The arterial system in the lower limbs feeds oxygenated blood to all the tissues in the legs.

Key to Illustration

1. Abdominal aorta

2. R. common iliac a. → 3. R. internal iliac a.

4. R. external iliac a.

5. R. deep femoral a. ← 6. R. femoral a.

7. R. popliteal a.

8. R. anterior tibial a.   9. R. posterior tibial a.

10. R. fibular a.

Note: Colored arrows indicate direction of blood flow.

Inguinal ligament

Femur

B
L
O
O
D

F
L
O
W

Tibia

Fibula

Dorsalis pedis a.

Right Lower Limb
(anterior view)

Dorsal arch
Metatarsal a.
Plantar arch
Digital a.

1
2
3
4
5
6
7 *
8
9 *
10 *

Label major branches

1. _____

2. _____

3. _____

4. _____

5. _____

6. _____

7. _____

8. _____

9. _____

10. _____

Note: *Small arrows indicate direction of blood flow.*

*only visible in posterior view

Key to Illustration

Head/Neck (H)

H1. External jugular v.

H2. Internal jugular v.

Shoulder (S)

S1. Axillary v.

Thorax (T)

T1. Superior vena cava

T2. Brachiocephalic v.

T3. Subclavian v.

Upper Limb (UL)

UL1. Cephalic v.

UL2. Brachial v.

UL3. Basilic v.

UL4. Ulnar v.

UL5. Radial v.

Abdomen/Pelvis (A)

A1. Hepatic v.

A2. Hepatic portal v.

A3. Gastric v.

A4. Splenic v.

A5. Inferior mesenteric v.

A6. Superior mesenteric v.

A7. Inferior vena cava

A8. Renal v.

A9. Gonadal v.
(testicular v. in male, ovarian v. in female)

A10. Common iliac v.

A11. Internal iliac v.

A12. External iliac v.

Lower Limb (LL)

LL1. Femoral v.

LL2. Great saphenous v.

LL3. Popliteal v.

LL4. Anterior tibial v.

LL5. Posterior tibial v.

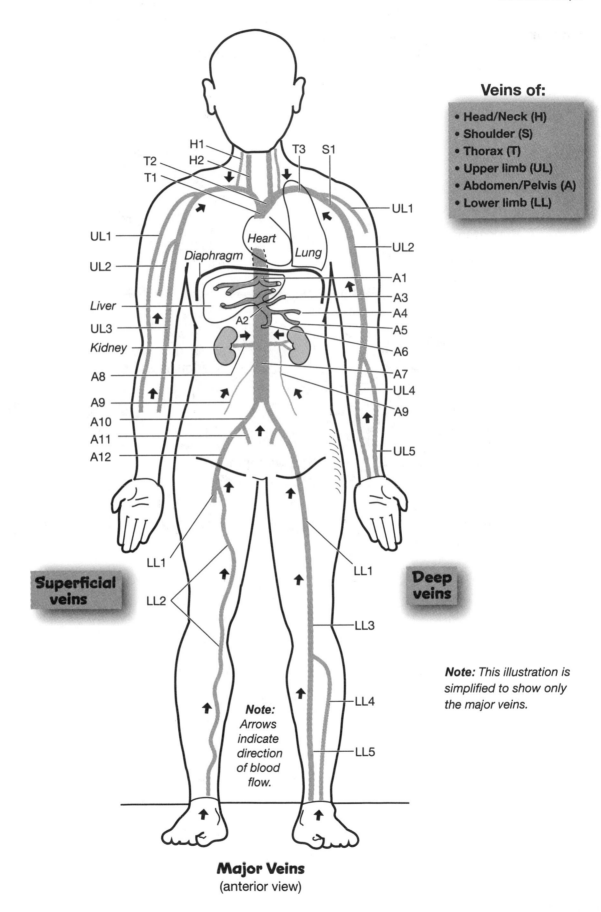

Veins of:

- Head/Neck (H)
- Shoulder (S)
- Thorax (T)
- Upper limb (UL)
- Abdomen/Pelvis (A)
- Lower limb (LL)

H1
T2
H2
T1
T3
S1

UL1
UL1
UL2
UL2

Heart
Diaphragm
Lung

A1
A3
A4
A5
A6
A7
UL4
A9
UL5

Liver
UL3
Kidney
A8
A9
A10
A11
A12
A2

Superficial veins

Deep veins

LL1
LL1
LL2
LL3

Note: This illustration is simplified to show only the major veins.

LL4

Note: Arrows indicate direction of blood flow.

LL5

Major Veins

(anterior view)

Description   The venous system in the head and neck drains deoxygenated blood from the brain, face, eyes, neck, larynx, pharynx, esophagus, and other important structures.

Key to Illustration

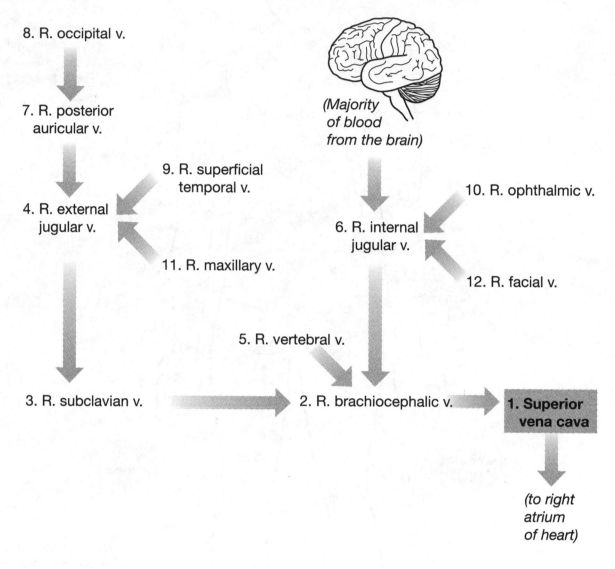

8. R. occipital v.

7. R. posterior auricular v.

9. R. superficial temporal v.

4. R. external jugular v.

11. R. maxillary v.

(Majority of blood from the brain)

10. R. ophthalmic v.

6. R. internal jugular v.

12. R. facial v.

5. R. vertebral v.

3. R. subclavian v.

2. R. brachiocephalic v.

1. Superior vena cava

(to right atrium of heart)

Note: *Gray arrows indicate direction of blood flow.*

Lateral View

Hyoid bone

Larynx

BLOOD FLOW

Note: Small arrows indicate direction of blood flow.

(Blood from upper limb)

1st rib (cut)

Sternum (cut)

1. _____
2. _____
3. _____
4. _____
5. _____
6. _____

7. _____
8. _____
9. _____
10. _____
11. _____
12. _____

CARDIOVASCULAR SYSTEM: Blood Vessels—Veins

Major Veins of the Abdominopelvic Cavity

Description The venous system in the abdominopelvic cavity drains deoxygenated blood into the inferior vena cava, which empties into the right atrium of the heart.

Study Tips
- Use diaphragm as a landmark
- Hepatic veins are the first vessels of the inferior vena cava below the diaphragm
- Renal veins are the widest vessels of the inferior vena cava
- Gonadal veins are very slender

Schematic of Major Veins of Abdomen

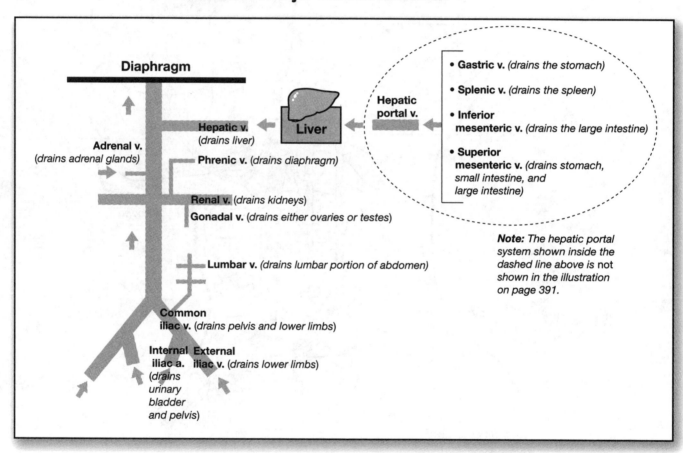

Key to Illustration

1. Inferior vena cava
2. Hepatic v.
3. Renal v.
4. Adrenal v.
5. Phrenic v.

6. Gonadal v.
   testicular v. in male
   ovarian v. in female
7. Lumbar v.
8. Common iliac v.

9. Internal iliac v.
10. External iliac v.

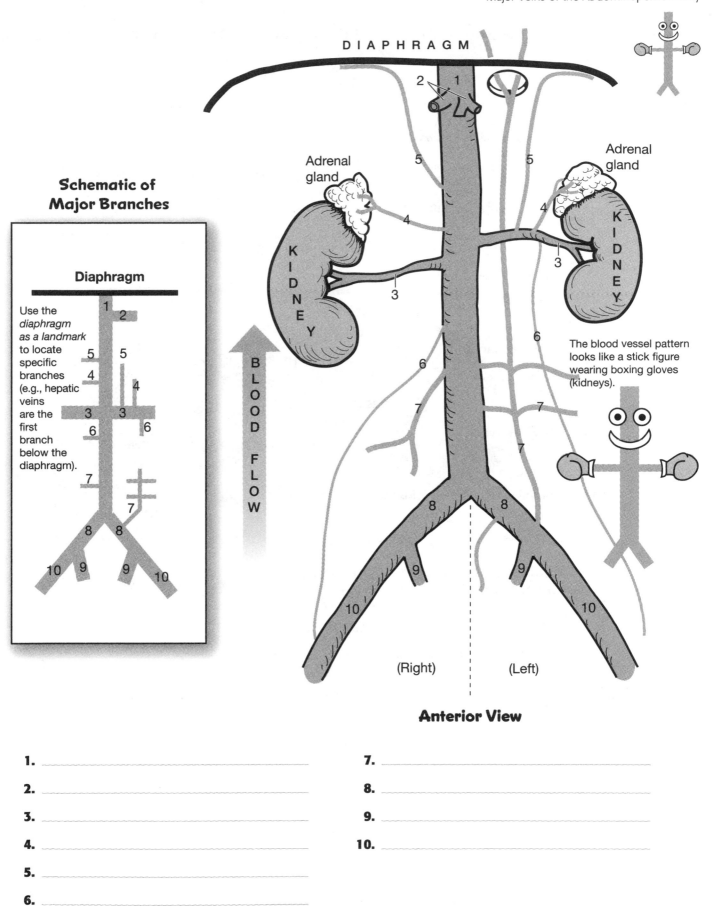

DIAPHRAGM

Adrenal gland

Adrenal gland

K I D N E Y

K I D N E Y

BLOOD FLOW

The blood vessel pattern looks like a stick figure wearing boxing gloves (kidneys).

(Right)   (Left)

Anterior View

Schematic of Major Branches

Diaphragm

Use the *diaphragm as a landmark* to locate specific branches (e.g., hepatic veins are the first branch below the diaphragm).

1. _____

2. _____

3. _____

4. _____

5. _____

6. _____

7. _____

8. _____

9. _____

10. _____

CARDIOVASCULAR SYSTEM: Blood Vessels—Veins

Veins of the Right Upper Limb and Shoulder

Description   The venous system in the upper limbs drains deoxygenated blood from all the tissues in the arms.

Key to Illustration

I. Superior vena cava

3. R. internal jugular v.

2. R. brachiocephalic v.

4. R. external jugular v.

5. R. subclavian v.

7. R. cephalic v.

6. R. axillary v.

10. R. median cubital v.

9. R. basilic v.

8. R. brachial v.

13. R. median antebrachial v.

11. R. radial v.   12. R. ulnar v.

R. Deep palmar venous arch   R. Superficial palmar venous arch   R. Digital veins

Note: Gray arrows indicate direction of blood flow.

3
4
5
6
2
1

Azygos v.

7

Humerus

8

9

This vessel (#10) is often used for taking blood.

10

BLOOD FLOW

R. posterior intercostal v.'s

L. ascending lumbar v.

Inferior vena cava

Note: *Small arrows indicate direction of blood flow.*

11   Ulna

   Radius

7

9

12

13

R. deep palmar venous arch

R. superficial palmar venous arch

R. digital v.

Right Upper Limb
(anterior view)

1. _____

2. _____

3. _____

4. _____

5. _____

6. _____

7. _____

8. _____

9. _____

10. _____

11. _____

12. _____

13. _____

Description   The venous system in the lower limbs drains deoxygenated blood from all the tissues in the legs.

Key to Illustration

1. Inferior vena cava
2. R. common iliac v.
3. R. internal iliac v.
4. R. external iliac v.
5. R. deep femoral v.
6. R. femoral v.
7. R. great saphenous v.
9. R. small saphenous v.
8. R. popliteal v.
10. R. anterior tibial v.
11. R. posterior tibial v.
12. R. fibular v.

Note: Gray arrows indicate direction of blood flow.

Inguinal ligament

BLOOD FLOW

Femur

Tibia

Fibula

Dorsal arch

Metatarsal v.

Plantar arch

Digital v.

Right Lower Limb
(anterior view)

Label major branches

1. _____

2. _____

3. _____

4. _____

5. _____

6. _____

7. _____

8. _____

9. _____

10. _____

11. _____

12. _____

Note: Small arrows indicate direction of blood flow.

* *only visible in posterior view*

Notes

LYMPHATIC SYSTEM

Description

The cardiovascular system has a close relationship with the lymphatic system. Like the veins running through the body, the lymphatic system consists of a network of thin-walled vessels called **lymphatic vessels**. Like veins, they contain one-way valves (semilunar valves) that assist in circulating the lymph, which is under very low pressure.

Instead of carrying blood, the lymphatic vessels carry **lymph**—tissue fluid originally filtered from the blood. In the illustration, the gray areas indicate this filtration process. The composition of lymph is similar to plasma—mostly water along with some solutes such as salts and small proteins.

Lymphatic vessels are connected to lymphatic capillaries and lymph nodes. **Lymphatic capillaries** are structurally similar to blood capillaries. Both are microscopic networks made of a single layer of simple squamous epithelium, but lymphatic capillaries contain flap-like structures that make them more permeable than blood capillaries.

Lymph nodes are pea-sized structures that act as tiny filters to clean the lymph. There are hundreds of nodes scattered within the body, but they tend to be most concentrated in the neck, armpits, and groin. Like an oil filter cleans the motor oil in your car's engine, lymph nodes filter pathogens such as bacteria and viruses out of your lymph before returning it to the plasma. These nodes contain macrophages that ingest and destroy pathogens. Lymphocytes are also present.

Lymph moves through lymph nodes in a one-way direction just as blood flows through blood capillaries. Lymph enters a node through multiple **afferent lymphatic vessels** and leaves through **efferent lymphatic vessels** (**note**: only a single vessel is shown in the illustration). Multiple afferents slow the flow of lymph through the node and allow time for an immune response. In short, the big picture is "unclean lymph in, clean lymph out." The cleansed lymph is finally returned to the cardiovascular system via the subclavian veins.

Flow of Lymph

Here is a summary of the flow of lymph through the lymphatic system:

Lymphatic capillaries ⟶ afferent lymphatic vessels ⟶ lymph nodes ⟶ efferent lymphatic vessels ⟶ subclavian veins

Lymph is really nothing more than filtered blood plasma. All blood capillaries constantly filter the blood as a result of the force of blood pressure (see p. 360). This fluid, called **interstitial fluid**, fills the interstitial spaces between body cells, bathing them in fluid. Filtration occurs at the higher-pressure arteriole end of the capillary. Although some of this fluid is reabsorbed at the venule end, there is still an excess amount. As interstitial fluid pressure builds in the interstitial spaces, it is shunted into the nearby lymphatic capillaries, which act as a drain for the excess fluid. Although this fluid has not changed its chemical composition in any way, once inside the lymphatic capillary, it now is called **lymph**. Think of this **fluid cycle** as recycling of our plasma. This helps maintain normal fluid levels in the blood and tissues. As shown in the illustration: **Plasma** is filtered to become **interstitial fluid**, which becomes **lymph**, which becomes plasma once again. The cycle is complete!

Study Tip

The terms "afferent" and "efferent" apply to multiple organ systems. Here is a way to distinguish them: **A**fferent as in **A**pproach; **E**fferent as in **E**xit.

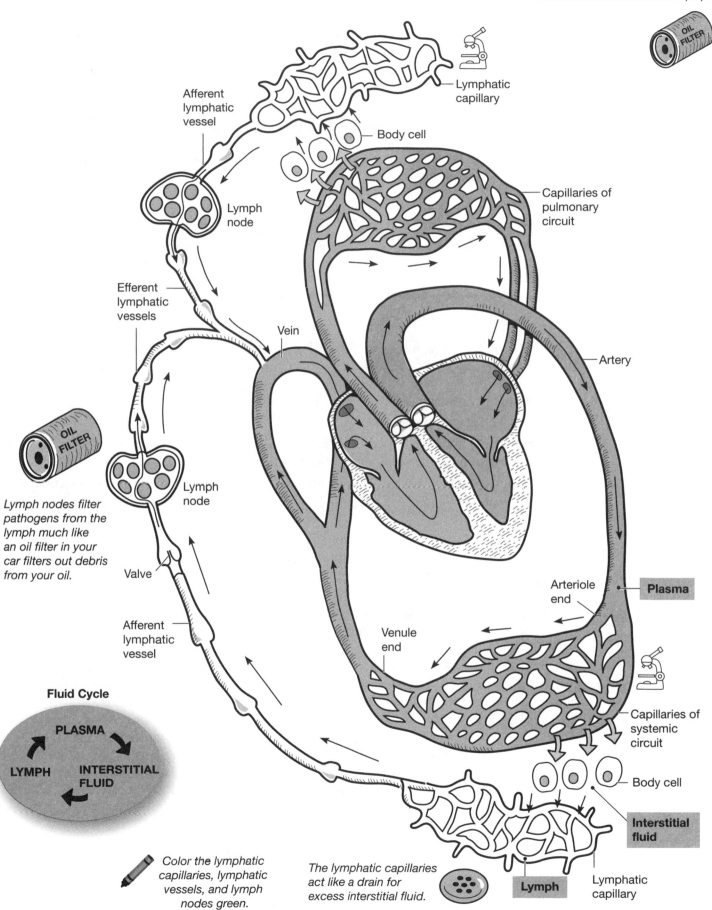

Afferent lymphatic vessel

Lymphatic capillary

Body cell

Capillaries of pulmonary circuit

Lymph node

Efferent lymphatic vessels

Vein

Artery

OIL FILTER

Lymph nodes filter pathogens from the lymph much like an oil filter in your car filters out debris from your oil.

Lymph node

Valve

Arteriole end

Plasma

Afferent lymphatic vessel

Venule end

Capillaries of systemic circuit

Fluid Cycle

PLASMA

LYMPH

INTERSTITIAL FLUID

Body cell

Interstitial fluid

Color the lymphatic capillaries, lymphatic vessels, and lymph nodes green.

The lymphatic capillaries act like a drain for excess interstitial fluid.

Lymph

Lymphatic capillary

Description

The **immune system** protects your body against foreign pathogens such as bacteria and viruses using several different lines of defense. The methods it uses can be grouped into two broad categories: (1) **Nonspecific resistance**, and (2) **Specific resistance**. Let's give an overview of each.

Nonspecific Resistance

Nonspecific resistance uses general defenses to prevent pathogenic invasion without targeting specific kinds of pathogens. Think of a pathogenic invasion as one army attacking another army inside a medieval castle. Your body is like the castle. The first line of defense is the wall around the castle. Similarly, your body has the following physical and chemical barriers as its **first line of defense:**

Physical barriers

- **Skin**—a thick layer of dead cells in the epidermis provides protection.
- **Mucous membranes**—the mucus covering these membrane surfaces traps microbes.

Chemical barriers

- **Lysozyme** in tears is an antibacterial agent.
- **Gastric juice** in the stomach is highly acidic (pH 2–3), which destroys bacteria.

Pathogens that pass through the first line of defense must deal with a second line of defense. Imagine archers along the top of the castle shooting arrows into the invading army or soldiers pouring vats of boiling oil on them. Both attack methods would kill clusters of enemy soldiers. Similarly, your body has its own **second line of defense,** and here are a few examples:

- **Phagocytic cells** ingest and destroy all microbes that pass into body tissues.
- **Inflammation** is a normal body response to tissue damage and other stimuli that brings more white blood cells to the site of pathogenic invasion.
- **Fever** inhibits bacterial growth and increases the rate of tissue repair during an infection.
- **Natural killer (NK) cells** destroy virus-infected cells and some tumor cells cells in body tissues.

Specific Resistance

The **third line of defense** deals with **specific resistance,** illustrated on the facing page. In modern warfare, these defenses would be like guided missiles that detect and destroy a specific target. In the medieval castle analogy, they might be specially trained soldiers who act as assassins to kill the enemy's general. In short, they have a specific mission with a very specific target. Your immune system also has "assassin" cells that attack specific microbes. Unlike in a war in which soldiers in different armies are wearing different uniforms, your immune system has a more difficult task in distinguishing its own tissues ("self") from foreign microbes ("nonself"). To make this important distinction, it relies on detecting foreign **antigens,** specific substances found on the surface of pathogens. Most are proteins that serve as the stimulus to induce an immune response. The term "antigen" is coined from *anti*body *gen*erator.

The illustration shows the immune response to an antigen. Once the antigen on a virus or bacterium is detected, a dual response is activated by two groups of specialized lymphocytes called **T cells** and **B cells.** These cells communicate with each other through chemical signaling. T cells typically are activated first. After activation, they can either directly destroy the microbes or use chemical secretions to destroy them. At the same time, T cells stimulate B cells to divide, forming other cells, called plasma cells, able to produce **antibodies.** These Y-shaped proteins circulate though the bloodstream and bind to specific antigens, thereby attacking microbes.

Details of the mechanisms T cells and B cells use to attack microbes are examined in subsequent modules. Together, the T cells and B cells provide specific resistance to specific antigens.

Immune System: Overview of Specific Resistance

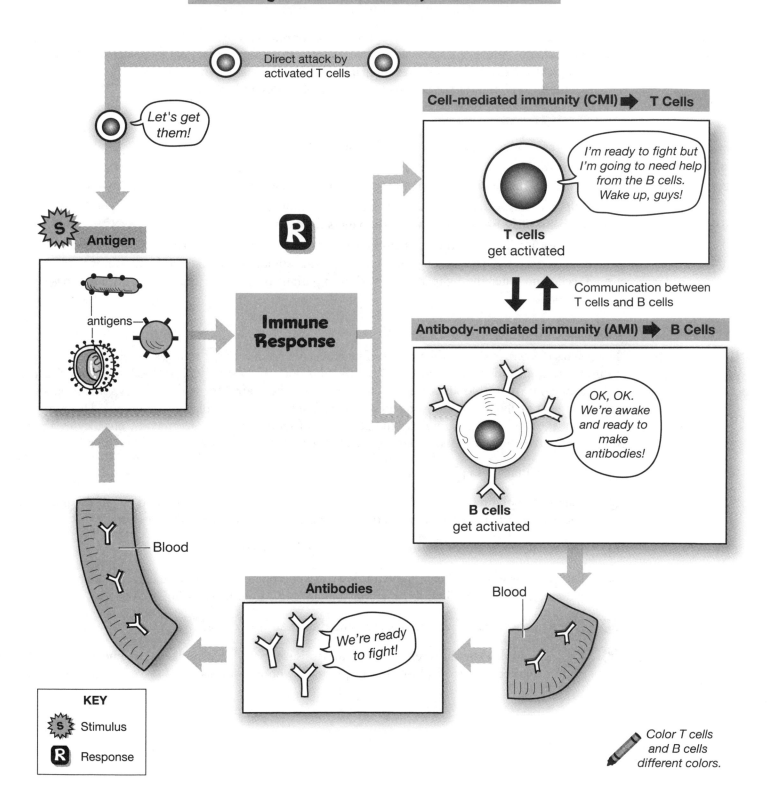

Description

Cell-mediated immunity (CMI) involves the activation of T cells (T lymphocytes) by a specific antigen. In total, the body contains millions of different T cells—each able to respond to one specific antigen.

Development of T cells

T cells are a special type of lymphocyte. **Immature lymphocytes** are produced from stem cells in the **red bone marrow**. Some of these cells are processed within the **thymus gland**—hence, T cell—during embryological development, then released into the blood. These mature T cells are located in the blood, lymph, and lymphoid organs such as the lymph nodes and spleen.

Common T cells and Their Functions

The three major types of T cells are as follows:

- **Cytotoxic T cells** secrete **lymphotoxin** and **perforin**. The former trigger destruction of the pathogen's DNA, and the latter create holes in the pathogen's plasma membrane, resulting in a **lysed cell**.

- **Helper T cells** secrete **interleukin 2** (I-2), which stimulates cell division of T cells and B cells. This can be thought of as recruiting more soldiers for the fight.

- **Memory T cells** remain dormant after the initial exposure to an antigen. If the same antigen presents itself again—even years later—the memory cells are stimulated to convert themselves into cytotoxic T cells and enter the fight. Memory is an important property of immunity and is the basis for vaccinations.

Phagocytosis and Antigen Presentation

Phagocytes ("eater cells") use the process of **phagocytosis** ("cell eating") to ingest foreign pathogens. An example is a macrophage ("big eater"), derived from the largest white blood cells—monocytes. Macrophages leave the bloodstream and enter body tissues to patrol for pathogens. As some phagocytic cells engage in phagocytosis, they present antigenic fragments on their plasma membrane. This is a very important mechanism that stimulates the activation of more T cells. Consequently, phagocytes are an important part of the CMI. A summary of the phagocytosis process shown in the illustration is:

(1) **Microbe** attaches to **phagocyte.**

(2) Phagocyte's **plasma membrane** forms armlike extensions that surround and engulf the microbe. The encapsulated microbe pinches off from the plasma membrane to form a **vesicle.**

(3) The vesicle merges with a **lysosome,** which contains **digestive enzymes.**

(4) The digestive enzymes begin to break down the microbe. The phagocyte extracts the nutrients it can use, leaving the **indigestible material** and **antigenic fragments** within the vesicle.

(5) The phagocyte makes **protein markers,** and they enter the vesicle.

(6) The indigestible material is removed by exocytosis (see p. 86). The antigenic fragments bind to the protein marker and are displayed on the plasma membrane surface. This serves to activate T cells.

Cell-Mediated Immunity (CMI)

Development of T Cells

Red bone marrow **Immature lymphocytes** **Thymus gland** (processing center) **T cell** (T lymphocyte)

Color the different T cells different colors.

Common Types of T Cells and Their Functions

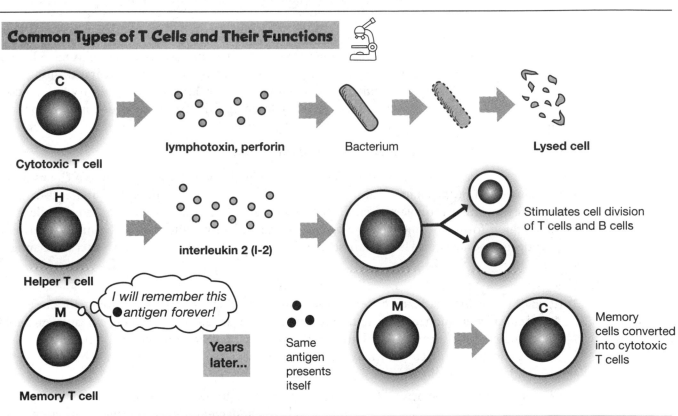

Cytotoxic T cell → **lymphotoxin, perforin** → **Bacterium** → **Lysed cell**

Helper T cell → **interleukin 2 (I-2)** → Stimulates cell division of T cells and B cells

Memory T cell — *I will remember this antigen forever!*

Years later... Same antigen presents itself **M** → **C** Memory cells converted into cytotoxic T cells

Phagocytosis and Antigen Presentation

Bacterium Plasma membrane Antigenic fragments Indigestible material T cell activation

Antigen ① ② Vesicle ③ ④ ⑤ ⑥ Protein marker + antigenic fragment

Lysosome — Digestive enzymes Protein marker Endoplasmic reticulum

Phagocyte

Description

Antibody-mediated (*humoral*) **immunity** (**AMI**) involves the activation of **B cells** (B lymphocytes) by a specific antigen. This triggers the B cells to transform into **plasma cells** able to secrete special proteins called **antibodies.** The antibodies are transported through the blood and the lymph to the pathogenic invasion site. In total, the body contains millions of different B cells—each able to respond to one specific **antigen** and produce one specific antibody. Amazing!

Development of B Cells

B cells are a special type of lymphocyte. **Immature B cells** are produced from stem cells in the **red bone marrow.** These immature cells are later processed within the red bone marrow—hence, B cell—during embryological development to become mature B cells, then are released into the blood. The mature B cells are located in the blood, lymph, and lymphoid organs such as the lymph nodes and spleen.

Antibody Production

B cells can be stimulated to divide, forming two types of numerous cells: (1) **plasma cells,** which secrete **antibodies,** and (2) **B memory cells,** which exist in the body for many years, ensuring a quick response to the same antigen. Antibodies (immunoglobulin, or Ig) are Y-shaped proteins subdivided into five classes: IgG, IgM, IgA, IgE, IgD. These are listed in order from the *most* common to the *least* common.

Study Tip: Mnemonic: *Get Me Another Excellent Donut!*

The basic structure on an antibody consists of four polypeptide chains—two **heavy chains** and two **light chains.** Both heavy chains and both light chains are identical to the other, and each contains a **constant region** and a **variable region.** The constant region forms the "trunk" of the molecule, and the variable region forms the **antigen-binding site** on the antibody. Note that each antibody has two of these antigen-binding sites. Think of these like claws on a lobster used to "grab" its specific antigen.

How Do Antibodies Work?

Antibodies work through many different mechanisms, of which the following are major ones:

1. **Neutralizing antigen:** The **antibody** can bind to an **antigen,** forming an **antigen–antibody complex.** This forms a shield around the antigen, preventing its normal function. In this way, a toxin from a bacterium may be neutralized or a viral antigen may not be able to bind to a body cell, thereby preventing infection.

2. **Activating complement:** "Complement" refers to a group of plasma proteins made by the liver that normally are inactive in the blood. An **antigen-antibody complex** triggers a cascade reaction that activates these proteins to induce beneficial responses. For example, some of these activated proteins can cluster together to form a pore or channel that inserts into a microbial plasma membrane. This results in a **lysed cell.** Other responses include **chemotaxis** and **inflammation.** Both of these mechanisms serve to increase the number of white blood cells at the site of invasion. (*Please note this mechanism is much more complex than what is presented here*).

3. **Precipitating antigens:** Numerous antibodies can bind to the same free antigens in solution to cross-link them. This cross-linked mass then precipitates out of solution, making it easier for phagocytic cells to ingest it by phagocytosis (see p. 86).

   Similarly, microbes (such as bacteria) can be clumped together by a process called *agglutination* (not illustrated). The antigens within the cell walls of the bacteria are cross-linked. As with precipitation, this is followed by phagocytosis.

4. **Facilitating phagocytosis:** An **antigen-antibody complex** acts like a warning sign to signal phagocytic cells to attack. In fact, the complex also binds to the surface of **macrophages** to further facilitate phagocytosis.

Development of B cells

Red Bone marrow

Immature lymphocytes

Red bone marrow (*processing center*)

B cell (B lymphocyte)

Color the B cells, plasma cells, and antibodies different colors.

Antibody production

antigen

S

Activated B Cell

R

Plasma cells

Antibodies

Memory B cells

Antigen

Antigen-binding site

Light chain (*gray*)

C C

Heavy chain (*color*)

C = Constant region

V = Variable region

Structure of an antibody

KEY

S Stimulus

R Response

How do antibodies work?

Color the antibodies all the same color.

1. Neutralizing antigen

● Antigen

Antibody

Antigen-antibody complex

3. Precipitating antigens

2. Activating complement

Activates complement

Bacterium

Lysed cell

• **Inflammation**
• **Chemotaxis**

4. Facilitating phagocytosis

Attack!

!

Macrophage

Antigen-antibody complex acts like a warning sign

Notes

RESPIRATORY SYSTEM

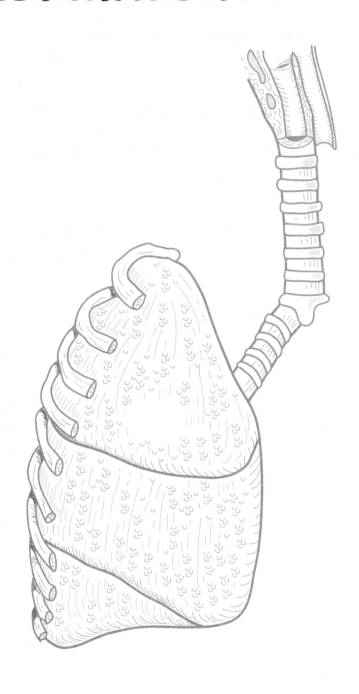

Description

The respiratory system is divided into two major divisions: *upper respiratory system* and *lower respiratory system*. The **upper respiratory system** consists of the nose, nasal cavity, paranasal sinuses, and pharynx. The **lower respiratory system** consists of the larynx, trachea, bronchi, and lungs.

Let's trace the pathway of a molecule of oxygen (O_2) through the respiratory system to its final destination at a body cell. The O_2 molecule enters the **nasal cavity** through the **external nares** *(nostrils)*. As it passes to the back of this moist chamber, it enters the **nasopharynx**, then the **oropharynx**, and finally the **laryngopharynx**.

After passing the rigid, flap-like structure called the **epiglottis**, it enters the **larynx**, then passes through the slit-like opening between the vocal cords called the **glottis**. Next it moves through the long, rigid tube of the **trachea** until it reaches a split in this passageway. Following the passageway branching into the left lung, it enters the **left primary bronchus**, then enters the next split in the passageway, the narrower **secondary bronchus**. The next branch is the even narrower **tertiary bronchus**. Finally, the O_2 molecule enters a microscopic tube called a **bronchiole**.

This continues to branch into a **terminal bronchiole**, then a **respiratory bronchiole**, and finally terminates in an air sac called an **alveolus**. This delicate air sac is the end of the bronchial tree in the lungs. The wall of each alveolus is made of **simple squamous epithelium**, which allows for easy diffusion of the O_2 molecule out of the alveolus and into the bloodstream, where it will be delivered to a body cell.

Analogies

- The **larynx** looks like the **head of a snapping turtle**. The **turtle's head** is the **thyroid cartilage**, and the **turtle's lower jaw** is the **cricoid cartilage**. The **neck of the turtle** is the **trachea**.

- The **clusters of alveoli** are like a **wad of bubble wrap** used in packaging. Both are small, saclike structures filled with air.

Study Tip

Palpate (*feel by touch*):

On average, the larynx is slightly larger in males than in females, but you can easily feel a portion of it in either gender. The main structure you feel beneath the skin is the thyroid cartilage, commonly called the *Adam's apple*. The portion of the Adam's apple that protrudes most anteriorly is called the laryngeal prominence.

Key to Illustration

Larynx (L)

L1. Thyrohyoid membrane
L2. Thyroid cartilage
L3. Laryngeal prominence
L4. Cricothyroid ligament
L5. Cricoid cartilage

Upper Respiratory Tract (UP)

UP1. Frontal sinus
UP2. Sphenoidal sinus
UP3. External nares (*nostrils*)
UP4. Nasal cavity
UP5. Superior concha
UP6. Middle concha
UP7. Inferior concha
UP8. Pharyngeal tonsil
UP9. Nasopharynx
UP10. Palatine tonsil
UP11. Oropharynx
UP12. Lingual tonsil
UP13. Epiglottis
UP14. Laryngopharynx

Lower Respiratory Tract (LO)

LO1. Vocal fold
LO2. Trachea
LO3. Tracheal rings
LO4. Parietal pleura
LO5. Visceral pleura
LO6. Location of carina (*internal ridge*)
LO7. Primary bronchus
LO8. Secondary bronchus
LO9. Tertiary bronchus
LO10. Terminal bronchiole
LO11. Respiratory bronchiole
LO12. Alveoli
LO13. Simple squamous epithelium

UP1
UP5
UP6
UP7
UP3
UP2
UP4
UP8
UP9
UP10
UP11
UP12
UP13
UP14

L2
LO1
LO2
L4

Esophagus

S S
M
I I

Color the lobes.

The left lung
has two lobes;
the right lung
has three lobes.

KEY
S = superior lobe
M = medial lobe
I = inferior lobe

LO3

**RIGHT
LUNG**

LO6

LO7

Ribs
(cut)

Diaphragm

Color the lobes.

Hyoid bone

L1

L2

L3
L4
L5

LO3

The larynx
looks like the
head of a
snapping
turtle.

Larynx (L)
(anterior view)

LO4 LO5

**LEFT
LUNG**

LO8

LO9

LO13

Alveolus
(cut in half)

LO11

LO10 LO12

The alveoli are
small, air-filled sacs
like the bubbles in
bubble wrap.

RESPIRATORY SYSTEM

Tracing the Pathway of an Oxygen Molecule

The following flowchart gives the pathway of oxygen through the respiratory system and into a body cell:

1. External nares *(nostrils)*

2. Nasal cavity

3. Nasopharynx

4. Oropharynx

5. Laryngopharynx

6. Larynx

7. Trachea

8. Primary bronchus

9. Secondary bronchus

10. Tertiary bronchus

11. Bronchiole

12. Terminal bronchiole

13. Respiratory bronchiole

14. Alveolus

15. Erythrocyte *(red blood cell)*

Tracing the Pathway of an Oxygen Molecule

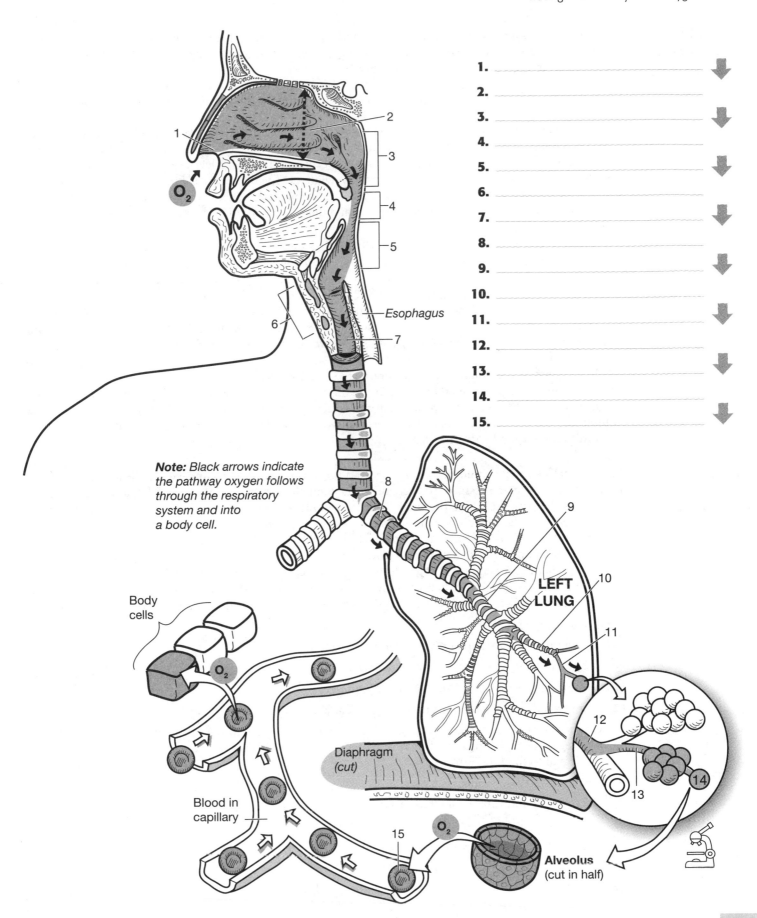

Note: Black arrows indicate the pathway oxygen follows through the respiratory system and into a body cell.

Esophagus

Body cells

Diaphragm *(cut)*

Blood in capillary

LEFT LUNG

Alveolus *(cut in half)*

1. _____
2. _____
3. _____
4. _____
5. _____
6. _____
7. _____
8. _____
9. _____
10. _____
11. _____
12. _____
13. _____
14. _____
15. _____

411

Concept 1: Atmospheric Pressure (AP)

Atmospheric pressure (AP) is a force that combines the weight of all the gases in the air we breathe. Of the many gases, the most abundant are: nitrogen (N_2), oxygen (O_2), carbon dioxide (CO_2), and water vapor (H_2O). Each of these gases has a different mass.

AP is the force on any surface that comes in contact with air. In the illustration, AP is indicated by the black arrows pushing against the external surface of the little boy's body. Even though he can't actually feel its presence, it is there.

The other illustration shows a barometer—a device used to measure the AP. A glass beaker is shown filled with liquid mercury (Hg). As the AP pushes down on the surface of this liquid, it forces the mercury into the open end of the glass tube so it rises to a specific height. At sea level, the standard measurement is 760 mm Hg. This number changes as the elevation changes. *The higher the elevation, the lower the AP*. For example, a mountain climber and a person strolling in the park at sea level experience different AP. The mountain climber has less atmosphere above him, so he has less AP than the person in the park.

Concept 2: Pressure and Volume Law (Boyle's Law)

The pressure and volume law (*Boyle's law*) is a gas law stating that *volume and pressure have an inversely proportional relationship for a gas held at a constant temperature*. The illustration shows a clear, hollow sphere containing five gas molecules. These molecules have kinetic energy, so they randomly bounce around against the wall of the sphere. This is the source of the force of pressure inside the sphere.

Consider what would happen if the small sphere were to increase in size. Has the number of gas molecules changed? No. Five gas molecules are still present. The only thing that has changed is the volume of the sphere. If the volume increases, what do you predict will happen to the pressure inside? That's right. It will decrease.

If we refer to the volume and pressure in the small sphere as V_1 and P_1 and use V_2 and P_2 for the enlarged sphere, the pressure and volume law can be expressed as this equation: $V_1 \times P_1 = V_2 \times P_2$. The product of the volume and the pressure in the larger sphere should be equivalent to the product of the volume and the pressure in the smaller sphere. Fill in the given numbers, and plug them into the formula to see if this relationship holds true.

Application: Mechanics of Breathing

In addition to knowing atmospheric pressure and Boyle's law, we need to understand the terms **interpleural pressure** and **alveolar pressure**, so we can understand the mechanics of normal breathing. Interpleural pressure is the pressure inside the small, liquid-filled, pleural space around the lungs. At rest, the normal value is 756 mm Hg. The pressure inside the lungs at any given moment is called the alveolar pressure. Its normal value at rest is the same as the normal AP, or 760 mm Hg. Normal, quiet breathing is a repeated cycle of **inhalations** and **exhalations**. Inhalation is an *active* process because it requires the contraction of muscles such as the **diaphragm** and the **external intercostal muscles** between the ribs. In contrast, an exhalation is a passive process because it mainly involves recoil of the elastic connective tissue in the lungs and thoracic wall. Let's summarize the steps involved in a normal inhalation and a normal exhalation.

Steps in a normal inhalation:
1. Diaphragm and external intercostal muscles contract
2. Thoracic cavity expands; lung volume increases
3. Alveolar pressure drops from 760 to 758 mm Hg
4. Air flows into lungs down its pressure gradient from 760 to 758 mm Hg

Note: There is no such thing as "suction." We do not "suck" air into the lungs. Instead, air flows from a region of higher pressure to a region of lower pressure. This continues until the alveolar pressure is equal to the AP (760).

Steps in a normal exhalation:
1. Diaphragm and external intercostals relax
2. Thoracic cavity contracts; lung volume decreases
3. Recoil effect from lungs and thoracic wall
4. Alveolar pressure increases from 760 to 762 mm Hg
5. Air flows out of the lungs down its pressure gradient from 762 to 760 mm Hg. This continues until the alveolar pressure is equal to the AP (760 mm Hg).

CONCEPT 1: ATMOSPHERIC PRESSURE (AP)

AP

I can't handle the pressure!

AP

AP

AP

AP

AP is a force exerted on any surface that comes in contact with air.

760 mm Hg = Atmospheric pressure (*at sea level*)

AP

AP

Open end of glass tube

Mercury (Hg)

Model of a Barometer

CONCEPT 2: PRESSURE AND VOLUME LAW

Gas molecule (*ex.*: O_2 gas)

Hollow sphere

(*increase size*)

(*decrease size*)

Volume (V_1) = 2
Pressure (P_1) = 2

Volume (V_2) = 4
Pressure (P_2) = 1

Formula: $V_1 \times P_1 = V_2 \times P_2$

Color the gas molecules.

___ ___ = ___ ___

APPLICATION: MECHANICS OF BREATHING

During inhalation, the ribs swing upward and outward like the handle on a pail.

Rib

Sternum

The recoil effect during exhalation is like the tension being released from a rubber band.

AP 760

Interpleural pressure

754

Thoracic wall

Alveolar pressure

758

Diaphragm

External intercostal muscles

Diaphragm moves downward

AP 760

Interpleural pressure

756

Alveolar pressure

762

Diaphragm

Diaphragm moves upward

Color the numbers for alveolar and interpleural pressures; color the regional areas for alveolar and intrapleural pressures.

INHALATION
(*active* process)

EXHALATION
(*passive* process)

413

Alveolar Structure and the Respiratory Membrane

The lungs contain about 300 million alveoli to increase surface area for exchange of respiratory gases. Like the bubbles in bubble wrap, the **alveoli** (sing. *alveolus*) are the delicate, microscopic air sacs where gas exchange occurs between the lungs and the blood. Each alveolus is composed of three different types of cells:

1. **Simple squamous epithelial cells** (*Type I cells*)
2. **Surfactant-secreting cells** (*Type II cells*)
3. **Alveolar macrophages** (*dust cells*)

The numerous simple squamous epithelial cells make up the wall of each alveolus, and their flat shape allows for better diffusion of respiratory gases (oxygen and carbon dioxide). The surfactant-secreting cells are fewer in number and are scattered within the alveoli. They secrete **surfactant**—an oily fluid film that lines the inside of the alveoli like a soap bubble and serves to reduce surface tension to prevent collapse. The alveolar macrophages are the "housekeepers" of the alveoli. They move around and engulf microorganisms, dust particles, and other debris to keep the lungs clean and free of disease.

Recall that the blood flows to each lung through the **pulmonary arteries** (left and right). Each of these vessels branches and becomes smaller until it finally reaches numerous capillary beds that surround the alveolar sacs—clusters of alveoli. This structural relationship is like a plastic mesh bag (blood capillaries) surrounding a cluster of grapes (alveolar sac) at the market. The **respiratory membrane** is called the "blood/air barrier" and is very thin to allow for easy diffusion of oxygen and carbon dioxide. This is the site where oxygen diffuses from alveolus to blood and carbon dioxide diffuses from blood to alveolus. This structure consists of the fusion of the alveolar and capillary walls. More specifically, it is where the **simple squamous epithelium** of the alveolus meets the **simple squamous epithelium** of the capillary with basement membranes in between.

Surfactant and Surface Tension

Surface tension is both a force and a property of water, attributed to the fact that water molecules are **polar**. Like a magnet, the oxygen end of any water molecule has a partially negative charge and the hydrogen end has a partially positive charge. When the negative end of one water molecule aligns itself with the positive end of another, a **hydrogen bond** is formed. In fact, water molecules are always rearranging themselves to maximize the number of hydrogen bonds because this is the most stable state for them. This gives water a high degree of surface tension and explains, by the way, why the insect called a **water strider** can stand on the surface of the water.

Surfactant is a fluid secreted by **surfactant-secreting cells** (Type II cells) that contains a mixture of phospholipids and lipoproteins. It functions to reduce the surface tension on alveoli to prevent them from collapsing after exhalation. If pure water were to line the inside of the alveoli, the surface tension in the water would tend to pull inward on the alveoli. During exhalation, the delicate alveoli would collapse because of the high surface tension on them. Surfactant reduces the amount of hydrogen bonds in the water normally found inside the alveoli, thereby reducing the surface tension. This keeps the alveoli partly inflated at all times. During normal breathing, the alveoli would look like slowly pulsating spheres that expand in size during inhalation and shrink in size during exhalation. This process is more energy-efficient than if we had to refill collapsed alveoli with every inhalation.

Lung Compliance

Note: Lung compliance is the only topic not illustrated on the facing page.

Lung compliance is a measure of the degree of effort needed to expand, or ease of expanding, the lungs and thoracic wall. This is determined primarily by two factors: (1) elasticity of lung tissue, and (2) alveolar surface tension. Two states of compliance are mentioned—high and low. Normal, healthy lungs have a *high* degree of compliance because of the elastic connective tissue within the lungs and the reduced surface tension in the lungs because of surfactant. States of *low* compliance may occur with some respiratory disorders.

For example, a premature baby may lack the normal surfactant found in a newborn. This increases surface tension on the alveoli. Emphysema also leads to lower compliance because of destruction of the walls of the alveoli. In short, any disease state that increases alveolar surface tension and/or damages normal elastic connective tissue in the lungs will result in lowered compliance.

Alveolar Structure and the Respiratory Membrane

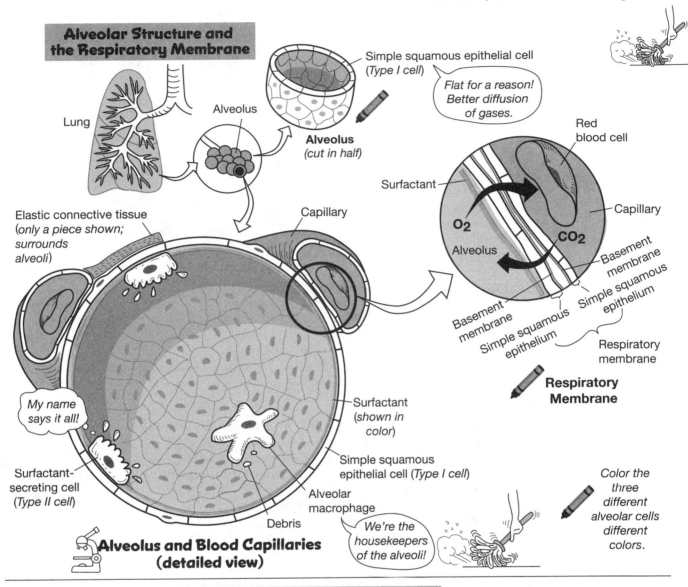

Lung

Alveolus

Alveolus
(cut in half)

Simple squamous epithelial cell
(Type I cell)

Flat for a reason! Better diffusion of gases.

Elastic connective tissue (*only a piece shown; surrounds alveoli*)

Capillary

Red blood cell

Surfactant

O_2

CO_2

Alveolus

Capillary

Basement membrane

Basement membrane

Simple squamous epithelium

Simple squamous epithelium

Respiratory membrane

Respiratory Membrane

My name says it all!

Surfactant (*shown in color*)

Simple squamous epithelial cell (*Type I cell*)

Alveolar macrophage

Surfactant-secreting cell (*Type II cell*)

Debris

We're the housekeepers of the alveoli!

**Alveolus and Blood Capillaries
(detailed view)**

Color the three different alveolar cells different colors.

Surfactant and Surface Tension

Alveolus

Notice that I never collapse during normal breathing!

Alveolus

Surfactant (*shown in color*)

Inhalation
(alveolus increases in size)

Exhalation
(alveolus decreases in size)

Water

Water (H_2O) molecules

Hydrogen bond

Water has a high degree of **surface tension** because of hydrogen bonding between water molecules. This allows insects called water striders to stand on water.

Surfactant *reduces* surface tension on alveoli, which prevents them from collapsing.

This module deals with the following concepts:

- Dalton's law
- Henry's law
- Diffusion of O_2 and CO_2 into and out of the blood
- External respiration and internal respiration

Model of a Barometer

Dalton's Law

Dalton's law is a gas law that deals with **partial pressures** of gases. It states that *in a mixture of gases (like the air we breathe), each gas exerts its own partial pressure*. Therefore, the sum of all the partial pressures is the total pressure exerted by the gas mixture.

Consider a barometer that measures atmospheric pressure (AP). At sea level, the normal AP is 760 mm Hg. Air contains the following gases: nitrogen (78%), oxygen (21%), water vapor (0.4%), carbon dioxide (0.04%), and other gases (0.06%). The partial pressure of nitrogen gas (P_{N_2}) is 78 percent of 760 or **597** mm Hg. The partial pressure of oxygen (P_{O_2}) is 21% of 760, or **159** mm Hg. Therefore, the combined partial pressures from nitrogen and oxygen account for **99** percent of atmospheric pressure (756 of the 760 total). In the body, we have to consider the partial pressures of physiologically important gases such as O_2 and CO_2 in the alveolar air and the blood.

Henry's Law

Another gas law, **Henry's law**, states that at a given temperature, the amount of gas dissolved in a liquid is directly proportional to the partial pressure of that gas. Gases normally move in and out of solution, with some more soluble in liquids than others. For example, carbon dioxide (CO_2) is highly soluble in water—much more so than oxygen (O_2).

As an example of a gas moving out of solution, let's consider soda pop. All sodas contain carbonated water—CO_2 dissolved in water. Although much of the CO_2 remains in solution, some is found in the air inside the bottle. If you open a soda and allow it to sit out at room temperature for a long time, all the CO_2 eventually will move out of solution, and the soda will go "flat." If you shake a fresh bottle, you can see the CO_2 bubble out of solution, which builds up pressure inside the bottle. This increase in pressure causes the soda to shoot out when the bottle is opened. Similarly, in the lungs, CO_2 moves out of solution from the blood capillaries and into the alveolar air.

Shake it up! CO_2 bubbles out of solution.

The increase in pressure causes the soda to burst out.

Diffusion of O_2 and CO_2 into and out of the Blood

The **blood P_{CO_2} level** is about 45 mm Hg as it approaches the lungs, and the P_{CO_2} in the alveoli is about 40 mm Hg. Because of the small gradient, CO_2 slowly diffuses out of the blood and into the alveoli. As the blood leaves the lungs, the P_{CO_2} level has dropped slightly, to about 40 mm Hg. Because tissues produce CO_2 as a by-product of normal metabolism, it gradually diffuses back into the blood causing the P_{CO_2} levels to rise back to 45 mm Hg. Therefore, the P_{CO_2} levels in the blood have a narrow range, between 40–45 mm Hg. This stable amount of CO_2 forms bicarbonate in the blood, which helps buffer the blood's pH to keep it in the normal range of 7.35–7.45.

The **blood P_{O_2} level** is about 40 mm Hg as it approaches the lungs, while the P_{O_2} of the alveolar air is about 105 mm Hg. Because of the large gradient, O_2 quickly diffuses into the blood. Because the hemoglobin in the red blood cells is especially efficient at binding O_2, the blood P_{O_2} levels rise to 100 mm Hg—the typical level for oxygenated blood. When the oxygenated blood approaches tissues low in oxygen, it rapidly diffuses into tissues. In deoxygenated blood, the P_{O_2} drops to about 40 mm Hg. In short, the blood P_{O_2} levels vary widely, from 40 mm Hg in deoxygenated blood to 100 mm Hg in oxygenated blood.

External Respiration and Internal Respiration

- **External respiration** occurs between the alveoli in the lungs and the blood capillaries; O_2 moves from the alveoli into the blood, and CO_2 moves from the blood into the alveoli.

- **Internal respiration** occurs between the blood capillaries and body cells; O_2 moves from the blood into body cells, and CO_2 moves from the body cells (where it is produced as a waste product) into the blood.

Once oxygen is inside body cells, it is used in the important process of **cellular respiration** (see p. 472).

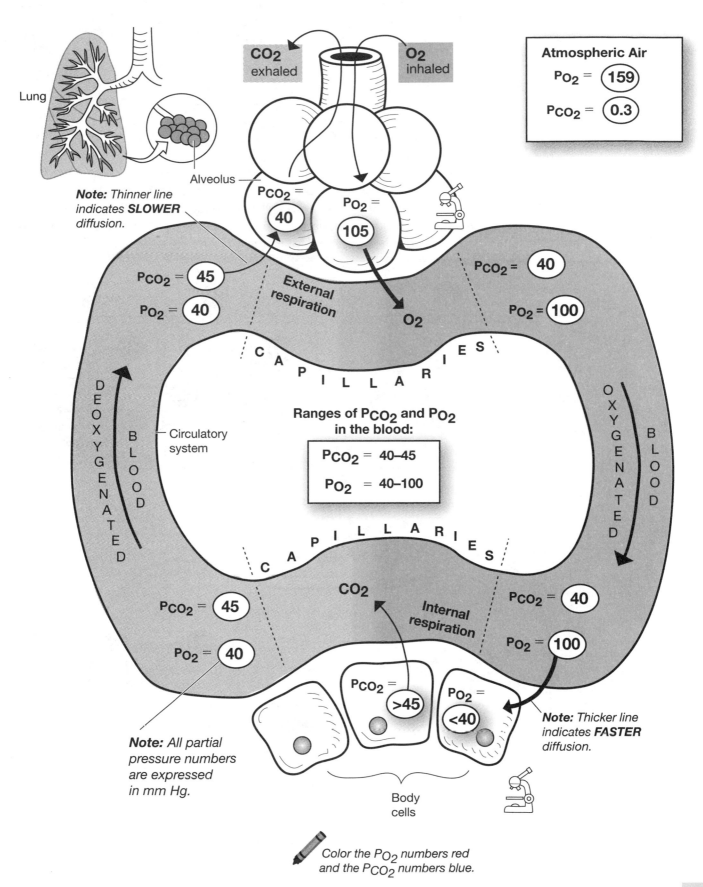

Lung

Note: Thinner line indicates **SLOWER** diffusion.

CO_2 exhaled

O_2 inhaled

Atmospheric Air

$P_{O_2} = 159$

$P_{CO_2} = 0.3$

Alveolus

$P_{CO_2} = 40$

$P_{O_2} = 105$

$P_{CO_2} = 45$

$P_{O_2} = 40$

External respiration

O_2

$P_{CO_2} = 40$

$P_{O_2} = 100$

CAPILLARIES

DEOXYGENATED BLOOD

Circulatory system

Ranges of P_{CO_2} and P_{O_2} in the blood:

$P_{CO_2} = 40-45$

$P_{O_2} = 40-100$

OXYGENATED BLOOD

CAPILLARIES

CO_2

$P_{CO_2} = 45$

$P_{O_2} = 40$

Internal respiration

$P_{CO_2} = 40$

$P_{O_2} = 100$

$P_{CO_2} = >45$

$P_{O_2} = <40$

Note: Thicker line indicates **FASTER** diffusion.

Note: All partial pressure numbers are expressed in mm Hg.

Body cells

Color the P_{O_2} numbers red and the P_{CO_2} numbers blue.

Description

Hemoglobin (Hb) is a protein pigment found in red blood cells. Each Hb molecule is composed of four subunits: alpha 1, alpha 2, beta 1, and beta 2. Each subunit has an iron-containing heme group where a molecule of oxygen is able to bind.

Function

The primary function of Hb is to transport oxygen from the lungs to the body cells, but it also transports some carbon dioxide from the body cells to the lungs. It is a pigment because its color changes depending on whether oxygen is bound to the molecule. It is bright red in its oxygenated form but changes to dark red in its deoxygenated form.

Two Stages for Hemoglobin

With respect to oxygen transport, hemoglobin exists in two states: **oxyhemoglobin** or **deoxyhemoglobin**. Deoxyhemoglobin is the state in which no oxygen is bound to Hb. Once inside the capillaries in the lungs, each Hb molecule binds one molecule of oxygen at a time to hold a maximum of four O_2 molecules. After binding its fourth oxygen molecule, it is saturated with oxygen and said to be in its oxyhemoglobin form.

Deoxyhemoglobin and Oxyhemoglobin

Capillary

Red blood cell

Red blood cell (RBC)
(surface view)

Hemoglobin—
inside RBC

(sectioned view)

A molecule of Hb is functionally like a taxicab. Just as a taxicab picks up passengers at the airport and drops them off at their hotel, hemoglobin picks up oxygen from the lungs and delivers it to tissues. It also picks up some of the carbon dioxide generated from body cells and delivers it to the lungs. Just like a taxicab has seats for its passengers, an Hb molecule has four binding sites for oxygen.

alpha 1

beta 1

heme group

alpha 2

beta 2

Hemoglobin (Hb) Molecule

H_3C

CH_2CH_2COOH

$H_2C=CH$

CH_2CH_2COOH

N

Fe

N

N

N

H_3C

CH_3

$H_2C=CH$

CH_3

Heme Group

Description

About 98 percent of all the oxygen (O_2) is transported through the blood on **hemoglobin** (**Hb**). The remaining 2 percent is dissolved in the plasma. Because all tissues need oxygen to survive, it is vital to understand factors that aid in oxygen binding to, and releasing from, hemoglobin. The binding of O_2 to Hb depends primarily on the *partial pressure created by oxygen* (P_{O_2}) in the blood (see p. 416). When the percent saturation of Hb is plotted versus the blood P_{O_2}, it yields a graph called the **oxygen-hemoglobin dissociation curve**.

Deoxyhemoglobin is like a taxicab without any oxygen "passengers." When the taxi is filled with four passengers, it is **oxyhemoglobin** (see p. 418). This cycle can be represented by the reversible reaction given below:

$$Hb \quad + \quad O_2 \quad \leftrightarrow \quad Hb\text{-}O_2$$
$$\text{deoxyhemoglobin} \qquad\qquad\qquad \text{oxyhemoglobin}$$

Examples

Let's interpret the curve according to the three different points numbered 1–3:

(1) At a P_{O_2} of **100**, hemoglobin is about 98 percent saturated with oxygen. An example of this would be the blood capillaries in the lungs. Because of the high P_{O_2} in the air sacs (alveoli) in the lungs, a large amount of oxygen diffuses rapidly into the blood capillaries surrounding the alveoli and immediately binds to hemoglobin.

(2) At a P_{O_2} of **40**, hemoglobin is about 75 percent saturated. On average, this is the typical blood P_{O_2} for tissues at rest. This means 25 percent of the available O_2 is released from hemoglobin to be used by tissues at rest.

(3) At a P_{O_2} of **20**, hemoglobin is about 35 percent saturated. This is a typical blood P_{O_2} for active tissues such as contracting muscle tissues during exercise. Notice that a large release of the available O_2—about 65 percent—comes with only a small change in P_{O_2} (20 mm Hg).

Other Factors

Other factors determine the **affinity** of hemoglobin for oxygen. These factors can *shift the normal curve to the left* (*higher affinity*) or *to the right* (*lower affinity*), as indicated by the arrows in the illustration. A shift to the *left* enhances the *binding* of oxygen to hemoglobin, and a shift to the *right* enhances the *release* of oxygen from hemoglobin.

Variables of Temperature, pH, and P_{CO_2}

Other variables also influence the binding of oxygen to hemoglobin. Exercise provides a practical example of three related variables we will examine: body temperature, pH, and P_{CO_2}. During exercise, skeletal muscle tissue becomes more metabolically active and changes the normal oxygen-hemoglobin dissociation curve in the following ways:

- **Effect of body temperature**—Contracting skeletal muscles generate more heat. This increase in temperature causes a shift to the right (lower affinity for oxygen), making it easier to release more oxygen from hemoglobin and deliver more oxygen to the active muscle tissue.

- **Effect of blood pH**—pH within muscle cells becomes more acidic because of the production of lactate and carbonic acid. As a result, free hydrogen ion (H^+) concentration increases. Some H^+ bind to hemoglobin. This induces a subtle shape change that causes another shift to the right.

- **Effect of P_{CO_2}**—All metabolically active tissues produce more CO_2 as a waste product. As with H^+, some of the CO_2 also binds to hemoglobin, inducing another shape change that causes a shift to the right.

Oxygen-Hemoglobin Dissociation Curve

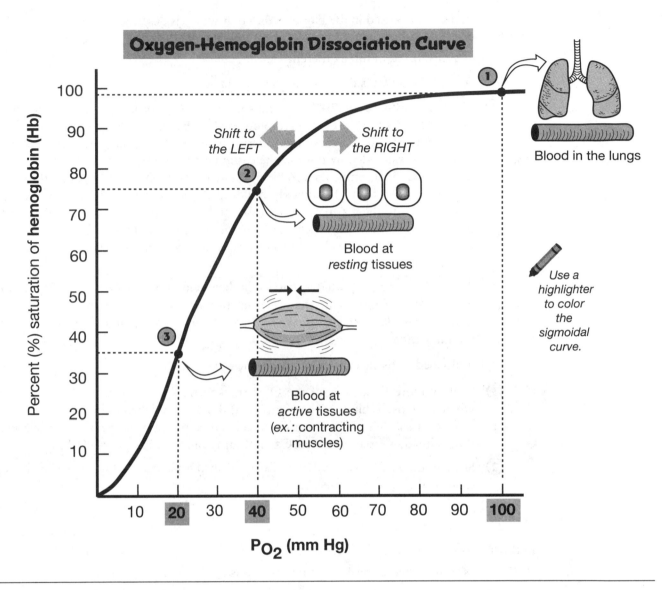

Blood in the lungs

Shift to the LEFT ← → *Shift to the RIGHT*

Blood at *resting* tissues

Blood at *active* tissues (*ex.:* contracting muscles)

Use a highlighter to color the sigmoidal curve.

Percent (%) saturation of **hemoglobin (Hb)**

P_{O_2} (mm Hg)

Effect of body temperature

Lower temp. (68° F)

Normal temp. (98.6° F)

Higher temp. (110° F)

Percent (%) saturation of hemoglobin

P_{O_2} (mm Hg)

Effect of blood pH

More basic (7.6)

Normal blood pH (7.4)

More acidic (7.2)

Percent (%) saturation of hemoglobin

P_{O_2} (mm Hg)

Effect of P_{CO_2}

Lower P_{CO_2}

Normal P_{CO_2}

Higher P_{CO_2}-

Percent (%) saturation of hemoglobin

P_{O_2} (mm Hg)

Description

CO_2 and O_2 are transported in the blood in different ways, as described below.

At the tissues: CO_2 is loaded into the blood, and O_2 is unloaded into tissue cells.

- **CO_2 is transported in three different ways:**

  (1) About 7% remains dissolved in the plasma as CO_2.

  (2) The majority—about **70%**—is transported in the form of bicarbonate ions (HCO_3^-). How? After CO_2 diffuses into **red blood cells** (**RBCs**), it undergoes a chemical reaction with water that produces an unstable intermediate product called carbonic acid (H_2CO_3). This reaction is fast, aided by the enzyme **carbonic anhydrase**. Because carbonic acid is unstable, it immediately decomposes into bicarbonate ions (HCO_3^-) and hydrogen ions (H^+).

  The levels of HCO_3^- quickly build up inside the RBC. The excess HCO_3^- has to be moved out into the plasma so more CO_2 can be converted into HCO_3^-. This is accomplished by the **chloride shift**, in which a membrane protein exchanges a HCO_3^- ion for a Cl^- ion. Each ion has the same charge, so this does not result in any change in total charge across the membrane.

  (3) About 23% is transported as **carbaminohemoglobin** (Hb-CO_2) when CO_2 loosely binds to hemoglobin (Hb). Because the main function of hemoglobin is to transport O_2, CO_2 *binds to a different binding site on the molecule than O_2 so the two gases are not competing for the same binding site.*

- **O_2 is unloaded to tissue cells in two different ways:**

  (1) Slightly more than 98% of the O_2 is transported in the form of **oxyhemoglobin** (Hb-O_2) when O_2 loosely binds to hemoglobin (Hb). Then oxyhemoglobin releases the O_2, and it diffuses tissue cells. Here it is used in cellular respiration. Some of the excess hydrogen ions (H^+) already present in the RBC temporarily bind to Hb to form H-Hb.

  (2) Slightly under 2% of the O_2 is dissolved in the plasma as O_2. Some of this diffuses into tissue cells.

In the lungs—CO_2 is unloaded into the alveoli, and O_2 is loaded into the blood.

- CO_2 is transported in reverse from how it was at the tissues.

  (1) Some of the CO_2 dissolved in the plasma diffuses into the alveoli.

  (2) The chemical reaction between CO_2 and water runs in reverse. In other words, the bicarbonate ions (HCO_3^-) and hydrogen ions recombine to form carbonic acid ($H_2CO_3^-$) again. Then, carbonic acid decomposes into CO_2 and water. The CO_2 liberated from this process diffuses into the **alveoli.**

The **chloride shift** is reversed in the lungs. In other words, a membrane protein transports a HCO_3^- into the RBC in exchange for moving a Cl^- into the plasma. This supplies lots of HCO_3^- for the purpose of continuing to run the reaction in the reverse direction.

  (3) CO_2 is released from **carbaminohemoglobin**, and it diffuses into the alveoli.

- O_2 is transported in reverse from how it was at the tissues.

  (1) More than 98% of the O_2 diffuses into red blood cells and loosely binds to **hemoglobin** (Hb) to form **oxyhemoglobin** (Hb-O_2). Some hydrogen ions (H^+) are produced as a by-product.

  (2) Less than 2% of the O_2 diffuses into the blood and remains dissolved in the plasma as O_2.

At the Tissues

CO_2 loading; O_2 unloading

Simple squamous epithelium

7%

CO_2 ① CO_2 dissolved in plasma

70%

② CO_2 + H_2O →[Carbonic anhydrase] H_2CO_3 → HCO_3^- + H^+

HCO_3^-

Cl^-

Membrane protein

Chloride shift

Cl^-

23%

③ CO_2 + Hb → Hb-CO_2 (Carbamino-hemoglobin)

H-Hb

Tissue cell

Interstitial fluid

Hb-O_2 (Oxyhemo-globin) → O_2 + Hb

Red blood cell

98%

① O_2

2%

② O_2 dissolved in plasma

Plasma

In the Lungs

CO_2 unloading; O_2 loading

7%

① CO_2 CO_2 dissolved in plasma

HCO_3^-

70%

② CO_2 CO_2 + H_2O ←[Carbonic anhydrase] H_2CO_3 ← HCO_3^- + H^+

Membrane protein

Chloride shift

Cl^-

Cl^-

23%

③ CO_2 CO_2 + Hb ← Hb-CO_2 (Carbamino-hemoglobin)

98%

O_2 ① O_2 + H-Hb → Hb-O_2 + H^+ (Oxyhemoglobin)

Red blood cell

2%

O_2 ② O_2 dissolved in plasma

Alveolus

Respiratory membrane

Plasma

Simple squamous epithelium

Note: The thickness of the lines indicates the relative amounts of CO_2 or O_2 in each pathway.

Color the CO_2 molecules blue and the O_2 molecules red.

Description

We take the rhythm and coordination of normal breathing for granted. The pattern of inhaling and exhaling constantly repeats itself. At rest, we take an average of 12–16 breaths per minute. This deceivingly simple process actually is complex and is regulated by the nervous system. The latest research has revealed that this mechanism is even more complex than previously thought.

Classic Explanation

In the classic explanation, clusters of specific neurons constitute respiratory control centers located in two regions of the brainstem—the **medulla oblongata** and the **pons**. The medulla oblongata sets the rate and rhythm of normal breathing. The pons regulates the rate and depth of breathing.

Medullary Respiratory Centers: DRG and VRG

- The **dorsal respiratory group (DRG)** is called the "inspiratory center" because it stimulates inhalations. Like a sparkplug, the neurons in the DRG fire in regular bursts. Each burst lasts about 2 seconds. As this happens, it simultaneously sends impulses along the **phrenic nerve** (to the diaphragm) and the **intercostal nerves** (to the **external intercostal muscles** of the ribs). This stimulates these muscles to contract, which increases the size of the thoracic cavity, and air rushes into the lungs. A 3-second delay between consecutive firings of the DRG allows time for a passive exhalation to follow each inhalation.

- The **ventral respiratory center (VRG)** is called the "expiratory center," but it is active *only* during a *forced* exhalation. During normal breathing, the VRG is inactive because exhalation is a passive, recoil process that does not require stimulation of muscles. During a *forced* exhalation, though, the DRG sends impulses to stimulate the VRG. The VRG responds by sending impulses to the internal intercostal muscles and abdominal muscles, which then contract, decreasing the size of the thoracic cavity for a forced exhalation.

Pons Respiratory Centers: Pneumotaxic Center (PC) and Apneustic Center (AC)

- The **pneumotaxic center (PC)** is the "regulator." It is like a traffic cop coordinating the flow of traffic to keep it running smoothly. Similarly, the pons coordinates the transition between inhalation and exhalation. It also prevents overinflation of the lungs by always sending inhibitory impulses to the inspiratory center (DRG).

- The **apneustic center (AC)** also coordinates the transition between inhalation and exhalation by fine-tuning the medullary respiratory centers. It accomplishes this by sending stimulatory impulses to the inspiratory center (DRG) that result in a slower, deeper inhalation. This is necessary when you choose to hold your breath. The pneumotaxic center is able to inhibit the apneustic center so a normal breath is not too slow or too deep.

Changes in Naming and Conceptual Understanding

The latest understanding brings a slight change in naming and conceptual understanding. Let's look at each, in turn. As for the naming, a new term has been introduced: the *respiration regulatory center*. This consists of all the aforementioned control centers (DRG, VRG, PC, and AC). But instead of considering them as individual units, we now think of them as being strung together in a loop or a circuit. This makes the classic definition of each control center a little more murky.

Conceptually, the latest thinking is that the respiration control center acts as a central pattern generator, or CPG. Like an electrical circuit, CPGs are neural circuits that generate periodic motor commands for rhythmic movements. The CPGs function automatically most of the time, but also allow voluntary control to override them.

Respiration Regulatory Center

PC = pneumotaxic center
AC = apneustic center
VRG = ventral respiratory group
DRG = dorsal respiratory group

Color the PC, AC, VRG, and DRG different colors.

Brain

Pons

Medulla

Abdominal muscles

External intercostal muscles

Internal intercostal muscles

Diaphragm

(via intercostal nerves)

(via phrenic nerve)

ANALOGIES

PC = Pneumotaxic Center

is like the traffic cop who keeps the flow of traffic (*breathing rhythm*) running smoothly.

DRG = Dorsal Respiratory Group

is like a sparkplug. It provides the "spark" for inspiration.

Color the positive (+) signs one color for stimulatory effects. Color the negative (−) signs a different color indicating inhibitory effects.

KEY

✚ Stimulatory

▭ Inhibitory

* Occurs only during a **forced exhalation**

Description

Even though the medulla oblongata and the pons control the typical rate and rhythm of normal breathing, they must be able to increase or decrease the breathing rate when needed. To do so, these respiratory centers in the brainstem receive various inputs from different parts of the body. This module summarizes the different types of input received and indicates whether it has a stimulatory or an inhibitory influence on normal breathing.

Functions

Chemoreceptors

Chemoreceptors are sensory neurons sensitive to shifts in specific chemicals in the blood, such as CO_2, H^+, and O_2. Changes in CO_2 levels are the *primary* stimulus, and changes in O_2 levels act as a *secondary* stimulus.

- **Peripheral chemoreceptors** are located in the aortic arch and the common carotid arteries. Consider what happens during exercise: P_{O_2} levels fall as skeletal muscle cells consume more oxygen; P_{CO_2} levels increase as skeletal muscle cells become more metabolically active; and H^+ levels increase, which makes the blood more acidic. Why? CO_2 diffuses into the blood. In the presence of the enzyme carbonic anhydrase, it chemically reacts with water (H_2O) to form carbonic acid (H_2CO_3), which dissociates into H^+ and bicarbonate (HCO_3^-). In this way, an increase in CO_2 results in an increase in H^+. All of these conditions stimulate an increase in respiration rate to supply the skeletal muscle cells with more O_2 and rid the body of excess CO_2.

- **Central chemoreceptors** are located in the medulla oblongata. They monitor P_{CO_2} levels and the pH of the cerebrospinal fluid (CSF). As mentioned above, as CO_2 levels increase, this also causes the pH of the CSF to become more acidic, which stimulates an increase in respiration rate.

Hypothalamic Controls

The **hypothalamus** has many functions. For example:

- The hypothalamus regulates body temperature. An increase in body temperature from a fever results in an increased breathing rate. A decrease in body temperature decreases breathing rate.

- The hypothalamus has connections with the limbic system, our "emotional brain." In times of emotional stress such as fear or anxiety, stimulatory impulses are sent to the inspiratory centers to increase breathing rate.

Pulmonary Controls

The lungs contain two important types of receptors: **irritant receptors** and **stretch receptors**.

- **Irritant receptors** are located in the bronchioles. They detect airborne particles such as dust and cigarette smoke, which results in stimulating reflex pathways to constrict bronchioles and decrease breathing rate. Other reflexes, such as coughing and sneezing, may be triggered also.

- **Stretch receptors** (*baroreceptors*) are located in the walls of the bronchi and bronchioles. As they detect stretching during inflation of the lungs, they send inhibitory impulses directly to the medullary inspiratory center. This results in an expiration, and the lungs relax. In this relaxed state, the stretch receptors no longer are stimulated, so the inhibition is lost and a new inspiration begins. This reflex is called the **inflation** (*Hering-Breuer*) **reflex**. The purpose seems to be a protective mechanism to prevent overinflation of the lungs.

Voluntary Controls: Cerebral Cortex

The **cerebral cortex** is the voluntary control over the body's activities. This can lead to either an increase or a decrease in respiration rate. For example, we may consciously choose to hold our breath. This stimulates motor neurons to send stimulatory impulses directly to our respiratory muscles for breathing. We may choose to go into a deep meditative state that indirectly leads to a decrease in respiration.

Proprioceptors

Proprioceptors are located in muscles, tendons, and joints. They monitor the tension in muscles and the movement and position of joints and deliver the information to the brain. For example, at the beginning of exercise, as tension in muscles and movement in joints increases, it stimulates an increase in respiration rate.

Other Influences on Respiration

- **Pain:** Sudden, acute pain may trigger **apnea** (absence of breathing), and extended somatic pain increases respiration rate.

- **Blood pressure (BP):** A quick increase in BP decreases respiration rate, and a drop in BP increases respiration rate.

Wedge of cerebral cortex

Voluntary control centers in cerebral cortex

Other influences on respiration (e.g., emotional stimuli such as fear or anxiety)

Hypothalamus

$CO_2 \uparrow$, $H^+ \uparrow$

Peripheral chemoreceptors

Irritant receptors

Pons

Medulla

$CO_2 \uparrow$, $H^+ \uparrow$

Central chemoreceptors

Respiratory centers in medulla and pons

Stretch receptors

Proprioceptors in muscles, tendons, and joints

Color the positive (+) signs one color for stimulatory effects; color the negative (−) signs a different color indicating inhibitory effects.

KEY

➕ Stimulatory

➖ Inhibitory

➕
➖ } Stimulatory OR inhibitory

Notes

DIGESTIVE SYSTEM

Overview of the Digestion Process

Let's look at the digestion process by examining what happens when someone eats a ham sandwich. The process begins in the mouth with **mechanical digestion**—the physical breakdown of food from chewing. The sandwich contains many different nutrients, including starch—the **complex carbohydrate** in the bread, **protein** in the ham, and **lipids** (*ex.:* triglycerides) in the mayonnaise. Each of these macromolecules has to be broken down by **chemical digestion** into its fundamental building blocks so they can be used by body cells:

Complex carbohydrates	→	Monosaccharides
Proteins	→	Amino acids
Lipids	→	Monoglycerides and fatty acids

Overview of Chemical Digestion

The process is a series of catabolic reactions that breaks chemical bonds in the macromolecules with the help of digestive enzymes. Different enzymes are required for different macromolecules, shown on the illustration.

Note: The enzymes are shown in *italics*, and the icons indicate the location in the digestive system.

= mouth   = stomach   = small intestine

Carbohydrates

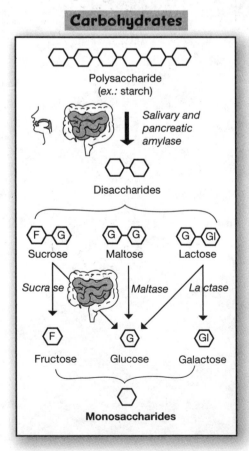

Polysaccharide
(*ex.:* starch)

Salivary and pancreatic amylase

Disaccharides

Sucrose   Maltose   Lactose

Sucrase   *Maltase*   *Lactase*

Fructose   Glucose   Galactose

Monosaccharides

Proteins

Protein

Pepsin

Large polypeptides

Pancreatic enzymes (trypsin, chymotrypsin, carboxypeptidase)

Small polypeptides

Intestinal enzymes (aminopeptidase, carboxypeptidase, dipeptidase)

Amino acids

Lipids

Unemulsified lipids

Bile

Triglyceride

Emulsified lipids

Pancreatic lipase

Monoglycerides   Fatty acids

NUTRIENTS IN A HAM SANDWICH

Starch

Protein

Triglycerides

Bread

Ham

Mayonnaise

I love ham sandwiches!

Oral cavity

Nasal cavity

Parotid salivary gland

Tongue

Epiglottis

Mandible

Sublingual salivary gland

Submandibular salivary gland

Trachea

Primary bronchus

Color different organs different colors.

THE DIGESTIVE SYSTEM: OVERVIEW

Bolus of food   1.

Accessory glands

(salivary glands, liver, gallbladder, and pancreas)

2.

Enzyme

(ex.: amylase)

Digestive tract

(tube from mouth to anus)

3.

4.

5.   Waste

Body cell

Circulatory system

Human body

1. Macromolecules in the bolus of food are ingested and mechanically broken down.

2. Enzymes are released from glands that aid in breaking down macromolecules into smaller products the cells can use.

3. Final products are absorbed into the bloodstream.

4. Final products are transported through the bloodstream and finally delivered to cells, where they can be used as nutrients.

5. Waste products are eliminated from the body.

Esophagus

Diaphragm

Pancreas   *Spleen*

Liver

Gallbladder

Stomach

Duodenum

Large intestine

Appendix

Ileum

Jejunum

Rectum   Large intestine

Description

The digestive tract [(*alimentary canal, gastrointenstinal (GI) tract*)] is the long, continuous, muscular tube between the mouth and the anus. Amazingly, it is nearly 30 feet in length. From innermost to outermost, it is divided into the following four layers: (1) *mucosa*, (2) *submucosa*, (3) *muscularis externa*, and (4) *serosa*.

The table below describes the four different layers of the digestive tract as listed from innermost to outermost.

Layer	Comments
Mucosa	• The innermost layer of the digestive tract. • The term mucosa refers to it being a mucous membrane, an epithelial tissue that secretes mucus and lines body cavities. • It is subdivided into three parts: epithelium, lamina propria, and muscularis mucosa. The type of epithelium changes in different parts of the tract. Depending on the type, it can function for either protection, absorption of nutrients, or secretion. The lamina propria is made of areolar connective tissue. It anchors the epithelium and houses many blood vessels and lymphatic vessels. The muscularis mucosa is a thin layer of smooth muscle that increases folds in the mucosa which increases surface area, thereby allowing increased absorption of nutrients.
Submucosa	• The term submucosa refers to its location as being "below the mucosa." • Composed of dense irregular connective tissue, it binds the mucosa to the muscularis externa. • It is a housing for blood vessels, lymphatic vessels, glands that secrete mucus, and a network of neurons.
Muscularis externa (muscularis)	• The term muscularis externa refers to the type of tissue it contains—smooth muscle. • It is divided into two sheets of smooth muscle: (1) the inner circular layer, and (2) the outer longitudinal layer. One exception to this rule is the stomach where there is a third layer of smooth muscle, the oblique layer. • Both layers contract for the purposes of mixing food with digestive secretions and moving contents through the digestive tract.
Serosa (visceral peritoneum)	• The outermost layer of the digestive tract. • The term serosa refers to it being a serous membrane. Like any serous membrane, it contains two layers: (1) an inner areolar connective tissue layer, and (2) an outer simple squamous epithelial layer. • The serosa is found on all the regions of the digestive tract except for the oral cavity, pharynx, esophagus, and rectum. These four regions contain a fibrous connective tissue layer instead of a serosa.

Analogy

Structurally, the layers of the digestive tracts are like tubes within tubes.

Study Tip

To recall the layers of the digestive tract in order from innermost to outermost, use the following mnemonic: *My Submarine Moved South.*

Key to Illustration			
1. Mucosa	1c. Epithelium	3a. Circular layer	4a. Areolar connective tissue
1a. Muscularis mucosa	2. Submucosa	3b. Longitudinal layer	
1b. Lamina propria	3. Muscularis externa	4. Serosa	4b. Simple squamous epithelium

Four layers of
the digestive tract

1

2

3

4

1a
1b
1c

3a

3b

4a

4b

Duct from gland outside the
digestive tract (ex: pancreas)

Gland in
submucosa

Artery
Vein

Notice that the layers
of the digestive tract
are like tubes within
tubes.

Mesentery

Artery
Vein

Description

The regulation of digestion is controlled simultaneously by the autonomic nervous system (ANS) and by hormones. This module focuses on the role of the nervous system in regulating digestion.

Phases

Control of digestion can be divided into three phases: (1) cephalic phase, (2) gastric phase, and (3) intestinal phase.

① Cephalic Phase

The **cephalic phase** begins before any food enters the oral cavity. The sight, smell, thought, or first taste of food stimulates reflexes to increase the production of saliva, gastric secretions, and pancreatic secretions. The collective purpose of this response is to prepare the digestive tract to receive food. This phase produces about 10% of all gastric secretions. Key digestive control areas in the nervous system include the **cerebral cortex, hypothalamus**, and **medulla oblongata**. Various stimuli (from food) are sent through nerve impulses to the cerebral cortex, where they are interpreted. Then the cerebral cortex responds by sending impulses to the medulla.

The **hypothalamus** contains the feeding center. Specific stimuli, such as smells and tastes of food, stimulate nerve impulses to be sent to the feeding center in the hypothalamus. This causes the hypothalamus to send nerve impulses to the medulla oblongata.

As stimulatory impulses are sent to the medulla oblongata from both the cerebral cortex and the hypothalamus, the medulla responds by sending out nerve impulses that produce three key responses: (1) increased saliva production, (2) increased gastric secretions, and (3) increased pancreatic secretions (small amount). More specifically, impulses sent from the medulla oblongata down the **vagus nerve** stimulate both gastric and pancreatic secretions. The gastric secretions contain pepsinogen, hydrochloric acid, and mucus. The pancreatic secretions are rich in pancreatic enzymes that aid in carbohydrate digestion, triglyceride digestion, protein digestion, and nucleic acid digestion. Cranial nerves other than the vagus stimulate the salivary glands to produce saliva.

② Gastric Phase

The **gastric phase** begins as food enters the stomach. This triggers reflex pathways that stimulate about 80% of all gastric secretions. The two types of sensory receptors in the stomach wall are stretch receptors and chemoreceptors. Upon entering the stomach, food stretches the stomach wall and triggers stretch receptors to send a nerve impulse along the vagus nerve to the medulla oblongata. The same response occurs when chemoreceptors detect a change in the pH of the gastric secretions.

After proteins in food enter the stomach, they buffer some of the normal acidity in the stomach. This leads to a more alkaline pH, which triggers the chemoreceptors to send out nerve impulses. These impulses are sent to the medulla oblongata, which responds by sending impulses back down the vagus nerve to the stomach. This further stimulates the gastric secretions of pepsinogen, hydrochloric acid, and mucus. It also stimulates the muscular contractions of peristalsis to move the chyme through the stomach and continue peristalsis in the esophagus. Finally, the hormone gastrin secreted during the gastric phase further stimulates gastric secretions.

③ Intestinal Phase

The **intestinal phase** begins when chyme enters the **duodenum** after passing through the pyloric valve. Within the wall of the duodenum are the same two important sensory receptors as in the stomach: stretch receptors and chemoreceptors. Activation of these receptors stimulates reflex pathways through the medulla that initially lead to more gastric secretion, gastric motility, and emptying of the stomach. This phase accounts for about 10 percent of total gastric secretion and ensures that the digestion process is fully completed. Later, when the duodenum becomes filled with fatty chyme, the hormones secretin and cholecystokinin (CCK) are produced to inhibit gastric secretion and stomach emptying. This inhibitory response is needed to signal the end of the passage of a meal.

Divisions

Parasympathetic (P) Divisions

The glands and organs of the digestive system are extensively innervated by both the sympathetic and the parasympathetic divisions of the ANS. Consequently, secretions and motility are under unconscious control by the nervous system. The **parasympathetic division (P)** controls stimulation of secretions and motility throughout the digestive tract primarily through action of the **vagus nerve** (cranial nerve X). The vagus nerve connects to all the major digestive system organs and structures except the salivary glands. It is a mixed nerve, meaning that it carries both sensory and motor information within it.

Sympathetic (S) Division

The **sympathetic division (S)** inhibits secretion and motility in the digestive system.

① CEPHALIC PHASE

The **cephalic phase** begins before any food enters the oral cavity. The sight, smell, thought, or first taste of food stimulates nerve impulses to increase the production of saliva, gastric secretions, and pancreatic secretions.

Yum!
I love spaghetti and meatballs!

THOUGHT

SIGHT

SMELL

TASTE

Cerebral cortex

Hypo-thalamus

Medulla oblongata

Salivary gland

Saliva

Use a highlighter to trace the neural pathways.

Vagus n. (cranial n. X)

Mucosal layer   Muscularis layer

Gastric secretions

Pancreatic secretions

SYMPATHETIC (S) and PARASYMPATHETIC (P) CONTROL

S

P

Ganglion

Parotid salivary gland

Vagus n.

Use a highlighter to trace the neural pathways. Color the different digestive organs different colors.

Vagus n. (cranial n. X)

② GASTRIC PHASE

The **gastric phase** begins as food enters the stomach.

Smooth muscle cell

Epithelial cell

Gastric secretions

Sensory n.

to medulla oblongata

Pyloric sphincter

Sensory receptor

Sensory n.

Duodenum

Sensory receptor

③ INTESTINAL PHASE

The **intestinal phase** begins when chyme enters the duodenum after passing through the pyloric sphincter.

General Description

The hormonal regulation of digestion is controlled mainly by three major hormones: (1) **gastrin**, (2) **secretin**, and (3) **cholecystokinin (CCK)**. The table below gives an overview of each hormone:

Hormone/Symbol	Stimulus	Site of Secretion	Major Actions
Gastrin ▽	Expansion of stomach wall by ingested materials; protein and caffeine in stomach; alkaline chyme	**G cells** within the mucosa of the stomach	Stimulates secretion of gastric juice (HCl, mucus, and pepsinogen)
Secretin ⊡	Acidic chyme in the duodenum	**S cells** within the mucosa of the duodenum	Stimulates secretion of pancreatic juice rich in bicarbonate ions (HCO_3^-) to help neutralize acid in chyme
Cholecystokinin (CCK) ⌒	Triclycerides, fatty acids, and amino acids in the duodenum	**CCK cells** in the mucosa of the duodenum	Stimulates production of pancreatic juice rich in digestive enzymes such as lipase; stimulates muscle contraction in wall of gallbladder to expel stored bile; stimulates relaxation of muscle to open hepatopancreatic sphincter

Regulation of Gastric ▽ Secretion

① A bolus moves through the lower esophageal sphincter of the esophagus and into the stomach.

② Partially digested proteins and caffeine from food stimulate G cells within the mucosal lining to produce gastrin and release it into the blood.

③ Gastrin travels in the blood and spreads to all body organs.

④ Gastrin binds only to target cells that contain a receptor for gastrin. Because the various secretory cells in the epithelial mucosa contain the gastrin receptor, gastrin binds to them and induces a response.

⑤ The binding of gastrin to the gastrin receptor causes a chemical chain reaction within the cell that leads to stimulation of the production of more gastric juices (hydrochloric acid, mucus, and pepsinogen).

Regulation of Pancreas ⊡ and Gallbladder ⌒ Secretions

① Acidic chyme stimulates S cells in the duodenal mucosa to manufacture secretin; fatty acids and triglycerides in chyme stimulate CCK cells in the duodenal mucosa to produce CCK.

② Secretin and CCK are released into the blood and travel to all body organs.

③ Secretin binds to secretin receptors in pancreatic cells. This stimulates the cells to make pancreatic juice rich in bicarbonate ions (HCO_3^-) to neutralize the acid in the chyme. CCK binds to CCK receptors in pancreatic cells to stimulate the production of pancreatic juice rich in enzymes such as lipase.

④ CCK also targets the gallbladder. CCK binds to its receptors in the smooth muscle in the gallbladder wall. This stimulates smooth muscle contraction that results in expelling bile from the gallbladder and eventually into the duodenum.

⑤ CCK also targets the hepatopancreatic sphincter, but with a response opposite from the gallbladder. As CCK binds to its receptors in the smooth muscle cells of the sphincter, it triggers a different chemical chain reaction within the target cells. This causes the smooth muscle to relax, opening the sphincter and allowing both pancreatic secretions and bile to be released into the duodenum.

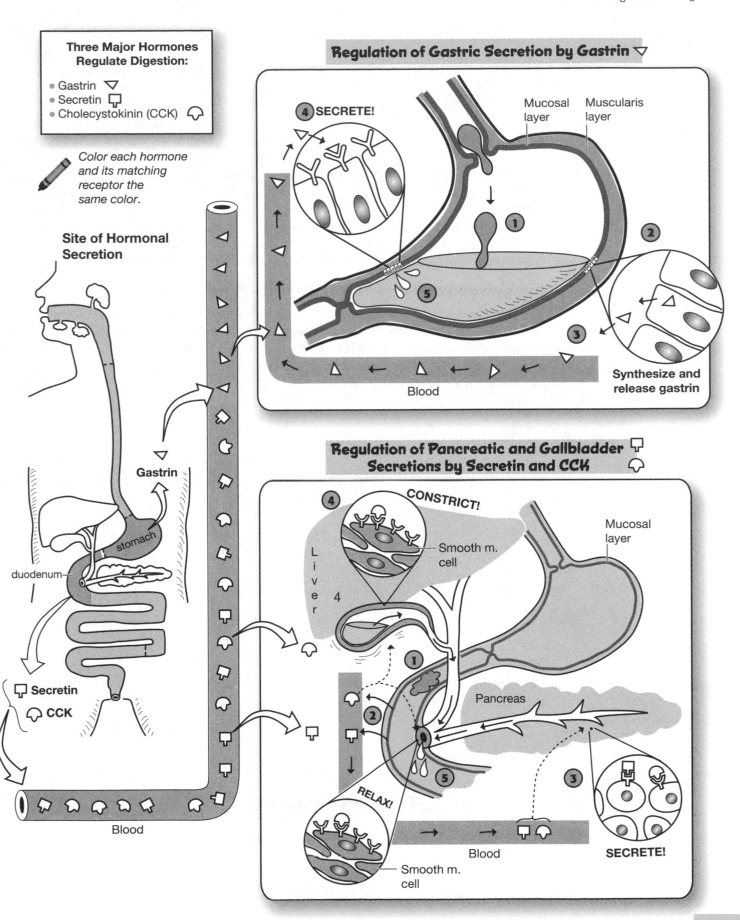

Three Major Hormones Regulate Digestion:

- Gastrin ▽
- Secretin ⊓
- Cholecystokinin (CCK) ◠

Color each hormone and its matching receptor the same color.

Site of Hormonal Secretion

Gastrin

stomach

duodenum

Secretin

CCK

Blood

Regulation of Gastric Secretion by Gastrin ▽

④ SECRETE!

Mucosal layer

Muscularis layer

①

②

⑤

③

Blood

Synthesize and release gastrin

Regulation of Pancreatic and Gallbladder Secretions by Secretin and CCK ⊓ ◠

④ CONSTRICT!

Smooth m. cell

Liver

4

①

②

Pancreas

Mucosal layer

③

⑤

RELAX!

Smooth m. cell

Blood

SECRETE!

Description

The innermost lining of the **oral cavity** (*buccal cavity*) is called the **mucosa**, and coats the roof and floor of the mouth, lines the inside of the cheeks, covers the tongue, and forms a ridge of tissue below the teeth called the gums, or **gingivae**. The mucosa is composed of two types of **stratified squamous epithelium**—*keratinized* and *nonkeratinized*. The keratinized form covers surfaces subjected to the most abrasion—the superior surface of the tongue, hard palate, and gingivae. The nonkeratinized form covers the inferior surface of the tongue, floor of the mouth, soft palate, cheeks, and lips. The mucosa curves anteriorly to form the upper and lower lips, or **labia**. Two separate, flattened plates of mucosa, each called a **frenulum,** anchor the upper and lower lips to the midline of the gingivae. Another similar structure called the **lingual frenulum** anchors the tongue to the floor of the oral cavity. The thick **tongue** covers the floor of the mouth and contains several different groups of muscles.

The **hard palate** is the most anterior portion of the roof of the oral cavity. It is composed of the palatine process of the maxilla and the palatine bone and functions to separate the oral cavity from the nasal cavity. Behind it lies a fleshy plate called the **soft palate** that contains no bone. It separates the oral cavity from the nasopharynx and covers the nasopharynx during swallowing to prevent food from entering it.

Dangling from the middle of the back of the soft palate is a small, flap-like structure called the **uvula**. It holds food in the oral cavity and prevents it from entering the **oropharynx** too soon. Two sets of arches are found at the back of the oral cavity. From the front to the back, the first set is called the **palatoglossal arches**. These curve to connect the soft palate with the base of the tongue. The second set is called the **palatopharyngeal arches**. These curve to connect the soft palate to the sides of the pharynx. Between these arches, on either side of the mouth, are two masses of lymphatic tissue called the **palatine tonsils**.

Key to Illustration

1. Frenulum of lower lip	5. Hard palate	9. Tongue
2. Palatine tonsil	6. Frenulum of upper lip	10. Gingivae
3. Uvula	7. Palatoglossal arch	
4. Soft palate	8. Palatopharyngeal arch	

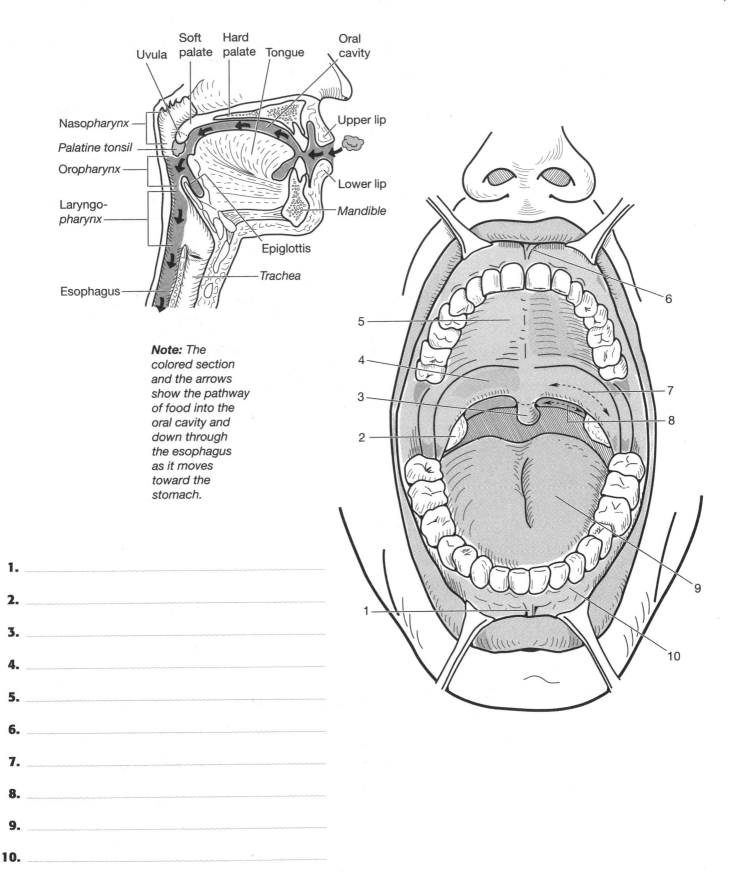

Uvula
Soft palate
Hard palate
Tongue
Oral cavity
Nasopharynx
Palatine tonsil
Oropharynx
Laryngo-pharynx
Esophagus
Upper lip
Lower lip
Mandible
Epiglottis
Trachea

Note: *The colored section and the arrows show the pathway of food into the oral cavity and down through the esophagus as it moves toward the stomach.*

1. _____

2. _____

3. _____

4. _____

5. _____

6. _____

7. _____

8. _____

9. _____

10. _____

Description

There are four different types of teeth—**incisors**, **canines** (*cuspids*), **premolars** (*bicuspids*), and **molars**. Incisors are for cutting, canines are for tearing, and premolars and molars are for crushing and grinding food. The adult jaws can accommodate a total of 32 permanent teeth.

Tooth Type	Upper Jaw	Lower Jaw	Total
Incisor	4	4	8
Canine	2	2	4
Premolar	4	4	8
Molar	6	6	12
		Total	**32**

Each tooth is divided into three regions: **crown**, **neck**, and **root**. The crown is visible above the gumline and is covered with a calcified coating of **enamel**—the hardest substance produced in the body. Deep to this covering is a bone-like substance called **dentin** that makes up the majority of the tooth. At the center of the tooth is a chamber called the **pulp cavity** that contains spongy tissue, blood vessels, and nerves. A bony substance called **cementum** covers the dentin in the root of the tooth. The **periodontal ligaments** anchor the root of the tooth to the bone in the jaw. At the tip of each root is an opening called the **apical foramen** that allows the **dental artery**, **dental vein**, and **dental nerve** to penetrate through the narrow **root canal** and up into the **pulp cavity**.

Key to Illustration

1. Incisors
2. Canines (*cuspids*)
3. Premolars (*bicuspids*)
4. Molars
5. Central incisors
6. Lateral incisor
7. Canine (*cuspid*)
8. 1st premolar
9. 2nd premolar
10. 1st molar
11. 2nd molar
12. 3rd molar (*wisdom tooth*)
13. Crown
14. Neck
15. Root
16. Enamel
17. Dentin
18. Pulp cavity
19. Gingiva
20. Periodontal ligaments
21. Cementum
22. Bone
23. Root canal
24. Apical foramen
25. Dental artery
26. Dental nerve
27. Dental vein

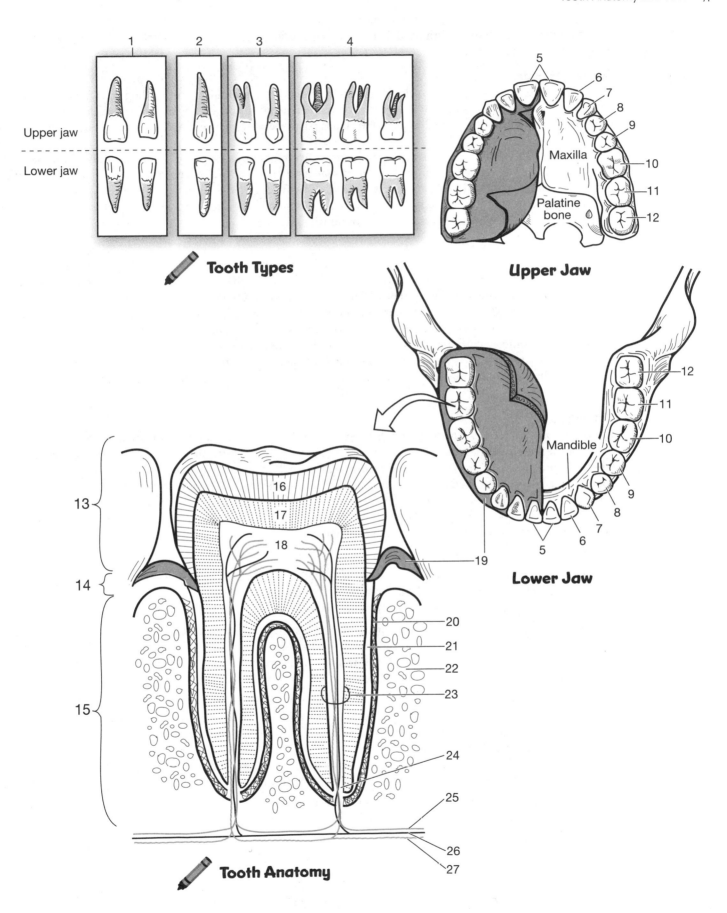

Tooth Types

Upper Jaw

Maxilla

Palatine bone

Lower Jaw

Mandible

Tooth Anatomy

Description   Digestion in the mouth deals with producing saliva from salivary glands, chewing, and swallowing. Each of these processes is summarized here.

Salivary Glands

The body contains three pairs of **salivary glands: parotid, submandibular,** and **sublingual.** Of these three, the parotid is the largest. The submandibular gland produces the most saliva, and the sublingual gland produces the least. Salivary glands are classified as exocrine glands partly because they all have ducts that connect them to the oral cavity.

At the microscopic level, salivary glands contain two types of secretory cells: **mucous cells** and **serous cells.** The mucous cells secrete a protein called **mucin** that forms mucus. The serous cells release a watery secretion containing electrolytes and salivary amylase. These collective secretions form saliva and release it into the oral cavity. Because secretion of saliva is controlled by the autonomic nervous system, salivary glands are innervated by the parasympathetic and the sympathetic divisions.

Saliva

Saliva is a mildly acidic (pH 6.4–6.8), watery secretion that has multiple functions. It contains mostly water (95%), as well as some important dissolved solutes (0.5%), including electrolytes such as sodium (Na^+), potassium (K^+), chloride (Cl^-), and others. **Mucus** is produced to lubricate the bolus to make swallowing easier. **Lysozyme** is an antibacterial agent that inhibits the growth of bacteria normally found in the oral cavity. **Antibodies** such as immunoglobulin A also control bacterial growth. The enzyme **salivary amylase** catalyzes the breakdown of complex carbohydrates into disaccharides. Like water spraying on cars in a carwash, saliva cleans oral cavity structures such as the teeth by constantly washing over them. Last, saliva serves as a fluid medium in which food molecules can dissolve to allow for chemical detection by taste buds. Without saliva, it is difficult to taste anything!

Chewing

Chewing (*mastication*) is the physical breakdown of food and the mixing of food with saliva. This is accomplished with the help of the teeth and the tongue. The four different types of teeth are incisors, canines (*cuspids*), premolars (*bicuspids*), and molars. Each type is best suited for a specific task. Incisors have a flat, chiseled edge for cutting food like biting into an apple. Cuspids have a pointed edge for puncturing, tearing, and shredding food. Bicuspids and molars have larger, flattened surfaces, best for grinding and crushing food. The tongue moves during chewing to mix the pulverized food with saliva to form a pasty, compressed mass called a **bolus.** This mixing prepares the **bolus** for swallowing.

Swallowing

Swallowing (*deglutition*) is a complex process that requires many different muscles to move ingested substances from the oral cavity to the stomach. It typically is divided into three stages: oral, pharyngeal, and esophageal. Note that the same process applies to ingested liquids and solids. The stages in the process are:

1. **Oral** (or *buccal*) **phase**—a voluntary phase that begins when substances are ingested. The tongue mixes these materials with saliva to form a moistened mass called a bolus. This bolus is compressed against the hard and soft palates in the roof of the mouth by upward movements of the tongue. Then the tongue pushes the bolus toward the oropharynx, which marks the end of the oral phase.

2. **Pharyngeal phase**—an involuntary process. It begins as the bolus pushes against the oropharynx, where it stimulates tactile receptors to send signals to the swallowing control center in the medulla oblongata. Three key events occur: (1) the soft palate and uvula move upward to prevent the bolus from entering the nasal cavity; (2) the bolus moves into the oropharynx; and (3) muscles raise the larynx forward and upward so the epiglottis closes over the glottis to prevent the bolus from entering the larynx or trachea. The pharyngeal phase ends after pharyngeal muscles constrict and force the bolus out of the pharynx and into the esophagus.

3. **Esophageal phase**—involuntary like the pharyngeal phase. The esophagus is a long, muscular tube that connects the pharynx to the stomach. This phase begins as the upper esophageal sphincter relaxes to allow the bolus to enter the esophagus. The rigid epiglottis springs back to its normal, upright position. Then the upper esophageal sphincter contracts to propel the bolus deeper into the esophagus. The muscular contractions of peristalsis (see p. 446) quickly move the bolus into the stomach. This entire phase takes less than 8 seconds!

> **DISCOVER for YOURSELF**
> Place your fingertips on the front of your larynx as you swallow. You can feel the larynx move forward and upward.

SALIVARY GLANDS

Oral cavity
Epiglottis
Nasal cavity
Tongue

SALIVARY GLANDS
Parotid
Submandibular
Sublingual

Trachea
Esophagus

Color the three pairs of salivary glands the same color.

SALIVA

Mucus
Mucous cell
salivary duct
Salivary gland
Serous cell
Fluid with electrolytes and amylase

Composition of Saliva	Functions of Saliva
99.5% H_2O	• **Lubricates** food
0.5% **SOLUTES**	• **Cleanses** oral cavity structures
• Electrolytes (Na^+, K^+, Cl^-, etc.)	• **Catalyzes** digestion of carbohydrates with AMYLASE
• Lysozyme	• **Inhibits bacterial growth** with LYSOZYME and ANTIBODIES
• Mucus	• **Dissolves** food molecules to allow for taste
• Salivary amylase	
• Immuno-globulin A	

amylase

CHEWING

Incisors → Cutting

Canines (*Cuspids*) → Tearing and shredding

Premolars and molars → Grinding and crushing

SWALLOWING

Color the bolus, larynx, uvula, and epiglottis different colors.

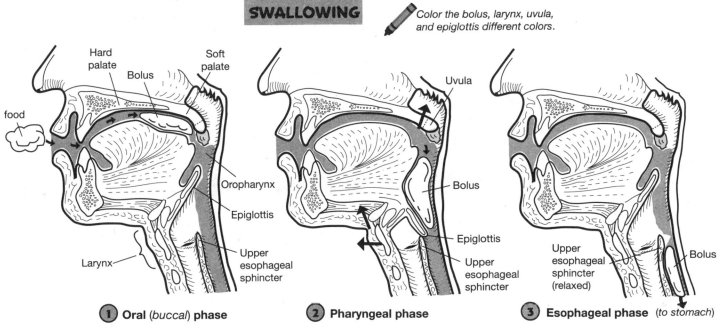

Hard palate
Bolus
Soft palate
food
Oropharynx
Epiglottis
Larynx
Upper esophageal sphincter

Uvula
Bolus
Epiglottis
Upper esophageal sphincter

Upper esophageal sphincter (relaxed)
Bolus

1 **Oral** (*buccal*) **phase**   2 **Pharyngeal phase**   3 **Esophageal phase** (*to stomach*)

Note: *The colored region shows the pathway ingested substances follow.*

Description

The stomach connects the esophagus to the first portion of the small intestine, called the duodenum. It is a specialized sac that contains three layers of smooth muscle within its wall. The innermost lining of the stomach, called the **mucosa**, is coated with a protective layer of alkaline mucus. Folds of mucosa called **rugae** allow the stomach to increase surface area to maximize **gastric juice**. This acidic mixture contains hydrochloric acid and enzymes. When this gastric juice combines with food, a mixture called **chyme** is formed.

Function

To release gastric juice for the purpose of chemical digestion of food, the mucosa manufactures an enzyme called **pepsin** that initiates protein digestion. Within the mucosal lining of the stomach are four different types of cells that have the following functions:

Mucosal Cell	Function
Mucous cells	Secrete an alkaline **mucus** to protect the stomach lining from acidic gastric juice
Chief cells	Secrete the inactive enzyme **pepsinogen**
Parietal cells	Secrete hydrochloric acid (HCl); this helps convert pepsinogen into the active enzyme **pepsin**
G cells	Produce and secrete the hormone **gastrin**, which increases secretion from parietal and chief cells. It also induces smooth muscle contraction in the stomach wall.

Study Tip

Use the following mnemonics to recall the mucosal cell types and the substances they produce:

CELL TYPE

Men		**M**ucous cells
Can	=	**C**hief cells
Pass		**P**arietal cells
Gas		**G** cells

SUBSTANCE PRODUCED

Men		**M**ucus
Pass	=	**P**epsinogen
Hot		**H**ydrochloric acid
Gas		**G**astrin

Key to Illustration

1. Esophagus
2. Cardia
3. Fundus
4. Serosa
5. Longitudinal muscle layer
6. Circular muscle layer
7. Oblique muscle layer

8. Mucosa
9. Lesser curvature
10. Body
11. Greater curvature
12. Rugae
13. Pyloric sphincter
14. Gastric pit

15. Gastric gland
16. Mucous cells
17. Parietal cells
18. Chief cells
19. G cells
20. Muscularis mucosa

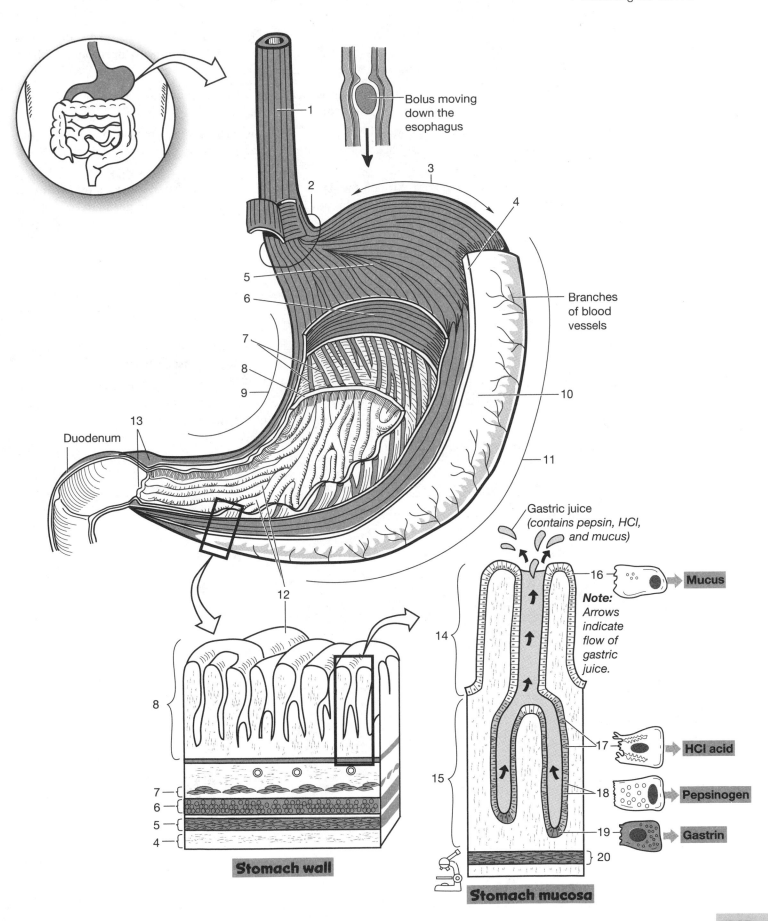

Bolus moving
down the
esophagus

1

2

3

4

Branches
of blood
vessels

5

6

7

8

9

10

11

13

Duodenum

12

Gastric juice
(contains pepsin, HCl,
and mucus)

16 → Mucus

Note:
*Arrows
indicate
flow of
gastric
juice.*

14

15

17 → HCl acid

18 → Pepsinogen

19 → Gastrin

20

8

7
6
5
4

Stomach wall

Stomach mucosa

Gastric Motility

Gastric **motility** refers to smooth muscle contractions in the wall of the stomach. This process is regulated by the autonomic nervous system and by the endocrine system with hormones such as **gastrin**. Waves of **peristalsis** move through the stomach regularly.

1. They begin as gentle muscular contractions near the **lower esophageal sphincter (LES)** and continue down the stomach toward the **pyloric sphincter**.

2. As the contractile waves near the distal end of the stomach, they become much stronger and more forceful. This results in thoroughly mixing **chyme** before it passes through the pyloric sphincter.

3. As the peristaltic wave passes through the partly opened pyloric sphincter, it causes the chyme to move through it in a back-and-forth fashion. This serves to break up the larger materials left in the chyme.

Gastric Emptying

Gastric **emptying** refers to the movement of **chyme** out of the stomach and into the duodenum.

1. Various stimuli, such as caffeine, alcohol, partially digested proteins, and distension of the stomach wall, all trigger the process of gastric emptying to begin.

2. These stimuli cause two things to occur: (1) an increase in parasympathetic impulses sent along the vagus nerve to the stomach, and (2) an increase in secretion of gastrin from the stomach.

3. Gastrin and vagus nerve impulses have the combined effects of closing the LES, opening the pyloric valve, and increasing gastric motility.

4. The final result is emptying of the chyme in the stomach into the duodenum.

GASTRIC MOTILITY

Lower esophageal sphincter (LES)

Pyloric sphincter

Muscularis layer

CHYME

Color the white arrows indicating peristaltic waves.

GASTRIC EMPTYING

① Stimuli

Food
(causes *stretching* of stomach wall)

Alcohol

Protein
(partially digested)

Caffeine

② Responses

Endocrine system response

• increase in **gastrin** secretion

Gastrin

Blood

Nervous system response

• stimulation of **vagus n.**

Vagus nerve

③ More responses

• LES **closes** and pyloric valve **opens**.

LES **CLOSES**

Pyloric valve **OPENS**

④ Gastric emptying

Open pyloric valve

Description

The **small intestine** is a hollow, muscular tube approximately 20 feet in length that links the stomach to the large intestine. It is divided into three parts—**duodenum** (10 inches), **jejunum** (8 feet), and **ileum** (12 feet). The jejunum and ileum are loosely held in place by a highly vascular serous membrane called **mesentery** that anchors the small intestine to the posterior wall of the abdominal cavity. From outermost to innermost, the layers in the wall of the small intestine are the **serosa**, **muscularis externa**, **submucosa**, and **mucosa**.

Folds are used to increase surface area anywhere in the body. In the small intestine there are three significant folded structures: (1) **circular folds**, (2) **villi** (sing. *villus*), and (3) **microvilli** (sing. *microvillus*). All these folds increase surface area to better facilitate absorption of nutrients. The large, macroscopic folds of the muscosa are the **circular folds**. Extending from these folds are the **villi**—microscopic fingerlike projections covered by a layer of **simple columnar epithelium**. Scattered within this tissue are **goblet cells** (they resemble a wine *goblet*) that secrete the **mucus** that coats the inside of the small intestine.

Each simple columnar epithelial cell has folds in its plasma membrane called **microvilli**. Nutrients must pass through the microvilli, then through the cell, in order to enter the villus. Within the villus is a **blood capillary** and a specialized lymphatic capillary called a **lacteal**. Depending on the type of nutrient, once inside the villus, it will enter either the blood capillary or the lacteal to complete its absorption.

Analogy

The **folds** on the inside of the **small intestine** are like a **folded carpet sample**. The **fold** is the **plica circularis**. The **carpet fibers** sticking out from this sample are the **villi**.

Study Tip

To recall the correct order of the three parts of the small intestine, use the mnemonic—Don't Jump In! (Duodenum, Jejunum, Ileum.)

Location

Abdominal cavity; surrounded by the large intestine; below the stomach.

Function

Absorption of nutrients

- **water, vitamins, minerals**
- **amino acids** (*from the digestion of proteins*)
- **fatty acids, glycerol** (*from the digestion of lipids*)
- **monosaccharides** (*from the digestion of complex carbohydrates*)
- **nucleotides** (*from the digestion of nucleic acids such as DNA*)

Key to Illustration

Wall of Small Intestine		
1. Serosa	4. Submucosa	9. Simple columnar epithelial cell
2. Muscularis externa (*longitudinal layer*)	5. Mucosa	10. Goblet cell
3. Muscularis interna (*circular layer*)	**Villus Structures**	11. Venule
	6. Arteriole	12. Plasma (*cell*) membrane
	7. Blood capillary	13. Nucleus of one cell
	8. Lacteal	

Increasing Surface Area

Duodenum

Jejunum

Ileum

A. Circular fold

C. Microvilli

Simple columnar epithelium

1

2

3

4

5 5 5 5 5 5

4 4

Layers of the wall of the small intestine

12

13

9

8

10

7 7

8

7 7

Note:
Arrows indicate direction of blood flow.

Villus = one carpet fiber

6

8

8

8

11

B. Villus

Imagine a folded carpet sample. The fold itself is like the large fold, or **circular fold**, in the small intestine, and each individual **carpet fiber** is like one **villus**.

Circular fold = folded carpet sample

Villus = one carpet fiber

Description

The small intestine is a 20-foot muscular tube that connects the stomach to the large intestine. It is the major site of **chemical digestion** and nutrient **absorption**. The wall of this organ contains a muscularis layer of smooth muscle subdivided into two thin layers: an inner circular layer and an outer longitudinal layer. When contracted, the circular layer pinches the small intestine, and the longitudinal layer shortens it. All the materials from the stomach must move through this tube by different types of coordinated muscle contractions.

The small intestine uses the processes of **peristalsis** and **segmentation** to move undigested materials along its length. These processes are not unique to the small intestine; they occur in other parts of the digestive tract as well. The sites of action for these muscular movements are summarized below:

Type of Muscular Movement	Site of Action
1. Peristalsis	Esophagus, stomach, small intestine, and large intestine
2. Segmentation	Small intestine and large intestine

Peristalsis

Peristalsis is a rhythmic wave of smooth muscle contraction that results in the propulsion of **chyme** through the digestive tract. More specifically, it is the alternating, coordinated stimulation of both the circular and the longitudinal layers of smooth muscle in the muscularis layer.

The following is the series of events illustrated on the facing page:

(1) The circular layer of smooth muscle contracts behind the mass of materials in the small intestine. The result is that the small intestine is pinched tighter, like rings becoming smaller in size.

(2) A section of the longitudinal layer in front of the pinched region contracts, which shortens part of the small intestine.

(3) This coordinated process of "pinching" and "shortening" continues to alternate at regular intervals along the length of the wall of the small intestine. Viewed in real time, it would look like a gradual, smooth undulating wave of muscle contraction.

Analogy

The action of peristalsis is like squeezing a tube of toothpaste so the toothpaste moves through the tube.

Segmentation

Segmentation is the pinching of the intestine into compartments and subsequent mixing of **chyme** with intestinal secretions. This ensures that the processes of chemical digestion and absorption are both completed. The circular layer of the muscularis is more active than the longitudinal layer. No net movement of materials results in this process as it did in peristalsis.

The following is the series of events illustrated on the facing page:

(1) The circular layer pinches off the small intestine into similar-sized compartments, or "segments."

(2) It further pinches to even smaller compartments as the undigested materials are moved back and forth by muscle contractions like an agitator in a washing machine.

(3) The undigested materials are mixed thoroughly with intestinal secretions.

Analogy

The action of segmentation is like the movement of a hand mixer combining all the ingredients in cake batter.

OVERVIEW

from stomach

Black arrows indicate movement of undigested materials through the small intestine.

Serosa

Muscularis
- **longitudinal layer**
- **circular layer**

Submucosa

Mucosa

Color each layer of the wall a different color.

Layers of the Wall of the Small Intestine
(midsaggital section)

PERISTALSIS

Muscularis layer
Longitudinal layer
Circular layer

direction of movement

Color the longitudinal and circular muscle layers red.

SEGMENTATION

The *p*urpose of *p*eristalsis is to **propel** undigested materials through the digestive tract like squeezing a tube of toothpaste moves toothpaste through the tube.

The purpose of segmentation is to **mix** undigested materials with enzymes and other secretions like a hand mixer combining all the ingredients in cake batter.

451

Description

Absorption is the movement of small, digested substances from the digestive tract into either the blood or the lymph. Almost all absorption in the digestive system occurs in the small intestine. Folds are used in organs and structures throughout the body to increase surface area for different purposes. The small intestine uses its many folds to maximize surface area for absorption of nutrients.

The three significant folded structures in the small intestine are the **plicae circulares** (sing. *plica circularis*), **villi** (sing. *villus*), and **microvilli**. The plicae circulares are folds of the inner mucosal layer of the small intestine. Extending out from the plica are numerous fingerlike projections called villi. Each villus is covered by a layer of **simple columnar epithelium**. The **goblet cells** scattered within this tissue secrete **mucus**. Each simple columnar epithelial cell has folds in its plasma membrane called microvilli.

Each cell also has two surfaces through which nutrients must pass: **apical** and **basal**. Nutrients first pass through the apical surface that contains the microvilli, then move through the cytosol, then cross the basal surface that lines the inside of the villi. In summary: nutrients follow this absorption pathway:

Lumen ⟶ Epithelial cell: apical surface ⟶ Cytosol ⟶ Basal surface ⟶ Blood or lymph (small intestine)

Process

The purpose of chemical digestion is to break down macromolecules into their simplest building blocks. These building blocks include **monosaccharides** from complex carbohydrates; **amino acids, dipeptides,** and **tripeptides** from **proteins;** and monoglycerides and **fatty acids** from **triglycerides** (lipids). These final products of digestion are small enough to be absorbed into the blood or the lymph and then pass directly into body cells where they can be used.

Some ingested substances already in their simplest form—such as the fructose in fruit juices—are partly absorbed through the wall of the stomach before they even reach the small intestine.

About 90% of all absorption occurs in the small intestine for the following reasons: (1) villi and other structures make it anatomically specialized for the process, (2) the final products of macromolecule catabolism are not even formed until they reach this site, and (3) the 20-plus-foot tube maximizes the time for absorption to take place.

Monosaccharides ⬡

Complex carbohydrates must be broken down into **monosaccharides** before they can be absorbed. The three types of monosaccharides are **glucose, galactose,** and **fructose.** Fructose, found in fruit and fruit juices, is transported across both the apical and basolateral surfaces by **facilitated diffusion.** Glucose and galactose are transported by a different method—secondary active transport with sodium (Na +). The sodium ion gradient is the driving force that helps move these two sugars across the apical surface and into the cell. Once inside the epithelial cell, both glucose and galactose cross the basolateral surface by facilitated diffusion. Fructose, glucose, and galactose all are absorbed directly into the blood capillary.

Amino acids ⬤   Dipeptides ⬤–▲   Tripeptides ⬤–▲–⬡

Most **amino acids** use **active transport** to cross the apical surface. Once inside the epithelial cell, the amino acids are actively transported again across the basolateral surface and then absorbed directly into the blood capillary.

Monoglycerides 〓   Fatty acids ⟋

Triglycerides are the most common type of lipids, so they will be considered here. When triglycerides are digested, two of their three fatty acid chains are broken off from their glycerol backbone, resulting in the formation of **monoglycerides** and **fatty acids.** In the chemical rule of "like goes into like," these products of triglyceride digestion easily diffuse through the plasma membrane of the apical surface because they are chemically similar to the phospholipids in the plasma membrane. Some of the fatty acid chains are short-chain fatty acids, and others are long-chain fatty acids. The **short-chain fatty acids** diffuse directly across both the apical and basolateral surfaces and are absorbed directly into the blood capillary.

The pathway for monoglycerides and **long-chain fatty acids** is a bit more complex. After diffusing through the apical surface, the monoglycerides are digested further into glycerol and a single fatty acid chain. The glycerol then is reassembled back into triglycerides by linking with free fatty acid chains. The newly formed triglycerides are packaged in a protein coat to form a structure called a **chylomicron.** Chylomicrons leave the cell by exocytosis (see p. 86). They are too large to enter the blood capillary, so they enter the more permeable lacteal instead. As they travel in the lymph of the **lacteal,** they eventually are deposited into the blood via the left subclavian vein.

Muscularis (*longitudinal layer*)

Serosa

Muscularis (*circular layer*)

Submucosa

Plica circularis

Mucosa

Submucosa

Layers of the wall of the small intestine

Simple columnar epithelial cell

Goblet cell

Lacteal

Blood capillary

Villus

Nutrient Microvilli

Apical surface

Plasma membrane

Nucleus

Basal surface

Simple columnar epithelium

SCHEMATIC OF ABSORPTION PROCESS

INGESTED FORM OF NUTRIENTS

LUMEN (of small intestine)

EPITHELIAL CELLS (in villus)

BLOOD CAPILLARY (in villus)

LACTEAL (in villus)

COMPLEX CARBOHYDRATES (*ex.*: starch in baked potato)

Glucose, galactose

Na^+

Na^+

Secondary active transport

Facilitated diffusion

MONOSACCHARIDES (*ex.*: sugar in fruit juice)

Facilitated diffusion

Fructose

Color the blood capillaries red, the lacteals green, the epithelial cells orange, and the ATP yellow.

Active transport

ATP

Active transport

ATP

Amino acids

PROTEIN (*ex.*: proteins in meat)

Simple diffusion

Simple diffusion

Short-chain fatty acids

Glycerol

Chylomicron

Simple diffusion

Simple diffusion

Monoglyceride

Exocytosis

TRIGLYCERIDES (*ex.*: lipids in creamy soups/sauces)

Long-chain fatty acids

Microvilli (of apical surface)

Triglyceride

Basolateral surface

Chylomicron

Description

The **pancreas** is an elongated, pinkish-colored gland 5 to 6 inches in length. It is divided into three main sections—**head, body,** and **tail**. The head is the widest portion; the body is the central region; and the tail marks the blunt, tapering end of the gland. The surface is bumpy and nodular. The individual nodes are called **lobules**. Scattered within the gland are many glandular epithelial cells. Also inside is a long tube called the **pancreatic duct** that runs through the middle of the gland and terminates in an opening to the duodenum. Smooth muscle surrounds this opening to form a valve called the **hepatopancreatic sphincter**.

The pancreas has a dual function as both an **endocrine** gland and an **exocrine** gland. Making it an exocrine gland, the vast majority of the epithelial cells (99 percent) form clusters called **acini**. The cells in the acini produce watery secretions and many different digestive enzymes to aid in the process of chemical digestion. These are released into the pancreatic duct and then into the **duodenum**. Making it an endocrine gland, the remaining 1 percent of its cells are arranged in separate cell clusters called **pancreatic islets** (*islets of Langerhans*). These various cells produce four different hormones that diffuse directly into the bloodstream, travel to various target organs, and induce responses in those organs in order to regulate a variety of different processes in the body.

Analogy

The general **gross anatomy of the pancreas** is like a **tadpole** with a head, body, and tail in one elongated shape. The **surface of the pancreas** has a nodular appearance like that of **alligator skin**. However, its texture is not as coarse.

Location

Abdominal cavity; the head of the pancreas is found near the duodenum and its tail is below and behind the stomach.

Function

- For the **digestive system**—Cells produce a wide variety of important enzymes to assist in chemical digestion in the following categories:
  - **Carbohydrases** (*digest carbohydrates*)
  - **Lipases** (*digest lipids*)
  - **Nucleases** (*digest nucleic acids such as DNA*)
  - **Proteases** (*digest proteins*)
- For the **endocrine system**—Produces the following hormones: **insulin, glucagon, somatostatin,** and **pancreatic polypeptide (PP)**

Key to Illustration

1. Common bile duct	5. Hepatopancreatic sphincter	7. Pancreatic duct
2. Duodenal papilla	6. Smooth muscle in wall of duodenum	8. Body of pancreas
3. Duodenum		9. Lobules
4. Droplet of either bile or pancreatic juice		

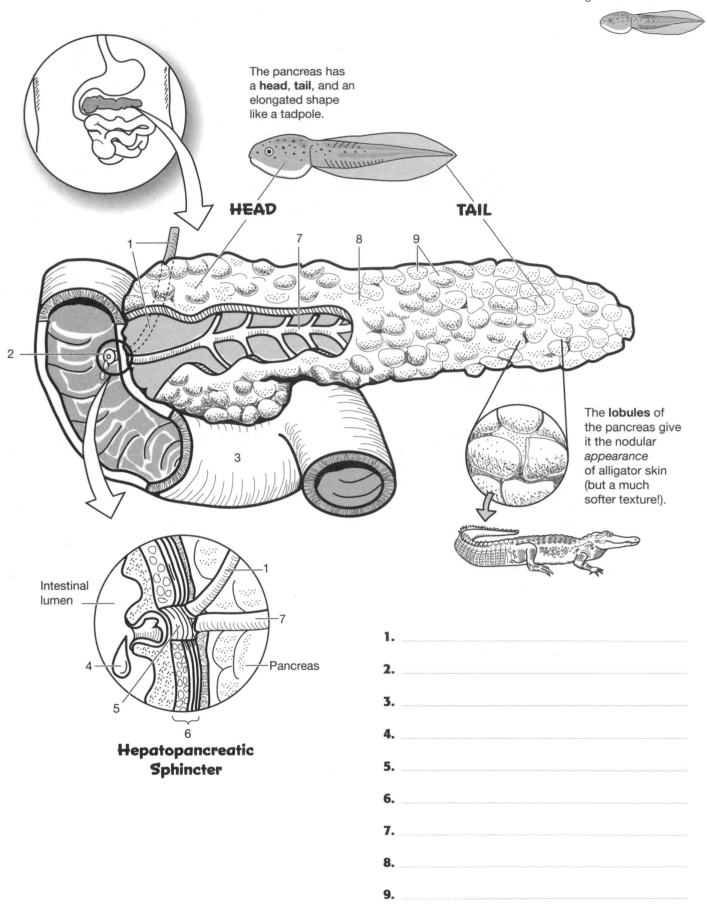

The pancreas has a **head**, **tail**, and an elongated shape like a tadpole.

HEAD

TAIL

The **lobules** of the pancreas give it the nodular *appearance* of alligator skin (but a much softer texture!).

Intestinal lumen

Pancreas

Hepatopancreatic Sphincter

1. _____

2. _____

3. _____

4. _____

5. _____

6. _____

7. _____

8. _____

9. _____

Description

The **pancreas** is vital to chemical digestion because it produces so many digestive enzymes. It is an elongated, pinkish-colored gland divided into three main sections: *head, body,* and *tail.* The head is the widest portion; the body is the central region; and the tail marks the blunt, tapering end of the gland. The surface is bumpy and nodular. The individual nodes are called **lobules**.

Scattered within the gland are many glandular epithelial cells. Also inside is a long tube called the pancreatic duct that runs through the middle of the gland and terminates in an opening to the duodenum. Smooth muscle surrounds this opening to form a valve called the **hepatopancreatic sphincter** (*sphincter of Oddi*).

The pancreas has a dual function as both an **endocrine** and an **exocrine** gland. Making it an exocrine gland, the vast majority of the epithelial cells (99%) form clusters called **acini**. The cells in the acini produce watery secretions including bicarbonate ions and many different digestive enzymes to aid in chemical digestion. These are released into the **pancreatic duct**, through the hepatopancreatic sphincter, and into the duodenum.

Making it an endocrine gland, the remaining 1 percent of cells are arranged in separate cell clusters called called **pancreatic islets** (*islets of Langerhans*). These various cells produce four different hormones: *insulin, glucagon, somatostatin,* and *pancreatic polypeptide* (PP). These hormones diffuse directly into the bloodstream, travel to various target organs, and induce responses in those organs to regulate a variety of different processes in the body.

Function

Pancreatic acini produce many different enzymes for chemical digestion. These include **amylase** (for complex carbohydrates), **lipase** (for lipids), and **proteases** (for proteins). In addition, the pancreas produces **nucleases** (for digesting nucleic acids such as DNA).

Let's examine the role of each of the first three enzyme types mentioned.

1. **Amylase** catalyzes the initial step in carbohydrate digestion, the conversion of a polysaccharide (*ex.:* starch) into disaccharides.

2. **Proteases:** The three different proteases—**trypsin, chymotrypsin,** and **carboxypeptidase**—all are used to speed up the middle step in protein digestion, conversion of large polypeptides into small polypeptides.

3. **Lipase** catalyzes the breakdown of lipids such as triglycerides into **monoglycerides** and **fatty acids**.

As chyme passes from the stomach into the duodenum, it stimulates the pancreas to make and release digestive enzymes. This process is regulated by the hormones **secretion** and **cholecystokinin** (CCK) (see p. 436). The hepatopancreatic sphincter normally is closed, but the hormone CCK causes it to open, which allows the release of pancreatic secretions into the duodenum. These secretions then mix with the chyme so the enzymes can do their jobs. As the hepatopancreatic sphincter opens, it also allows for the release of bile from the common bile duct. **Bile** mediates the first step in lipid digestion. It is needed to emulsify the lipids in the chyme so the lipases can work.

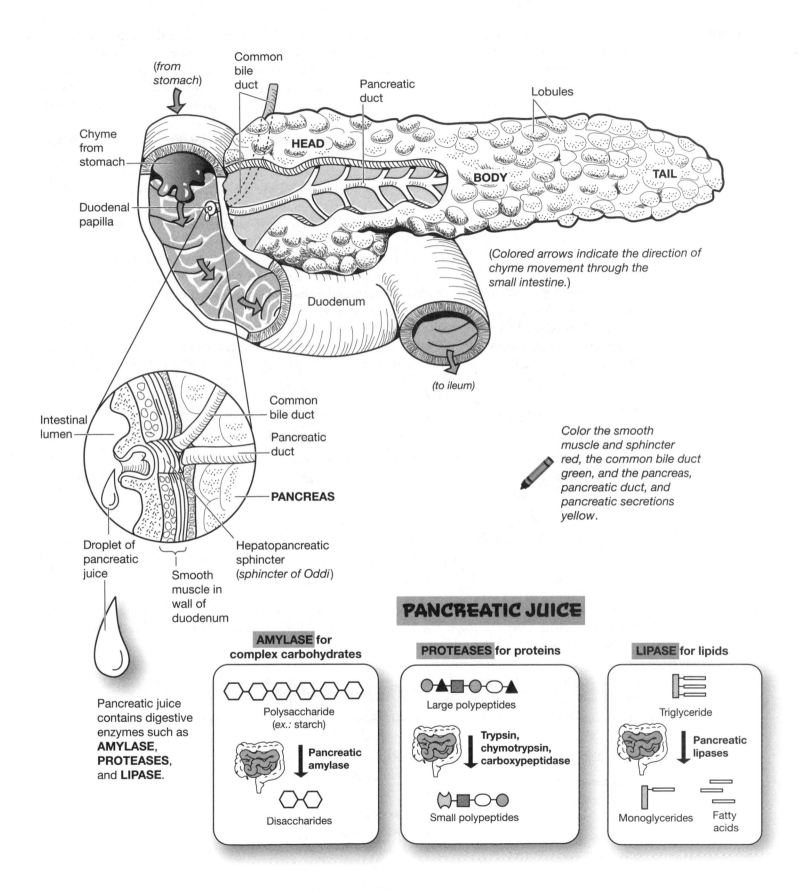

(from stomach)

Common bile duct

Pancreatic duct

Lobules

Chyme from stomach

HEAD

BODY

TAIL

Duodenal papilla

(Colored arrows indicate the direction of chyme movement through the small intestine.)

Duodenum

(to ileum)

Intestinal lumen

Common bile duct

Pancreatic duct

PANCREAS

Color the smooth muscle and sphincter red, the common bile duct green, and the pancreas, pancreatic duct, and pancreatic secretions yellow.

Droplet of pancreatic juice

Smooth muscle in wall of duodenum

Hepatopancreatic sphincter (sphincter of Oddi)

Pancreatic juice contains digestive enzymes such as **AMYLASE**, **PROTEASES**, and **LIPASE**.

PANCREATIC JUICE

AMYLASE for complex carbohydrates

Polysaccharide (ex.: starch)

↓ **Pancreatic amylase**

Disaccharides

PROTEASES for proteins

Large polypeptides

↓ **Trypsin, chymotrypsin, carboxypeptidase**

Small polypeptides

LIPASE for lipids

Triglyceride

↓ **Pancreatic lipases**

Monoglycerides

Fatty acids

Description

The **liver** is the largest abdominal organ and is located below the diaphragm in the abdominal cavity. It is divided into two major lobes—left and right—separated by a sheet of connective tissue called the **falciform ligament**. Along the bottom edge of the falciform ligament is a fibrous, cable-like structure called the **round ligament** (*ligamentum teres*) derived from the fetal **umbilical vein**. The liver is anchored to the diaphragm by the **coronary ligament**. The liver also has two minor lobes—**caudate** and **quadrate**—visible in the posterior view.

Microscopically, each lobe of the liver is composed of more than 100,000 individual hexagon-shaped units called **hepatic lobules**. These structures are the fundamental functional units of the liver. At each corner of a lobule is a cluster of three vessels called a **hepatic triad** that consists of a hepatic artery, a branch of the hepatic portal vein, and a bile duct. At the center of each lobule is a long vessel called the **central vein**.

Radiating out from the central vein like spokes from a wheel are the very permeable, blood-filled capillaries called **sinusoids**. Fixed to the inner lining of the sinusoids are macrophages, also called Kupffer's cells. The **hepatic artery** and **hepatic portal vein** deliver blood into the sinusoids, and it travels toward the central vein, then into the hepatic veins, and into the **inferior vena cava**. Most **hepatocytes** (*liver cells*) are located close to the blood in the sinusoids, which makes it easier to deposit products into the blood or screen materials out of it.

Bile is produced by hepatocytes and drains into vessels called **bile canaliculi**, then into the larger **bile ducts**, which transport the bile away from the liver and into the **gallbladder** for storage. Note that the direction of bile flow in the bile canaliculi is opposite to that of the blood in the sinusoids.

Functions

The liver has more than 200 different functions. Major functions of the liver are:

- **Synthesis**   *ex.:* plasma proteins, clotting factors, bile, cholesterol
- **Storage**   *ex.:* iron (Fe), glycogen, blood, fat-soluble vitamins
- **Metabolism**   *ex.:* convert glucose to glycogen and glycogen to glucose; convert carbohydrates to lipids, maintain normal blood glucose levels
- **Detoxification** *ex.:* alcohol and other drugs
- **Conjugation**—this means making substances water soluble

    ex.: bilirubin is converted to a water soluble form to make it easier to be excreted

Analogy

In cross section, each **hepatic lobule** is like a simple **Ferris wheel** with six swinging chairs. Each **swinging chair** is a **hepatic triad**. The **hub** of the Ferris wheel is the **central vein**, and the **spokes** of the Ferris wheel are like the **sinusoids**.

Key to Illustration		
Liver Lobes (L)	**Lobule Structures**	**Other**
L1 Left major	CV Central vein	GB Gallbladder
L2 Right major	HT Hepatic triad	
Ligaments (L)	BC Bile canaliculus	
RL Round ligament	BD Bile duct	
(*ligamentum teres*)	HA Branch of hepatic artery	
FL Falciform ligament	HPV Branch of hepatic portal vein	
CL Coronary ligament	S Sinusoid	
	H Hepatocyte	
	K Kupffer's cell (macrophage)	

LIVER
(anterior view)

Diaphragm

CL

FL

L2

L1

GB

RL

to inferior vena cava

Wedge of a hepatic lobule

CV

S

H   K   HPV (branch)

BD

HA

oxygenated blood

deoxygenated blood

to gallbladder

H

Hepatic lobule

CV

Schematic of a Lobule:

CV

S

HT

Color the bile canals green, the hepatic cells yellow, the sinusoids blue, the hepatic arteries red, and all the veins blue.

Ferris Wheel Analogy

The schematic of a hepatic lobule resembles a Ferris wheel.

CV

Interlobular vein

BC

H

BC

S

BD

CV

Hepatic Lobule
(detailed view)

H

HA

BD

HPV (branch)

Bile Production, Storage, and Release

Description

Bile is a thick, alkaline solution produced by **hepatocytes** (*liver cells*), transported through bile ducts, stored and concentrated in the **gallbladder**, and released into the lumen of the **duodenum**. It is composed mostly of water, various ions, bile pigments such as **bilirubin** (orange or yellow pigment derived from hemoglobin catabolism), and various lipids such as **cholesterol, phospholipids,** and **triglycerides**. In the gallbladder the color of bile varies from yellow to brown or green.

After being produced, it drains away from the hepatic lobule and enters a microscopic **bile duct**, which drains into a larger **hepatic duct**. The **left hepatic duct** drains bile from the left lobe of the liver, and the **right hepatic duct** drains bile from the right lobe of the liver. The two hepatic ducts then fuse to form a **common hepatic duct**.

Bile flows down the common hepatic duct and into the **common bile duct**, which terminates in an opening leading to the duodenum. Around this opening is a thickened band of smooth muscle called the **hepatopancreatic sphincter**. Normally the smooth muscle around this sphincter is contracted, causing it to be closed.

Bile is stored in the **gallbladder**, located under the right lobe of the liver. To fill the gallbladder with bile, the hepatopancreatic sphincter must remain closed. As bile drains from the liver, it backs up into the common bile duct and enters the **cystic duct**, which leads to the gallbladder. Hard, solid deposits called **gallstones** can form in either the gallbladder or bile ducts. There are two types based on which component in the bile became too concentrated—either cholesterol or bilirubin. Gallstones vary widely in size and can become too large to exit the gallbladder, which can lead to *cholecystitis*—inflammation of the gallbladder. If the gallstones damage the lining of the gallbladder, a surgical procedure called a *cholecystectomy*—removal of the gallbladder—may need to be performed.

Emptying of the gallbladder is regulated by a hormone called **cholecystokinin**, or **CCK**. As lipid-rich **chyme** (food and gastric juice) passes through the **mucosa** (inner lining) of the duodenum, it stimulates specific cells in the duodenum to release CCK. This hormone then travels through the bloodstream and targets the smooth muscle around the gallbladder.

These smooth muscle cells have receptors for CCK. Once CCK binds to these receptors, it induces the smooth muscle to contract. The result is that bile is expelled forcefully from the gallbladder. Simultaneously, CCK targets the smooth muscle of the hepatopancreatic sphincter and causes it to relax, opening the sphincter. With the sphincter open, bile moves out of the gallbladder, down the cystic duct, down the common bile duct, past the hepatopancreatic sphincter, and into the duodenum.

Function

Emulsification of lipids (fats) is the first step in the chemical digestion of lipids. Rather than breaking chemical bonds, bile physically breaks large masses of lipids into smaller lipid droplets. This makes it easier for enzymes such as **lipase** to facilitate chemical digestion.

Key to Illustration		
1. Right hepatic duct	**3.** Common hepatic duct	**5.** Common bile duct
2. Left hepatic duct	**4.** Cystic duct	**6.** Pancreatic duct

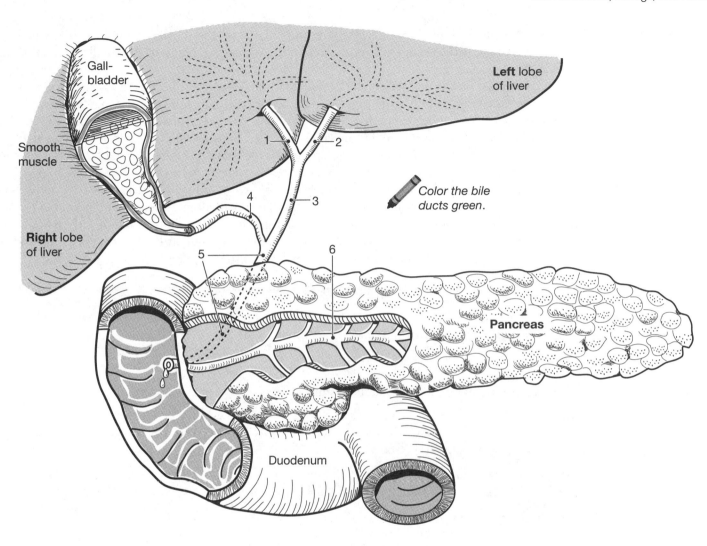

Gall-
bladder

Left lobe
of liver

Smooth
muscle

Right lobe
of liver

1

2

3

4

5

6

Pancreas

Duodenum

*Color the bile
ducts green.*

FILLING the gallbladder:

Hepatic
lobule

Smooth
muscle

Gall-
bladder

Pancreas

Duodenum

**Sphincter
CLOSED**

EMPTYING the gallbladder:

Hepatic
lobule

Smooth
muscle

Gall-
bladder

Pancreas

Duodenum

**Sphincter
OPEN**

*Use green to color the pathways
indicated by the black arrows.*

Description   The **large intestine**, or large bowel, is a hollow, muscular tube about 5 feet long. It is subdivided into three different parts—the **cecum**, **colon**, and **rectum**. It connects the end of the small intestine (*ileum*) to the **anus**.

The **cecum** is a pouch that marks the beginning of the large intestine. On the posterior side of this structure, the **appendix** (*vermiform appendix*) can be located. This is a long, slender, hollow, tubelike structure that opens into the cecum. It is classified as part of the lymphatic system because it contains lymphatic tissue. Inflammation of the appendix is called *appendicitis*. The surgical removal of the appendix is called an *appendectomy*. The next portion is the **colon**, the longest part of the large intestine. It is subdivided into four segments—*ascending, transverse, descending,* and *sigmoid* colon.

Along the length of the colon are three separate bands of smooth muscle called the **teniae coli**. They constrict the colon into pouches called **haustra** that run all along its length and permit expansion and elongation of the colon. At regular intervals along the taenia coli are flaps of fatty tissue called **fatty appendices** (*epiploic appendages*). The **rectum** is the last segment of the large intestine and the digestive tract and allows for temporary storage of fecal waste.

Analogy   The **appendix** is also called the *vermiform appendix*. The term "vermiform" means "wormlike" so the appendix is compared to a **worm** because of its general shape.

Location   Abdominal cavity; surrounds the small intestine.

Functions
- **Reabsorption** of water and electrolytes
- **Absorption** of some vitamins (e.g., vitamin K, B-complex vitamins) produced by the bacteria *Escherichia coli (E. coli)*, which naturally live within the colon
- **Compaction and temporary storage** of fecal waste

Key to Illustration

1. Cecum	**4.** Left colic (*splenic*) flexure	**7.** Ileocecal valve
2. Right colic (*hepatic*) flexure	**5.** Teniae coli	**8.** Rectum
3. Haustra	**6.** Fatty appendages	**9.** Anus

Visualizing the General Structure

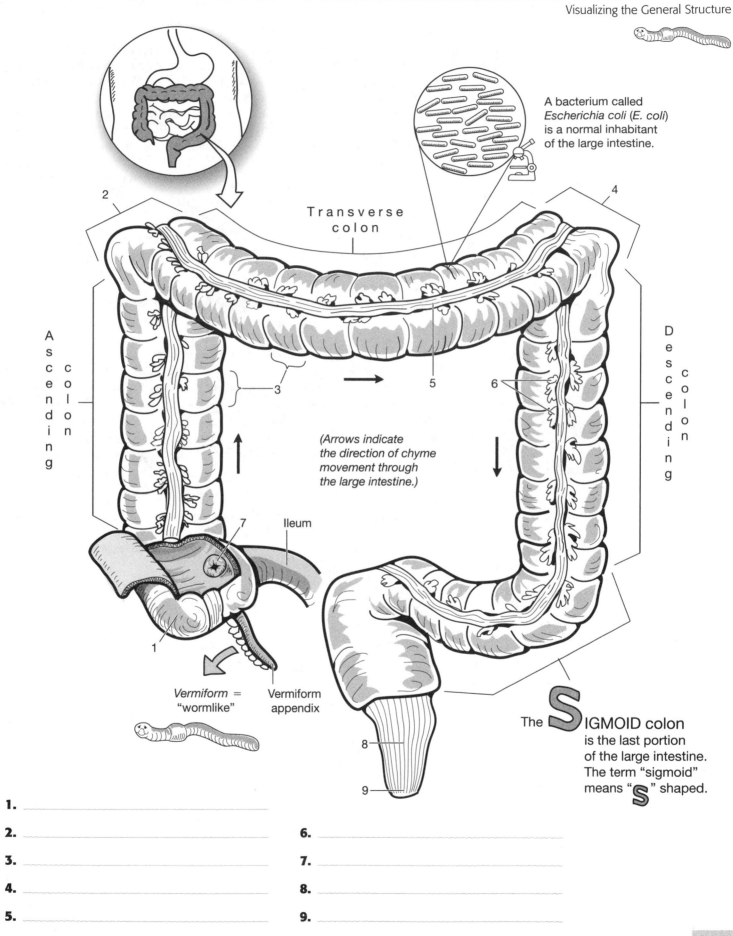

Transverse colon

A bacterium called *Escherichia coli* (*E. coli*) is a normal inhabitant of the large intestine.

Ascending colon

Descending colon

(Arrows indicate the direction of chyme movement through the large intestine.)

Ileum

Vermiform = "wormlike"

Vermiform appendix

The **S**IGMOID colon is the last portion of the large intestine. The term "sigmoid" means "**S**" shaped.

1. _____

2. _____

3. _____

4. _____

5. _____

6. _____

7. _____

8. _____

9. _____

DIGESTIVE SYSTEM: Large Intestine

Function of the Large Intestine

Description

The **large intestine** (*large bowel*) is a hollow, muscular tube about 5 feet long that connects the end of the small intestine (*ileum*) to the **anus**. It is subdivided into three different parts—the **cecum, colon,** and **rectum**. The cecum is a pouch that marks the beginning of the large intestine. On the posterior surface of this structure, the **appendix** is located. This is a long, slender, hollow tube that opens into the cecum.

The next portion is the colon, the longest part of the intestine. It is subdivided into four segments—the *ascending, transverse, descending,* and *sigmoid* colon. Along the length of the colon is a band of smooth muscle called the **taenia coli**. It bulges the colon into pouches called **haustra** that run along its length and permit expansion and elongation of the colon. At regular intervals along the taenia coli are flaps of fatty tissue called **fatty appendices**. The last segment of the large intestine and the digestive tract is the rectum, which allows for temporary storage of fecal waste.

Process

Most nutrients have been digested and absorbed into the blood by the time they reach the end of the small intestine. The remaining undigested chyme passes through the **ileocecal sphincter** and into the large intestine. Two opposing processes—absorption and secretion—occur. **Absorption** is the movement of nutrients from the large intestine into the blood and then into cells. **Secretion** is the movement of substances from the blood to the large intestine for eventual excretion. Remaining to be absorbed in the large intestine are some electrolytes, water, and vitamins. Absorption and secretion take place in the ascending and transverse colon. The remaining undigestible material is compacted into a semisolid mixture called **feces**, stored temporarily in the descending and sigmoid colon, and finally ejected from the body. The process of eliminating the feces is controlled by a defecation reflex.

We share a mutually beneficial relationship with the various **bacteria**—such as *Escherichia coli* (*E. coli*)—that live inside the large intestine. We provide them a warm environment and food to live and reproduce, and they make vitamins for us that we can't make on our own, such as vitamin K and B-complex vitamins. But bacteria also contribute to *flatus*, or gas. As they metabolize any remaining carbohydrates, they produce hydrogen, carbon dioxide, and methane gases. Another gas produced by bacteria—hydrogen sulfide—gives off a "rotten egg" odor. But the bacteria are not entirely to blame because the major source of intestinal gas is air swallowed during the rapid ingestion of food.

Motility

The **muscularis layer** of the wall of the large intestine contains two layers of **smooth muscle**—an inner, circular layer and an outer, longitudinal layer. Contraction of this smooth muscle in different ways allows for movement of undigested materials inside the large intestine. The three types of motility through the large intestine are:

1. **Segmentation** (*haustral churning*). As undigested material enters the haustra, the expansion triggers muscularis layer contractions that pinch the haustra into separate compartments and turn the contents inside. This serves to enhance absorption and secretion of water and electrolytes as well as to propel the contents to the next haustrum.

2. **Peristalsis.** Stimulation of the muscularis layer results in a weak, rhythmic wave of contraction, a combination of pinching the circular layer and shortening the longitudinal layer. This contraction combination results in the gradual propulsion of undigested material through the tract.

3. **Mass movement.** This strong muscle contraction forces fecal material into the rectum. It is regulated by a reflex and occurs two to three times a day on average.

Absorption and Secretion

Absorption follows a predictable sequence, illustrated on the facing page, as follows:

1. Levels of **sodium** (Na^+) and **potassium** (K^+) are controlled by the **sodium-potassium pump** (see p. 90), which transports these ions in opposite directions. Sodium is actively transported out of the lumen of the large intestine and into the blood, and potassium is transported into the large intestine. The net result is that sodium is reabsorbed into the blood, and potassium is secreted into the large intestine to be removed in the feces. The result is that it causes a higher solute concentration in the blood relative to the lumen of the large intestine.

2. Because **water** (H_2O) always moves passively across a plasma membrane toward the higher solute concentration via osmosis (see p. 80), this is what occurs next: The body forces osmosis by actively transporting sodium out first.

GROSS ANATOMY

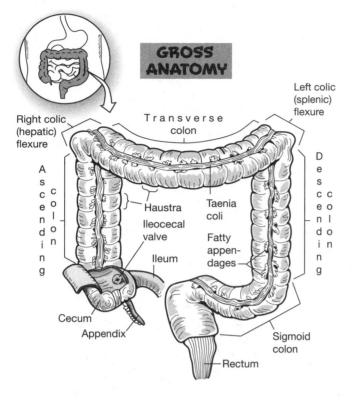

Right colic (hepatic) flexure

Transverse colon

Left colic (splenic) flexure

Ascending colon

Descending colon

Haustra

Ileocecal valve

Taenia coli

Ileum

Fatty appendages

Cecum

Appendix

Sigmoid colon

Rectum

FUNCTIONAL OVERVIEW

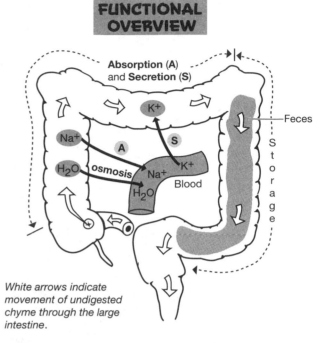

Absorption (A) and Secretion (S)

K^+

Na^+

A

S

Feces

H_2O osmosis

Na^+ K^+

H_2O Blood

Storage

White arrows indicate movement of undigested chyme through the large intestine.

MOTILITY

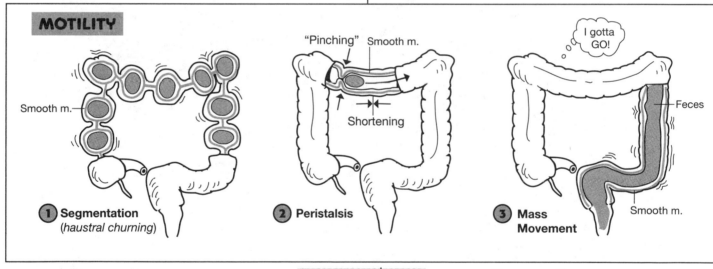

Smooth m.

"Pinching" Smooth m.

Shortening

I gotta GO!

Feces

Smooth m.

(1) **Segmentation** (*haustral churning*)

(2) **Peristalsis**

(3) **Mass Movement**

ABSORPTION (A) AND SECRETION (S)

Bacteria are normal inhabitants of the large intestine and manufacture vitamins for the body.

vit. K
vit. B complex

A

vit. K
vit. B complex

(1) Na^+

A

Na^+

K^+

K^+

S

(2) H_2O

A

H_2O

(osmosis)

Large intestine

Blood capillary

Color the large intestine brown and the blood capillary red.

465

Notes

METABOLIC PHYSIOLOGY

Description

Metabolism is the sum total of all the chemical reactions that occur in the body. It consists of two different parts—catabolic reactions and anabolic reactions. **Catabolic reactions** break down biomolecules into simpler building blocks (e.g., triglycerides into fatty acids and glycerol), and **anabolic reactions** use these building blocks to synthesize more complex biomolecules (e.g., amino acids into proteins). Energy metabolism focuses on the elements of metabolism that generate ATP. As a general overview of energy metabolism, let's examine its three main phases:

PHASE 1:
The Release of Nutrients in the Digestive Tract Lumen

The digestion of three major nutrients—triglycerides, complex carbohydrates, and proteins—is shown below. As these macromolecules are ingested, they enter the digestive tract and gradually are broken down into their fundamental building blocks:

- Proteins → → → → → → **amino acids**
- Complex carbohydrates → → **glucose** (and other simple sugars)
- Triglycerides → → → → **monoglycerides** and **fatty acids**

This phase of catabolism is catalyzed by digestive enzymes in the digestive tract. These building blocks now are small enough to be used by cells. After being absorbed into the bloodstream through microscopic structures called **villi** on the inner lining of the small intestine, they are transported to various cells throughout the body.

PHASE 2:
The Fate of Nutrients within Cells of Various Tissues

Glucose has three possible fates:
- **Catabolized:** When the body requires energy, glucose is catabolized to produce ATP molecules. The three stages of glucose catabolism are glycolysis, the citric acid cycle, and the electron transport system.
- **Stored:** This takes place in the form of **glycogen**, a long chain of glucose molecules. Skeletal muscle cells and liver cells store lots of glycogen when excess glucose is present in the blood.
- **Converted:** Glucose can be converted into several **amino acids** that may be used to make proteins. Alternatively, when storage of glycogen is maximized, liver cells can convert glucose into **glycerol** and **fatty acids** to form **triglycerides**. The triglycerides then are deposited in fatty tissue.

Fatty acids and **glycerol** have one of two general fates:
- **Converted/catabolized:**
  Glycerol can be converted into glyceraldehyde-3-phosphate, one of the compounds formed during the process of glycolysis. Then, depending on the ATP levels in the cell, it is either converted into **pyruvate** (via glycolysis) or **glucose** (via gluconeogenesis).
  Fatty acids follow a different pathway than glycerol. They can be converted into **acetyl CoA**, which then enters the citric acid cycle to produce ATP.
- **Stored:** Glycerol and fatty acids recombine to form triglycerides within liver cells and fat cells.

Amino acids have three different fates:
- **Anabolized:** Amino acids may be used to produce **proteins**.
- **Converted:** Some amino acids can be converted into **acetyl CoA** and enter the citric acid cycle to produce ATP. When there are excess amino acids, they can be converted into either **glucose** (via gluconeogenesis) or **fatty acids**.
- **Deaminated:** Liver cells can remove the amino group from some amino acids to produce carboxylic acids and ammonia (NH_3) in a process called deamination. Then the carboxylic acids can enter different stages of the citric acid cycle.

PHASE 3:
Aerobic Catabolism of Nutrients in the Mitochondria of Cells

Aerobic respiration refers to the two parts of cellular respiration that occur inside the mitochondrion: the citric acid cycle (see p. 474) and the electron transport system (see p. 476). **Acetyl CoA**, a key chemical intermediate in metabolism, is generated from fatty acids, pyruvate, and some amino acids within the mitochondrion. Once acetyl CoA is produced, it may enter the citric acid cycle as part of normal catabolism.

TRIGLYCERIDES

COMPLEX CARBOHYDRATES

PROTEINS

PHASE 1: Release of nutrients in digestive tract lumen

Fatty acids

Glycerol

Glucose, other simple sugars

Amino acids

Glucose

GLYCOLYSIS

GLUCONEOGENESIS

GLYCOGEN

PROTEIN

LIPIDS

e−

ATP

Pyruvate

PHASE 2: Fate of nutrients within cells of various tissues

NH_3

e−

CO_2

Acetyl CoA

CO_2

CITRIC ACID CYCLE

e−

ELECTRON TRANSPORT SYSTEM

O_2

H_2O

ATP

ATP

PHASE 3: Aerobic catabolism of nutrients in mitochondria of cells

Mitochondrion

KEY

Catabolic reactions

Anabolic reactions

METABOLIC PHYSIOLOGY

Overall Equation for Oxidation of Glucose

In the process of cellular respiration, 1 molecule of glucose reacts with 6 oxygen molecules through a series of chemical reactions to produce 6 carbon dioxide molecules, 6 water molecules, 36 molecules of ATP, and heat energy lost to the environment.

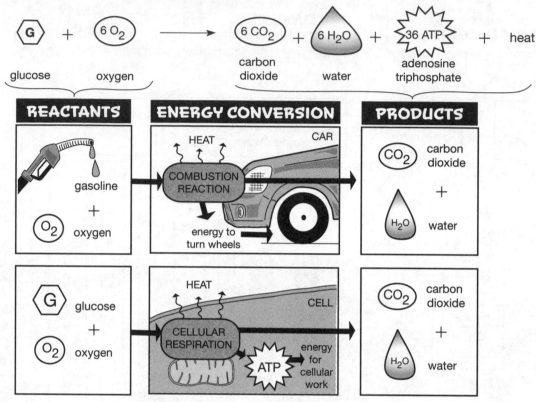

Description

The process of cellular respiration is like the combustion of gasoline in a car engine. Both require a fuel that must be chemically reacted with oxygen. Glucose is the fuel for the cell, whereas gasoline is the fuel for the car engine. The car engine uses a spark plug to ignite the combustible oxygen/gasoline mixture, causing it to produce a small explosion that moves a piston up and down inside a metal cylinder. This kinetic energy is transferred to an axle, causing it to spin. Because the wheels are attached to the axle, they spin as the axle spins. The direct products of this reaction, carbon dioxide and water vapor, are released in the exhaust gases, and heat is lost to the external environment. In summary, the potential energy in gasoline is converted into kinetic energy that propels the car.

Cellular respiration works in a similar manner. Instead of having all the energy released at once, glucose is gradually broken down through a long series of chemical reactions. The overall purpose is to trap as much energy as possible from glucose in the form of an energy molecule called ATP. Then the ATP can be used to do cellular work such as contracting a muscle cell. During the process, some heat is lost to the environment. As in the combustion of gasoline, the final products of this chemical process are carbon dioxide and water. In summary, some of the chemical-bond energy in glucose is transferred slowly into ATP molecules that can be used to do cellular work.

Cellular respiration is a series of oxidation-reduction reactions, or redox reactions. These involve the transfer of electrons from one substance to another. In this case, glucose is oxidized because it loses electrons to oxygen. Oxygen is reduced because it gains the electrons lost from glucose. To distinguish oxidation from reduction remember the words OIL RIG = Oxidation Is Loss, Reduction Is Gain.

Efficiency Rating of Cellular Respiration

Cellular respiration incorporates three processes: (1) glycolysis, (2) the citric acid cycle, and (3) the electron transport system (ETS). Each of these processes will be discussed in more detail in separate modules.

The process of cellular respiration is more efficient than a car engine. During the oxidation of glucose, 686 kilocalories of energy are released. Of this total, 278 are captured in the bonds of ATP molecules, giving it an efficiency rating of 41%. The remaining 59% is lost as heat. Compared to a typical car engine, which is between 10–30% efficient, this rating is looking pretty good!

Road map

KEY

- = electron
- NAD+ = nicotinamide adenine dinucleotide (coenzyme)
- = hydrogen ion

FUN FACT: The carbon atom in each molecule of carbon dioxide gas that you exhale was originally one of the carbon atoms in glucose.

METABOLIC PHYSIOLOGY

Overview

Location within cell:	cytosol
Aerobic or anaerobic?	anaerobic (doesn't require O_2)
Initial reactant:	glucose
Final product(s):	• 2 pyruvates
Side products:	• 2 NADH (reduced coenzymes)
Net yield of energy:	• 2 ATP molecules (4 created – 2 used)

Description

The term **glycolysis** means the "splitting of glucose," which accurately describes the process. It begins with one glucose molecule containing 6 carbon atoms and ends with the formation of 2 new 3-carbon molecules called **pyruvate**. This process is catalyzed by various enzymes in the **cytosol** of cells and yields a net gain of 2 ATP molecules and 2 NADH. There are numerous intermediate steps.

Instead of presenting the various chemical reactions that occur in each of the intermediate steps, the process will be presented conceptually in two phases:

Phase 1:   **Energy Input Phase** (splitting of glucose)

Key idea: **Free energy from the hydrolysis of 2 ATP molecules is invested to help split the original glucose molecule.**

- Glucose enters the cell and remains in the **cytosol**.
- Step ① With the help of an enzyme, 2 phosphate groups from 2 ATP molecules are transferred onto glucose, thereby changing it into **fructose-1,6-bisphosphate**. These phosphate molecules have a net negative charge and repel each other. The repulsion makes the molecule more unstable in preparation for splitting it.
- Step ② With the help of an enzyme, the covalent bond between carbons number 3 and 4 in fructose-1,6-bisphosphate is broken, yielding 2 new 3-carbon molecules: **dihydroxyacetone phosphate** and **glyceraldehyde-3-phosphate**. These two molecules are isomers of each other.

Phase 2:   **Energy Capture Stage**

- Step ③ The 2 new 3–carbon molecules from phase 1 each have an additional phosphate group transferred onto them with the help of yet another enzyme. At the same time, a carrier molecule called **NAD$^+$** picks up 2 electrons and 1 proton from each of the molecules and is reduced to **NADH**. These electrons can be transported into the electron transport system for later use.
- Step ④ The 2 phosphates from each 3-carbon intermediate are transferred onto ADP to form a total of 4 new ATP molecules. The final result is that 2 new molecules of pyruvate are created.
- The net gain in ATP for glycolysis is 2 ATP (**4 created** in phase 2 minus **2 used** in phase 1).

Analogy

NAD$^+$ is an ion called **nicotinamide adenine dinucleotide**. It is derived from the vitamin niacin and functions as a carrier ion that transports electrons to the electron transport system (ETS). When NAD$^+$ is reduced (gains electrons), it becomes **NADH**. The NAD$^+$ ion is like a **car pulling a trailer with no cargo**, and **NADH** is like a **car filled with its cargo of 2 electrons (and 1 proton)**.

Road map

GLYCOLYSIS: CONCEPTUAL OVERVIEW

Energy input phase

Glucose

2 ATP

2 ADP

(Covalent bond to be broken)

Inorganic phosphate

Fructose-1,6-bisphosphate

Energy capture phase

Dihydroxyacetone phosphate

Glyceraldehyde-3-phosphate

NAD^+

NAD^+

NADH

NADH

to electron transport system

2 ADP

2 ATP

2 ADP

2 ATP

to electron transport system (see p. 476)

1 Pyruvate

1 Pyruvate

2 Pyruvates

If oxygen present

If oxygen not present

CITRIC ACID CYCLE

2 lactates (see p. 478)

(see p. 474)

Overview

Location within cell:	mitochondrion
Aerobic or anaerobic?	aerobic (indirectly)
Initial reactants/substrates:	oxaloacetate and acetyl CoA
Final product:	oxaloacetate (which condenses with acetyl CoA to form citrate)
Important side products: (for **both** pyruvates from glycolysis or two rounds of the citric acid cycle)	• 6 NADH, 2 $FADH_2$ • 6 CO_2 molecules (4 during the cycle; 2 in formation of acetyl CoA)
Net yield of energy:	• 2 ATP molecules

Three Key Events:

1. Conversion of pyruvate (from glycolysis) into **acetyl CoA**. This is what links glycolysis to the citric acid cycle.

2. Formation of CO_2 as a side product. The carbon in carbon dioxide can be traced back to the carbon in glucose. Think about that the next time you exhale some CO_2!

3. Formation of many **reduced coenzymes** (**NADH, FADH₂**). These reduced coenzymes link the citric acid cycle to the electron transport system.

Before Citric Acid: Cycle Begins

Pyruvate is first converted into **acetic acid** and then into **acetyl coenzyme A** (CoA).
—CO_2 is released in the process.
—**CoA** is used to carry the 2-carbon remnant of pyruvate oxidation.
—**NAD⁺** is reduced to **NADH** by picking up 2 electrons.

Citric Acid Cycle: Step by Step

NOTE: Each of the steps in the citric acid cycle requires the help of a different enzyme.

Step 1: Oxaloacetate condenses with **acetyl coenzyme A** (CoA) to form **citrate**.

Step 2: **Citrate** is converted into **isocitrate**.

Step 3: **Isocitrate** is converted into **alpha (α)-ketoglutarate**.
 • NAD⁺ is reduced to NADH
 • CO_2 is released in the process

Step 4: **Alpha (α)-ketoglutarate** is converted into **succinyl CoA**.
 • CoA enters this step to carry the succinyl group
 • NAD⁺ is reduced to NADH
 • CO_2 is released in the process

Step 5: **Succinyl CoA** is converted into **succinate**.
 • Phosphate group is transferred to GDP to create GTP
 • Hydrolysis of GTP transfers phosphate group to ADP to form ATP

Step 6: **Succinate** is converted into **fumarate**.
 • FAD is reduced to FADH₂

Step 7: **Fumarate** is converted into **malate**.

Step 8: **Malate** is converted into **oxaloacetate**.
 • NAD⁺ is reduced to NADH

Analogy:

NAD⁺ and **FAD** are compounds that act like **shuttle buses**. **NAD⁺** is called nicotinamide adenine dinucleotide, and **FAD** is called flavin adenine dinucleotide. **FAD** is derived from vitamin B_2 (riboflavin). Both are classified as carrier molecules, represented as different types of vehicles carrying a cargo. By picking up 2 electrons, **NAD⁺** gets reduced to **NADH**, and **FAD** gets reduced to **FADH₂**. The **NAD⁺** or **FAD** molecule is like a **car pulling a trailer with no cargo**, and **NADH** or **FADH₂** is like a **car filled with its cargo of electrons (and protons)**.

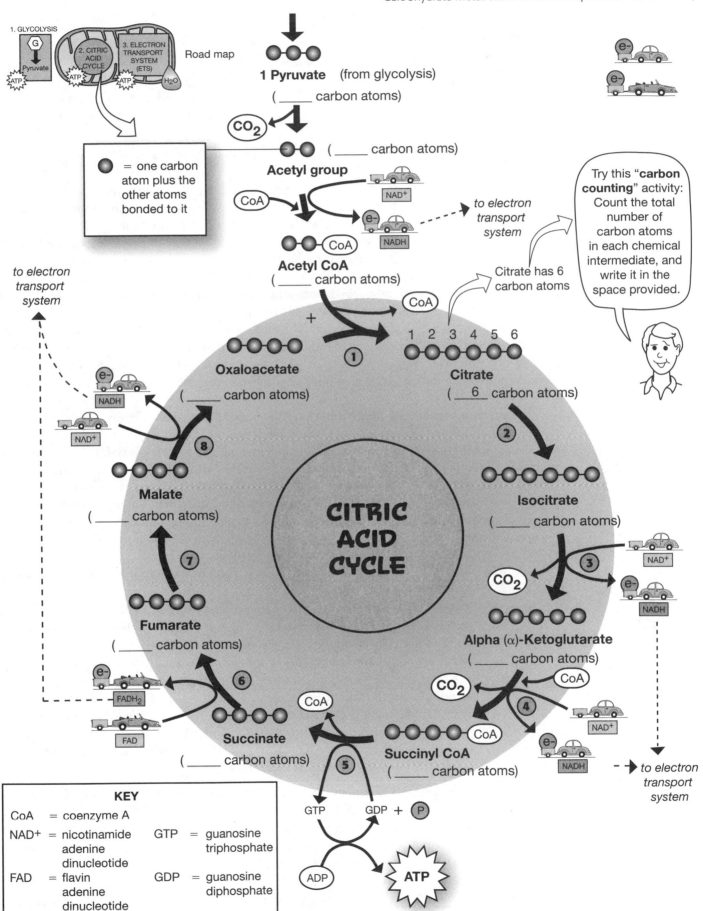

1. GLYCOLYSIS

G

Pyruvate

2. CITRIC ACID CYCLE

3. ELECTRON TRANSPORT SYSTEM (ETS)

H₂O

ATP

Road map

1 Pyruvate (from glycolysis)

(_____ carbon atoms)

CO_2

= one carbon atom plus the other atoms bonded to it

Acetyl group

(_____ carbon atoms)

CoA

NAD⁺

e⁻

NADH

to electron transport system

Acetyl CoA

(_____ carbon atoms)

Try this "**carbon counting**" activity: Count the total number of carbon atoms in each chemical intermediate, and write it in the space provided.

Citrate has 6 carbon atoms

to electron transport system

CoA

+

Oxaloacetate

(_____ carbon atoms)

①

Citrate

1 2 3 4 5 6

(_6_ carbon atoms)

e⁻

NADH

NAD⁺

⑧

Malate

(_____ carbon atoms)

CITRIC ACID CYCLE

②

Isocitrate

(_____ carbon atoms)

⑦

CO_2

③

NAD⁺

e⁻

NADH

to electron transport system

Fumarate

(_____ carbon atoms)

⑥

Alpha (α)-Ketoglutarate

(_____ carbon atoms)

e⁻

FADH₂

FAD

Succinate

(_____ carbon atoms)

CoA

⑤

CO_2

④

CoA

CoA

Succinyl CoA

(_____ carbon atoms)

NAD⁺

e⁻

NADH

to electron transport system

GTP

GDP + P

ADP

ATP

e⁻

e⁻

KEY

CoA	= coenzyme A	
NAD⁺	= nicotinamide adenine dinucleotide	GTP = guanosine triphosphate
FAD	= flavin adenine dinucleotide	GDP = guanosine diphosphate

METABOLIC PHYSIOLOGY

Carbohydrate Metabolism: Cellular Respiration—Electron Transport System (ETS)

Overview

Location within cell:	mitochondrion
Aerobic or anaerobic?	aerobic (O_2 used **directly**)
Important side products:	• 10 NADH and 2 $FADH_2$ (total of 12 reduced coenzymes)
Final electron acceptor:	O_2 (oxygen)
Final product:	H_2O (water)
Net yield of energy:	**32 ATP** molecules (maximum)

Description

The **electron transport system** (**ETS**) is the last step in cellular respiration, yet it produces far more ATP—**32**—than either of the first two steps. ETS accomplishes three major things: (1) uses the energy from stored electrons (e–) to create a potential energy gradient in the form of hydrogen ions (protons), (2) converts this potential energy into kinetic energy to produce lots of ATP, and (3) regenerates NAD^+ and FAD to allow glycolysis and the citric acid cycle to continue.

The reduced coenzymes (**NADH** and **$FADH_2$**) directly link the citric acid cycle to the ETS. They are carrying electrons from the original glucose molecule that started the whole process. These electrons are delivered to membrane proteins and other electron carriers in the inner mitochondrial membrane where the ETS process occurs. The electrons are shuttled through a series of electron carriers, where they go from a high-energy state to a low-energy state. Oxygen serves as the final electron acceptor. As oxygen bonds with the electron, it also bonds to hydrogen ions (H^+) to create a water molecule in the matrix. Where do the hydrogen ions come from? They are in the solution of the matrix.

The electron carriers use the energy from the shuttled electrons to pump hydrogen ions from the mitochondrial matrix into the intermembrane space. The result is that a gradient of hydrogen ions is created that serves as a source of potential energy. The primary way for the hydrogen ions to move quickly out of the intermembrane space is to flow through a pore in a membrane protein called an **ATP synthase**. As the hydrogen ions diffuse down their gradient by flowing through this pore, the potential energy is converted into kinetic energy. This kinetic energy is used to covalently bond a phosphate group (P_i) to an ADP molecule to produce an ATP molecule. Note that both the phosphate group and the ADP molecule are present already in the solution of the matrix.

Two Key Concepts and Their Analogies:

• **Concept 1:** **The high-energy electron provides energy to the electron carriers to create a hydrogen ion gradient.**

• **Analogy 1:** The electron being transported between the proteins in the inner mitochondrial membrane is like a **hot baked potato**. The hot potato has **thermal energy** (heat) just like the electron has **energy**. The **electron carriers in the inner mitochondrial membrane** are like a **row of people** standing next to each other. Imagine that the first person tosses the baked potato to the next in line, who then tosses it to the next, and so on. The result is that each person's hands absorb a little bit of the heat from the potato. Similarly, the electron transfers some energy to the electron carriers that is then used to transport a hydrogen ion (or proton) from the mitochondrial matrix to the inner membrane space. The net result is that a hydrogen ion gradient is formed across the inner mitochondrial membrane.

• **Concept 2:** **The potential energy in the hydrogen ion gradient is converted into kinetic energy and used to create ATP.**

• **Analogy 2:** The **potential energy** in the hydrogen ion gradient is like **water behind a dam**. The **dam** is the **inner mitochondrial membrane**, and the **water behind the dam** is like the **gradient of hydrogen ions** in the intermembrane space. The **floodgate** in the dam is like the **ATP synthase**, and the kinetic energy from the flow of water through the dam is like the **flow of hydrogen ions** through the pore in the ATP synthase protein.

Road map

CYTOSOL

OUTER MITOCHONDRIAL MEMBRANE

INTER-MEMBRANE SPACE

INNER MITOCHONDRIAL MEMBRANE

Electron

Electron carriers

Electron carriers

$\frac{1}{2}$ O_2 + 2 H^+ → H_2O

NADH

My job is done here!

NAD+

MITOCHONDRIAL MATRIX

First STEP: Electron Transport Chain

ADP + P_i

H^+

ATP

Second STEP: ATP Synthase

Electron Transport Chain: HOT POTATO ANALOGY

Ouch!

Electron carriers

O_2

H_2O ← + H^+

ATP Synthase: DAM ANALOGY

The hydrogen ion gradient is like water behind a dam in the sense that it has potential energy to do work.

Hydrogen ions moving through the ATP synthase protein are like water moving through the dam's floodgate. Kinetic energy drives a turbine to generate hydroelectric power and also is used by the cell to bond a phosphate group to ADP to form ATP.

Water (hydrogen ion gradient)

Dam (cell membrane)

Floodgate (closed)

Floodgate (open)

Description

Fermentation is the production of 2 **lactate** molecules from the breakdown of a single **glucose** molecule when no oxygen is present in the cell. This process occurs in the cytosol and yields 2 ATP. Recall that in glycolysis (see p. 472), glucose is broken down into 2 pyruvate molecules. When oxygen is present—**aerobic respiration**—the pyruvic acid is converted into acetyl coenzyme A, enters the citric acid cycle, and then enters the electron transport chain. In this case, the number of ATP produced is about 36. But when no oxygen is present, such as in skeletal muscle cells during vigorous exercise, the pyruvate is converted into lactate. This reduces ATP production to a total of only 2.

The conversion of pyruvate into lactate is a type of chemical reaction called a reduction reaction. Oxidation-reduction reactions are common in metabolic pathways and are always paired. These reactions involve the transfer of electrons from one substance to another. To distinguish between these two reactions, recall OIL RIG = Oxidation Is Loss, Reduction Is Gain. The NADH is oxidized to become NAD^+, and pyruvate is reduced to become lactate. NAD^+ is a carrier molecule with electrons as its cargo. It acts like a shuttle bus in that it is picking up and dropping off electrons constantly. The production of NAD^+ in the formation of lactate allows glycolysis to continue because it is needed for that process.

Lactate accumulates in skeletal muscle cells but doesn't remain there. This accumulation is a problem because it tends to lower the local pH, which interferes with normal chemical reactions that have to occur. Consequently, cells have to get rid of it. As its levels increase, lactate diffuses out of cells and into the blood, from which it is taken up by liver cells. Unlike most body cells, liver cells contain special enzymes that can convert lactate back into pyruvate in the presence of oxygen. After this occurs, liver cells can convert pyruvate back into glucose. The glucose then can diffuse back into the blood, where it can be used as an energy source for other muscle cells. This chemical cycling of lactate back into glucose between muscle cells and liver cells is called the **Cori cycle**.

FERMENTATION

KEY

e- = electron	NADH = nicotinamide adenine dinucleotide (reduced)
NAD^+ = nicotinamide adenine dinucleotide (oxidized)	

Glucose

Glycolysis

2 ATP

2 NADH

2 Pyruvate

(in absence of O_2)

e-

2 NADH

2 NAD$^+$

2 Lactate

Liver cells

Blood

CORI CYCLE

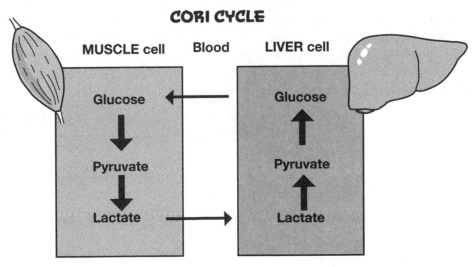

MUSCLE cell   Blood   LIVER cell

Glucose   Glucose

Pyruvate   Pyruvate

Lactate   Lactate

METABOLIC PHYSIOLOGY

Lipid Metabolism

Description

Many different types of lipids exist in the body, including **triglycerides, glycerophospholipids,** and **cholesterol.** Some of the stored lipids can be removed and either oxidized to be used as fuel or redeposited in other fat cells. Although many cells in your body prefer to burn a carbohydrate-based fuel such as glucose, they also can burn other fuels such as triglycerides or amino acids when necessary. Alternatively, new lipids also may be synthesized from other compounds. Lipids do not dissolve in water, which makes it difficult to transport them in the blood. To solve this problem, lipids are combined with proteins to form spheres called **lipoproteins** that are more easily transported through the blood and delivered to body cells. Triglycerides are the most common form of lipid in the body, so we will refer to this type when discussing how lipids are metabolized.

Lipolysis

(*lipo* = lipid; *–lysis* = breaking down)

Lipolysis is the process of breaking down lipids to produce ATP.

- **Triglycerides** are broken down via hydrolysis in the small intestine into their component parts: **glycerol** and **fatty acids.** Once inside cells, glycerol can be converted into a glycolysis intermediate, **glyceraldehyde-3-phosphate,** and then into **pyruvate.** Pyruvate is oxidized through the citric acid cycle, producing reduced coenzymes (NADH and $FADH_2$). As these coenzymes are oxidized by the electron transport system, ATP is produced.

- **Fatty acids** follow a different pathway through a process called **beta oxidation.** In this series of chemical reactions, which occur in mitochondria, fatty acids are broken down into 2-carbon fragments. Each of these fragments then is converted into **acetyl CoA,** which is oxidized through the citric acid cycle, producing reduced coenzymes. As these coenzymes are oxidized by the electron transport system, ATP is produced. In total, this accounts for a large production of ATP. For example, some long-chained fatty acids produce about four times more ATP compared to the complete catabolism of a single glucose molecule. In liver cells, fatty acids are catabolized to produce a group of substances called **ketone bodies.** This process is called **ketogenesis,** and it follows this pathway:

Fatty acids ⟶ acetyl CoA ⟶ **ketone bodies**

These ketone bodies enter the bloodstream and are delivered to cells. Some cells, such as cardiac muscle cells, prefer ketone bodies rather than glucose to produce ATP. They accomplish this by converting ketone bodies back into acetyl CoA, which enters the citric acid cycle and then the electron transport system.

Lipogenesis

(*lipo* = lipid; *–genesis* = generation, birth)

The process of synthesizing lipids is called **lipogenesis.** This occurs in liver cells and adipose cells, which can convert carbohydrate, proteins, and fats into triglycerides. For example, drinking lots of soda pop on a regular basis gives the body more simple sugars than it needs, so they are converted into triglycerides and stored in fat cells. Excess simple sugars such as glucose lead to the production of glycerol or fatty acids by the following pathways:

Glucose ⟶ glyceraldehyde-3-phosphate ⟶ **glycerol** ...OR...
Glucose ⟶ pyruvate ⟶ acetyl CoA ⟶ **fatty acids**

The excess **glycerol + fatty acids** ⟶ **triglycerides.**
Proteins are digested in the small intestine to form their component amino acids. Certain types of amino acids can be converted into triglycerides. Here is the pathway for proteins:

Protein ⟶ amino acids (*certain types*) ⟶ acetyl coA ⟶ fatty acids ⟶ **triglycerides**

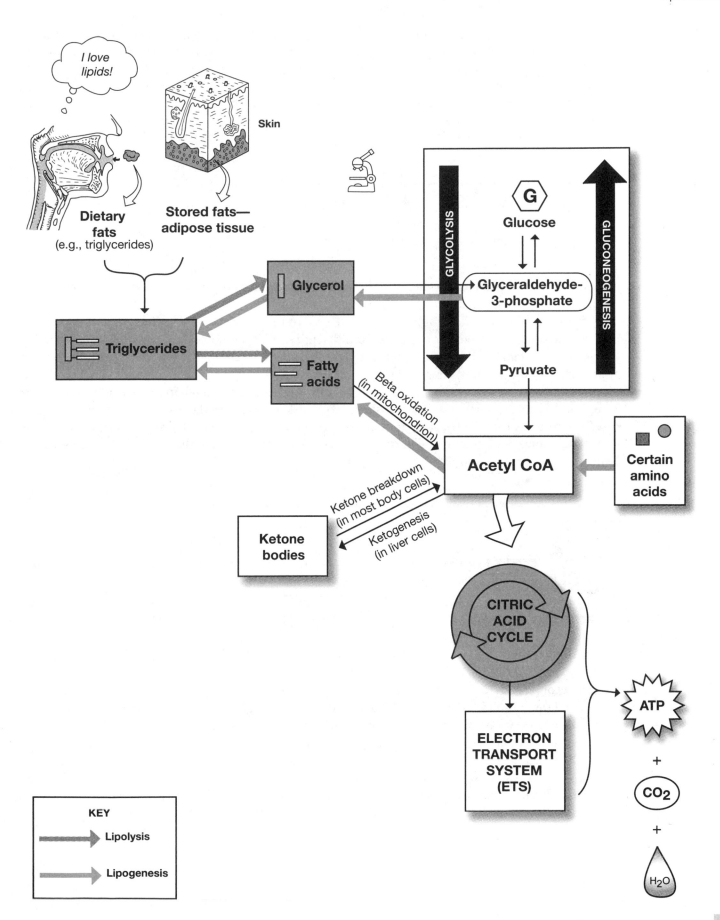

Description

Your body is made primarily of water, fat, and **protein**. A healthy adult body may contain as much as 18% protein. If you eat a chicken sandwich, the protein in the chicken will be broken down gradually into **amino acids** in the digestive tract. The amino acids then are transported into the bloodstream and delivered to body cells. All body cells need amino acids for two purposes: (1) **protein synthesis**—to make new proteins within the cells, or (2) to use as a fuel source to provide energy (to make ATP). If you ingest more protein than your body needs daily, the excess amino acids are converted into either glucose $\langle G \rangle$ or triglycerides ▤.

Amino Acids for Protein Synthesis

amino acids   protein

The process of protein synthesis occurs at a ribosome (see p. 72). During human growth and development, proteins have to be produced regularly and rapidly. In the adult, protein synthesis serves the purpose of replacing worn-out proteins and repairing damaged tissues.

Just as the different letters in the alphabet are used to form various words, different amino acids are used to make various proteins. There are 20 different types of amino acids. Your body can synthesize roughly half of them, so the others must come from our diet. Some cells, such as liver cells, are more active at protein synthesis than others. Sometimes a cell has to convert one amino acid into another to aid the process of protein synthesis. This is achieved by a process called **transamination**, in which an α-**amino group** (NH_3^+) from an amino acid is transferred to an α-**keto acid** with the help of the enzyme transaminase. The result is that the α-keto acid is converted into a new amino acid that can be used in protein synthesis. Examples of other tissues active in protein synthesis are cells in the brain, skeletal muscle, and heart.

Amino Acids Used as Fuel Source

amino acids

Alternatively, amino acids can be used by body cells to produce ATP by entering the citric acid cycle. For this to occur, amino acids must go through a **deamination** process whereby they lose their α-**amino group** (NH_3^+) and a hydrogen atom with the help of the enzyme deaminase. The result is that an **ammonium ion** (NH_4^+) is produced along with an α-**keto acid**. The keto acid can enter the citric acid cycle and electron transport system to produce ATP for the cell. Because ammonium ions (NH_4^+) in high quantities are toxic to cells, they are further converted into a waste product called urea, a harmless substance. This occurs in liver cells when ammonium ions and carbon dioxide enter a biochemical pathway called the urea cycle, in which urea is the primary side product. The liver is the major site for these deaminations because the liver cells have the proper enzymes to do the job. Urea normally travels through the blood and into the kidneys, where it is filtered out and excreted from the body as part of the urine.

PROTEIN DIGESTION

Protein

Amino acid

Used in protein synthesis

TRANSAMINATION

Glutamate (amino acid)

+

α-Keto acid 1

Transaminase

α-Keto acid 2

+

Tyrosine (new amino acid)

OR

DEAMINATION

Glutamate (amino acid)

Deaminase

H_2O

NAD^+ → NADH

α-Keto acid 2

+

NH_4^+

Ammonium ion

Color the NH_3^+ group and the NH_4^+ different colors.

Liver

UREA CYCLE

NH_4^+ + CO_2

Ammonium ion

Carbon dioxide

Urea cycle

$H_2N-C-NH_2$

Urea

CITRIC ACID CYCLE

Kidney

$H_2N-C-NH_2$

Urea

BLOOD

Urea

(excreted in urine)

Notes

URINARY SYSTEM

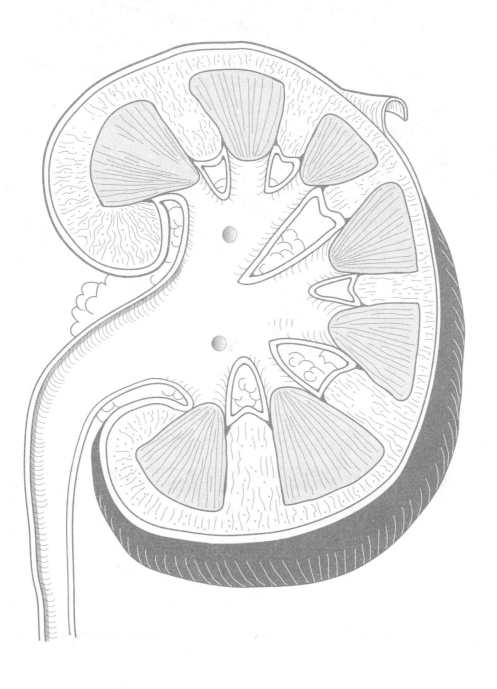

Description

The urinary system is composed of the **kidneys, ureters, urinary bladder**, and **urethra**. The kidneys are located on either side of the upper lumbar region of the vertebral column. They lie between the dorsal body wall and the **parietal peritoneum** in a retroperitoneal position. They are tightly covered by the parietal peritoneum on their anterior surface and further protected by layers of fatty tissue. Major abdominal organs such as the small and large intestines are all positioned anterior to the kidneys.

The **renal arteries** branch off the **abdominal aorta** and supply oxygenated blood to the kidneys, and the **renal veins** drain blood from the kidneys and empty into the **inferior vena cava**. The blood that enters the kidney via the renal artery contains various waste products that must be removed from the blood. The functional units within the kidney are microscopic structures called **nephrons**. They filter waste products and nutrients out of the blood and transport them through a tubular system. The fluid that enters the tubular system is called the **filtrate**. The filtrate is processed in a manner that transports nutrients back to the bloodstream and retains waste products in the tubular system.

When the processing is completed, the liquid inside the tube is called **urine**. In reality, urine is nothing more than processed blood plasma. The urine then moves out of the nephrons and into the pelvis of the kidneys. Slender, muscular tubes called **ureters** transport urine from the kidney to the urinary bladder. The muscular **urinary bladder** can expand to hold as much as 600–800 mL of urine. Finally, urine is removed from the body by the process of **micturition** (*urination*), which uses forceful muscular contractions in the wall of the urinary bladder to transport urine into the urethra and out of the body.

Key to Illustration

Major Organs/Structures	Blood Vessels (B)	
1. Kidney	B1. Inferior vena cava	B6. Internal iliac a.
2. Ureter	B2. Abdominal aorta	B7. External iliac a.
3. Urinary bladder	B3. Renal a.	B8. Gonadal v. (testicular v. in male; ovarian v. in female)
	B4. Renal v.	
	B5. Common iliac a.	B9. Gonadal a. (testicular a. in male; ovarian a. in female)

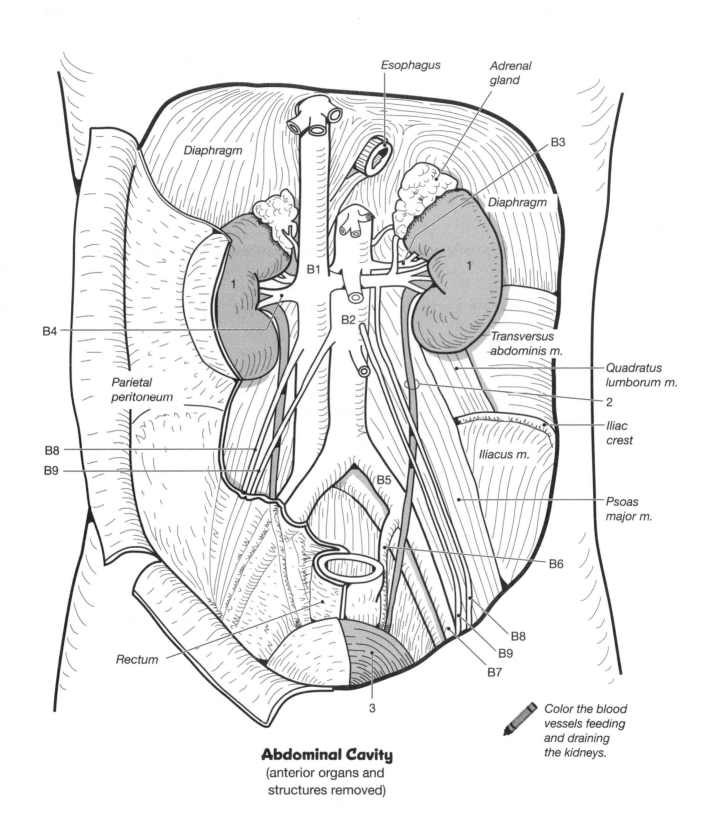

Esophagus

Adrenal gland

Diaphragm

B3

Diaphragm

B1

B2

B4

Transversus abdominis m.

Parietal peritoneum

Quadratus lumborum m.

2

Iliac crest

Iliacus m.

B8

B9

B5

Psoas major m.

B6

B8

Rectum

B9

B7

3

Color the blood vessels feeding and draining the kidneys.

Abdominal Cavity
(anterior organs and structures removed)

URINARY SYSTEM: Kidneys—Structure and Function

Description

To begin, let's examine the gross anatomy of the kidneys—macroscopic structures that are visible to the naked eye. The kidneys are covered by a fibrous connective tissue called the **renal capsule**. It is good to note that the term *renal* means *relating to the kidneys* and is used repeatedly for various anatomical structures here. The **renal hilum**, a cleft on the medial surface of the kidneys is a handy landmark for locating the ureters and blood vessels (renal a., renal v.) that serve these organs. Each kidney is divided into two regions: (1) **Renal cortex**, and (2) **Renal medulla**. Like the bark around a tree, the renal cortex is the outermost region and appears granular. The middle region—the renal medulla—contains many cone-shaped structures called **renal pyramids**. When the kidney is sectioned, as in the illustration on the facing page, these structures appear triangular in shape. At the tip of each renal pyramid is a nipple-like structure called a **renal papilla**. Separating the pyramids are projections of cortical tissue called **renal columns**.

Now let's consider structures involved in the collection and transport of urine. Urine is essentially filtered and processed blood plasma. The filtering of the blood occurs in microscopic structures called nephrons found mostly in the renal cortex. Portions of these nephrons and related tubular structures extend into the renal medulla and through the renal pyramids. Urine exits the papilla of each pyramid and enters a **minor calyx**—a cup-like structure that surrounds the papilla and extends into a tube. Several minor calyces (sing., *calyx*) join together to form a larger collecting chamber called a **major calyx**. Note that the terms *minor* and *major* are used to describe the relative sizes of these chambers. The **major calyces** collectively drain urine into a larger, funnel-shaped structure called the **renal pelvis**. The pelvis channels urine into the long, slender tube called the **ureter** which leads to the temporary storage area for urine—the urinary bladder.

Analogy

The **calyces** (sing., *calyx*) are functionally like a **plumbing system** of smaller-diameter pipes leading to larger ones. Each **smaller pipe** is a **minor calyx**, and each **larger pipe** is a **major calyx**. The flow of water through the pipes follows a path similar to the flow of urine through the calyces.

Key to Illustration

Regional areas	Structures
1. Cortex	3. Pelvis
2. Medulla	4. Renal papillus
	5. Renal pyramids
	6. Renal capsule
	7. Renal column
	8. Minor calyx
	9. Major calyx
	10. Adipose tissue in renal sinus
	11. Renal hilum
	12. Ureter

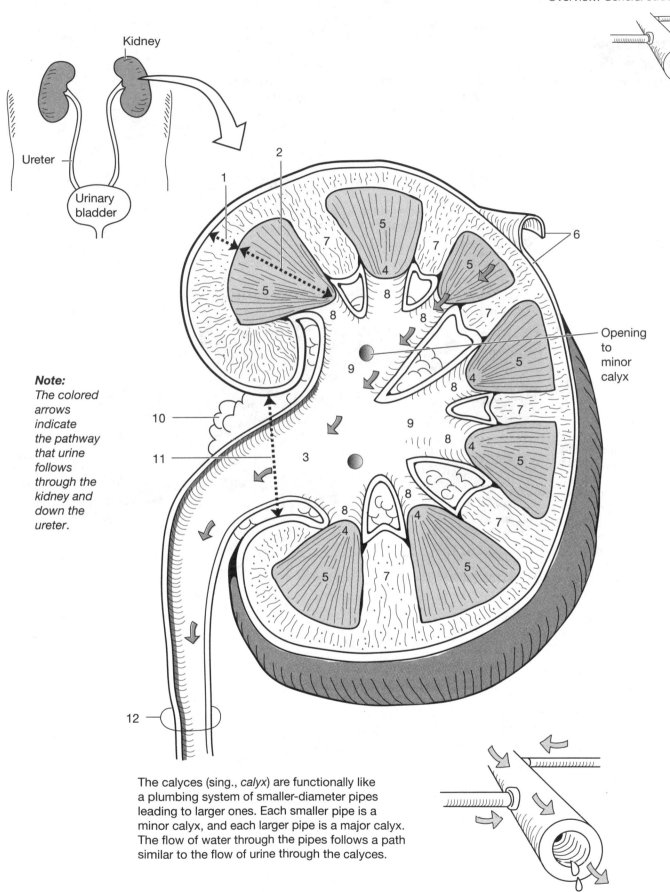

Kidney

Ureter

Urinary
bladder

Note:
*The colored
arrows
indicate
the pathway
that urine
follows
through the
kidney and
down the
ureter.*

Opening
to
minor
calyx

The calyces (sing., *calyx*) are functionally like
a plumbing system of smaller-diameter pipes
leading to larger ones. Each smaller pipe is a
minor calyx, and each larger pipe is a major calyx.
The flow of water through the pipes follows a path
similar to the flow of urine through the calyces.

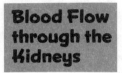

Blood Flow through the Kidneys

The renal artery feeds oxygenated blood to the kidneys, and the **renal vein** drains deoxygenated blood from them.

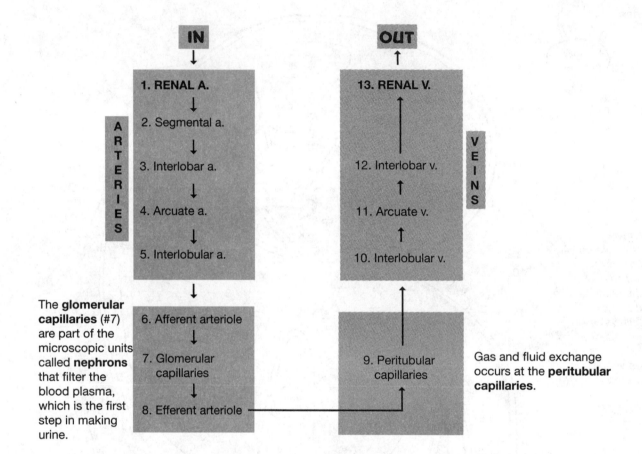

The **glomerular capillaries** (#7) are part of the microscopic units called **nephrons** that filter the blood plasma, which is the first step in making urine.

Gas and fluid exchange occurs at the **peritubular capillaries**.

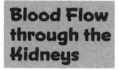

Blood Flow through the Kidneys

Trace the pathway of blood flowing into and out of the kidneys.

Cortex

Medulla

Collecting tubule

Arcuate vessels

Description

The **nephron** is the microscopic functional unit of the kidney. Each kidney has more than a million of these units. Each nephron is divided into five distinct parts: *glomerulus, glomerular capsule, proximal convoluted tubule, nephron loop,* and *distal convoluted tubule.* As blood enters the kidney via the **renal artery,** it branches into smaller vessels until it becomes the interlobular artery in the **renal cortex.** The afferent arteriole branches off the interlobular artery and connects to a special coiled ball of capillaries called the **glomerulus.** Completely surrounding the glomerulus is a cuplike structure called the **glomerular capsule.**

Connected to the capsule is a long tubular system. The glomerulus is highly permeable because of pores in its walls called **fenestrae.** Wrapped around the glomerular capillaries are cells called **podocytes**—modified simple squamous epithelial cells. The blood in the glomerulus is under high pressure so the plasma is passively filtered into the glomerular capsule. The fluid inside the glomerular capsule is referred to as *filtrate.* This filtrate flows from the glomerular capsule into the first part of the tubular system called the **proximal convoluted tubule (PCT).** Note that the term *convoluted* simply means *coiled* which correctly describes this structure. Next it flows into the **nephron loop,** which consists of a descending limb where the filtrate flows downward and an ascending limb where the filtrate moves upward. Then the filtrate enters the **distal convoluted tubule (DCT)** and finally moves into the **collecting duct.**

The general function of the nephron is to collect the filtrate (*filtered blood plasma*) in a separate tubular system and process it. The filtrate in the glomerular capsule is a solution that contains both nutrients and waste products mixed together. The nephron processes the filtrate by returning the nutrients to the bloodstream and retaining the waste products in the tubular system. By the time this processing job is completed, the resulting fluid is referred to as **urine.** The urine then follows this pathway: *minor calyx, major calyx, ureter, urinary bladder, urethra.* Then it exits the body.

Analogy

The tubular system in the nephron is **structurally** like the sewer lines in private homes that connect to a larger public sewer drain under a street. The **sewer drain** is like the **collecting duct.**

Key to Illustration			
Blood Vessels (B)	**Nephron Structures**	5. Descending limb	**Collecting systems**
B1. Interlobular artery	1. Glomerular (*Bowman's*) capsule	6. Ascending limb	9. Collecting duct
B2. Afferent arteriole	2. Glomerulus	7. Nephron loop (*loop of Henle*)	
B3. Efferent arteriole	3. Renal corpuscle	8. Distal convoluted tubule (DCT)	
	4. Proximal convoluted tubule (PCT)		

Podocyte

Fenestrae

Filtrate

Renal capsule

3

2

1

Filtrate

B3

B1

B2

B3

B2

3

4

8

3

Renal Corpuscle
(cross section)

B3

B2

Renal cortex

9

5

7

6

Renal medulla

Renal capsule

Wedge of cortex and medulla

Papilla

Minor calyx

Kidney
(sagittal section)

Urine

Note: Black arrows show the direction the filtrate flows through the nephron and the direction urine flows through the kidney.

Many nephrons empty into a single collecting tubule. This is like the sewer lines from individual homes in a city emptying into a public sewer drain.

Public sewer drain

Sewer line

Description

The major function of the **kidneys** is to filter and process the blood plasma to produce **urine**. This occurs within the kidney's millions of microscopic filter units called **nephrons**. The first step in this process is **filtration** of the blood (see pp. 364-365), which occurs in the nephron's coiled ball of capillaries called the **glomerulus**. During this process, the blood pressure in the glomerulus forces a liquid mixture of nutrients and waste products to move into the cuplike structure surrounding the glomerulus, called the glomerular capsule.

This mixture is referred to as the filtrate. It contains substances such as water, amino acids, glucose, sodium, chloride, urea, uric acid, and many others. Working together, both kidneys filter about 48 gallons (182 liters) of plasma per day. This is equivalent to 91 two-liter bottles of soda pop! In total, about 99 percent of this large volume with all its nutrients will be reabsorbed back into the blood, and the remaining 1 percent will be released from the body as urine. To better understand the process of glomerular filtration, we will examine the forces and counterforces involved.

Force

Glomerular hydrostatic pressure (GHP): This is the force of the blood pressure against the walls of the glomerulus. Filtration is completely dependent on pressure. If the blood pressure drops too low, filtration stops. The blood pressure in the glomerulus averages about 50 mm Hg and cannot be measured directly, so it must be determined mathematically. As a comparison, recall that the typical pressure on the ateriole end of a capillary bed is 35 mm Hg (see pp. 364-365). Think of the GHP like the force exerted by a sumo wrestler as he thrusts himself forward against his opponent(s).

Counterforces

Two forces work against the GHP: the CHP and the BCOP. Each is like a boy sumo wrestler pushing against the adult sumo wrestler.

1. **Capsular hydrostatic pressure (CHP):** Like the force of water in a water balloon, this is the force of the filtrate fluid against the wall of the glomerular capsule. It averages about 15 mm Hg and works against filtration.

2. **Blood colloidal osmotic pressure (BCOP):** This is the only force that is not a hydrostatic pressure of fluid against a wall. Instead, it is an osmotic pressure, which refers to the tendency of water to move toward the solution with the higher concentration of nonpenetrating solutes. Because plasma proteins like albumin remain in the blood in the glomerulus, the water in the filtrate tends to move back into the glomerulus. This osmotic pressure is a force working against filtration, and it averages about 25 mm Hg.

Net Filtration Pressure Formula

Net filtration pressure = GHP − (CHP + BCOP)   or "forces = sum of all counterforces"

$$= 50 - (15 + 25)$$
$$= 50 - (40)$$
$$= \textbf{10 mm Hg}$$

In other words, a net positive pressure is maintained so the process of filtration can continue. If this were not the case, renal failure could result.

Nephron Function 1: Filtration

KEY

GHP = glomerular hydrostatic pressure
BCOP = blood colloidal osmotic pressure
CHP = capsular hydrostatic pressure
PCT = proximal convoluted tubule

Note: To review the process of filtration, see p. 82.

Kidney

Ureter

Urinary bladder

Renal corpuscle

Nephron

Afferent arteriole

Efferent arteriole

Afferent arteriole

Efferent arteriole

Glomerulus

Glomerular capsule

GHP

BCOP

CHP

(to PCT)

Renal Corpuscle
(cross section)

(to PCT)

Schematic of Forces and Counterforces

Color the glomerulus red to represent the blood that runs through it.
Color the space inside the glomerular capsule yellow to represent the filtrate inside it.

GHP

CHP   BCOP

The forces and counterforces involved in filtration within the renal corpuscle are like an adult sumo wrestler pushing against the two child sumo wrestlers. The force from the adult sumo wrestler is like the GHP, and the combined push from the two child wrestlers is like the CHP and BCOP.

Description

The **juxtaglomerular apparatus** is vital to the **glomerular filtration rate (GFR)** and consists of the following two structures:

① **Juxtaglomerular (JG) cells:** a group of modified smooth muscle cells around the afferent arteriole. They secrete the enzyme **renin** (see pp. 512–513).

② **Macula densa:** a group of modified epithelial cells within the wall of the distal convoluted tubule adjacent to the afferent arteriole. They act as *chemoreceptors* that sense changes in solute concentration of the filtrate.

Each glomerulus in every nephron filters the blood to produce filtrate. The GFR is the volume of filtrate produced every minute by all the nephrons in both kidneys and is equal to about 125 mL filtrate/min. The blood pressure (BP) in each glomerulus must be kept relatively constant to ensure that filtration continues. If the GFR rises or falls too far from this normal level, **nephron** function is impaired.

As shown in the illustration, a general scheme is used to either increase or decrease the GFR as needed. This involves controlling the blood flow through the glomerulus.

Analogy

Let's compare the glomerulus to a loop of garden hose that has some holes for filtration. If we expand the incoming end of the hose, this allows more water in and increases the water pressure. If we squeeze the outgoing end, this causes a backup of fluid in the loop and increases water pressure. Similarly, the glomerulus uses vasodilation and vasoconstriction of its afferent and efferent arterioles to change the BP in the glomerulus. By doing so, this directly changes the GFR.

Mechanisms

To control the GFR, three major mechanisms are used: (1) autoregulation, (2) neural regulation, and (3) hormonal regulation. Let's summarize each, in turn.

- *Autoregulation* is the dominant control mechanism at rest. It is subdivided into two mechanisms: (1) *smooth muscle mechanism*, and (2) *tubular mechanism*. The first uses stretching of the wall of the afferent arteriole as a stimulus to induce vasoconstriction of itself. This leads to a decrease in blood flow to the glomerulus and decreases GFR back to normal levels. The second mechanism involves the **macula densa**. For example, when BP increases, the GFR also increases. This results in an increase in filtrate flow in the renal tubules. Because this also increases the level of solute (such as Na^+ and Cl^-) in the filtrate, the macula densa detects this change and responds by constricting the afferent arteriole. This reduces blood flow to the glomerulus and decreases the GFR back to normal levels.

- *Neural regulation* occurs during periods of physical activity and stressful situations. The idea here is you don't need to make urine if you are running or fighting for your life. Because it is the sympathetic division of the autonomic nervous system (ANS) that helps us respond to these situations, it also controls the GFR. Neurons are linked to the afferent arteriole through a reflex pathway. In an effort to deliver more blood to vital organs, less blood should be going to the kidneys. To accomplish this, nerve impulses stimulate the smooth muscle around the afferent arteriole to constrict, which decreases the blood flow to the glomerulus and decreases GFR.

- *Hormonal regulation* involves the hormone angiotensin II and is triggered by a decrease in blood volume or BP. This is part of the normal renin-angiotensin aldosterone system (see p. 512). Juxtaglomerular cells release the enzyme renin into the bloodstream, which leads to the production of angiotensin II, a powerful vasoconstrictor. By causing peripheral blood vessels to constrict, it increases systemic BP, which increases GFR.

Regulation of the Glomerular Filtration Rate

Juxtaglomerular apparatus

Renal Corpuscle
(cross section)

Nephron

General Scheme for Controlling the GFR

By increasing or decreasing the BP in the glomerulus

There are two ways to change the BP in the glomerulus	There are two ways to change the water pressure in the garden hose

Renal Corpuscle
(cross section)

This loop of garden hose is like the glomerulus.

URINARY SYSTEM: Kidneys—Structure and Function

Nephron Function 2: Tubular Reabsorption

Description

Reabsorption is the transport of nutrients such as sodium ions or water from the renal tubule into the blood of the peritubular capillaries.

Key

R = reabsorption

Reabsorption by Nephron Region

Some of the major substances reabsorbed include ions such as sodium (Na^+), chloride (Cl^-), potassium (K^+), calcium (Ca^{+2}), magnesium (Mg^{+2}), and bicarbonate (HCO^{3-}), along with water, glucose, and amino acids. The table shows what substances are reabsorbed in each part of the nephron. Both active and passive transport processes are used in reabsorption (see p. 76). Some substances, such as glucose, are cotransported.

Nephron Region	Major Substances Reabsorbed		
Proximal convoluted tubule (PCT)	• Na^+ • Cl^- • K^+ H_2O	• Ⓖ • amino acids	• HCO_3^- • Ca^{+2} • MG^{+2}
Nephron loop	• Na^+ • Cl^- • K^+ H_2O	• HCO_3^- • Ca^{+2} • MG^{+2}	
Distal convoluted tubule (DCT)	• Na^+ • Cl^- H_2O	• Ca^{+2}	
Collecting tubule (CT)	• Na^+ H_2O	• HCO_3^-	

General Scheme for Reabsorption

(1) Sodium ions (Na^+) typically are transported actively into the **interstitial space**. These ions then diffuse into the blood. This creates a gradient of positive (+) charge in the interstitial space.

(2) The gradient of *positive* charge in the interstitial space draws *negatively* charged anions such as chloride ions (Cl^-) passively out of the **tubule** into the interstitial space and then into the **blood**.

(3) Water (H_2O) follows the salt. Because of the higher solute concentration in the interstitial space relative to the tubule, water leaves the tubule via osmosis (see p. 80). Therefore, by creating this gradient of sodium and chloride ions, the nephron has really forced osmosis to occur.

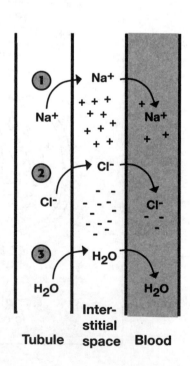

Reabsorption in the Nephron

PCT

DCT

Peritubular capillaries

Nephron loop

Collecting tubule

Amount of Filtrate Reabsorbed by Nephron Region

NEPHRON REGION	AMOUNT REABSORBED (out of 125 mL filtrate)
Proximal convoluted tubule (PCT)	100 mL
Nephron loop	7 mL
Distal convoluted tubule (DCT)	12 mL
Collecting tubule (CT)	5 mL

TOTAL REABSORBED: 124 mL

This is based on a rate of filtrate being formed at 125 mL/min. by both kidneys. Therefore, 124 of the 125 mL is reabsorbed into the blood. About 80 percent of this total occurs in the PCT.

Functional Analogy

The reabsorption process in the nephron is like workers sorting different products on a conveyor belt. The substances filtered from the blood in the glomerulus contain a mixture of nutrients (*ex.*: water, glucose, amino acids) and waste products (*ex.*: urea, uric acid). The waste products remain in the tubules (PCT, nephron loop, DCT, and collecting tubule), and the nutrients must be reabsorbed back into the blood. The workers are like the membrane proteins in the tubules that transport specific substances into the blood. The tubule conveyor represents the filtrate moving through the renal tubules, and the blood conveyor represents the blood in the peritubular capillaries that surround the tubules. In the end, the waste products are collected in the urine, stored in the urinary bladder, and excreted from the body, and the nutrients travel through the blood to finally be used by body cells.

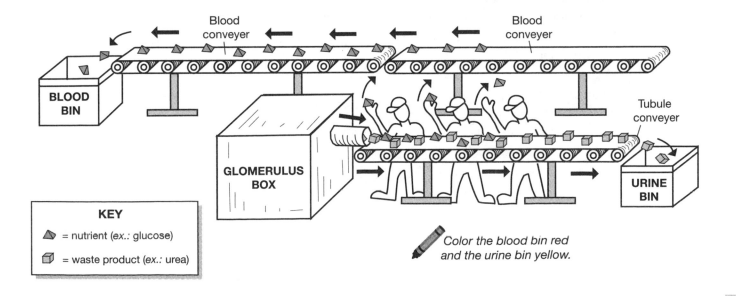

Blood conveyer

Blood conveyer

BLOOD BIN

GLOMERULUS BOX

Tubule conveyer

URINE BIN

KEY

△ = nutrient (*ex.*: glucose)

▱ = waste product (*ex.*: urea)

Color the blood bin red and the urine bin yellow.

499

Description

Secretion is the transport of substances such as hydrogen ions from the blood of the **peritubular capillaries** into the **renal tubules**. The purpose is to remove these substances from the body in the urine.

Key

S = secretion

Secretion by Nephron Region

The table shows the substances secreted in each part of the nephron.

Key

H_3O^+ = hydronium ions*

K^+ = potassium ions

NH_4^+ = ammonium ions

*** Note:** In solution, hydrogen ions (H^+) bind to water molecules to form hydronium ions (H_3O^+).

Nephron Region	Substance(s) Secreted
Proximal convoluted tubule (PCT)	H_3O^+ NH_4^+ Urea Creatine
Nephron loop	Urea
Distal convoluted tubule (DCT)	H_3O^+ K^+
Collecting tubule (CT)	H_3O^+ K^+

Outcomes of Secretion

Tubular secretion has three important outcomes:

- to control blood pH: achieved by the secretion of hydronium ions;
- to rid the body of nitrogen wastes; substances such as urea, creatinine, and ammonium ions are all products of metabolic processes that must be released in the urine; and
- to rid the body of unwanted drugs; drugs such as penicillin also are secreted and released in the urine.

Functional Analogy

The secretion process in the nephron is like workers moving products from one conveyor belt to another. When certain substances such as potassium ions and hydrogen ions reach high levels in the blood, they are transported into the renal tubules and eventually released from the body in the urine. The workers are like the membrane proteins in the blood capillaries that transport substances into the renal tubules. The tubule conveyor represents the filtrate moving through the renal tubules, and the blood conveyor represents the blood in the peritubular capillaries that surround the tubules.

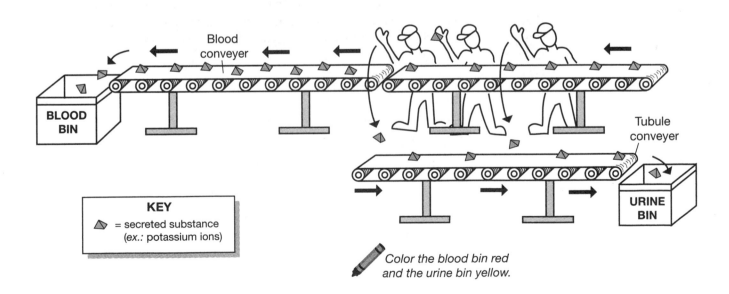

KEY

△ = secreted substance
(*ex.*: potassium ions)

Color the blood bin red and the urine bin yellow.

Description

This module covers the three general mechanisms used to regulate acid-base balance in the body. These are buffer systems, exhalation of CO_2, and kidney excretion of hydrogen ions (H^+).

pH Scale

See pages 40–41.

Major Mechanisms

It is vital that body fluids remain in a normal pH range. The **normal blood pH** is 7.35–7.45. Enzymes and other proteins in the blood are highly sensitive to pH changes and can lose their function if the blood becomes either too acidic (<7.35) or too alkaline (>7.45). By-products from normal metabolism make the blood more acidic. The body uses different mechanisms to raise the pH back to normal. Here is a summary of the three major mechanisms:

1 Buffer systems

Buffers allow a pH to maintain a stable pH at a given pH value. It's like a vise that clamps down on a pH value and prevents it from shifting. The most important buffering system is called the **carbonic acid-bicarbonate buffering system**. When carbon dioxide (CO_2) diffuses into the blood it reacts with water (H_2O) in the plasma to produce carbonic acid (H_2CO_3). This weak acid then dissociates into hydronium ions (H_3O^+) and bicarbonate ions (HCO_3^-). When CO_2 levels are high, as during exercise, more carbonic acid is produced, and the blood becomes more acidic. But when CO_2 levels are low, the reaction can run in reverse. In short, the HCO_3^- binds the free H_3O^+ to produce carbonic acid and convert it into CO_2 and H_2O. This makes the blood pH more alkaline.

2 Exhalation of CO_2

Respiration centers in the brainstem control the rate and depth of breathing. Increasing these factors results in a faster, more forceful exhalation, which removes CO_2 from the body. As explained about the **carbonic acid-bicarbonate buffering system** above, less CO_2 in the blood means less carbonic acid being formed, making the blood pH more alkaline.

3 Kidney excretion of hydronium ions (H_3O^+)

Unlike buffering systems, the kidneys actually *remove* the excess H_3O^+ from the body and excrete them in the urine. This makes the urine *more* acidic and the blood *less* acidic. For the details of this process, see p. 506. (**Note:** In solution, hydrogen ions [H^+] bind to water molecules to form hydronium ions [H_3O^+].)

pH Scale

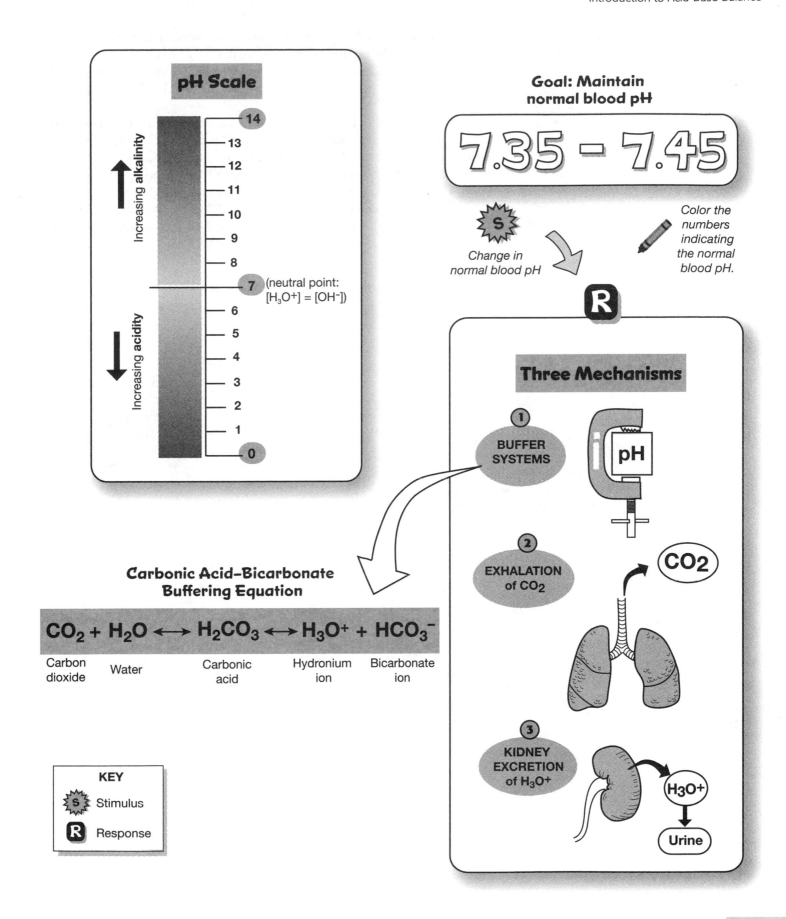

Increasing alkalinity

Increasing acidity

14
13
12
11
10
9
8
7 (neutral point: $[H_3O^+] = [OH^-]$)
6
5
4
3
2
1
0

Goal: Maintain normal blood pH

7.35 - 7.45

(S) Change in normal blood pH

Color the numbers indicating the normal blood pH.

R

Three Mechanisms

1. **BUFFER SYSTEMS**

   pH

2. **EXHALATION of CO_2**

   CO_2

3. **KIDNEY EXCRETION of H_3O^+**

   H_3O^+

   Urine

Carbonic Acid–Bicarbonate Buffering Equation

$$CO_2 + H_2O \longleftrightarrow H_2CO_3 \longleftrightarrow H_3O^+ + HCO_3^-$$

| Carbon dioxide | Water | Carbonic acid | Hydronium ion | Bicarbonate ion |

KEY

(S) Stimulus

(R) Response

Description

The kidneys regulate acid-base balance in the body. Normal metabolic reactions in body cells produce acids, such as sulfuric acid, deposited in the blood. As with any acid, this increases the amount of free hydronium ions (H_3O^+) in solution and makes the blood pH more **acidic** than normal. Because these acids are produced day after day, they must be eliminated from the body. This is accomplished by excreting the H_3O^+ in the urine.

Let's explain this idea a bit further. Within the kidneys are millions of microscopic functional units called **nephrons** that constantly filter the blood plasma and process it into urine. Two specific parts of the nephron—the **proximal convoluted tubule** (**PCT**) and the **collecting tubule** (**CT**)—are involved in excreting H_3O^+ into the urine. Any time a substance is moved out of the blood and into the renal tubules, it is called *secretion*. Consequently, it is more accurate to say that H_3O^+ is secreted into the **renal tubules** (PCT and CT). At the same time H_3O^+ is being secreted, bicarbonate ions (HCO_3^-) are being reabsorbed back into the blood so they are not lost in the urine. This constant removal of metabolic acids ensures that the blood pH remains in its normal range of 7.35–7.45.

Renal Regulation of Acid-Base Balance

Body cells

ACIDS

H_3O^+

H_3O^+

H_3O^+

H_3O^+

H_3O^+

H_3O^+

H_3O^+

H_3O^+

H_3O^+

Blood

Proximal convoluted tubule (PCT)

Collecting tubule (CT)

Nephron

Renal tubule

R

Urine becomes more **acidic**

KEY

S Stimulus

R Response

Color the hydrogen ions.

Note: *In solution, hydrogen ions (H^+) bind water molecules to form hydronium ions (H_3O^+).*

Water Conservation and Antidiuretic Hormone (ADH)

Description

Antidiuretic hormone (ADH) helps conserve water. Imagine a hot day where you are working outdoors, sweating, and getting dehydrated. This triggers the following series of events:

① Sweating increases the **blood osmotic pressure.** This serves as a stimulus (S).

② This increase in osmotic pressure is detected by **osmoreceptors** within the **hypothalamus** that constantly monitor the osmolarity ("saltiness") of the blood.

③ Osmoreceptors stimulate groups of neurons within the **hypothalamus** to release **ADH** from the **posterior lobe** of the **pituitary gland.**

④ Like all hormones, ADH travels through the bloodstream to its various target organs.*

⑤ One of its targets, the **renal tubules** in the **kidney**, contains receptors for ADH. Most of these receptors are in the **collecting tubules** (**CT**). When ADH binds to its receptor, it makes the membrane more permeable to water.

⑥ The final response (R) is an increase in water **reabsorption**, which leads to a decrease in **urine output.**

Now imagine that you drink lots of water to get rehydrated after working outdoors. What happens? The excess water decreases the blood osmotic pressure. This inhibits osmoreceptors, which, in turn, stops secretion of ADH. If no ADH is released, the renal tubules remain impermeable to water. Therefore, the excess water is retained by the renal tubules, and the result is an increase in urine output. In short, although ADH helps the body conserve water, it also controls whether the urine is diluted or concentrated. That's one important little hormone!

Have you ever wondered why alcohol consumption makes you urinate more frequently? It's because alcohol inhibits the release of ADH, thus increasing urine production. Contrary to what many people think, drinking alcohol actually *dehydrates* you.

* Other targets that have receptors for ADH are sweat glands and arterioles. In sweat glands, ADH decreases perspiration to conserve water. In arterioles, it causes the smooth muscle in the wall to constrict, narrowing the vessel diameter and increasing blood pressure.

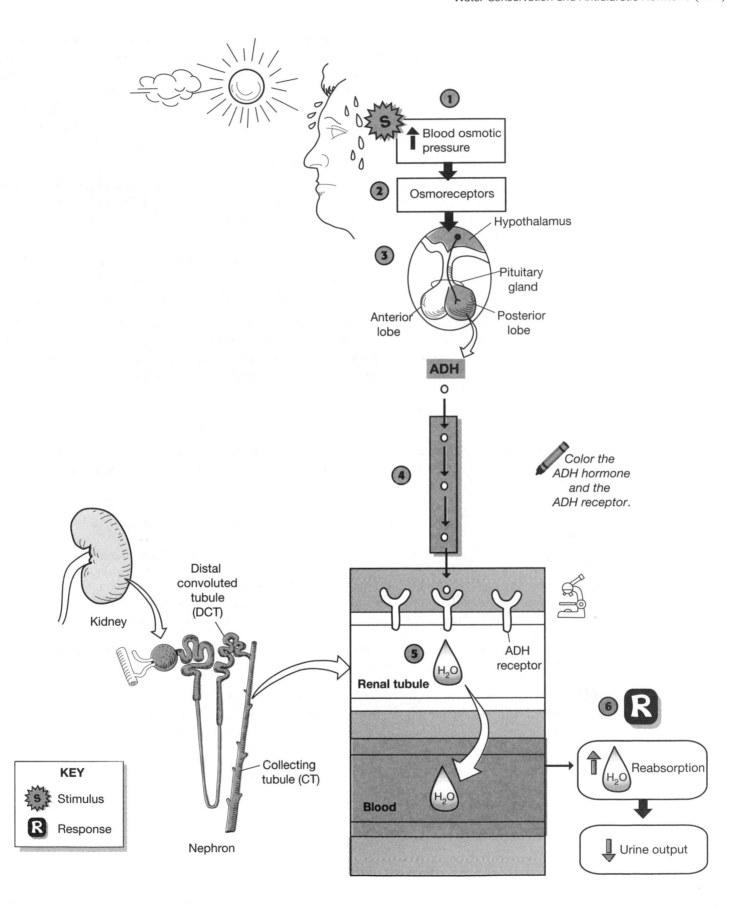

1 ↑ Blood osmotic pressure

2 Osmoreceptors

Hypothalamus

3

Pituitary gland

Anterior lobe

Posterior lobe

ADH

4

Color the ADH hormone and the ADH receptor.

Kidney

Distal convoluted tubule (DCT)

ADH receptor

5

Renal tubule

H₂O

Collecting tubule (CT)

Blood

H₂O

6 **R**

↑ H₂O Reabsorption

↓ Urine output

Nephron

KEY

S Stimulus

R Response

Description

This module covers the **countercurrent multiplier** mechanism in the **nephron loop** of the **nephrons**. First, though, let's summarize a few key things about the nephron loop. It is composed of a **descending limb (DL)** and an **ascending limb (AL)**. Filtrate from the glomerular capsule flows down the DL and up the AL before entering the **distal convoluted tubule (DCT)**, followed by the **collecting tubule (CT)**. The DL has a thinner wall permeable to water but not salt, sodium ions (Na^+) and chloride ions (Cl^-). In contrast, the wall of the AL becomes thicker (as indicated by a heavier line in the illustration). This thicker portion is impermeable to water and contains protein pumps used to actively transport Na^+ and Cl^- out of the AL and into the **extracellular fluid (ECF)**.

- Name? Why is it called the **countercurrent multiplier?** The term "countercurrent" refers to the flow of filtrate in opposite directions in the nephron loop—down the DL and up the AL. The term "multiplier" refers to the fact that it increases (or multiplies) the salt concentration in the renal medulla.

- **Osmolarity?** This is a measure of the amount of dissolved particles in 1 liter of solution. In living systems, smaller units called **milliosmoles (mOsm)** are used. For our purposes, this is a measure of the saltiness of the solution. The higher the number, the saltier the solution, and vice versa.

- Purpose? Notice from the illustration that the ECF deep in the medulla is four times saltier than the ECF near the cortex (1,200 mOsm versus 300 mOsm). The general purpose of the countercurrent multiplier mechanism is to maintain a constant gradient of salt deep within the renal medulla. Why? *This allows the collecting tubule (CT) to reabsorb water via osmosis, thereby concentrating the urine.*

Analogy

The salt in your ECF is the same as that in table salt—sodium chloride (NaCl). In solution, sodium chloride ionizes into Na^+ and Cl^-. A salt shaker will remind you of this connection to everyday life.

Mechanism Step by Step

For the step-by-step process of this cyclical mechanism, refer to 1–5 in the illustration:

1. Salty filtrate (300 mOsm) enters the nephron loop from the **proximal convoluted tubule (PCT)**.

2. Because the ECF is saltier than the filtrate, water leaves the descending limb via osmosis. Then, it is reabsorbed into the blood. The saltier the ECF becomes, the more water is removed from the descending limb. This loss of water decreases the volume and makes the filtrate saltier.

3. As water leaves the descending limb, the filtrate becomes saltier until it reaches a peak of about 1,200 mOsm at the bottom of the nephron loop.

4. As the salty filtrate moves up the ascending limb, salt (Na^+ and Cl^-) is actively transported into the ECF. The saltier the filtrate, the more salt is pumped out. As this occurs, the filtrate becomes less salty but the volume remains the same, moving into the DCT at 100 mOsm.

5. As salt is constantly pumped out of the ascending limb, the ECF becomes saltier in the renal medulla.

This ensures that the medulla is always saltier than the cortex.

Cycle repeats. Go back to step 1.

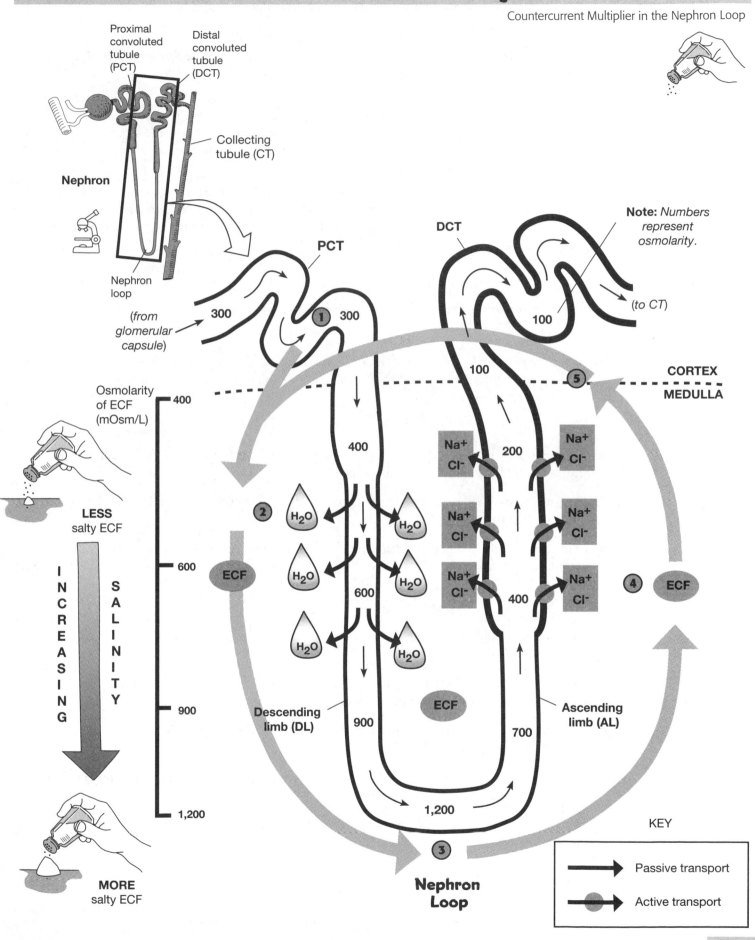

Proximal convoluted tubule (PCT)

Distal convoluted tubule (DCT)

Collecting tubule (CT)

Nephron

Nephron loop

(from glomerular capsule)

PCT

300   ①   300

DCT

(to CT)

100

Note: *Numbers represent osmolarity.*

Osmolarity of ECF (mOsm/L)

400

300

400

100   100   ⑤

CORTEX

MEDULLA

LESS salty ECF

②   H₂O   H₂O   Na⁺ Cl⁻   200   Na⁺ Cl⁻

600

ECF   H₂O   H₂O   Na⁺ Cl⁻   Na⁺ Cl⁻

600

④   ECF

I N C R E A S I N G   S A L I N I T Y

H₂O   H₂O   Na⁺ Cl⁻   400   Na⁺ Cl⁻

900

Descending limb (DL)   900   ECF   **Ascending limb (AL)**   700

MORE salty ECF

1,200

1,200

③

Nephron Loop

KEY

Passive transport

Active transport

Description

This module covers the **countercurrent exchanger** mechanism in the **nephron loop** of the nephrons. First, let's go over a few fundamentals:

- Name? Why is it called the countercurrent exchanger? The term "countercurrent" refers to the flow of blood in opposite directions in the **vasa recta**—down one side and up the other. The vasa recta is blood capillary branches located next to the nephron loop in the same looping pattern. The term "exchanger" refers to the fact that water is exchanged for salt along the length of the vasa recta.

- **Osmolarity**? This is a measure of the amount of dissolved particles in 1 liter of solution. In living systems, smaller units called **milliosmoles** (**mOsm**) are used. For our purposes, this is a measure of the saltiness of the solution. The higher the number, the saltier the solution, and vice versa.

- Purpose? The general purpose of the countercurrent exchanger mechanism is to maintain the gradient of salt deep within the renal medulla established by the countercurrent multiplier mechanism. *Without the countercurrent exchanger, the salt would be carried away by the blood in the vasa recta, and the gradient would be lost.*

Analogy

The salt in your extracellular fluid (ECF) is the same as that in table salt—sodium chloride. In solution, it ionizes into sodium ions (Na^+) and chloride ions (Cl^-). A salt shaker will remind you of this connection to everyday life.

Mechanism

The **countercurrent multiplier mechanism** (see p. 508) created a salt gradient in the extracellular fluid (**ECF**) of the renal medulla. Here are the steps:

1. The gradient allows water to be reabsorbed from the **collecting tubule** (**CT**) by osmosis, thereby concentrating the urine.

2. **Urea** is one of the waste products in the nephron tubules. Although most of the urea remains in the tubules, some of it is recycled by diffusing out of the permeable lower end of the CT and into the permeable descending limb of the nephron loop. Because the thickened portions of the nephron (as indicated by a heavier line) in the ascending limb and DCT are impermeable to urea, it can't diffuse out of that portion of the tubule. The net result of this recycling is that urea adds substantially to the total solute concentration in the ECF, which increases the ECF's osmolarity.

3. The **vasa recta** is adjacent to the nephron loop and follows the same looping pattern with opposite flow. As mentioned previously, if the salt were to diffuse into the blood in the vasa recta, the salt gradient could be lost. The countercurrent exchanger stops this problem.

4. As blood flows downward in the vasa recta, salt (Na^+ and Cl^-) diffuses into the blood, and water leaves the blood by osmosis.

5. In contrast, as blood flows upward in the vasa recta, water in the ECF moves into the blood by osmosis, and salt diffuses out of the blood.

In short, the blood carries away more water than salt. The salt is recycled in the vasa recta, so most of it remains in the ECF, which stabilizes the salt gradient.

Countercurrent Exchanger in the Nephron Loop

Regulation of Extracellular Volume: Renin-Angiotensin-Aldosterone (RAA) System

Description

The **renin-angiotensin-aldosterone (RAA)** system is one of the body's mechanisms to detect falling blood pressure (BP) and bring it back to normal. There are serious consequences for allowing BP to drop below normal levels. For example, the kidneys depend on a constant, normal BP to filter the blood and remove waste products. If the BP drops too low, shock and even renal failure could result. Not surprisingly, the body has numerous mechanisms to detect falling blood pressure and bring it back to normal.

RAA System: Overview

Let's say you were in a car accident and suffered a bad laceration that caused you to lose some blood. This reduction in blood volume causes a decrease in BP, which, in turn, sets a cascade of events in motion. First, this decrease in BP would be detected by juxtaglomerular cells within the kidney, and they would respond by secreting the enzyme renin, which through a chain of events leads to the activation of the hormone angiotensin II. This hormone functions to increase BP in two important ways: (1) vasoconstriction of blood vessels, and (2) stimulation of the adrenal cortex to secrete **aldosterone,** which increases blood volume. Aldosterone does this by increasing reabsorption of sodium ions. Because the general rule is that "water follows the salt," water also is reabsorbed.

Formation of Angiotensin II: Detail

How does renin lead to the formation of angiotensin II? This process actually involves two steps:

1. **Renin** helps convert **angiotensinogen** into **angiotensin I**

2. **ACE** helps convert **angiotensin I** into **angiotensin II.**

As shown in the illustration, angiotensinogen is a plasma protein made by the liver. It normally travels through the bloodstream without doing much of anything. But when the enzyme renin is released into the blood from the kidney, it specifically catalyzes the conversion of angiotensinogen into a slightly different protein called angiotensin I. This conversion involves the cutting off of one part of the protein. As angiotensin I travels through the capillaries of the lung and other tissues, it gets exposed to **angiotensin-converting enzyme (ACE),** which catalyzes the conversion of angiotensin I into **angiotensin II.** This conversion also involves cutting off one part of the protein and completes the process. Now angiotensin II is ready to do its important work.

RAA System: Overview

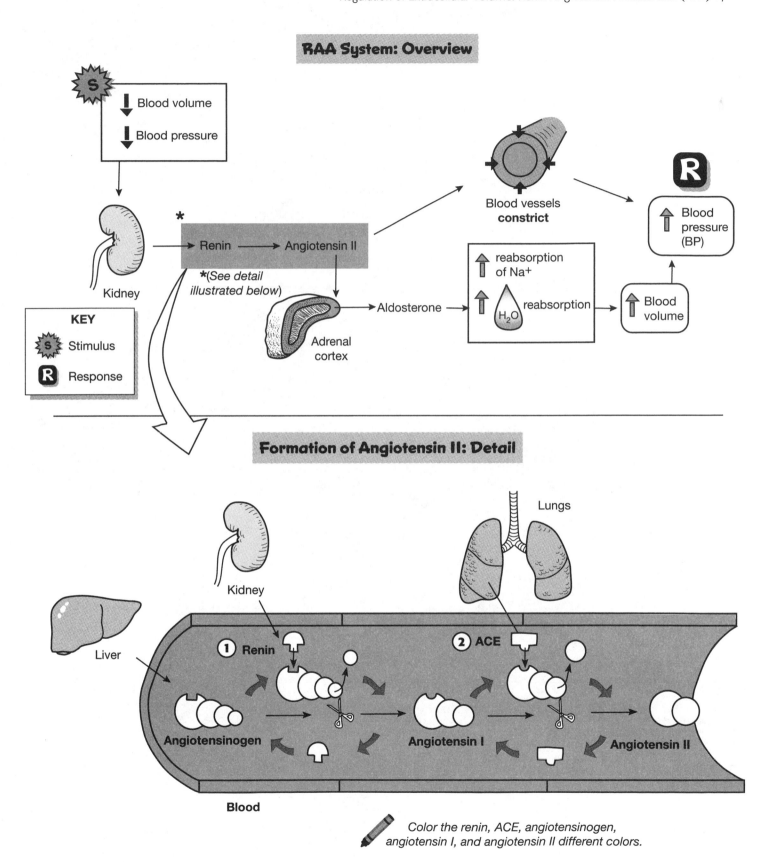

Formation of Angiotensin II: Detail

Color the renin, ACE, angiotensinogen, angiotensin I, and angiotensin II different colors.

Notes

REPRODUCTIVE SYSTEMS

Overview of General Structures, Functions

Description

The male and female reproductive systems are introduced and compared here. Both reproductive systems are derived from the same general tissues and are composed of the following three general structures:

- **Gonads:** produce the gametes (*sex cells*)
- **Ducts:** transport the gametes to the site of fertilization or outside the body (the "plumbing" of the reproductive system)
- **Accessory glands:** add liquid secretions to the reproductive tract to act as a lubricant or a medium in which the gametes are transported

For the external genitalia, similar structures between the sexes are:

- **Penis** and **clitoris:** both contain erectile tissues
- **Scrotal sac** and **labial folds**

Comparison

The table below gives a simplified, not comprehensive, structural comparison of male and female reproductive systems.

Structure/Product	Male	Female
Gonads (*glands that produce gametes*)	Testes	Ovaries
Gametes (*sex cells*)	Spermatozoa (*sperm cells*)	Ova (*egg cells*)
Ducts (*transport gametes*)	Epididymis, ductus deferens, and urethra	Uterine tube
Accessory glands (*add liquid secretions to the reproductive tract*)	Seminal vesicles, prostate gland, bulbourethral glands	Greater vestibular glands

Both sexes have a pair of gonads: in the male, two **testes** within the scrotal sac and, in the female, two ovaries in the pelvic cavity. The testes produce **sperm cells,** and the **ovaries** produce **ova.** The duct system in the male is longer and more complex than in the female. It connects the testes to the external urethral orifice, where sperm cells are released from the body. The female has only one type of main duct, the uterine tubes, which connect each ovary to the uterus. If fertilization occurs, the embryo implants in the **endometrium,** the innermost lining of the uterus.

The three accessory glands in the male produce a collective secretion called the **seminal fluid.** A pair of **seminal vesicles** secrete most of this, the **prostate gland** adds a milky secretion, and the **bulbourethral glands** contribute the smallest amount. In the female, the **greater vestibular glands** secrete a mucus that lines the inside of the **vagina.**

Both sexes use many of the same hormones to regulate reproductive processes. Both use the anterior pituitary hormones—**follicle-stimulating hormone** (FSH) and **luteinizing hormone** (LH)—to regulate gonadal activity. The details are explained in other modules. In the testes, hormone-producing cells called **interstitial cells** secrete the "masculinizing" hormone **testosterone.** During puberty, it stimulates the development of all the secondary sexual characteristics in the male, such as deepening of the voice, enlargement of the genitals, and distribution of body hair. Similarly, the ovaries produce the "feminizing" hormone **estrogen,** responsible for the secondary sexual characteristics in the female, such as enlargement of the breasts, widening of the pelvis, and development of body hair. Although it is easy to focus on gender differences, the similarities are numerous.

Schematic of a Reproductive System

(to site of fertilization)

The **gonads** produce a **gamete**(s) that travels through a **duct system** to the site of fertilization or outside the body. **Accessory glands** add secretions to either the duct system or some other part of the reproductive tract.

Testes

Testosterone

(masculinizing hormone)

Nipple

Ovaries

Estrogen

(feminizing hormone)

Lactating mammary glands

Adipose tissue

Nipple

Pituitary gland

Anterior lobe

Posterior lobe

FSH

LH

KEY

Gonads

Duct system

Accessory glands

Male Reproductive Structures
(posterior view)

Ureter

Urinary bladder

Seminal vesicle

Prostate gland

Bulbo-urethral gland

Ductus deferens

Epididymis

Testis

Urethra

Penis

Spermatozoon
(sperm cell)

Female Reproductive Structures
(posterior view)

Ovary

Uterus

Uterine tube

Endometrium

Greater vestibular gland

Vagina

Ovum
(egg cell)

Detailed Pathway

1. Lumen of seminiferous tubule

2. Rete testis

3. Efferent ductules

4. Epididymis

5. Ductus (*vas*) deferens

6. Ampulla of ductus deferens

7. Ejaculatory duct

8. Prostatic urethra

9. Membranous urethra

10. Penile urethra

11. External urethral orifice

Simplified Pathway

Seminiferous tubule

Rete Testis

Efferent ductules

Epididymis

Ductus deferens

Ejaculatory duct

Urethra

Study Tip

Use this mnemonic for the simplified pathway shown above:

"*Some Really Elderly Elephants Don't Even Urinate!*"

Key to Illustration		
A. Spermatogonium	**C.** Secondary spermatocyte	**E.** Spermatozoon (*sperm cell*)
B. Primary spermatocyte	**D.** Spermatid	

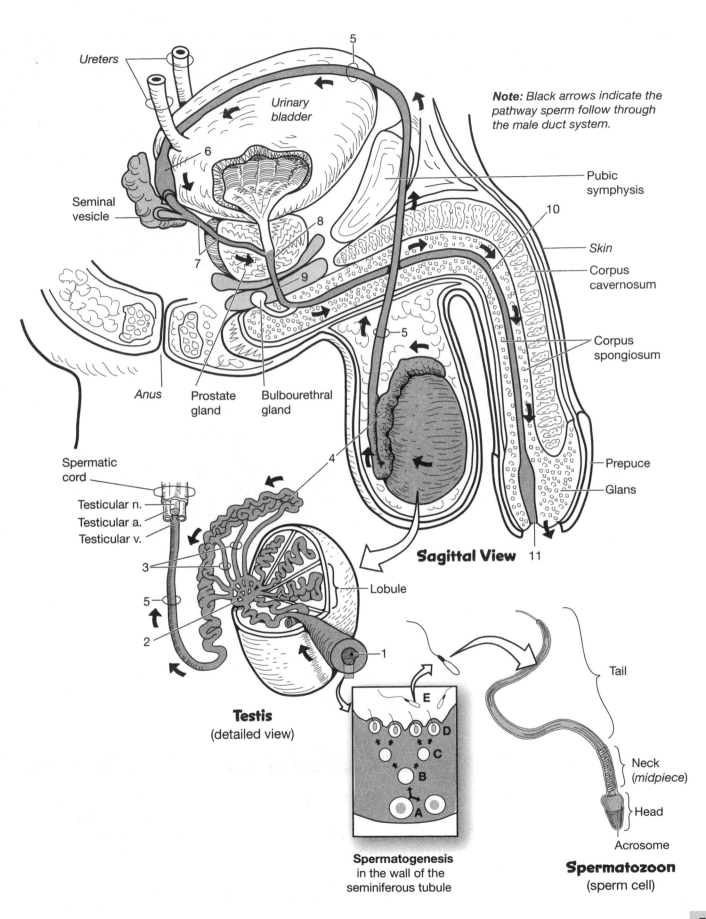

Ureters

Urinary bladder

5

6

Note: *Black arrows indicate the pathway sperm follow through the male duct system.*

Pubic symphysis

Seminal vesicle

8

7

9

Skin

10

Corpus cavernosum

Anus

Prostate gland

Bulbourethral gland

5

Corpus spongiosum

Spermatic cord

Testicular n.

Testicular a.

Testicular v.

4

Prepuce

Glans

3

5

2

Lobule

Sagittal View 11

1

Tail

Testis
(detailed view)

E

D

C

B

A

Neck (*midpiece*)

Head

Acrosome

Spermatogenesis
in the wall of the
seminiferous tubule

Spermatozoon
(sperm cell)

Description

The penis, the male sex organ, is composed of three tubes of spongy connective tissue: two corpora cavernosa and one corpus spongiosum. The two **corpora cavernosa** constitute the bulk of the penis and rest on top of the **corpus spongiosum**, which surrounds the urethra. Sexual stimulation causes these tubes to fill with blood during a normal erection. The corpus spongiosum becomes the **glans** at its distal end and the **bulb** of the penis at its proximal end. A loose sleeve of skin called the foreskin or prepuce normally covers the glans. This is removed during a surgical procedure called a circumcision. Each corpus cavernosum (*plural, corpora cavernosa*) becomes the **crus** (*plural crura*) of the penis at its proximal end. These crura are part of an attachment for the root of the penis to the pubic arch in the pelvis.

Analogy

In cross section, the penis is like a monkey's face. The **corpus cavernosa** are like the **mask around the eyes** of the monkey's face. The **central arteries** are like the **eyes**. The **corpus spongiosum** is like the **area around the mouth,** and the **male urethra** is like the **monkey's mouth.**

Key to Illustration

1. Bulbourethral (*Cowper's glands*) glands
2. Membranous urethra
3. Bulb of penis
4. Crus of penis
5. Subcutaneous dorsal v.

6. Deep dorsal a.
7. Deep fascia
8. Corpora cavernosa
9. Central a.
10. Corpus spongiosum
11. Glans

12. External urethral orifice
13. Tunica albuginea
14. Superficial fascia
15. Urethra

Root of penis

Shaft (body) of penis

Skin

Structural View of the Penis

Skin

Septum

The penis in cross section looks like...

...a monkey's face.

Description   Spermatogenesis is the process of developing sperm cells in the **testes** of the male. It begins for the first time in **puberty** and continues throughout adulthood. The process occurs within the walls of the **seminiferous tubules** in the testes and takes about 2 to 2.5 months to complete. It begins near the outer portion of the seminiferous tubules, proceeds toward the center, and finally ends when the sperm cells are released into the **lumen** of the seminiferous tubules. One of the hormones that regulates spermatogenesis is testosterone. It is secreted by **interstitial cells** (*cells of Leydig*), located between the seminiferous tubules.

Sustentacular cells (*Sertoli cells*) compose the walls of the seminiferous tubules. They physically support and nourish the developing sperm cells (DSCs), control the movement of the DSCs, remove wastes, produce the fluid that fills the lumen of the seminiferous tubules, and are involved in other functions. Each sustentacular cell is joined tightly to another by **tight junctions** like rivets that anchor cell membranes together (see p. 94). This tight seal helps maintain the **blood testis barrier**, which has two specific functions:

1. It protects developing sperm cells from potentially harmful substances by preventing proteins and other large molecules in the blood from coming in direct contact with developing sperm cells.

2. It prevents the immune system from being exposed to sperm cell antigens not found on any other body cells. If this were to occur, the immune system would view sperm cells mistakenly as foreign and respond by making antisperm antibodies in an effort to destroy the sperm cells.

Stages   This flowchart shows the **stages in spermatogenesis**:

| Spermatogonium (SG) ⟶ primary spermatocyte (PS) ⟶ secondary spermatocyte (SS) ⟶ |
| early spermatids (EST) ⟶ late spermatids (LST) ⟶ sperm cells (S) |

The **spermatogonium** is a **stem cell** found in the outer wall of the seminiferous tubules within the testes. It undergoes cell division to produce two new copies of itself. One of these new cells remains in the outer wall as a new spermatogonium, and the other differentiates into a slightly larger cell called a **primary spermatocyte.** This cell contains the total number or **diploid (2n)** number of chromosomes found in all human body cells: 46.

The primary spermatocyte enters a special process called **meiosis**, in which a cell undergoes two cell divisions to produce four final cells, *each containing half the number of chromosomes as the original cell*. This half-set of chromosomes is called the **haploid (n) number**. The only purpose of meiosis is to produce **gametes** (sperm, ova).

Because the total chromosome number is reduced by half, meiosis is called "reduction division." The primary spermatocyte divides to produce two **secondary spermatocytes**. Each of these cells contains a set of 23 replicated chromosomes. Each secondary spermatocyte then divides to produce two new **spermatids**. In total, four spermatids are produced, each containing the haploid (n) number, or 23 chromosomes. Last, the spermatids develop into spermatozoa (sperm cells).

Spermiogenesis is the last part of the spermatogenesis process, in which a spermatid is transformed into a sperm cell. Highlights of this metamorphosis are:

1. **Acrosome formation:** A sac covers the head of the sperm cell and contains the digestive enzymes used to penetrate the outer portion of the ova in the fertilization process.

2. **Mitochondrial reproduction:** The mitochondria replicate themselves and cluster to form a tightly coiled spiral around the neck of the sperm cell. They will be used to supply large amounts of ATP to propel the sperm cell by moving its flagellum.

3. **Flagellum formation:** The centrioles form the microtubules that make up the flagellum.

The **spermatozoon** or **sperm cell** has three main parts:

1. **Head:** has a nucleus that contains the haploid (n) number of chromosomes (23).

2. **Midpiece:** contains many mitochondria that produce large amounts of ATP to fuel the whipping of the tail.

3. **Tail** (*flagellum*): contains microtubules and is used to propel the sperm cell.

Testes: Production of Sperm Cells

SPERMATOGENESIS

Spermatogonium (SG)

Primary spermatocyte (PS)

Secondary spermatocytes (SS)

Early spermatids (EST)

Late spermatids (LST)

Spermatozoa (S) (sperm cells)

Color the 2n cells one color and the n cells another.

SPERMIOGENESIS

Nucleus
Mitochondrion
Centriole
Plasma membrane

Acrosomal vesicle

Microtubule
Plasma membrane

Mitochondria
Acrosome

Spermatozoon (sperm cell)
Head
Acrosome
Midpiece
Tail

Testis
Seminiferous tubule

Interstitial cells (cells of Leydig)

Sustentacular cell
Sustentacular cell nucleus

Tight junction

Lumen of seminiferous tubule

Wall of seminiferous tubule

Interstitial space

SG
SS
EST
LST
PS
S

Description

Spermatogenesis, the development of sperm cells, occurs within the seminiferous tubules in the male testes. This process begins at puberty and continues throughout adult life. The group of hormones that regulates this process is the same basic set that regulates the ovarian cycle in the female (see p. 536).

Stages

A step-by-step description of this process is given below:

1. It all begins when the hypothalamus in the brain releases into the blood a **peptide** called **gonadotropin-releasing hormone (GnRH)**. In adult males, GnRH is released in regular bursts every 60–90 minutes. It travels through the blood to target cells within the anterior lobe of the pituitary gland, which has receptors for GnRH.

2. The binding of GnRH induces a response in these cells to make two **gonadotropin** hormones: **follicle-stimulating hormone (FSH)** and **luteinizing hormone (LH)**. Like any hormone, these chemical messengers are released into the bloodstream and travel to their respective targets, which contain receptors for these hormones.

3. In the male, **FSH** targets the **sustentacular cells** within the walls of the seminiferous tubules in the testes. These cells support and nourish the spermatogenetic cells. The binding of FSH also induces these cells to secrete two proteins: **androgen-binding protein (ABP)** and **inhibin**.

4. At the same time, **LH** binds to receptors on the **interstitial cells**, located between the seminiferous tubules. After binding, LH induces these cells to produce the hormone **testosterone**, which has many different targets. It is responsible for producing the secondary sex characteristics in the male (deepening of voice, enlargement of the genitals, growth of body hair, among other things). In this case, the testosterone binds to the protein **ABP** to form a complex. Once formed, this complex induces the spermatogenic cells to develop into sperm cells. The newly developed sperm cells are released into the lumen of the seminiferous tubule, where they travel through the male duct system until they are released from the body through the external orifice of the penis.

5. Negative feedback is used to shut down the pituitary gland. As the levels of **testosterone** rise, they inhibit the hypothalamus and the pituitary gland. The result is that production of GnRH ceases, which, in turn, leads to no production of either FSH or LH. This inhibition is enhanced by production of the hormone **inhibin** by the sustentacular cells. It has the same effect on the hypothalamus and pituitary as did testosterone. Once the testosterone and inhibin levels fall, inhibition is lost, and the cycle begins again. In the adult male, this cycle results in a steady level of testosterone production that fluctuates within a normal range.

Brain

Hypothalamus

Pituitary gland

① GnRH

Posterior lobe

Anterior lobe

KEY

→ Stimulates

--→ Inhibits

Note: *Compare with female regulation of ovarian cycle, p. 536.*

Testosterone — ⑤ — Inhibin

② LH ☐

③ ● FSH

④

LH receptor

Interstitial cells

Testosterone

+

(ABP = Androgen-binding protein)

ABP

to other targets

Spermatogenic cells

FSH receptor

ABP

Sustentacular cell

Direction of spermatogenesis

Sperm cell

Testis

Seminiferous tubule

Lumen of seminiferous tubule

Wall of seminiferous tubule

Description

The female **external genitalia** (*vulva*) are illustrated on the facing page. Please understand that this is an idealized rendering and does not account for the many variations of normal. In the developing embryo, the external genitalia of males and females appear very similar. Then they differentiate over a period of about eight weeks. The table will point out *homologous structures*—features structurally similar between male and female genitalia.

Structure	Description
1. Mons pubis	A relatively large mound of skin and fatty tissue located anterior to the pubic symphysis. It's covered with pubic hair in adult females
2. Prepuce of clitoris	A hoodlike structural extension of the labia minora that covers the glans of the clitoris; homologous to the male prepuce (foreskin)
3. Glans of clitoris	The tip of the clitoris, which contains many sensory nerve endings for sexual pleasure in the female; homologous to the glans of the penis; a small, erectile body that engorges with blood during sexual excitation
4. Urethral orifice (*opening*)	Opening from the urethra located between the glands of the clitoris and the vaginal opening; a narrow tube that connects the urinary bladder to the outside of the body; urine collects in the urinary bladder, passes through the urethra, and is expelled from the body
5. Labia minora ("smaller lips")	Smaller, hairless folds located inside the larger labia majora that may have increased pigmentation because of the abundance of melanocytes; homologous to the ventral shaft of the penis
6. Vestibule	The space between the labia minora; contains the urethral opening, the vaginal orifice, and openings to the greater vestibular glands
7. Vaginal orifice (*opening*)	The opening into the vagina; a thick, muscular tube that connects the uterus to the outside of the body; acts as the organ to receive the penis during sexual intercourse; functions as the birth canal and passageway for menstruation
8. Openings for the greater vestibular glands	The openings that lead to the pair of greater vestibular glands; during sexual arousal, produce a secretion that serves as a vaginal lubricant and increases during sexual intercourse; homologous to the bulbourethral glands in males
9. Labia majora ("larger lips")	Thick, protruding folds of fatty skin homologous to the male scrotum; outer margins are covered with coarse pubic hair in the adult female

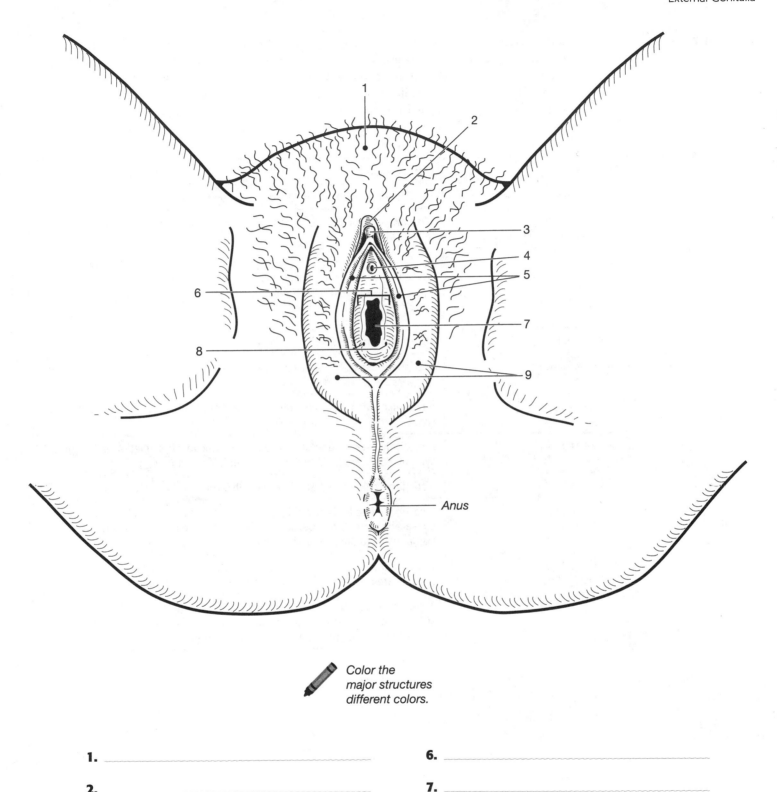

Color the major structures different colors.

1. _____   6. _____

2. _____   7. _____

3. _____   8. _____

4. _____   9. _____

5. _____

Description

The midsagittal view of the female reproductive system allows you to see its relationship to other organs/structures in the body. For example, notice that the uterus is superior to the urinary bladder and anterior to the rectum. Moreover, it allows you to see structures in their full length like the clitoris and the vagina. It also helps illustrate how structures connect one structure to another, such as the vagina connecting the uterus to outside the body.

A description of each of the structures numbered in the illustration is given in the table below.

Structure	Description
1. Fimbriae of the uterine tube	The fingerlike extensions of the uterine tube nearest the ovary
2. Uterine tube (*fallopian tube, oviduct*)	The hollow, muscular tubes that connect the ovaries to the uterus; the site of fertilization
3. Ovary	The female gonad that produces an ovum or egg cell; two ovaries are suspended in the pelvic cavity, one on each side of the uterus
4. Perimetrium	The serous membrane that extends from the peritoneal lining that covers most of the outside of the uterus
5. Myometrium	The thick, muscular layer that forms the wall of the uterus
6. Endometrium	The innermost, glandular layer of the uterus; site where the developing embryo implants
7. Cervix	A neck-like structure at the inferior portion of the uterus that projects into the vagina
8. Mons pubis	A relatively large mound of skin and fatty tissue located anterior to the pubic symphysis; covered with pubic hair in adult females
9. Clitoris	A small, erectile body that engorges with blood during sexual excitation
10. Labia minora ("smaller lips")	Smaller, hairless folds located inside the larger labia majora that may have increased pigmentation because of the abundance of melanocytes; homologous to the ventral shaft of the penis
11. Labia majora ("larger lips")	Thick, protruding folds of fatty skin homologous to the male scrotum; outer margins are covered with coarse pubic hair in the adult female
12. Vagina	A thick muscular tube that connects the uterus to the outside of the body; acts as the organ to receive the penis during sexual intercourse; functions as the birth canal and the passageway for menstruation

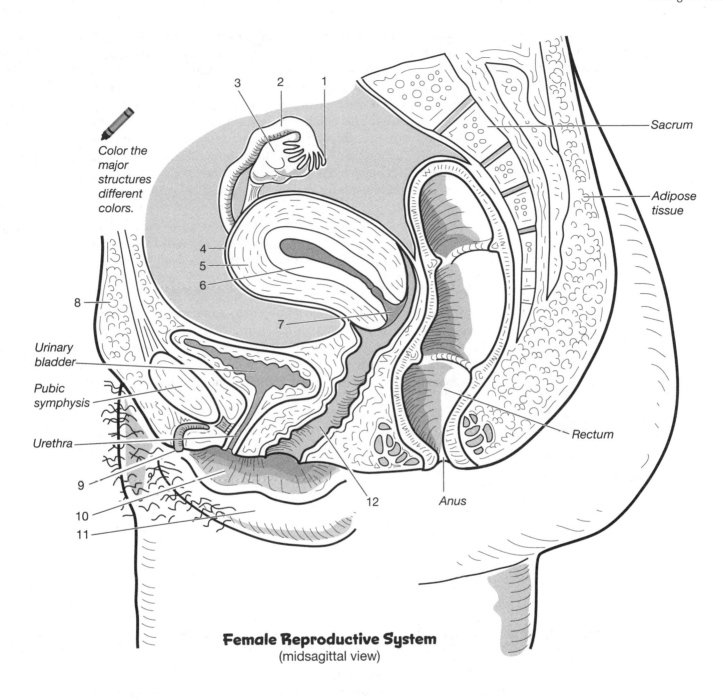

Color the major structures different colors.

Female Reproductive System
(midsagittal view)

1. _____   7. _____

2. _____   8. _____

3. _____   9. _____

4. _____   10. _____

5. _____   11. _____

6. _____   12. _____

Description

The **uterus** (*womb*) is a hollow, muscular organ divided into three regional areas: (1) **fundus**, (2) **body**, and (3) **cervix**. The **fundus** is the most superior portion of the uterus, the main portion is called the **body**, and the narrowed, neck-like portion that extends into the vagina is called the **cervix**. The wall of the uterus is made of three layers, from innermost to outermost: the endometrium, myometrium, and perimetrium. The **endometrium**, or mucosal lining, is made of simple columnar epithelium and an underlying vascular connective tissue. This entire layer thickens during a normal menstrual cycle in preparation for the implantation of an embryo. If no implantation occurs, this layer is sloughed off at the end of the menstrual cycle. The thick middle layer, the **myometrium**, is made of multiple layers of smooth muscle. The myometrium is hormonally stimulated to contract during childbirth to help move the baby out of the uterus. The outermost layer, the fibrous connective **perimetrium,** is the same as the **visceral peritoneum.**

The **vagina** is a thick, muscular tube that extends from the cervix to outside the body. It is lined with stratified squamous epithelium. The penis enters this passageway during sexual intercourse.

The small, lumpy **ovaries** are loosely held in place within the abdomen by various connective tissue ligaments. The **ovarian ligament** anchors the medial side of each ovary to the lateral side of the uterus, the **suspensory ligament** anchors the ovaries to the pelvic wall, and the wide, flat, **broad ligament** extends like a tarp over the uterus and ovaries and cradles the vagina, uterus, and uterine tubes.

The **uterine tubes** (consisting of the *fallopian tubes* and *oviducts*) serve to transport a female gamete (*egg cell; ova*) from the ovary to the uterus. The proximal end (near the uterus) is a long, narrow tube called the **isthmus**. The distal end widens and curves around the ovary to form the **ampulla**. The ampulla becomes the funnel-shaped **infundibulum,** which has fingerlike extensions called **fimbriae.**

Key to Illustration

Uterus	Vagina (V)	Ovary (O)
1. Fundus of uterus	Blood Vessels (B)	Uterine Tubes (U)
2. Body of uterus	B1. Ovarian artery	U1. Ampulla
3. Cervix	B2. Ovarian vein	U2. Isthmus
4. Lumen of uterus	B3. Uterine artery	U3. Infundibulum
5. Endometrium	B4. Uterine vein	U4. Fimbriae
6. Myometrium		
7. Perimetrium	Ligaments (L)	
8. Internal os	L1. Round ligament	
9. Cervical canal	L2. Suspensory ligament	
10. External os	L3. Ovarian ligament	
	L4. Broad ligament	
	L5. Uterosacral ligament	

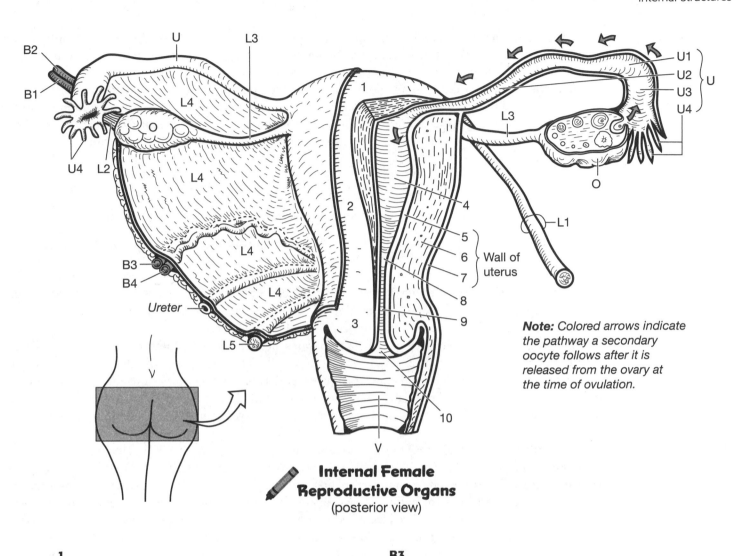

Note: *Colored arrows indicate the pathway a secondary oocyte follows after it is released from the ovary at the time of ovulation.*

Internal Female Reproductive Organs
(posterior view)

1. _____

2. _____

3. _____

4. _____

5. _____

6. _____

7. _____

8. _____

9. _____

10. _____

V _____

B1 _____

B2 _____

B3 _____

B4 _____

L1 _____

L2 _____

L3 _____

L4 _____

L5 _____

O _____

U1 _____

U2 _____

U3 _____

U4 _____

Description

Oogenesis is the process of developing **ova** (*egg cells*) within the ovaries of the female. Unlike sperm production in the male, this process begins during fetal development instead of at puberty. It involves two processes occurring together: **meiosis** and **follicle maturation**.

Imagine putting a marble in a balloon and then filling that balloon with water. In this analogy, the marble is the cell that has to go through meiosis to produce an ovum, and the water balloon is like the follicle that has to grow and develop by filling itself with fluid. It is important to remember that both of these processes occur together. In other words, meiosis occurs inside the maturing follicles. Let's consider each process separately.

Meiosis

This flowchart shows the stages of **meiosis** to produce an **ovum** within a follicle:

Oogonium (OG) ⟶ primary oocyte (PO) ⟶ secondary oocyte (SO) ⟶ zygote (if fertilized)

Meiosis is a type of cell division used solely to produce **gametes** (*sperm, ova*). It is called "reduction division" because during this process a cell undergoes two cell divisions to produce four final cells that each contain half the number of chromosomes as the original cell. This half-set of chromosomes, called the **haploid (n) number**, is equal to 23 chromosomes in humans.

The **oogonium** (OG) is a stem cell that contains the full set of chromosomes called the **diploid (2n) number**, or 46 ($2 \times n$ or 2×23). It is found in the outer region of the ovaries and undergoes cell division during fetal development to produce millions of cloned copies of itself. Most of these are degenerated, but some are stimulated to grow into a slightly larger cell called a **primary oocyte** (**PO**). Though they begin their first cell division of meiosis (meiosis I) in fetal development, *they do not complete it*.

At birth, about 400,000 to 4,000,000 oogonia (sing., *oogonium*) and POs remain in both ovaries. By puberty, about 40,000 remain, and the meiosis process is now strictly regulated by hormones. At this time, one PO is triggered to complete its first cell division of meiosis (meiosis I) every month. This cell division results in the formation of two unequally sized cells—a large, viable **secondary oocyte** (**SO**) and a small, nonfunctional cell called a **polar body**. The SO contains 23 replicated chromosomes. It will undergo its second cell division (meiosis II) only if it is fertilized by a sperm cell. If this occurs, the result is two new cells: a fertilized egg called a **zygote** and a small, nonfunctional **polar body**.

Follicle Maturation

This flowchart shows the stages in **follicle maturation**:

Primordial follicle ⟶ primary follicle ⟶ secondary follicle ⟶ tertiary follicle

Follicle maturation is like a water balloon filling with water until it bursts. The primitive **primordial follicles** are located in the outer region of the ovaries. They consist of a primary oocyte surrounded by a single layer of follicular cells. During the fetal stage, the primordial follicles are transformed into **primary follicles** by gradually thickening the follicle wall with multiple layers of cells around the primary oocyte.

As maturation continues, cells in the wall of the follicle, called granulosa cells, begin to secrete **follicular fluid**, which fills a potential space called the **antrum**. This marks the formation of the **secondary follicles** during adulthood. Like the expansion of a water balloon filling with fluid, the secondary follicle increases in size to become a **tertiary follicle** (*Graafian follicle; mature follicle; vesicular follicle*). At the same time, the primary oocyte has undergone its first meiotic division and produced a secondary oocyte and first polar body.

The tertiary follicle soon ruptures because of a hormonal surge and releases the **secondary oocyte** from the ovary in an event called **ovulation**. Last, the damaged follicle forms a temporary gland called a **corpus luteum** that produces the hormones progesterone and estrogen.

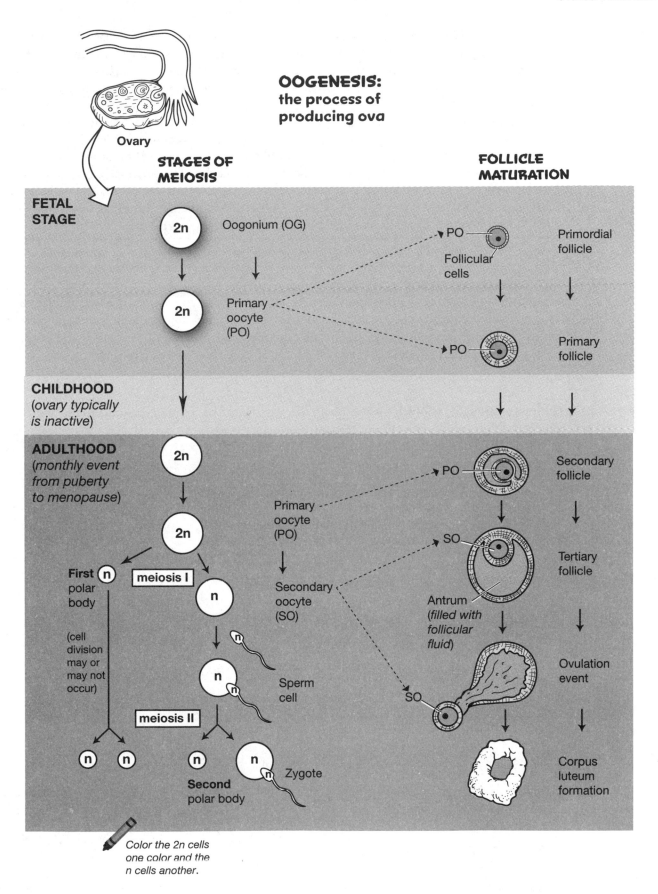

Ovary

OOGENESIS:
the process of producing ova

STAGES OF MEIOSIS

FOLLICLE MATURATION

FETAL STAGE

2n — Oogonium (OG)

2n — Primary oocyte (PO)

PO
Follicular cells — Primordial follicle

PO — Primary follicle

CHILDHOOD (*ovary typically is inactive*)

ADULTHOOD (*monthly event from puberty to menopause*)

2n

2n

First **n** polar body — meiosis I — n

(cell division may or may not occur)

n — Sperm cell

meiosis II

n   n   n   **n** Zygote

Second polar body

Primary oocyte (PO) — PO — Secondary follicle

Secondary oocyte (SO) — SO — Tertiary follicle

Antrum (*filled with follicular fluid*)

SO — Ovulation event

Corpus luteum formation

Color the 2n cells one color and the n cells another.

Description

The female has two lumpy, oval-shaped structures called **ovaries**, each measuring about 5 cm. in length. Blood is brought to the ovary by an **ovarian artery** and is drained by an **ovarian vein**. The surface of the ovary is covered by a simple cuboidal epithelial layer called the **germinal epithelium**. The two regional areas are the outer **cortex** and the inner **medulla**. Production of **gametes** (*sex cells*) occurs in the **cortex**.

Ideally, the ovaries in a sexually mature female alternate to produce one ovum each month. The specific process for developing an **ovum** (*egg cell*) is called **oogenesis**. This is a part of the ovarian cycle, which refers to all the processes that occur in the ovary during this monthly event.

The ova are produced in special chambers called **follicles**. The process begins when a hormone stimulates an immature follicle called a **primordial follicle** to begin to mature. The wall of the follicle thickens, and cells within it begin to produce a fluid called **follicular fluid** that fills a space called the **antrum**. As the follicle expands, it goes through the following progression:

> Primordial follicle ⟶ primary follicle ⟶ secondary follicle ⟶ tertiary follicle

Once a **tertiary follicle** has been formed, a hormonal surge causes it to rupture, and an ovum is released in an event called **ovulation**. The specific name for this ovum is a **secondary oocyte**. The broken tissue within the tertiary follicle thickens and becomes a temporary endocrine gland called a **corpus luteum** that produces a mixture of **estrogens** and **progestins**. Gradually the corpus luteum shrinks and degrades into a small mass of scar tissue called the **corpus albicans**.

Analogy

The **maturation of a follicle** in the ovary is comparable to a **water balloon filling with water**. As follicular fluid accumulates inside the antrum, it increases pressure just like water inside the water balloon. When the pressure becomes too great, the follicle ruptures at the **moment of ovulation** just like the **bursting of the water balloon**.

Location

Ovaries are located along the lateral edges of the pelvic cavity.

Function

The ovaries have two basic functions:

1. Produce **gametes** (sex cells)
2. Manufacture and release **hormones**
   - Follicles (primary, secondary, and tertiary) → **estrogen**
   - Corpus luteum → **estrogen, progesterone**

Key to Illustration	
1. Primary follicle	5. Early corpus luteum (still forming)
2. Secondary follicle	6. Corpus luteum
3. Tertiary follicle	7. Corpus albicans
4. Ovulation	

Ovary
(frontal section)

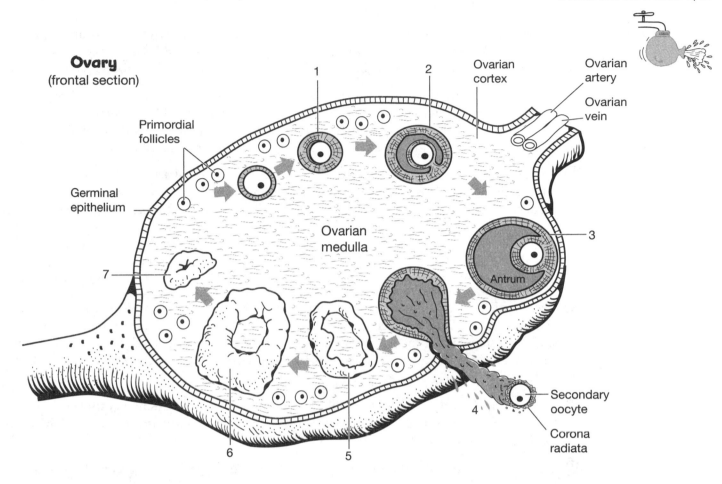

The maturation of a follicle is like...

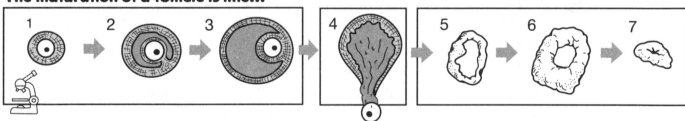

...a water balloon filling up with water and then bursting.

Write the correct term in each box:

1	2	3	4	5	6	7

Description

The **ovarian cycle** refers to the regular, monthly events that occur within the ovary of a sexually mature female. The group of hormones that regulates this process is the same basic set that regulates **spermatogenesis** in the male (see p. 524). A step-by-step description of this process is given below:

1. It all begins when the hypothalamus in the brain releases into the blood a **peptide** called **gonadotropin-releasing hormone (GnRH)**. It travels through the blood to target cells within the anterior lobe of the pituitary gland because they have receptors for GnRH. Unlike in the male, GnRH is not released in regular pulses. Instead, the amount released fluctuates in the female.

2. The binding of GnRH induces a response in these cells to make two **gonadotropin** hormones: **follicle-stimulating hormone (FSH)** and **luteinizing hormone (LH)**. Like any hormone, these chemical messengers are released into the bloodstream and travel to their respective targets, which contain receptors for these hormones. In the female, FSH targets the follicles within the ovary. The binding of FSH induces these follicles to mature and increase in size as a result of the production of a fluid called follicular fluid. As the fluid accumulates within a chamber called the antrum, it increases fluid pressure within the follicle. Meanwhile, LH causes the follicles to produce **estrogen**.

3. These initial lower levels of estrogen inhibit the anterior pituitary from *releasing* any hormones while *stimulating* it to produce mostly LH and some FSH within the cells of the anterior lobe. It is similar to stepping on the gas pedal of a car placed in neutral. Though you may be revving the engine, the car is not moving. Similarly, though the cells may be actively producing gonadotropins, they are not releasing any yet.

4. As the **tertiary** (*Graafian, vesicular*) **follicle** is established, it produces more estrogen. These higher levels have the opposite effect of the initial lower levels: they *stimulate* the anterior lobe to release its accumulated LH (and some FSH) in one big surge.

5. This one- to two-day spike in LH levels in the blood triggers many events: First, it stimulates formation of the secondary oocyte within the tertiary follicle.

6. **LH** triggers rupturing of the tertiary follicle and release of the secondary oocyte in an event called **ovulation**. Estrogen levels fall slightly after ovulation because of damage to the tertiary follicle.

7. The LH surge also changes the damaged follicle into a **corpus luteum**, which functions as a temporary endocrine gland, producing **estrogen**, **progesterone**, and **inhibin**.

8. Working together, when these three hormones reach a critical level, they serve to inhibit the hypothalamus and pituitary gland. This results in no release of GnRH, which translates into no release of FSH or LH. As a result, the corpus luteum shrinks to become a small, inactive mass of scar tissue, called the **corpus albicans**. With the loss of the corpus luteum, the estrogen, progesterone, and inhibin levels fall. As a result, inhibition is lost, and the hypothalamus again begins to release GnRH to start another ovarian cycle.

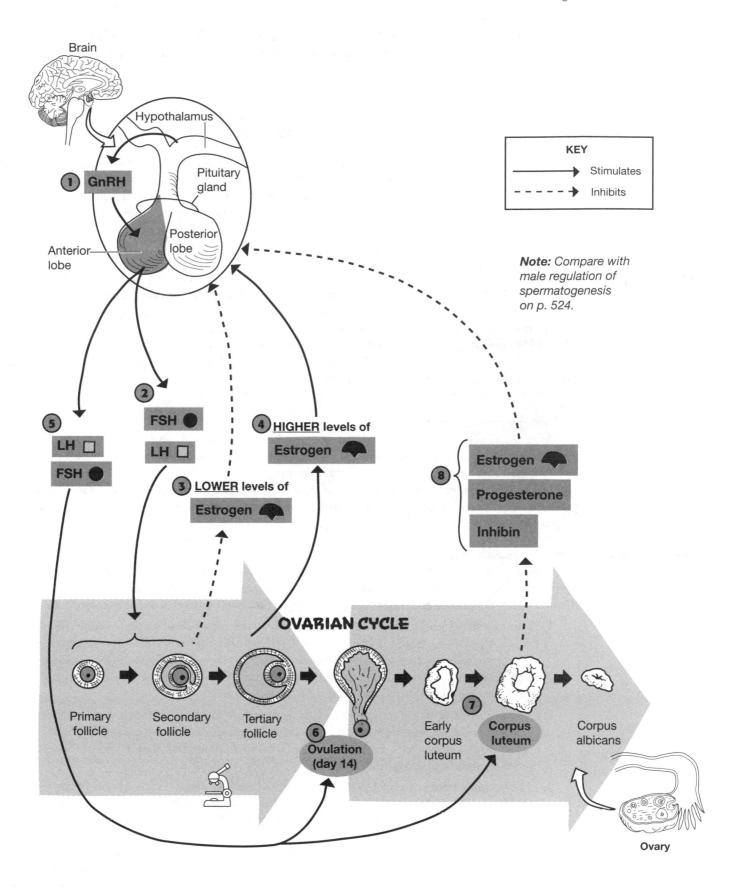

Brain

Hypothalamus

Pituitary gland

1 **GnRH**

Posterior lobe

Anterior lobe

KEY

→ Stimulates

--→ Inhibits

Note: Compare with male regulation of spermatogenesis on p. 524.

2

5

FSH ●

LH ◻

LH ◻

4 **HIGHER** levels of

Estrogen ◗

FSH ●

3 **LOWER** levels of

Estrogen ◗

8 **Estrogen** ◗

Progesterone

Inhibin

OVARIAN CYCLE

Primary follicle

Secondary follicle

Tertiary follicle

6 **Ovulation (day 14)**

Early corpus luteum

7 **Corpus luteum**

Corpus albicans

Ovary

Description

In a sexually mature female, a predictable series of changes occurs every month within the ovaries and the uterus. These processes are referred to separately as the **ovarian cycle** and the **uterine cycle** (or menstrual cycle) and are regulated by specific hormones. These processes are linked by a series of cascading events. Here is an overview:

> Hypothalamus ⟶ GnRH ⟶ anterior lobe of pituitary gland ⟶
> FSH, LH ⟶ ovary ⟶ estrogen, progesterone ⟶ uterus

It all begins when the hypothalamus in the brain releases a **peptide** into the blood, called **gonadotropin-releasing hormone (GnRH)**. It travels through the blood to target cells within the anterior lobe of the pituitary gland because they have receptors for GnRH. The binding of GnRH to its receptor induces a response in these cells to make the hormones **follicle-stimulating hormone (FSH)** and **leuteinizing hormone (LH)**. Like any hormone, these chemical messengers are released into the bloodstream and travel to their respective targets. Target sites always contain receptors for these hormones.

Ovarian Cycle

The **ovarian cycle** refers to all the changes that occur within the ovary. Ideally, the ovaries in a sexually mature female alternate to produce one ovum each month. The process for developing an **ovum** (*egg cell*) is called **oogenesis**. This is a part of the ovarian cycle, which refers to all the processes within the ovary during this monthly event. The ova are produced in special chambers called **follicles**. The process begins when **FSH** stimulates a **primary follicle** to begin to mature. The wall of the follicle thickens, and cells within it begin to produce a fluid called **follicular fluid** that fills a space called the **antrum**. As the follicle expands, it goes through the following progression:

> Primary follicle ⟶ secondary follicle ⟶ tertiary follicle

After a **tertiary** (*Graafian, vesicular, mature*) **follicle** is established, an **LH surge** occurs, lasting about one to two days and causing the follicle to rupture and release an ovum in an event called called **ovulation**. The specific name for this ovum is a **secondary oocyte**. The broken tissue within the tertiary follicle thickens and becomes a temporary endocrine gland called a **corpus luteum** that produces a mixture of **estrogens** and **progestins**. Gradually the corpus luteum shrinks and degrades into a small mass of scar tissue called the **corpus albicans**.

Uterine Cycle

The **uterine cycle** refers to changes within the **endometrium**, the innermost lining of the uterus. The endometrium is subdivided into two layers—the **functional zone**, which lines the lumen of the uterine cavity, and the **basal zone**, which lies beneath it. The entire cycle averages about 28 days. This cycle is divided into three phases: the **menstrual phase**, the **proliferative phase**, and the **secretory phase**.

1. The **menstrual phase** (days 1–5). During this phase, blood flow to the endometrium is reduced, which starves the tissues in the functional zone of their oxygen and other nutrients. As a result, these tissues deteriorate. When they break away from the uterine lining, it causes some blood loss. The sloughing off of these tissues and the associated blood, called **menstruation**, lasts one to seven days. During menstruation the basal zone remains intact.

2. The **proliferative phase** (days 6–13). Estrogens induce all the changes throughout this phase. The major event is to reestablish the functional zone lost during menstruation. Epithelial tissues and blood vessels grow back into the functional zone. **Endometrial glands** appear, which produce a mucus containing glycogen.

3. The **secretory phase** (days 15–28). This last and longest phase covers about 14 days. The tissues within the functional zone thicken and increase in vascularization, while endometrial glands enlarge and produce more mucus. The combination of estrogen and progesterone working together are responsible for these changes. The purpose of this thickening of the endometrium is to prepare for possible implantation of a human embryo into the functional layer of the endometrium. The glycogen in the mucus from the endometrial glands will serve to nourish the developing embryo. If no implantation occurs, both the ovarian and uterine cycles begin once again with release of GnRH from the pituitary gland.

Description

Each of us began as a single cell—a fertilized egg called a **zygote**. Through numerous cell divisions, this single cell gave rise to many more cells. This eventually leads to the establishment of an **embryo** and then a **fetus**. The first stage of human development occurs over about six days and is referred to as **pre-embryonic development**. It begins at conception when a sperm cell fertilizes an egg cell and ends when a blastocyst implants itself into the mother's uterus. Each stage of pre-embryonic development is described in the table below.

Structure	Description
1 **Zygote** (*zygotos = union*)	The result of a single sperm cell fertilizing an oocyte; occurs in the uterine tube near the ovary and marks the beginning of a genetically unique organism
2 **2-cell stage**	The zygote's first cell division resulting in two identical cells; occurs about one-and-a-half days after fertilization
3 **4-cell stage**	The second cell division creating a total of four identical cells; occurs about two days after fertilization
4 **Morula** (*moros = mulberry*)	A solid ball of cells; formed about three days after fertilization
5 **Blastocyst** (*blastos = germ, kystis = bag*)	A hollow sphere of cells with a fluid-filled central cavity; first forms about four days after fertilization and contains two different groups of cells: the trophoblast and the inner cell mass; the flat trophoblast cells form the outer shell of the blastocyst and will become part of the placenta; the inner cell mass will eventually develop into the embryo; the blastocyst travels through the uterine cavity and implants into the endometrium—the innermost lining of the uterus; this implantation event begins about six days after fertilization

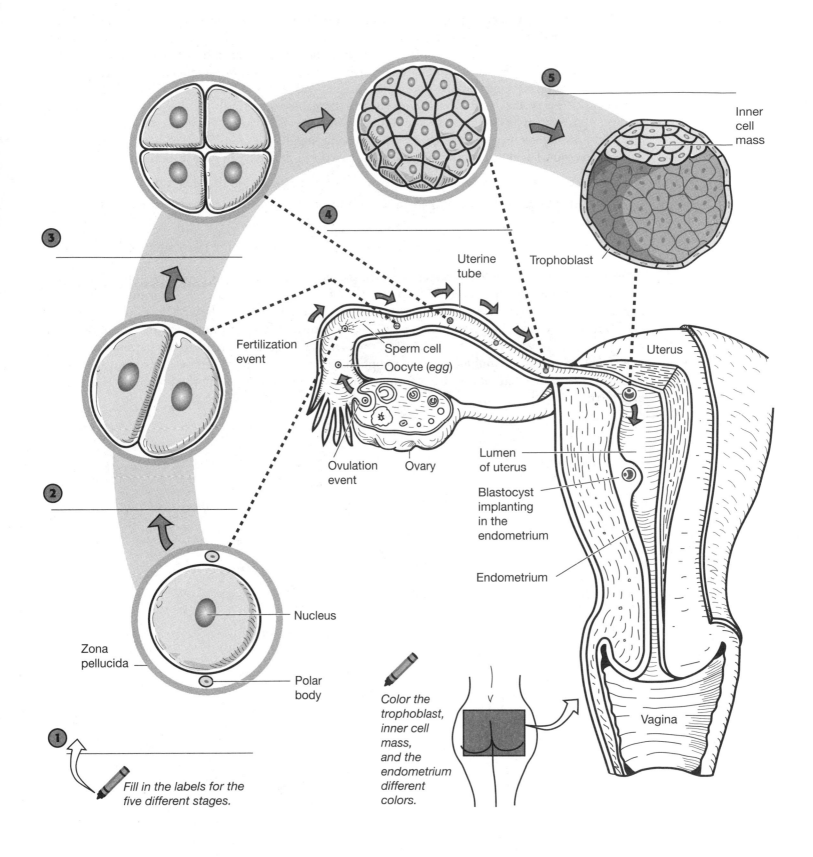

⑤ _____

Inner cell mass

③ _____

④ _____

Trophoblast

Uterine tube

Uterus

Fertilization event

Sperm cell

Oocyte (egg)

② _____

Ovulation event

Ovary

Lumen of uterus

Blastocyst implanting in the endometrium

Endometrium

Nucleus

Zona pellucida

Polar body

① _____

Fill in the labels for the five different stages.

Color the trophoblast, inner cell mass, and the endometrium different colors.

Vagina

Description

The illustration on the facing page shows a woman nine months pregnant with a full-term fetus. Many of the anatomical changes that occur during pregnancy can be seen in the illustration. For example, the normally fist-sized **uterus** has expanded to an enormous size, filling most of the abdominal cavity. Notice how it compresses the **urinary bladder**, resulting in increased frequency of urination. This uterine expansion also pushes the abdominal organs and diaphragm upward. This, in turn, causes pressure on the stomach, often resulting in heartburn. Another result of the swollen abdomen is lordosis—an increased curvature of the lumbar vertebrae. This causes backaches. Hormonal stimulation causes enlargement of the breasts. Around the nipple, the **areola** often increases in size and darkens in color. The **pubic symphysis** widens and increases flexibility in preparation for birth.

Here are some other anatomical changes that cannot be viewed in the illustration:

- **Heart enlarges**—this leads to an increase in cardiac output to support the growing fetus
- **Weight gain**—a woman can gain as much as 30 lbs.
- **Vagina changes**—increased elasticity and vascularity in preparation for serving as the birth canal

Let's examine the **full-term fetus**. It floats in the **amniotic sac** filled with **amniotic fluid**. This liquid serves as a protective cushion and also regulates fetal body temperature. When a pregnant woman says her "*water breaks*" it refers to the rupturing of the amniotic sac, followed by a flow of amniotic fluid out of the body. This often occurs prior to childbirth. Exchange of nutrients and wastes between mother and fetus occurs at a structure called the **placenta**. Linking the fetus to the placenta is a cable-like structure called the **umbilical cord**. It contains the usual umbilical arteries (two) and vein (one). Following birth, the placenta detaches from the uterus and is expelled from the body. This is called the **afterbirth**.

Lactating mammary glands

Areola

Nipple

Adipose tissue

Intestine

Placenta

Uterus

Amniotic sac

Umbilical cord

Amniotic fluid

Urinary bladder

Pubic symphysis

Urethra

Clitoris

Labia minora

Labia majora

Liver

Diaphragm

Spleen

Aorta

Liver

Stomach

Muscle tissue

Adipose tissue

Cervix

Mucous plug in cervical canal

Rectum

Anus

Vagina

Color the uterus, cervix, and amniotic sac.

Pregnant Female with Full-Term Fetus

Description

The **placenta** is a temporary pancake-shaped organ attached to the mother's inner uterine wall that is the site for the exchange of nutrients and wastes between the mother and the fetus. In terms of development, it is unique because it forms by combining maternal tissue with fetal tissue. Even though the mother and the fetus have separate circulatory systems and their blood does not mix together, the placenta is the site where the blood of both mother and fetus come in closest contact with one another. Through the process of **diffusion**, nutrients such as oxygen move from the mother's blood to the fetal blood, while waste products such as carbon dioxide diffuse in the opposite direction. It should be noted that drugs such as alcohol and cocaine also pass between the mother's blood and the fetal blood. *After* the *birth* of the baby, the placenta detaches from the uterus and is removed and called the **afterbirth**.

Let's examine the structural and functional details of the placenta with the illustration on the facing page. The upper half of the illustration (#6) shows the vascular **decidua basalis**. This is the maternal portion of the placenta. Notice that it contains **maternal arteries** (#4) that supply the fetus with oxygenated blood and **maternal veins** (#5) that carry away the mother's deoxygenated blood. The lower half of the illustration (#7) shows the **chorion**, or the fetal portion of the placenta. Within the chorion are fingerlike structures called **villi** (#1) that contain **fetal capillaries** (#2). At the bottom of the illustration, the **umbilical cord** (#8) is shown to contain two different blood vessels. There are **two umbilical arteries** (#9) and one **umbilical vein** (#10). The umbilical arteries carry deoxygenated blood from the fetus to the fetal capillaries, the specific site for gas exchange. The mother's oxygenated blood pools in the **intervillous space** (#3) where oxygen diffuses into the fetal capillaries to oxygenate the fetal blood. This oxygenated fetal blood is then returned to the fetus through the single umbilical vein. Note that the black arrows indicate the direction of blood flow. The maternal blood is constantly feeding and draining the intervillous space, while the fetal blood is constantly feeding and draining the fetal capillaries. Lastly, notice that the barrier that is the chorion prevents the maternal and fetal blood from mixing together.

Key to Illustration

1. Chorionic villi (containing fetal capillaries)	6. Decidua basalis (maternal portion of placenta)
2. Fetal capillaries	7. Chorion (fetal portion of placenta)
3. Intervillous space (filled with maternal blood)	8. Umbilical cord
4. Maternal artery	9. Umbilical arteries
5. Maternal vein	10. Umbilical vein

Placenta

Uterus

Umbilical cord

Fetus

1. _____

2. _____

3. _____

4. _____

5. _____

6. _____

7. _____

Note: Black arrows indicate the direction of blood flow.

Color the fetal capillaries to match the others to the left; red for oxygenated blood and blue for deoxygenated blood.

8. _____

9. _____

10. _____

KEY

▨	Oxygenated blood
▨	Deoxygenated blood

Notes

Glossary of Prefixes and Suffixes

Element	Definition and Example	Element	Definition and Example	Element	Definition and Example
a-	absent, deficient, or without: atrophy	chondr-	cartilage: chondrocyte	extra-	outside of, beyond, in addition: extracellular
ab-	off, away from: abduct	chrom-	color: chromosome		
abdomin-	abdomen	-cid(e)	destroy: germicide	fasci-	band: fascia
-able	capable of: viable	circum-	around: circumduct	febr-	fever: febrile
ac-	toward, to: actin	-cis	cut, kill: excision	-ferent	bear, carry: efferent arteriole
acou-	hear: acoustic	co-	together: copulation	fiss-	split: fissure
ad-	denoting to, toward: adduct	coel-	hollow cavity: coelom	for-	opening: foramen
af-	movement toward a central point: afferent artery	-coel	swelling, and enlarged space or cavity: blastocoel	-form	shape: fusiform
alba-	pale or white: linea alba	con-	with, together: congenital	gastro-	stomach: gastrointestinal
-alg	pain: neuralgia	contra-	against, opposite: contraception	-gen	an agent that produces or originates: pathogen
ambi-	both: ambidextrous	corn-	denoting hardness: cornified		
angi-	pertaining to vessel: angiogram	corp-	body: corpus	-genic	produced from, producing: carcinogenic
ante-	before: antebrachium	crypt-	hidden: cryptorchism		
anti-	against: anticoagulant	cyan-	blue color: cyanosis	gloss-	tongue: glossopharyngcal
aqua-	water: aqueous	cysto-	sac or bladder: cystoscope	glyco-	sugar: glycogen
archi-	to be first: archeteron	cyto-	cell: cytology	-gram	a record, recording: myogram
arthri-	joint: arthritis			gran-	grain, particle: agranulocyte
-asis	condition or state of: homeostasis	de-	down, from: descent	-graph	instrument for recording: electrocardiograph
aud-	pertaining to ear: auditory	derm-	skin: dermatology		
auto-	self: autolysis	di-	two: diarthrotic	grav-	heavy: gravid
		dipl-	double: diploid	gyn-	female sex: gynocology
bi-	two: biceps	dis-	apart, away from: disarticulate		
bio-	life: biology	duct-	lead, conduct: ductus deferens	haplo-	simple or single: haploid
blast-	generative or germ bud: osteoblast	dur-	hard: dura mater	hema(o)-	blood: hematology
		-dynia	pain: gastrodynia	hemi-	half: hemisphere
brachi-	arm: brachialis	dys-	bad, difficult, painful: dysentery	hepat-	liver: hepatic portal
brachy-	short: brachydont			hetero-	other, different: heterosexual
brady-	slow: bradycardia	e-	out, from: eccrine	histo-	tissue: histology
bucc-	cheek: buccal cavity	ecto-	outside, outer, external: ectoderm	holo-	whole, entire: holocrine
		-ectomy	surgical removal: tonsillectomy	homo-	same, alike: homologous
cac-	bad, ill: cachexia	ede-	swelling: edema	hydro-	water: hydrophilic
calci-	stone: calcification	-emia	pertaining to a condition of the blood: lipemia	hyper-	beyond, above, excessive: hypertension
capit-	head: capitate				
carcin-	cancer: carcinogenic	end-	within: endoderm	hypo-	under, below: hypoglycemia
cardi-	heart: cardiac	entero-	intestine: enteritis		
cata-	lower, under, against: catabolism	epi-	upon, in addition: epidermis	-ia	state or condition: hypoglycemia
caud-	tail: cauda equina	erythro-	red: erythrocyte	-iatrics	medical specialties: pediatrics
cephal-	head: cephalis	ex-	out of: excise	idio-	self, separate, distinct: idiopathic
cerebro-	brain: cerebrospinal fluid	exo-	outside: exocrine	ilio-	ilium: iliosacral
chol-	bile: cholic			infra-	beneath: infraspinatus

GLOSSARY OF PREFIXES AND SUFFIXES

Element	Definition and Example	Element	Definition and Example	Element	Definition and Example
inter-	among, between: intercellular	par-	give birth to, bear: parturition	-sis	state or condition: dialysis
intra-	inside, within: intracellular	para-	near, beyond, beside: paranasal	steno-	narrow: stenohaline
-ion	process: acromion	path-	disease, that which undergoes sickness: pathology	-stomy	surgical opening: tracheostomy
iso-	equal, like: isotonic			sub-	under, beneath, below: subcutaneous
-ism	condition or state: rheumatism	-pathy	abnormality, disease: neuropathy		
-itis	inflammation: meningitis	ped-	children: pediatrician	super-	above, beyond, upper: superficial
		pen-	need, lack: penicillin		
labi-	lip: labium major	-penia	deficiency: thrombocytopenia	supra-	above, over: suprarenal
lacri-	tears: lacrimal	per-	through: percutaneous	syn (sym)	together, joined, with: synapse
later-	side: lateral	peri-	near, around: pericardium		
leuc-	white: leucocyte	phag-	to eat: phagocyte	tachy-	swift, rapid: tachometer
lip-	fat: lipid	-phil	have an affinity for: neutrophil	tele-	far: telencephalon
-logy	science of: cytology	phlebo-	vein: phlebitis	tens-	stretch: tensor fasciae latae
-lysis	solution, dissolve: hemolysis	-phobe	abnormal fear, dread: hydrophobe	tetra-	four: tetrad
				therm-	heat: thermogram
macro-	large, great: macrophage	-plasty	reconstruction of: rhinoplasty	thorac-	chest: thoracic cavity
mal-	bad, abnormal, disorder: malignant	platy-	flat, side: platysma	thrombo-	lump, clot: thrombocyte
		-plegia	stroke, paralysis: paraplegia	-tomy	cut: appendectomy
medi-	middle: medial	-pnea	to breathe: apnea	tox-	poison: toxic
mega-	great, large: megakaryocyte	pneumato-	breathing: pneumonia	tract-	draw, drag: traction
meso-	middle or moderate: mesoderm	pod-	foot: podiatry	trans-	across, over: transfuse
meta-	after, beyond: metatarsal	-poiesis	formation of: hematopoiesis	tri-	three: trigone
micro-	small: microscope	poly-	many, much: polyploid	trich-	hair: trichology
mito-	thread: mitosis	post-	after, behind: postnatal	-trophy	a state relating to nutrition: hypertrophy
mono-	alone, one, single: monocyte	pre-	before in time or place: prenatal		
mons-	mountain: mons pubis	prim-	first: primitive	-tropic	turning toward, changing: gonadotropic
morph-	form, shape: morphology	pro-	before in time or place: prosect		
multi-	many, much: multicellular	proct-	anus: proctology	ultra-	beyond, excess: ultrasonic
myo-	muscle: myofibril	pseudo-	false: pseudostratified	uni-	one: unicellular
		psycho-	mental: psychology	-uria	urine: polyuria
narc-	numbness, stupor: narcotic	pyo-	pus: pyoculture	uro-	urine, urinary organs or tract: uroscope
necro-	corpse, dead: necrosis				
neo-	new, young: neonatal	quad-	fourfold: quadriceps femoris		
nephro-	kidney: nephron			vas-	vessel: vasoconstriction
neuro-	nerve: neurolemma	re-	back, again: repolarization	vermi-	worm: vermiform
noto-	back: notochord	rect-	straight: rectus abdominis	viscer-	organ: visceral
		reno-	kidney: renal	vit-	life: vitamin
ob-	against, toward, in front of: obturator	rete-	network: retina		
		retro-	backward: retroperitoneal	zoo-	animal: zoology
oc-	against: occlusion	rhin-	nose: rhinitis	zygo-	union, join: zygote
-oid	resembling, likeness: sigmoid	-rrhage	excessive flow: hemorrhage		
oligo-	few, small: oligodendrocyte	-rrhea	flow, or discharge: diarrhea		
-oma	tumor: lymphoma				
oo-	egg: oocyte	sanguin-	blood: sanguiferous		
or-	mouth: oral	sarc-	flesh: sarcoplasm		
orchi-	testicles: cryptorchidism	-scope	instrument for examination of a part: stethoscope		
-ory	pertaining to: sensory				
-ose	full of: adipose	-sect	cut: dissect		
osteo-	bone: osteocyte	semi-	half: semilunar		
oto-	ear: otolith	serrate-	saw-edged: serratus anterior		
ovo-	egg: ovum				

Glossary

A

abdomen (AB-doh-men) The region between the diaphragm and the pelvis.

absorption (ab-ZORP-shun) The passage of gases, liquids, or solutes through a membrane.

acetylcholine (ACh) (ass-ee-til-KOH-leen) A neurotransmitter secreted by various neurons into synapses; may have an excitatory effect (neuromuscular junction of skeletal muscles) or an inhibitory effect (neuromuscular junction of cardiac muscle tissue).

acetylcholinesterase (AChase) (ass-ee-til-koe-lin-ESS-ter-ase) An enzyme that rapidly inactivates the acetylcholine bound to postsynaptic receptors.

acidic (ah-SID-ik) Describes a solution in which the pH is less than 7; having a relatively high concentration of hydrogen ions.

acromial (ah-KROH-mee-al) The outer end of the scapula; site where the clavicle is attached.

acrosome (AK-roh-sohm) A cap-like structure on the head of a sperm cell; produces enzymes for penetrating eggs.

actin (AK-tin) A thin protein filament found in skeletal muscle cells; protein component of microfilaments.

action potential an electrical signal that advances along the plasma membrane of an excitable cell, such as a neuron or a muscle cell; used to communicate over long or short distances.

active transport A carrier-mediated process in which cellular energy is used to move molecules against their concentration gradient through a plasma membrane.

adenosine diphosphate (ah-DEN-oh-seen dye-FAHS-fate) An energy molecule composed of an adenine base, a ribose sugar, and two phosphate groups.

adenosine triphosphatase (ATPase) (ah-DEN-oh-seen try-FAHS fate-ays) (AY-tee-pee-ase) An enzyme that catalyzes the hydrolysis of ATP to form ADP, a phosphate group, and a net release of free energy.

adenosine triphosphate (ATP) (ah-DEN-oh-seen try-FAHS-fate) The energy currency of a cell; composed of an adenine base, a ribose sugar, and three phosphate groups.

adenylate cyclase (ah-DEN-il-ayt SYE-klayz) An enzyme that catalyzes the conversion of ATP to cyclic AMP (cAMP).

adipocyte (AD-i-poh-syte) A fat cell.

adrenergic neuron (ad-ren-ER-jik NOO-ron) A neuron that secretes norepinephrine (*noradrenaline*) or epinephrine (*adrenaline*) as its neurotransmitter.

aerobic Any process that requires oxygen.

afferent (AF-fer-ent) Toward; opposite of efferent.

aldosterone (al-DAH-stair-ohn) A mineralocorticoid; a hormone produced by the adrenal cortex that acts on the kidneys to increase sodium levels in the blood and increase excretion of potassium in urine.

alkaline (AL-kah-lin) Refers to a base, a pH greater than 7; having a relatively low concentration of hydrogen ions.

alveolus/alveoli (al-VEE-oh-luss/al-VEE-oh-lye) Delicate air sacs in the lungs where gas exchange occurs.

amino acid (ah-MEE-no ASS-id) The structural unit of a protein.

amnion (AM-nee-on) One of the extraembryonic membranes; develops around the embryo/fetus, forming the amniotic cavity.

amniotic fluid (am-nee-OT-ik fluid) The fluid that surrounds and cushions the developing embryo and fetus.

amniotic sac (am-nee-OT-ik sak) Fluid-filled chamber in which the embryo floats during development.

amylase (AM-eh-layz) An enzyme that breaks down poly-saccharides; produced by salivary glands and pancreas.

anabolism (ah-NAB-oh-liz-em) Synthesis reactions that require energy to join small molecules to form more complex molecules; for example, amino acids bonding to make proteins.

anaerobic (an-air-OH-bik) Any process that does not require oxygen.

anaphase (AN-nah-fayz) Stage of mitosis when the chromatid pairs separate and the daughter chromosomes move toward the opposite ends of the cell.

antagonist (an-TA-gun-ist) Something that has an opposing action on something else, such as a muscle, hormone, or drug.

antebrachial (an-tee-BRAY-kee-al) Pertaining to the forearm.

antecubital (an-tee-KYOO-bi-tal) Pertaining to the anterior side of the elbow.

anterior (an-TEER-ee-or) Toward the front or ventral; opposite of posterior.

antibody Proteins produced by plasma cells in response to specific antigens; antibodies combine with antigens to render a pathogen harmless; also known as immunoglobulin (Ig).

antigen (AN-tih-jen) Means "*anti*body *gen*erating substances"; macromolecules that induce the immune system to make antibodies.

antrum (AN-trum) Central chamber.

anus (AY-nus) External opening at the end of the rectum.

aorta (ay-OR-tah) Largest artery in the body; carries oxygenated blood from left ventricle and into the systemic circuit.

apex (AY-peks) Pointed tip of a structure.

apical (AY-pik-al) Located near or relating to the apex or pointed structure; opposite of basal.

apnea (APP-nee-ah) Temporary cessation of breathing at the end of a normal expiration.

aponeurosis (ap-oh-nyoo-ROH-sis) Broad, flat collagenous sheets that may serve as anchor points for skeletal muscle.

appendix (ah-PEND-diks) Small organ connected to the cecum of the large intestine.

aqueous humor (AY-kwee-us HYOO-mor) A fluid that fills the anterior chamber of the eye.

arachnoid (ah-RAK-noyd) Middle meninges that surround the CSF and protect the brain and spinal cord.

arbor vitae (AR-bor VYE-tay) Central area of white matter in the cerebellum.

arteries (AR-ter-eez) Blood vessels that carry blood away from the heart.

arteriole (ar-TEER-ee-ohl) Microscopic blood vessel that connects small arteries to capillaries.

articulation (ar-tik-yoo-LAY-shun) Joint; point of contact between bones.

atom (AT-om) The smallest particle of a chemical element that displays the properties of that element.

atrioventricular (AV) (ay-tree-oh-ven-TRIK-yoo-lar) **bundle** Group of specialized cardiac muscle cells that extends from the AV node to the Purkinje fibers; coordinates contraction of the heart muscle.

atrioventricular (ay-tree-oh-ven-TRIK-yoo-lar) (**AV**) **node** Secondary pacemaker of the heart; a small mass of specialized cardiac muscle cells that plays a role in the electrical conduction system within the heart.

atrioventricular (**AV**) (ay-tree-oh-ven-TRIK-yoo-lar) **valves** Pair of heart valves located between the atria and the ventricles; tricuspid and bicuspid; prevent backflow of blood into atria during ventricular contraction.

atrium (AY-tree-um) An upper chamber of the heart; receives blood from the pulmonary or systemic circuit.

auditory (AW-di-toh-ree) Pertaining to the sense of hearing.

auditory tube (AW-di-toh-ree tube) Passageway that connects the nasopharynx with the middle ear.

auricle (AW-ri-kul) Curved, flexible upper portion of the ear; also, expandable flap-like structure of an atrium in the heart.

autonomic (aw-toh-NAHM-ik) **nervous system (ANS)** The part of the nervous system that governs itself without our conscious knowledge; divided into sympathetic and parasympathetic divisions; controls organs through reflex pathways that link to smooth muscle, cardiac muscle, and glands.

axilla (AK-sil-ah) The armpit.

axon hillock (AK-son HILL-ok) Funnel-shaped portion of neural cell body from which the axon extends.

axons (AK-sonz) Long, single processes of neurons that conduct nerve impulses toward the synaptic knobs.

B

basal (BAY-sal) Located at or near the base of a structure; opposite of apical.

basement membrane (BAYSE-ment MEM-brayne) A layer of protein fibers that connects the epithelium to the underlying connective tissue.

basic *See alkaline.*

basophil (BAY-so-fil) White blood cell; releases histamine into damaged tissue.

bicuspid valve (bye-KUSS-pid valve) The left atrioventricular (AV) valve in the heart; also known as the mitral valve.

bicuspids (bye-KUSS-pids) Teeth used for crushing, mashing, and grinding; premolars.

bile A secretion of the liver stored in the gallbladder and released into the duodenum; physically breaks down fat into smaller droplets.

bilirubin (BILL-y-roo-bin) Orange or yellow pigment found in bile and derived from hemoglobin catabolism.

bipolar neuron (bye-POH-lar NOO-ron) Nerve cell with two distinct processes; one dendrite, one axon.

blastocyst (BLASS-toh-sist) Early stage in embryonic development; a hollow ball of cells consisting of an inner cell mass and an outer cell mass.

blastomere (BLASS-toh-meer) The first cleavage division, which produces a pre-embryo consisting of two identical cells.

bolus (BOH-luss) A ball or mass of chewed food that passes from the mouth to the stomach.

Bowman's capsule *See glomerular capsule.*

Boyle's law A physics law stating that the volume of a gas is inversely proportional to its pressure.

brachial (BRAY-kee-al) Pertaining to the upper limb between shoulder and elbow.

brainstem (brayn-stem) Part of the brain that contains important processing centers; consists of the medulla oblongata, pons, and midbrain.

bronchiole (BRONG-kee-ohl) A small, tubelike branch of a bronchus; lacks cartilaginous supports, but wall contains smooth muscle.

bronchus (BRONG-kuss) A branch of the bronchial tree between the trachea and the bronchioles.

buccal (BUK-al) Pertaining to the cheek.

bulbourethral glands (BUL-boh-yoo-REE-thral glands) Small mucus glands located at base of the penis; secretions lubricate the urethra.

bundle branches Two bands of specialized cardiac muscle cells that transmit impulses from the AV bundle to the Purkinje fibers.

bundle of HIS *See atrioventricular (AV) bundle.*

C

calcification (kal-sih-fih-KAY-shun) Process of hardening a tissue with deposits of calcium salts.

calcitonin (CT) (kal-sih-TOE-nin) A hormone produced by the thyroid gland that serves to decrease calcium levels in the blood.

canaliculus/canaliculi (kan-ah-LIK-yoo-luss/kan-ah-LIK-yoo-lye) Microscopic channels between cells, found in compact bone and liver; in compact bone, canaliculi allow diffusion of nutrients and wastes; in liver, bile canaliculi transport bile to bile ducts.

canine tooth *See cuspids.*

capillary (KAP-i-lair-ee) Smallest, simplest blood vessel; microscopic; connects arterioles and venules.

carbohydrate (kar-boh-HYE-drayt) Organic compound containing carbon, oxygen, and hydrogen; sugars, starches, and cellulose.

cardiac (KAR-dee-ak) **cycle** A complete heartbeat or pumping cycle consisting of systole and diastole.

cardiac (KAR-dee-ak) **output** The volume of blood pumped by one heart ventricle each minute.

carotid artery (kah-ROT-id AR-ter-ee) The large artery of the neck that provides a major blood supply to the brain.

carpals (KAR-puls) Wrist bones.

carpus/carpal (KAR-pus/KAR-pul) The wrist.

catabolism (kah-TAB-oh-liz-em) Chemical reactions that break down complex organic compounds into simpler ones with a net release of energy; for example, proteins broken down into amino acids.

catecholamines (kat-eh-KOLE-ah-meenz) A class of neurotransmitters synthesized from the amino acid tyrosine; include norepinephrine, epinephrine, and dopamine; may play a role in sleep, motor function, regulating mood, and pleasure recognition.

caudal (KAW-dal) The tail.

CCK *See cholecystokinin.*

cecum (SEE-kum) The pouch located at the beginning of the large intestine.

cell (sell) The basic unit of life.

cellular respiration Metabolic process by which carbohydrates, fatty acids, and amino acids are broken down to produce ATP.

cementum (see-MEN-tum) Bone-like material covering the root of the tooth.

centriole (SEN-tree-ohl) Tiny, cylindrical organelle of a cell; involved with spindle formation during mitosis.

centromere (SEN-troh-meer) The region where two chromatids are connected together during the early stages of cell division.

centrosome (SEN-troh-sohm) The region of cytoplasm that coordinates the activities of centrioles.

cerebellum (sair-eh-BELL-um) Second largest part of the brain; coordinates and refines learned movement patterns.

cerebrospinal (SAIR-eh-broh-SPY-nal) **fluid (CSF)** Liquid found in the central nervous system by filtering and processing blood plasma; circulates in the ventricles, central canal, and subarachnoid space of the brain and spinal cord.

cerebrum (SAIR-eh-brum) Largest region of the brain; origin of conscious thoughts and all intellectual functions; controls sensory and motor integration.

cervical (SER-vih-kal) Pertaining to the neck.

cervix (SER-viks) Neck-like structure at the inferior portion of the uterus; projects into vagina.

chief cells (CHEEF sells) Cells in the gastric mucosa that primarily secrete pepsinogen, which later is converted into the protein-digesting enzyme called pepsin.

cholecystokinin (CCK) (koh-lee-sis-toh-KYE-nin) A hormone produced by the duodenal mucosa that stimulates contraction of the gallbladder and secretion of pancreatic juice rich in digestive enzymes.

cholesterol (ko-LES-ter-all) Steroid compound produced by animals; found in their plasma membranes.

cholinergic neuron (koh-leh-NER-jik NOO-ron) Neuron that secretes acetylcholine as its neurotransmitter.

chondrocyte (KON-droh-syte) A cartilage cell.

chorion (KOH-ree-on) A membrane consisting of the mesoderm and trophoblast; develops into a membrane of the placenta.

choroid (KOH-royd) Middle vascular layer of the eye.

chromatid (KROH-mah-tid) Either of two daughter strands of chromosomes joined by a single centromere.

chromatin (KROH-mah-tin) Chromosomal material that is loosely coiled, forming a tangle of fine filaments with a grainy appearance.

chromosomes (KROH-moh-sohms) Tightly compacted structures that contain coiled DNA wrapped around histone proteins; normal human body cells contain 46 chromosomes; term means "colored body."

chyme (kyme) A soupy, viscous mixture of ingested substances and gastric juices leaving the stomach.

cilia (SIL-ee-ah) Long folds of the plasma membrane of an epithelial cell that contains microtubules; move back and forth to propel fluids or secretions across cell surfaces.

ciliary body (SIL-ee-air-ee body) A muscular structure that surrounds the perimeter of the lens of the eye and attaches to it through the suspensory ligaments.

citric acid cycle (SIT-rik ASS-id SYE-kul) An aerobic chemical cycle that begins with the formation of citric acid and results in the formation of oxaloacetic acid; in the process, ATP is produced and carbon dioxide is released; occurs in the mitochondrion.

clitoris (KLIT-oh-ris) Small female organ composed of erectile tissue located behind the junction of the labia majora.

coccygeal (KOKS-ih-jee-al) Relating to or near the coccyx.

coccyx (KOKS-siks) The tailbone; most inferior portion of the vertebral column.

cochlea (KOHK-lee-ah) An inner ear structure that resembles a snail shell; contains nerve endings essential to hearing.

collagen (KAHL-ah-jen) The most common type of protein fiber found in connective tissues; serves to strengthen tissues.

colon (KOH-lon) The large intestine.

common bile duct (KOM-mon byle dukt) Formed by the union of the cystic duct and the hepatic duct; carries bile from the liver and gallbladder to the duodenum.

compact bone (KOM-pak bone) Dense bone; contains osteons.

compound (KOM-pound) A substance formed by the union of two or more elements.

connective tissue (koh-NEK-tiv TISH-yoo) One of the four major tissue types; serves to give structural support to other tissues and organs in the body; most contains cells, protein fibers, and ground substance.

convolution See gyrus.

cornea (KOHR-nee-ah) The transparent, anterior region of the sclera.

corona radiata (koh-ROHN-ah ray-dee-AY-tah) Follicular cells that surround the oocyte.

coronoid (KOHR-oh-noyd) Pertaining to certain processes of the bone; shaped like a crow's beak.

corpus/corpora (KOHR-pus/KOHR-pohr-ah) Body.

corpus albicans (KOHR-pus AL-bi-kans) Pale scar tissue in the ovaries that replaces the nonfunctional corpus luteum.

corpus callosum (KOHR-pus kah-LOH-sum) Area of the brain that links the right and left cerebral hemispheres.

corpus cavernosum (KOHR-pus kav-er-NO-sum) Two columns of erectile tissue that extend along the length of the penis.

corpus luteum (KOHR-pus LOO-tee-um) An ovarian structure transformed from a ruptured follicle; secretes estrogen and progesterone.

corpus spongiosum (KOHR-pus spun-jee-OH-sum) Erectile body that surrounds the urethra.

cortex (KOHR-teks) Outer part of an organ; adrenal cortex.

cortisol (KOHR-tih-sawl) A glucocorticoid; a hormone produced by the adrenal cortex that helps regulate responses to stress; also known as hydrocortisone.

Cowper's glands See bulbourethral glands.

coxa/coxae (KOKS-ah/KOKS-ee) Pelvic bone or hip bone.

crest (as in bone marking) Slightly raised narrow ridge on a bone; site for muscle attachments.

crural (KROOR-al) Refers to the anterior portion of the leg (below the knee).

cubital (KYOO-bi-tal) Pertaining to the elbow.

cuspids (KUS-pids) Sharp, pointed teeth; canines.

cyclic adenosine monophosphate (SIK-lik ah-DEN-oh-seen mon-oh-FAHS-fate) (**cAMP**) A second messenger formed from ATP; composed of an adenine base, a ribose sugar, and one phosphate group; used for signal transduction.

cystic duct (SIS-tik dukt) A tube that leads from the gallbladder toward the liver; unites with the common hepatic duct to form the common bile duct.

cytokinesis (SYE-toe-kih-nee-siss) Physical division of the cytoplasm during cell division.

cytoplasm (SYE-toh-plaz-em) Gel-like material between the nucleus and the plasma membrane; includes cytosol and organelles.

cytoskeleton (sye-toh-SKEL-eh-ton) An integrated network of microtubules and microfilaments in the cytoplasm that gives shape and provides support to the cell and its organelles.

cytosol (SYE-toh-sawl) Fluid portion of the cytoplasm.

D

decidua basalis (dih-SID-yoo-ah bah-SAY-lis) Area of the endometrium that develops into the maternal part of the placenta.

deep Away from the surface; opposite of superficial.

deltoid (DEL-toyd) Triangular shape.

dendrite (DEN-dryte) A long cellular extension of a neuron that responds directly to stimuli.

dentin (DEN-tin) The mineralized matrix found in teeth.

dermis (DER-mis) Layer of connective tissue that lies beneath the epidermis.

diaphragm (DYE-ah-fram) Dome-shaped muscle that separates the thoracic and abdominal cavities; a major muscle involved with respiration.

diaphysis (dye-AF-i-sis) Tubular shaft of a long bone.

diastole (dye-ASS-toh-lee) Relaxation of both atria and both ventricles in the heart; opposite of systole.

diffusion (dih-FYOO-shun) The net movement of substances from an area of high concentration to an area of low concentration.

disaccharide (dye-SAK-ah-ryde) A double-unit sugar formed by bonding a pair of monosaccharides; *ex:* lactose, sucrose.

distal (DIS-tal) Refers to the region or reference away from an attached base; opposite of proximal.

distal convoluted tubule (DCT) (DIS-tal KON-voh-loo-ted TOOB-yool) The part of the nephron distal to the ascending limb of the nephron loop; site for active secretion and selective reabsorption of ions and other substances.

dorsal (DOR-sal) Pertaining to the back; posterior; opposite of ventral.

ductus deferens (DUK-tus DEF-er-ens) A smooth muscular tube that propels sperm from the epididymis to the ejaculatory duct; also known as vas deferens.

duodenum (doo-oh-DEE-num) First and shortest segment of the small intestine—about 10 inches long; receives chyme from the stomach and digestive secretions from the pancreas.

dura mater (DOO-rah MAH-ter) Outermost layer of the meninges.

E

efferent (EF-fer-ent) Away from; opposite of afferent.

ejaculatory duct (ee-JAK-yoo-lah-toh-ree dukt) Short passageway that allows sperm to enter the urethra.

electrocardiogram (ECG or EKG) (ee-lek-troh-KAR-dee-oh-gram) A graphic record of the heart's electrical activity or conduction of impulses; used to evaluate the heart's action potential.

electroencephalogram (EEG) (ee-lek-troh-en-SEF-ah-loe-gram) A graphic record of brain electrical potentials; used to evaluate nerve tissue function and diagnose specific disorders (such as epilepsy).

electrolytes (ee-LEK-troh-lytes) Inorganic substances that break up in solution to form ions and conduct electricity; include acids, bases, and salts.

electrons (ee-LEK-trons) Negatively charged subatomic particles in atoms; located in energy shells outside the nucleus of the atom.

electron transport system (ETS) (ee-LEK-tron TRANS-port SIS-tem) Transfer of electrons along a series of membrane-bound electron carrier molecules in the mitochondria; energy from this process is used to synthesize ATP; aerobic process that produces water.

embolus (EM-boe-luss) A clot that dislodges and circulates through the bloodstream.

embryo (EM-bree-oh) A stage of human development beginning at fertilization and ending at the start of the eighth developmental week.

emulsification (ee-mul-sih-fih-KAY-shun) The process by which bile breaks up fats.

enamel (ee-NAM-el) Hardest manufactured substance in the body; covers the crown of the tooth.

endocardium (en-doh-KAR-dee-um) Simple squamous epithelial inner layer of the heart.

endocytosis (en-doh-sye-TOH-sis) An active transport mechanism that allows extracellular substances to enter the cell by phagocytosis, pinocytosis, or receptor-mediated endocytosis.

endometrium (en-doh-MEE-tree-um) Innermost, glandular layer of the uterine wall.

endomysium (en-doh-MISH-ee-um) Inner layer of connective tissue that surrounds each skeletal muscle fiber.

endoneurium (en-doh-NOO-ree-um) A layer of connective tissue that surrounds individual axons.

endoplasmic reticulum (en-doh-PLAS-mik re-TIK-yoo-lum) The network of intracellular membranes that synthesizes and manufactures membrane-bound proteins.

endosteum (en-DOS-tee-um) A membrane that lines the marrow cavity inside a bone.

enzyme (EN-zyme) A biological catalyst.

eosinophil (ee-oh-SIN-oh-fil) A phagocytic white blood cell; numbers increase during allergic reaction.

epicardium (ep-i-KAR-dee-um) Outer covering of the heart; also called the visceral pericardium.

epidermis (ep-i-DER-mis) Outermost layer of the skin.

epididymis (ep-i-DID-i-miss) A long, coiled and twisted tubule that lies along the posterior border of the testis; stores sperm and facilitates their maturation.

epiglottis (ep-i-GLOT-iss) Flap of elastic cartilage that folds back over the larynx during swallowing.

epimysium (ep-i-MISH-ee-um) Connective tissue layer that surrounds the entire skeletal muscle.

epineurium (ep-i-NOO-ree-um) Outermost fibrous connective tissue sheath that surrounds a peripheral nerve.

epiphysis (eh-PIF-i-sis) Expanded end of a long bone.

epithelial tissue (ep-i-THEE-lee-al TISH-yoo) One of the four major tissue types; serves to cover exposed body surfaces and lines internal cavities and passageways.

erythrocyte (e-RITH-roh-syte) Red blood cell.

esophagus (eh-SOF-ah-gus) Hollow, muscular tube that transports food and liquid from the pharynx to the stomach.

estrogens (ES-troh-jens) Steroid hormones produced by the ovaries; dominant sex hormones in females.

exocytosis (eks-oh-sye-TOH-sis) An active transport mechanism that involves movement of vesicle-bound substances out of the cell; vesicles fuse with the plasma membrane and contents are released outside the cell.

extracelluar fluid (ECF) Liquid located outside body cells, such as plasma, lymph, and interstitial fluid.

F

facet (FASS-et) Flat surface of a bone that forms a joint with another bone.

facilitated diffusion A passive transport process wherein a substance binds to a carrier molecule in the plasma membrane and is moved from an area of higher concentration to an area of lower concentration.

fallopian tubes *See uterine tubes.*

fascia (FAY-sha) A connective tissue sheath consisting of fibrous tissue and fat; unites skin to underlying tissue.

fascicle (FASS-i-kul) A single bundle.

feces (FEE-seez) The waste material eliminated from the large intestine; residue of digestion; also known as *stool.*

femoral (FEM-or-al) Pertaining to the thigh.

fetus (FEE-tus) The name given to the unborn young from the eighth week of pregnancy to birth.

fibroblasts (FYE-broh-blasts) The most abundant fixed cells in connective tissue proper.

fibular (FIB-yoo-lar) Pertaining to the fibula.

fight-or-flight response A group of reactions that ready the body to engage in maximum muscular exertion needed to deal with a perceived threat.

filtrate Fluid produced by the process of filtration, such as that made by the glomeruli in the kidneys.

filtration A process that uses hydrostatic pressure to move water and solutes through a membrane.

fimbria/fimbriae (FIM-bree-ah/FIM-bree-ee) Fingerlike projections, such as those found on the ends of the uterine tubes nearer the ovaries.

first messenger Typically, a nonsteroid hormone that binds to its plasma membrane receptor, thereby inducing a response in the target cell.

fissure (FISH-ur) Deep groove.

follicle (FOL-lih-kul)-**stimulating hormone (FSH)** A hormone that stimulates structures within the ovaries and primary follicles to grow toward maturity; stimulates follicle cells to synthesize and secrete estrogen in the female and stimulates development of the seminiferous tubules in the male.

fontanel (FON-tah-nel) Fibrous area between the cranial bones; "soft spot."

foramen/foramina (foh-RAY-men/foh-RAM-i-nah) A hole or passageway in bone for blood vessels and/or nerves to pass through.

forearm (FORE-arm) The part of the arm between the elbow and the wrist.

fossa (FOSS-ah) Depression.

frenulum (FREN-yoo-lum) Small bridle; thin, flat fold of mucous membrane between two structures, such as the lingual frenulum that anchors the body of the tongue to the floor of the oral cavity.

frontal (FRUN-tal) Toward the anterior.

frontal plane (FRUN-tal playne) Runs parallel to the long axis of the body; divides the body into anterior and posterior sections.

G

gallbladder Pear-shaped organ attached to the liver that stores and concentrates bile.

gametes (GAM-eets) Mature male or female reproductive cells—sperm or secondary oocyte.

ganglion (GANG-lyun) A collection of neuronal cell bodies outside the CNS.

gastric (GAS-trik) Pertaining to the stomach.

gastric juice (GAS-trik joos) A mixture of secretions that contain water, mucus, hydrochloric acid, and the enzyme pepsin; secreted by exocrine gastric glands.

gastrin (GAS-trin) The hormone that stimulates the secretion of both parietal and chief cells in the stomach.

G cells (jee sells) Cells found in the stomach; secrete the hormone gastrin.

genitalia (jen-i-TAYL-yah) The reproductive organs.

glomerular (gloh-MARE-yoo-lar) **capsule** Cup-shaped mouth of the nephron; surrounds the glomerulus; receives the glomerular filtrate; also known as Bowman's capsule.

glomerulus/glomeruli (gloh-MARE-yoo-lus/gloh-MARE-yoo-lye) Cluster of capillaries tucked into the glomerular capsule; forms the glomerular filtrate.

glucagon (GLOO-kah-gon) A hormone produced by alpha cells in the pancreas; increases blood glucose levels.

glucocorticoids (gloo-koe-KORE-tih-koyds) A group of hormones produced by the adrenal cortex that help regulate normal glucose levels and increase resistance to stress; *ex:* cortisol.

gluconeogenesis (gloo-koh-nee-oh-JEN-eh-sis) The synthesis of glucose from lipid or protein.

glucose (GLOO-kohs) A monosaccharide; principal source of energy for the cells.

gluteal (GLOO-tee-al) Pertaining to the buttocks.

glycogen (GLYE-koh-jen) A polysaccharide consisting of a long chain of glucose units; found mainly in liver and skeletal muscle cells.

glycolysis (glye-KOL-ih-sis) An anaerobic process that breaks down a glucose molecule into two pyruvic acid molecules; ATP is produced; occurs in the cytoplasm.

Golgi complex (GOL-jee KOM-pleks) Membranous cellular organelle that stores, alters, and packages secretory products such as proteins; gives rise to lysosomes and secretory vesicles; also known as Golgi apparatus or Golgi body.

G protein Membrane protein that reacts with guanosine triphosphate (GTP); activates the membrane-bound enzyme adenylate cyclase.

Graafian follicle *See tertiary follicle.*

graded potential (GRAY-ded poh-TEN-shul) A small, local change in the membrane potential; used to communicate *only* over short distances.

gray matter (gray MAT-er) Regions in the brain and spinal cord containing nerve cell bodies, glial cells, and unmyelinated axons.

growth hormone Hormone produced by the anterior pituitary; promotes bodily growth indirectly by stimulating the liver to produce certain growth factors; also known as somatotropin.

gyrus/gyri (JYE-russ/JYE-rye) Elevated ridge.

H

hard palate (hard PAL-let) Bony roof of the mouth.

haustrum/haustra (HAWS-trum/HAWS-trah) Series of pouches in the wall of the colon; permits distention and elongation.

Haversian system *See osteon.*

head Rounded process at the end of a long bone.

hemoglobin (hee-mo-GLOH-bin) A protein containing iron in red blood cells; functions to transport oxygen and some carbon dioxide in the blood.

hepatic duct (heh-PAT-ik dukt) Collects bile from all of the bile ducts of the liver lobes, which unite to form common bile duct.

hepatic portal vein (heh-PAT-ik POR-tal vane) Delivers blood to the liver.

hepatic triads (heh-PAT-ik TRYE-ads) Portal areas located at each of the six corners of the hepatic lobule; each triad consists of the following three structures: (1) branch of hepatic portal vein, (2) branch of hepatic artery proper, and (3) branch of the bile duct.

hepatocytes (heh-PAT-oh-sytes) Liver cells.

hepatopancreatic sphincter (heh-PAT-oh-pan-kree-AT-ik SFINGK-ter) Muscular ring that surrounds the lumen of the common bile duct and the duodenal ampulla; seals off the passageway and prevents bile from entering the small intestine; also known as sphincter of Oddi.

hilus (HYE-luss) A depression on an organ that serves as an entrance site for blood vessels, lymphatic vessels, and/or nerves; *ex.:* renal hilus.

homeostasis (hoh-mee-oh-STAY-sis) The process of maintaining the relatively stable state in the body's internal environment despite regular changes in the external environment.

hormone (HOR-mohn) Chemical messenger; secreted by an endocrine gland and transported through the blood.

hydroxyapatite (hye-DROK-see-ap-ah-tyte) The chief structural component of bone; formed by the interaction of calcium phosphate and calcium hydroxide.

hyperpolarization (hi-per-pohl-er-iz-A-shun) An increase in the membrane potential causing the measured voltage to become more negative.

hypertonic (hye-per-TON-ik) **solution** Solution that contains a higher concentration of nonpenetrating solutes than the adjacent cell; causes cell to shrink.

hypodermis (hye-poh-DER-mis) Layer of loose connective tissue that separates the skin from underlying tissue and organs; also known as subcutaneous layer.

hypothalamus (hye-poh-THAL-ah-mus) Part of the diencephalon in the brain; contains control centers regulating autonomic functions, emotions, and hormone production.

hypotonic (hye-poh-TON-ik) **solution** Solution that contains a lower concentration of nonpenetrating solutes than the adjacent cell; causes cell to swell.

I

ileum (IL-ee-um) Final 12-foot segment of the small intestine.

ilium (IL-ee-um) Largest coxal bone.

incisors (in-SYE-zors) Blade-shaped teeth found at the front of the mouth; used for clipping or cutting.

incus (IN-kuss) Middle ear bone; also called the anvil.

inferior (in-FEER-ee-or) Below; opposite of superior.

inferior vena cava (inferior VEE-nah KAY-vah) One of the great veins; carries blood to the right atrium from the trunk and lower extremities.

inguinal (ING-gwih-nal) Pertaining to the groin.

inhibin (in-HIB-in) A hormone secreted by the testes and ovaries that inhibits the release of follicle-stimulating hormone (FSH) from the anterior pituitary.

inner ear Region of the ear that contains the receptors of equilibrium and hearing.

inorganic (in-or-GAN-ik) Pertaining to compounds that lack hydrocarbons (carbon atoms bonded to hydrogen atoms); *ex.:* water, salts.

insulin A hormone produced by the pancreas; promotes the movement of glucose, amino acids, and fatty acids out of the blood and into body cells.

intercalated disks (in-TER-kah-lay-ted disks) Specialized regions that form connections between cardiac muscle cells.

interdigitate (in-ter-DIJ-ih-tayt) To interlock.

intermediate filaments (in-ter-MEE-dee-it FIL-ah-ments) Part of the cell's cytoskeleton; provide strength, stabilize the position of the organelles, and transport material within cells.

interneuron (in-ter-NOO-ron) Short neurons in the central nervous system located between sensory and motor neurons; association neuron.

interphase (IN-ter-fayz) Longest stage in the cell life cycle; when the cell prepares for division; DNA is replicated in this stage.

interstitial (in-ter-STISH-al) **fluid** Solution that fills the spaces between cells.

intervertebral disk (in-ter-VER-tee-bral disk) Pad of fibrocartilage that separates and cushions the vertebrae.

intra-alveolar (in-trah-al-VEE-oh-lar) **pressure** Air pressure inside the alveoli of the lungs.

intracellular fluid (ICF) Liquid located within the cells.

ion (EYE-on) A charged particle; *ex.:* Na^+.

iris (EYE-riss) Part of the eye that contains blood vessels, pigment cells, and smooth muscle.

ischium/ischia (IS-kee-um/IS-kee-ah) One of the three bones that fuse to create a coxal bone in the hip.

isotonic (eye-soh-TON-ik) **solution** Solution that contains the same concentration of nonpenetrating solutes as the adjacent cell; causes cells to neither shrink nor swell.

J

jejunum (jeh-JOO-num) 8-foot middle segment of the small intestine between the duodenum and the ileum.

K

kidney (KID-nee) Major organ of the urinary system; filters waste products from the blood.

Krebs cycle *See citric acid cycle.*

L

labia (LAY-bee-ah) "Lips"; skin folds on the sides of the opening to the vagina.

lacrimal gland (LAK-rih-mal gland) Almond-shaped tear gland.

lacteal (LAK-tee-al) Lymphatic capillary within a villus of the small intestine.

lactic acid A waste product of glucose metabolism as a result of insufficient oxygen supply; produced in muscles after an intense workout.

lacuna (lah-KOO-nah) A small space where bone or cartilage cells are found.

lamella/lamellae (lah-MEL-ah/lah-MEL-ee) Layer of calcified bone matrix.

lamina propria (LAY-min-ah PRO-pree-ah) The loose connective tissue component of a mucous membrane.

laryngopharynx (lair-ring-go-FAIR-inks) Portion of the pharynx lying between the hyoid bone and the entrance to the esophagus.

larynx (LAIR-inks) Cartilaginous structure that surrounds and protects the glottis; voice box.

lateral (LAT-er-al) Away from the midline of the body; opposite of medial.

lens A part of the eye that lies posterior to the cornea; focuses the visual image on the retinal photoreceptors.

leukocyte (LOO-koh-syte) White blood cell.

ligament (LIG-ah-ment) A type of connective tissue that connects bone to bone.

line A long, narrow strip or mark; in the skeletal system, similar to a crest.

lipase (LYE-payse) A pancreatic enzyme that breaks down lipids.

lipids (LIP-idz) Organic molecules containing carbon, hydrogen, and oxygen, such as in fats and oils.

lobule (LOB-yool) A small lobe.

loop of Henle *See nephron loop.*

lumbar (LUM-bar) Lower back.

lumen (LOO-men) Hollow area within a tube.

lungs The pair of respiratory organs that exchanges oxygen and carbon dioxide in the blood.

luteinizing (loo-tee-in-EYE-zing) **hormone (LH)** A hormone produced by the anterior pituitary; in females, acts on ovaries to stimulate ovulation and progesterone secretion from corpus luteum; in males, acts on testes to stimulate the synthesis and secretion of testosterone.

lymph (limf) Fluid transported by the lymphatic system; similar to plasma but contains lower concentration of proteins.

lymph node (limf node) Small oval organ that filters and purifies lymph before it reaches the venous system.

lymphocyte (LIM-foh-syte) Primary cell of the lymphatic system; a type of white blood cell.

lysosome (LYE-so-sohm) Cell organelle that contains digestive enzymes.

M

macromolecules (mak-roh-MOL-eh-kyools) Large, complex molecules made from simpler molecules; *ex.*: proteins, nucleic acids.

malleus (MAL-ee-us) One of three ear bones; also called the hammer.

marrow (MAIR-roh) Soft vascular tissue that fills the cavities of most bones.

matrix (MAY-triks) Extracellular substance of a connective tissue.

meatus (mee-AY-tus) Tubelike opening or channel.

medial (MEE-dee-al) Toward the midline longitudinal axis of the body; opposite of lateral.

median plane (MEE-dee-an plane) A section that passes along the midline and divides the body into right and left halves.

mediastinum (mee-dee-as-STYE-num) Region of the thoracic cavity between the lungs.

medulla (meh-DUL-ah) Inner portion of an organ; *ex.*: adrenal medulla.

medulla oblongata (meh-DUL-ah ob-long-GAH-tah) Most inferior portion of the brainstem; site of vital control centers for respiratory and cardiovascular systems.

medullary cavity (MED-oo-lair-ee cavity) Hollow space inside the shaft of the bone; marrow cavity.

meiosis (my-OH-sis) Type of cell division that results in the formation of sperm or ova that contain half the number of chromosomes as the parent cell; occurs in the testes and ovaries.

melanin (MEL-ah-nin) Brown skin pigment that absorbs ultraviolet radiation.

melanocytes (MEL-ah-noh-sytes) A fixed cell that synthesizes and stores melanin.

melatonin (mel-ah-TOE-nin) A hormone produced by the pineal gland; involved in regulating the body's sleep-wake cycles.

membrane potential (MEM-brayn po-TEN-shul) A separation of positive and negative charges across a plasma membrane that results in an electrical potential or a voltage.

meninx/meninges (ME-ninks/meh-NIN-jeez) One of the three membranes surrounding the spinal cord or brain.

menstrual phase (MEN-stroo-al fayz) The first phase of the uterine cycle, characterized by menstrual bleeding as a result of shedding the uterine lining called the endometrium.

metabolism (meh-TAB-oh-liz-em) Sum of all chemical activities in the body.

metacarpals (met-ah-KAR-puls) The bones in the palm of each hand.

metaphase (MET-ah-fayz) Stage of mitosis evident when chromatids line up at the middle of the cell.

microfilaments (my-KROH-FIL-ah-ments) Slender protein strands that make up the framework of the cytoskeleton; also called actin.

microtubules (my-kroh-TOOB-yools) Hollow protein tubes found in all body cells; framework of the cytoskeleton.

microvillus/microvilli (my-kroh-VIL-luss/my-kroh-VIL-eye) Small, finger-shaped projections of the plasma membrane of an epithelial cell that contain microtubules; function to increase surface area for better absorption.

midbrain Portion of the brainstem; provides pathways between brainstem and cerebrum.

middle ear Small cavity between the external ear and inner ear that houses the three small ear bones.

mineraloccorticoids (min-er-al-oh-KOR-tih-koyds) A group of hormones produced by the adrenal cortex that regulate mineral homeostasis, such as sodium and potassium levels in the blood; *ex.*: aldosterone.

mitochondrion/mitochondria (my-toh-KON-dree-ohn/my-toh-KON-dree-ah) Cellular organelle that is the site of aerobic cellular respiration; produces ATP for the cell.

mitosis (my-TOH-sis) Division of the cell nucleus that results in the formation of two new daughter cells that contain the same type and number of chromosomes as the parent cell; divided into four stages: prophase, metaphase, anaphase, and telophase.

molars (MOHL-larz) Teeth with flattened crowns with prominent ridges; used for grinding and crushing.

molecules (MOL-eh-kyools) Compound consisting of two or more atoms held together by chemical bonds.

monocyte (MON-oh-syte) A phagocytic white blood cell.

monosaccharide (mon-oh-SAK-ah-ryde) A simple, single-unit sugar such as glucose or fructose.

monosynaptic reflex (mon-oh-sin-AP-tic RE-flex) A reflex pathway that does not contain any interneurons between the sensory and motor neurons.

morula (MOR-yoo-lah) A solid ball of cells formed from mitotic divisions of the blastomeres.

motility (moh-TIL-i-tee) Contraction of smooth muscle in the wall of the digestive tract for the purpose of mixing and propelling digested substances; includes the processes of peristalsis and segmentation.

mucosa (myoo-KOH-sah) Mucous membrane.

mucous (adj.) (MYOO-kuss) Describes lubricating secretions along the digestive, respiratory, urinary, and reproductive tracts.

mucus (n.) (MYOO-kuss) A thick, gel-like substance secreted by glands in the digestive, respiratory, urinary, and reproductive tracts.

multipolar neuron (mul-tee-PO-lar NOO-ron) A neuron that has several dendrites and a single axon.

myelin (MY-eh-lin) A membranous insulation consisting of many layers of glial cell membrane around the axon of a nerve cell; improves the speed of impulse conduction.

myelination (my-eh-lih-NAY-shun) The process of forming a myelin sheath.

myocardium (my-oh-KAR-dee-um) The cardiac muscle in the wall of the heart.

myofibril (my-oh-FYE-brill) Slender fibers found in skeletal muscle and extending the length of the cell; composed of thick and thin filaments.

myometrium (my-oh-MEE-tree-um) The middle muscular layer of the uterine wall.

myosin (MY-oh-syn) Thick protein filament found in skeletal muscle cells.

N

nasal (NAY-zal) Pertaining to the nose.

nasopharynx (nay-zoh-FAIR-inks) The superior portion of the larynx.

neck A slender region, usually connected to the head of a bone.

nephron (NEF-ron) The basic functional unit of the kidney.

nephron loop (NEF-ron loop) A subdivision of the nephron; located between the proximal convoluted tubule and the distal convoluted tubule; also known as the loop of Henle.

nerves Long, cable-like structures that extend throughout the body; bundles of peripheral neurons held together by several layers of connective tissues.

neurilemma (noo-rih-LEM-mah) The cytoplasmic covering around the axon provided by the Schwann cells; found only in the peripheral nervous system.

neuroglia (noo-ROG-lee-ah) A class of cell types in neural tissue that provides nutrients and a supporting framework to neural tissue; does not transmit impulses.

neurolemmocyte (noo-ROH-lem-moh-syte) A neuroglial cell that wraps itself around peripheral axons to form a protective covering.

neurons (NOO-rons) Conducting cells of the nervous system; each consists of a cell body, one or more dendrites, and a single axon.

neurotransmitters (noo-roh-trans-MIH-terz) Chemical messengers released by neurons for the purpose of stimulating or inhibiting target cells.

neutron (NOO-tron) Subatomic particle in the nucleus of an atom; no electrical charge.

neutrophil (NOO-troh-fil) Phagocytic white blood cell; produced in bone marrow.

Nissl bodies (NISS-ul bodies) Ribosomal clusters found in neuron cell bodies.

node of Ranvier (node of rahn-vee-AY) Gaps of exposed axon not covered by a myelin sheath; appear at regular intervals in some myelinated axons.

norepinephrine (nor-ep-ih-NEF-rin) **(NE)** A neurotransmitter and a hormone; involved in regulating autonomic nervous system activity; also known as noradrenaline.

nuclease (NOO-klee-ayz) An enzyme that catalyzes the hydrolysis of nucleic acids.

nucleic acid (noo-KLAY-ik ASS-id) Organic molecule found in the nucleus; made up of functional units called nucleotides; *ex.*: RNA, DNA.

nucleus Prominent organelle of the cell; contains DNA molecules.

O

oblique (oh-BLEEK) Diagonal.

olecranon (oh-LEK-rah-nohn) The point of the elbow.

olfaction (ohl-FAK-shun) The sense of smell.

oocyte (OH-oh-syte) Immature stage of the female gamete.

oogenesis (oh-oh-JEN-eh-sis) Production of female gametes or ova; occurs in the ovaries.

optic chiasma (OP-tik kye-AS-mah) Area near the diencephalon where the optic nerves cross over each other to eventually connect to opposite sides of the brain.

optic nerve (OP-tik nerve) Carries visual information from the retina of the eyes to the brain.

oral (OR-al) Pertaining to the mouth.

oral cavity (OR-al KAV-i-tee) Space within the mouth; includes lips, cheeks, tongue and its muscles, and hard and soft palates.

orbital (OR-bi-tal) Pertaining to the eye.

organ (OR-gan) A group of tissues that performs a specific function.

organelle (or-gah-NELL) Small internal structures in cells; divided into membranous and nonmembranous types; *ex.*: ribosome.

organic (or-GAN-ik) Pertaining to compounds that contain hydrocarbons (carbon atoms bonded to hydrogen atoms); *ex.*: carbohydrates, proteins.

organism (OR-gan-ism) An individual living thing.

organ system (OR-gan SISS-tem) Varying number and kinds of organs arranged in a way to perform complex functions; most complex organizational unit of the body; *ex.*: respiratory system.

oropharynx (oh-roh-FAIR-inks) Portion of the pharynx that extends between the soft palate and the base of the tongue at the level of the hyoid bone.

osmosis (os-MOH-sis) Movement of water across a semi-permeable membrane toward the solution containing the higher solute concentration.

osteoblasts (OS-tee-oh-blasts) Cells that lay down the specialized matrix of bone; responsible for the production of new bone.

osteoclasts (OS-tee-oh-clasts) Cells that decompose the matrix of bone.

osteocytes (OS-tee-oh-sytes) Mature bone cells.

osteon (OS-tee-ohn) The basic functional unit of mature compact bone.

oval window (OH-val WIN-doh) Membranous structure in the inner ear; separates the middle and inner ear; stapes connects to it.

ovary (OH-vah-ree) Female gonad; produces ova.

ovulation (ov-yoo-LAY-shun) Release of an egg cell called a secondary oocyte from a tertiary follicle in the ovary.

ovum/ova (OH-vum/OH-vah) Female sex cell; egg cell.

oxidation Loss of one or more electrons from an atom or molecule.

P

palpate (PAL-payt) To examine by feeling part of the body.

pancreas (PAN-kree-ass) Slender, elongated organ that lies in the abdominopelvic cavity; has the dual function of manufacturing digestive enzymes and making hormones such as insulin.

pancreatic juice A mixture of secretions that contains mostly water, salts, sodium bicarbonate, and various digestive enzymes; secreted by the exocrine acinar cells of the pancreas.

papilla (pah-PILL-ah) Nipple-shaped mounds.

parathyroid hormone (PTH) Hormone secreted by the parathyroid glands that is the primary regulator of calcium homeostasis; increases calcium levels in the blood.

parietal (pah-RYE-i-tal) Relating to or forming the wall of an organ or cavity.

parietal cells (pah-RYE-i-tal sells) Cells located in the stomach; secrete hydrochloric acid.

pectoral (PEK-toh-ral) Pertaining to the chest; between sternal and axillary regions.

pedicle (PED-i-kul) Bony structure that attaches the vertebral arch to the body of a vertebra; means "foot."

pelvic (PEL-vik) Relating to the pelvis.

pepsin (pep-SYN) Enzyme made by chief cells in the stomach; functions to digest proteins.

pepsinogen (pep-SYN-oh-jen) An inactive enzyme secreted by chief cells in the stomach; converted to the active enzyme pepsin during gastric digestion.

perimysium (pair-i-MISH-ee-um) A connective tissue layer that divides the skeletal muscle into a series of compartments or bundles called fascicles.

perineal (pair-i-NEE-al) The region between the scrotum and the anus in males and between the posterior vulva junction and the anus in females.

perineurium (pair-i-NOO-ree-um) Connective tissue sheath that surrounds a bundle of nerve fibers within a nerve and holds them together.

periosteum (pair-ee-OS-tee-um) A fibrous layer that covers the bone.

peristalsis (pair-ih-STAWL-siss) Wavelike ripple of muscular contractions along the wall layer of a hollow organ, such as the small intestine.

peroxisomes (per-AHK-si-sohms) Cell organelles that absorb and neutralize toxins.

pH Means "potential Hydrogen scale"; a measure of the concentration of hydrogen ions in a solution; scale ranges in value from 0–14 where 7 is the neutral point; values below 7 are acidic and those above 7 are alkaline.

phagocytosis (fag-oh-sye-TOH-siss) Type of endocytosis; process in which microorganisms or other large particles are engulfed by the plasma membrane and enter the cell.

phalanx/phalanges (FAH-lanks/fah-LAN-jeez) The finger bones and toe bones.

pharynx (FAIR-inks) Passageway that connects the nose, mouth, and throat; commonly called the throat.

physiology (fiz-ee-AHL-oh-jee) Science that examines the function of the living organism and its parts.

pia mater (PEE-ah MAH-ter) The highly vascular, innermost layer of the meninges.

pinocytosis (pin-oh-sye-TOH-sis) Type of endocytosis; process in which a plasma membrane invaginates to capture some extracellular fluid and brings it into the cytoplasm in a membranous vesicle.

pituitary (pih-TOO-i-tair-ee) **gland** Small endocrine gland located near the base of the brain that releases many different hormones; consists of two separate glands: anterior pituitary and posterior pituitary; referred to as the "master gland" because many of its hormones control other endocrine glands in the body.

placenta (plah-SEN-tah) The specialized organ within the uterus that supports embryonic and fetal development; also called the afterbirth.

plasma Fluid portion of the blood.

polysaccharide (pol-ee-SAK-ah-ryde) An organic macro-molucule; a complex sugar formed by bonding many mono-saccharides in a long chain, such as starch or cellulose.

polysynaptic reflex (pahl-y-sin-AP-tic RE-flex) A reflex pathway that contains interneurons between the sensory neurons and motor neurons.

pons (ponz) Part of the brainstem; connects the brainstem to the cerebellum; contains relay centers and is involved with somatic and visceral motor control.

positive feedback A physiological mechanism that tends to amplify or reinforce the change in internal environment.

posterior (pos-TEER-ee-or) Back; behind; opposite of anterior.

postganglionic neuron (post-gang-glee-ON-ik NOO-ron) The second neuron in the motor output portion of an autonomic reflex pathway; connects a ganglion to smooth muscle or cardiac muscle, or a gland.

preganglionic neuron (pree-gang-glee-ON-ik NOO-ron) The first neuron in the motor output portion of an autonomic reflex pathway; connects the central nervous system to a ganglion.

premolars *See bicuspids.*

prepuce (PREE-pus) The fold of skin that surrounds the glans penis in males and clitoris in females; foreskin.

primary active transport Refers to the transport of a specific substance across the plasma membrane, against its concentration gradient; typically involves a membrane protein, referred to as a "pump," that uses ATP.

progesterone (pro-JES-ter-ohn) Female hormone produced by the ovaries; prepares the endometrium for possible implantation of the embryo and stimulates milk secretion in mammary glands.

prolactin (pro-LAHK-tin) **(PRL)** Hormone secreted by the anterior pituitary gland; stimulates milk secretion in mammary glands.

proliferative phase (PROH-lif-eh-rah-tiv fayz) Second phase of the uterine cycle; involves reestablishing a portion of the endometrium.

prophase (PRO-fayz) Initial stage of mitosis; evident when the chromosomes become visible.

prostate gland (PROSS-tayt) Small, muscular, rounded organ; produces a weak acidic secretion that contributes about 30% of the volume of semen.

protease (PRO-tee-ayz) Enzyme that catalyzes the breakdown of proteins into intermediate compounds.

protein (PRO-teen) Organic macromolecule with a long, complex polymer structure and a variety of functions; structural units are amino acids.

proton (PRO-ton) Positively charged subatomic particle; located in the nucleus of an atom.

proximal (PROK-sih-mal) Refers to the region or reference toward an attached base; opposite of distal.

proximal convoluted tubule (PROK-sih-mal kon-voh-LOO-ted TOOB-yool) Part of the nephron in the kidney; located between the glomerular capsule and the nephron loop; primary site of reabsorption of nutrients from the filtrate.

pubic (PYOO-bik) Relating to the region of the pubis.

pubis (PYOO-biss) The articulation between the two coxal bones.

pudendum (pyoo-DEN-dum) Region enclosing the female external genitalia; usually called the vulva.

pulp cavity (pulp KAV-i-tee) A spongy, highly vascular cavity in the tooth.

pupil (PYOO-pill) Central opening in the iris of the eye that allows light to enter the eye.

Purkinje (pur-KIN-jee) **fibers** Specialized cardiac muscle cells that extend out to the lateral walls of the ventricles and papillary muscles; conduct impulses from the AV bundle, stimulating heart contraction.

pyloric sphincter (pye-LOR-ik SFINGK-ter) A muscular structure that regulates the release of chyme from the stomach to the duodenum.

R

ramus (RAY-mus) A branch, such as the vertical section of bone in the mandible that connects it to the skull.

reabsorption Second step in urine formation; movement of substances out of the renal tubules and into the blood.

rectum (REK-tum) Last segment of the large intestine; end of the digestive tract.

rectus (REK-tus) Straight.

reduction Gain of one or more electrons by an atom or molecule.

reflex A fast, automatic response to a stimulus.

renal (REE-nal) Pertaining to the kidney.

renin (REH-nin) An enzyme produced by the nephrons in the kidneys; converts angiotensinogen to angiotensin I in the plasma.

resting potential *See membrane potential.*

rete testis (REE-tee TES-tis) A maze of passageways formed from the seminiferous tubules and located within the mediastinum of the testis.

retina (RET-i-nah) The inner layer of the eye; contains the photoreceptor cells of the eye.

ribosomes (RYE-boh-sohms) Cellular organelles that are the sites of protein synthesis; composed of two subunits.

round window (round WIN-doh) A thin, membranous partition that separates the cochlear chambers from the air spaces of the middle ear.

rugae (ROO-gee) Longitudinal folds that line the inner wall of the stomach.

S

sacral (SAY-kral) Pertaining to the sacrum.

saggital (SAJ-i-tal) Runs along with the long axis of the body.

sagittal plane (SAJ-i-tal plane) Extends from anterior to posterior and divides the body into right and left sections.

sarcolemma (sar-koh-LEM-ah) Plasma membrane of a skeletal muscle cell.

sarcomere (SAR-koh-meer) Contractile unit of a muscle cell.

scapula (SKAP-yoo-lah) Shoulder blade.

Schwann cell *See neurolemmocyte.*

sclera (SKLAIR-ah) The dense fibrous, white outer covering of the eye.

scrotum (SKRO-tum) Pouch-like sac, divided into two chambers; contains the testes.

sebaceous gland (seh-BAY-shus gland) Holocrine gland that secretes waxy, oily sebum into the hair follicles.

sebum (SEE-bum) An oily secretion produced by sebaceous glands that lubricates hair shafts.

secondary active transport Typically involves cotransport of substances across the plasma membrane; does not use ATP; relies on the concentration gradients established by primary active transport.

secondary oocyte (SEK-on-dair-ee OH-oh-syte) Ovum released at ovulation.

second messenger An intracellular molecule, such as cyclic AMP, produced in response to a first messenger (*ex.*: hormone) binding to its plasma membrane receptor; triggers a chemical chain reaction within the target cell, causing cellular changes.

secretin (seh-KREE-tin) A digestive hormone produced by duodenal mucosa; stimulates secretion of pancreatic juice rich in bicarbonate ions.

secretion Production and release of a chemical substance from a cell; *ex.*: neurotransmitters from a neuron.

secretory phase (SEEK-reh-toh-ree fayz) The last and longest phase of the uterine cycle; characterized by thickening of the endometrium in preparation for possible implantation of a human embryo.

segmentation Mixing movement; occurs when digestive reflexes cause a forward-and-backward movement within a single region of the digestive tract.

semen (SEE-men) Fluid released from the penis during ejaculation; contains sperm and seminal fluid produced by various glands.

semicircular canals (sem-i-SIR-kyoo-lar kah-NALS) Three bony rings located in the inner ear; contain fluid-filled ducts that, in turn, contain receptors that function to maintain dynamic equilibrium.

semilunar valve (sem-i-LOO-nar valv) Pair of heart valves located at the exit point of each ventricle; each valve has three pouch-like flaps; consists of the pulmonary and aortic valves.

seminiferous tubules (seh-mih-NIF-er-us TOOB-yools) A tightly coiled structure located in each lobule of the testis; site where spermatozoa develops.

sensory transduction (SEN-sohr-y trans-DUCK-shun) The conversion of energy from a stimulus into an action potential in a sensory neuron.

septum/septae (SEP-tum/SEP-tee) A terminal partition that divides an organ.

serosa (seh-ROH-sah) A serous membrane that covers most of the digestive tract.

simple diffusion *See diffusion.*

sinoatrial (sye-NOH-AY-tree-al) **(SA) node** Primary pacemaker of the heart; specialized mass of cardiac muscle cells located in the wall of the right atrium that plays a role in electrical conduction system within the heart.

sinus (SYE-nus) A cavity or space in a tissue; a large dilated vein.

sinusoid (SYE-nu-soyd) Large, permeable blood capillaries found in organs such as the liver, spleen, and bone marrow; specialized to allow large proteins to smoothly move in or out of the blood.

sodium-potassium pump Protein pump that uses active transport; located in the plasma membrane of most cells; exchanges three sodium ions moved out of the cell for two potassium ions brought into the cell; also known as the sodium-potassium ATPase.

soft palate (soft PAL-let) The portion of the roof of the mouth that lies posterior to the hard palate.

solute Dissolved substance(s) in a solution; *ex:* salt.

solution (suh-LOO-shun) Mixture in which a solute is dissolved in a solvent; *ex.:* salt water.

solvent Liquid portion of a solution; dissolves the solute; *ex:* water.

soma (SO-mah) Cell body.

sperm *See spermatozoon.*

spermatic cord (sper-MAT-ik cord) A layer of fascia, tough connective tissue, and muscle surrounding the ductus deferens and the blood vessels and nerves of the testes.

spermatogenesis (sper-mah-toh-JEN-eh-sis) Production of sperm cells; occurs in the testes.

spermatozoon/spermatozoa (sper-mah-tah-ZOH-ohn/sper-mah-tah-ZOH-ah) Sperm cell; male gamete or sex cell.

spindle fiber (SPIN-dul FYE-ber) Network of tubules in the cell that extends between the centriole pairs.

spine Slender, pointed process on a bone; site for muscle attachment.

spongy bone (SPUN-jee bone) Type of bone that has a porous network of bony plates; also known as cancellous bone.

stapes (STAY-peez) The inner ear bone; also called the stirrup.

stem cells Nonspecialized cells that have the ability to divide continually to form new types of more specialized cells.

steroids A large class of lipids characterized by a carbon ring chemical structure; act as components for cells and regulate body function; *ex.:* cholesterol.

stimulus Any agent that induces a response; causes a change in an excitable tissue.

striation (STRYE-ay-shun) A striped or banded appearance such as that found in skeletal muscle cells and cardiac muscle cells.

submucosa (sub-myoo-KOH-sah) A layer of loose connective tissue in the wall of the digestive tract; large blood vessels, lymphatic vessels, and nerve fibers are found here.

sulcus (SUL-kus) A shallow depression.

superficial (soo-per-FISH-al) On the surface; opposite of deep.

superior (soo-PEER-ee-or) Above; opposite of inferior.

surface tension The force of attraction between water molecules as a result of hydrogen bonding; creates a thin surface film on water.

surfactant (sur-FAK-tant) Chemical mixture that contains lipids and proteins and coats the inner surface of each alveolus in the lungs; reduces surface tension.

suture (SOO-chur) The boundary between the skull bones; immovable joint.

synapse (SIN-aps) Specialized junction where a neuron communicates with another cell such as another neuron, a muscle cell, or a glandular cell.

synaptic cleft (sin-AP-tic kleft) Within the synapse, the extracellular space between the membrane of the neuron and the membrane of the other cell with which it is communicating.

synaptic knob (sin-AP-tic nahb) The bulbous structure at the end of a neuron's axon where neurotransmitters are secreted; also known as axon terminal, synaptic terminal, and synaptic end bulb.

synovial fluid (sih-NO-vee-all FLOO-id) Thick and colorless lubricating fluid secreted by synovial membranes.

synovial joint (sih-NO-vee-all joynt) Freely movable joint; most numerous and anatomically most complex joint in the body.

synovial membrane (sih-NO-vee-all MEM-brayne) Connective tissue membrane that lines the spaces between bones and joints; secretes synovial fluid.

systole (SISS-toh-lee) Contraction of both atria and both ventricles in the heart; opposite of diastole.

T

target cell A general term for any cell affected by a hormone because it contains a receptor for that hormone.

tarsals (TAR-sals) Ankle bones.

tarsus (TAR-sus) Ankle.

telophase (TEL-oh-fayz) Stage of mitosis; nuclear membrane forms, nuclei enlarge, and chromosomes gradually uncoil and disappear.

temporal (TEM-poh-ral) Pertaining to the temples on the sides of the head, above the zygomatic arch.

tendon (TEN-don) A cable-like, fibrous connective tissue that connects skeletal muscle to bone.

tertiary follicle (TER-shee-air-ee FOL-lih-kul) Mature ovum in ovary; also known as Graafian follicle or vesicular follicle.

testis/testes (TES-tiss/TES-teez) The male gonad that produces sperm.

testosterone Principal male sex hormone; produced by interstitial cells in the testes; responsible for growth and maintenance of male sexual characteristics and for sperm production.

thalamus (THAL-ah-mus) Located in diencephalon of the brain; final relay point for ascending sensory information that will be sent to the cerebrum.

thoracic (tho-RASS-ik) Pertaining to the chest.

threshold stimulus Minimum stimulus needed to produce a response in an excitable cell.

thrombocyte (THROM-boh-syte) A blood cell that has a role in clotting: also known as a platelet.

thrombus (THROM-bus) Stationary blood clot.

thyroid gland (THY-royd gland) Endocrine gland located near the trachea in the neck; produces the hormones T_3, T_4 and calcitonin.

thyroid-stimulating hormone (TSH) Hormone secreted by anterior pituitary gland; stimulates thyroid gland to produce the hormones T_3 and T_4.

thyroxin (T_4) (thy-ROK-syn) Hormone produced by the thyroid gland; influences metabolic rate, growth, and development.

tissue (TISH-yoo) Group of similar cells that perform a common function.

tonsil (TAHN-sil) A large nodule containing masses of lymphoid tissue; located in the walls of the pharynx.

trachea (TRAY-kee-ah) Long tube between the larynx and the primary bronchi that serves as air passageway; windpipe.

transverse plane (TRANS-vers plane) A division that lies at right angles to the long axis of the body; divides the body into superior and inferior sections.

tricuspid valve (try-KUS-pid valv) The right atrioventricular (AV) valve in the heart; located between the right atrium and right ventricle.

triiodothyronine (T_3) (try-eye-oh-doh-THY-roe-neen) Hormone produced by the thyroid gland; regulates metabolic rate, growth, and development.

trochanter (troh-KAN-ter) Large bump on a bone; site for muscle attachment.

trochlea (TROHK-lee-ah) A bone process that resembles a pulley.

tubercle (TOO-ber-kal) Small bump on a bone.

tuberosity (too-bah-ROS-i-tee) Bump on a bone; smaller than a trochanter.

tympanic membrane (tim-PAN-ik membrane) The eardrum.

U

umbilical cord (um-BIL-i-kul cord) The vascular structure that connects the fetus to the placenta.

unipolar neuron (YOO-nee-POH-lar NOO-ron) A neuron in which the dendrite and axonal processes are continuous, and the cell body lies off to the side.

ureters (YOOR-eh-ters) Pair of long, slender, muscular tubes that transports urine from the kidneys to the urinary bladder.

urethra (yoo-REE-thrah) A passageway that transports urine from the neck of the urinary bladder to the exterior; in males, also acts as a passageway for spermatozoa to the exterior.

uterine tubes (YOO-ter-in tubes) Hollow muscular tubes that transport a secondary oocyte to the uterus; also known as oviducts or fallopian tubes.

uterus (YOO-ter-us) Hollow muscular organ that provides mechanical protection, nutritional support, and waste removal for a developing embryo.

V

vagina (vah-JYE-nah) Muscular passageway in the female that connects the uterus with the exterior genitalia.

vascular (vas-KYOO-ler) Pertaining to the blood vessels.

vas deferens *See ductus deferens.*

veins (vanes) Vessels that collect blood from tissues and organs and return it to the heart.

ventral (VEN-tral) Anterior or belly side in humans; opposite of dorsal.

ventricle (VEN-tri-kul) Cavity or chamber.

venule (VEN-yool) Microscopic blood vessel that connects capillaries to small veins.

villus (VIL-us) Fingerlike extension of mucous membrane of small intestine.

vitreous humor (VIT-ree-us humor) A gel-like mass that fills the posterior chamber of the eye; helps maintain shape of the eye and give support to the retina.

W

white matter (whyte MAT-ter) Nerve fibers covered with the myelin sheath.

Y

yolk sac (yohk sak) The first of the extraembryonic membranes to appear; an important site for blood cell production.

Z

zona pellucida (ZOH-nah pel-LOO-sih-dah) A region formed between the innermost follicular cells and the developing oocyte.

zygote (ZYE-goht) Fertilized egg cell.

Index

INDEX